Vol. 29. **The Analytical Chemistry of Sulfur and Its Compounds** (*in three parts*). By J. H. Karchmer

Vol. 30. **Ultramicro Elemental Analysis.** By C

Vol. 31. **Photometric Organic Analysis** (*in tw*

Vol. 32. **Determination of Organic Compoun** ck T. Weiss

Vol. 33. **Masking and Demasking of Chemic**

Vol. 34. **Neutron Activation Analysis.** By D. De Soete, R. Gijbels, and J.

Vol. 35. **Laser Raman Spectroscopy.** By Marvin C. Tobin

Vol. 36. **Emission Spectrochemical Analysis.** By Morris Slavin

Vol. 37. **Analytical Chemistry of Phosphorus Compounds.** Edited by M. Halmann

Vol. 38. **Luminescence Spectrometry in Analytical Chemistry.** By J. D. Winefordner, S. G. Schulman and T. C. O'Haver

Vol. 39. **Activation Analysis with Neutron Generators.** By Sam S. Nargolwalla and Edwin P. Przybylowicz

Vol. 40. **Determination of Gaseous Elements in Metals.** Edited by Lynn L. Lewis, Laben M. Melnick, and Ben D. Holt

Vol. 41. **Analysis of Silicones.** Edited by A. Lee Smith

Vol. 42. **Foundations of Ultracentrifugal Analysis.** By H. Fujita

Vol. 43. **Chemical Infrared Fourier Transform Spectroscopy.** By Peter R. Griffiths

Vol. 44. **Microscale Manipulations in Chemistry.** By T. S. Ma and V. Horak

Vol. 45. **Thermometric Titrations.** By J. Barthel

Vol. 46. **Trace Analysis: Spectroscopic Methods for Elements.** Edited by J. D. Winefordner

Vol. 47. **Contamination Control in Trace Element Analysis.** By Morris Zief and James W. Mitchell

Vol. 48. **Analytical Applications of NMR.** By D. E. Leyden and R. H. Cox

Vol. 49. **Measurement of Dissolved Oxygen.** By Michael L. Hitchman

Vol. 50. **Analytical Laser Spectroscopy.** Edited by Nicolo Omenetto

Vol. 51. **Trace Element Analysis of Geological Materials.** By Roger D. Reeves and Robert R. Brooks

Vol. 52. **Chemical Analysis by Microwave Rotational Spectroscopy.** By Ravi Varma and Lawrence W. Hrubesh

Vol. 53. **Information Theory As Applied to Chemical Analysis.** By Karel Eckschlager and Vladimir Štěpánek

Vol. 54. **Applied Infrared Spectroscopy: Fundamentals, Techniques, and Analytical Problem-solving.** By A. Lee Smith

Vol. 55. **Archaeological Chemistry.** By Zvi Goffer

Vol. 56. **Immobilized Enzymes in Analytical and Clinical Chemistry.** By P. W. Carr and L. D. Bowers

Vol. 57. **Photoacoustics and Photoacoustic Spectroscopy.** By Allan Rosencwaig

Vol. 58. **Analysis of Pesticide Residues.** Edited by H. Anson Moye

Vol. 59. **Affinity Chromatography.** By William H. Scouten

Vol. 60. **Quality Control in Analytical Chemistry.** By G. Kateman and F. W. Pijpers

Vol. 61. **Direct Characterization of Fineparticles.** By Brian H. Kaye

Vol. 62. **Flow Injection Analysis.** By J. Ruzicka and E. H. Hansen

(*continued on back*)

Modern Methods of Polymer Characterization

Edited by

HOWARD G. BARTH

E. I. du Pont de Nemours & Company
Experimental Station
Wilmington, Delaware

JIMMY W. MAYS

Department of Chemistry
University of Alabama at Birmingham
Birmingham, Alabama

A WILEY-INTERSCIENCE PUBLICATION

JOHN WILEY & SONS

New York / Chichester / Brisbane / Toronto / Singapore

In recognition of the importance of preserving what has been written, it is a policy of John Wiley & Sons, Inc. to have books of enduring value published in the United States printed on acid-free paper, and we exert our best efforts to that end.

Library of Congress Cataloging in Publication Data:

Modern methods of polymer characterization / edited by Howard G. Barth, Jimmy W. Mays.
p. cm.—(Chemical analysis, ISSN 0069-2883; v. 113)
"A Wiley-Interscience publication."
Includes bibliographical references.
ISBN 0-471-82814-9
1. Polymers—Analysis. I. Barth, Howard G. II. Mays, Jimmy W. III. Series.
QD139.P6M63 1991 90-20906
547.7′046—dc20 CIP

Printed in the United States of America

10 9 8 7 6 5 4 3 2 1

CONTRIBUTORS

Stephen T. Balke, Department of Chemical Engineering and Applied Chemistry, University of Toronto, Toronto, Ontario, Canada M5S 1A4

Howard G. Barth, E. I. du Pont de Nemours & Co., Experimental Station, P.O. Box 80228, Wilmington, DE 19880-0228

Karin D. Caldwell, Departments of Chemistry and Bioengineering, University of Utah, 2232 Merrill Engineering Bldg., Salt Lake City, UT 84112

H. N. Cheng, Hercules Inc., Research Center, Wilmington, DE 19894

Richard B. Flippen, E. I. du Pont de Nemours & Co., Experimental Station, P.O. Box 80228, Wilmington, DE 19880-0228

Nikos Hadjichristidis, Division of Chemistry, University of Athens, Athens 157 71, Greece

Shyhchang S. Huang, B. F. Goodrich Company, Research and Development Center, 9921 Brecksville Rd., Brecksville, OH 44141

Jeffrey S. Lindner, Diagnostic Instrumentation and Analysis Laboratory, Mississippi State University, P. O. Drawer MM, Mississippi State, MS 39762

Jimmy W. Mays, Department of Chemistry, University of Alabama at Birmingham, Birmingham, AL 35294

Gregorio R. Meira, INTEC (CONICET and Universidad Nacional del Litoral) Guemes 3450, 3000 Santa Fe, Argentina

Petr Munk, Department of Chemistry and Biochemistry and Center for Polymer Research, University of Texas at Austin, Austin, TX 78712

Alfred Rudin, Guelph-Waterloo Centre for Graduate Work in Chemistry, Department of Chemistry, University of Waterloo, Waterloo, Ontario, Canada N2L 3G1

Paul Vouros, Department of Chemistry and Barnett Institute of Chemical Analysis, Northeastern University, 360 Huntington Ave., Boston, MA 02115

John W. Wronka, Department of Chemistry and Barnett Institute of Chemical Analysis, Northeastern University, 360 Huntington Ave., Boston, MA 02115

PREFACE

Polymer characterization is an area usually overlooked or given minimal treatment in most polymer science books and courses. Because of the complexity of polymeric materials in terms of molecular weight distribution, and compositional and microstructural heterogeneity, polymer characterization offers unique challenges to both analytical chemists and polymer scientists. Methodology and instrumentation used for polymer characterization continue to grow at a steady pace. To this end, we have recognized the need for a book containing state-of-the-art advances and developments in major technique areas of polymer characterization.

This book is written for two audiences: those with little background in a given area who would like to learn more about a specific technique and its applicability, and those who have some expertise in the area but require an updated review of recent advances. The main focus of this book is on synthetic polymers, although applications in the area of biopolymers are also included. Each chapter contains theory and fundamental principles, instrumentation, and applications. Adequate reference coverage is given to direct the reader to more detailed studies. Complete titles of cited references are given as an aid in selecting pertinent papers.

The book is divided into the following major areas: separations, inverse gas chromatography, osmometry, viscometry, ultracentrifugation, light scattering, and spectroscopy. The first two chapters discuss the use of size exclusion chromatography (SEC) to characterize complex polymers, while Chapter 3 deals with the measurement of long-chain branching with SEC. Principles and applications of field-flow fractionation as applied to polymer analysis are treated in Chapter 4. The determination of thermodynamic parameters and diffusion properties of polymers using inverse gas chromatography is covered in Chapter 5.

Molecular weight determinations using osmometry and dilute solution viscometry are reviewed in Chapters 6 and 7, respectively. In addition, the use of viscometry to obtain fundamental molecular parameters is also discussed. The importance of ultracentrifugation in polymer characterization is presented in Chapter 8.

The theory and applications of low-angle laser light scattering are examined in Chapter 9. In Chapter 10, photon correlation spectroscopy (dynamic light scattering) is reviewed. Characterization of polymers using nuclear magnetic resonance and mass spectrometry are treated in detail in Chapters 11 and 12, respectively.

Although the terms "characterization" and "analysis" are often times used interchangeably, there is a subtle distinction between the two terms: analysis is used to identify the composition of a polymer, while characterization is used to study the kinetic or thermodynamic interactions of a polymer with its environment. Many technique discussed in this book can be used for both analysis and characterization. For example, ultracentrifugation is a valuable method for determining the molecular weight distribution of a polymer (analysis) and also for studying thermodynamic properties (characterization). Inverse gas chromatography is strictly used for polymer characterization, while mass spectrometry is used for polymer analysis.

The editors wish to express their gratitude and appreciation to all the contributors for their cooperation and patience during the writing of this volume. Special thanks are given to the Central Research & Development Department of the Du Pont Company for supporting this project.

HOWARD G. BARTH
JIMMY W. MAYS

Wilmington, Delaware
Birmingham, Alabama
February 1991

CONTENTS

Modern Methods of Polymer Characterization

CHAPTER

1

CHARACTERIZATION OF COMPLEX POLYMERS BY SIZE EXCLUSION CHROMATOGRAPHY AND HIGH-PERFORMANCE LIQUID CHROMATOGRAPHY

STEPHEN T. BALKE

Department of Chemical Engineering and Applied Chemistry
University of Toronto
Toronto, Ontario, Canada

Modern Methods of Polymer Characterization, Edited by Howard G. Barth and Jimmy W. Mays
ISBN 0-471-82814-9

1. INTRODUCTION

Size exclusion chromatography (SEC), also referred to as gel permeation chromatography (GPC), is a well-recognized technique for the determination of polymer molecular weight distributions. However, it is now recognized that some polymers are much more difficult to analyze by SEC than polystyrene and other linear homopolymers. The topic of this chapter is analysis of these "complex" polymers by both SEC, and by the closely related method of high-performance liquid chromatography (HPLC).

There have been a number of revolutionary changes in chromatographic technology that have occurred in recent years (1–14). In particular

- Size exclusion chromatography is no longer one simple technique. Many options are available in both how the analysis is conducted and how the data are interpreted.
- Size exclusion chromatography is often now used for analysis of properties other than molecular weight distribution, for example, copolymer composition.
- Microprocessors are now commonly installed in SEC instruments to control conditions and to process data.
- New column technology now permits analyses to be conducted in 20 min rather than 3 h. However, strong reliance is placed upon strict control of chromatographic conditions by microprocessors.
- "Chemometrics," the application of computer implemented mathematics, particularly statistics, in chemistry, is now a major, well-recognized area of research, which is strongly affecting data processing.

- High-performance liquid chromatography, a method traditionally used for small molecule analysis, is now being proposed as an alternative, which can be superior to SEC for polymer analysis in some cases.

Selecting analytical options that provide needed results for complex polymers is a critical need in this dynamic situation. The objective of this chapter is to assist the analyst in this selection.

In the next section, a brief description of SEC and HPLC is provided. Following this description, polymer properties are reviewed and classified. This provides the basis for defining what is meant by "monodisperse," "polydisperse," and "complex" polymers in an SEC–HPLC context. After these definitions are formulated, a plan for systematically examining the alternative methods is presented.

2. DESCRIPTION OF SEC AND HPLC

Figure 1 shows a schematic of a typical instrument that is used for both SEC and HPLC. The sample is dissolved in a solvent, injected into a flowing stream of the same solvent at the top of the chromatographic columns, and carried through the columns at a constant flow rate. When the instrument is used for SEC, the columns are filled with a porous packing (e.g., porous glass or a cross-linked gel). The smaller molecules in the sample go into and out of more pores in the packing than do the larger molecules and exit later. The fractionation is on the basis of molecular size in solution with the largest molecules leaving the columns first followed by the smaller molecules. Thus polymer exits before monomer, and larger molecules in the polymer exit before the smaller ones. Upon leaving the columns, the sample molecules enter one or more sequentially attached detectors. A typical detector is a fixed wavelength ultraviolet (UV)–visible (vis) spectrophotometer. However, as we discuss later in this chapter, many different detectors may be used.

High-performance liquid chromatography utilizes basically the same instrumentation as SEC. However, HPLC utilizes columns that are intended to encourage adsorption and partition mechanisms. Size exclusion is considered not to be present. Currently there is much debate regarding the actual mechanism of separation. A generalized model allowing for the possibility of simultaneous adsorption, partition, and size exclusion has been published (15). In HPLC, the columns used most frequently employ packing of small porosity with a hydrophobic coating. They are termed "reversed-phase" columns because the packings are less polar than the mobile phase (10). Because of the different mechanistic basis for the separation, the solvents used are usually different than employed in SEC. Binary, ternary, and even quaternary

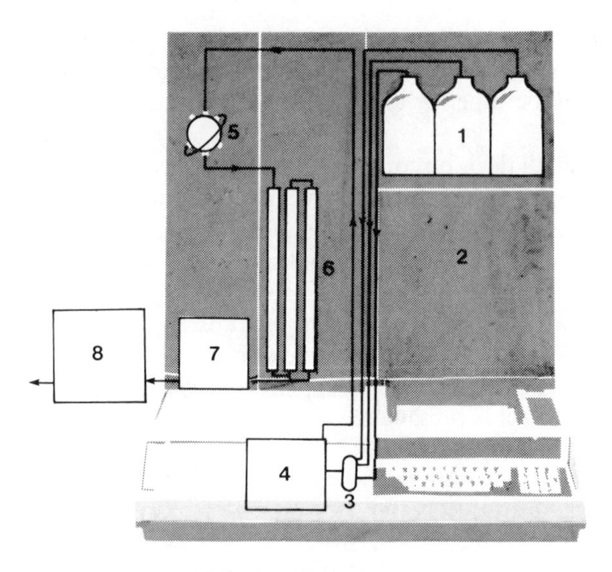

1. SOLVENT RESERVOIR
2. MICROPROCESSOR CONTROL MODULE
3. MIXING VALVE
4. PUMP
5. INJECTION VALVE
6. CHROMATOGRAPHY COLUMNS
7,8. DETECTORS

Figure 1. A typical size exclusion chromatograph (SEC)–high performance liquid chromatograph (HPLC).

miscible mixtures of organic solvents and water are sometimes employed. Furthermore, the ratio of these solvents may be constant during the run (isocratic elution) or may be programmed to vary with time (gradient elution). Until recently, HPLC was used only to analyze small molecules.

3. POLYMER MOLECULAR PROPERTIES

Polymers, especially those used in industry, are highly complex multicomponent materials. Properties typically considered important to polymer performance in products include the following (16):

Residual small molecules (e.g., monomers and catalysts)
Total weight of microgel or molecular aggregates present
Total weight of polymer present

Molecular weight

Tacticity (e.g., isotactic, syndiotactic, and atactic)

Composition

Sequence length (In a copolymer, this is the number of one type of monomer unit linked together in a chain, uninterrupted by a monomer unit of a different type.)

Sequence frequency (e.g., alternating, random, or block)

Branch length

Branch frequency

The residual small molecule content, total weight of microgel or aggregates present, total weight of polymer present, and properties that depend on only these measures are termed *simple properties.* They include conversion of monomer to polymer in a polymerization, monomer composition in a copolymer, and average copolymer composition.

Conversion (symbolized by X) can be obtained from the total weight of polymer and the weight of the sample [assuming the sample contains only polymer and monomer(s)].

$$X = \text{weight of polymer in sample/weight of sample} \quad (1)$$

Alternatively, it can be obtained from measuring monomer concentrations

$$X = (W_0 - W)/W_0 \quad (2)$$

where W_0 = weight of monomer in sample before reaction
W = present weight of monomer

For a copolymer, monomer composition can be obtained from

$$f_1 = [\mathrm{A}]/([\mathrm{A}] + [\mathrm{B}]) \quad (3)$$

where f_1 = mole fraction of monomer A in the monomer mixture
[A] = moles of A per liter of reaction mixture
[B] = moles of B per liter of reaction mixture

For other properties, such as molecular weight, different molecules in the same polymer will have different values of the property. Therefore, for example, we generally speak of a "distribution of molecular weights." In plotting such a distribution in its "differential form," the abscissa of the distribution is molecular weight and the ordinate is $W(M)$, where $W(M)dM$ is the weight of polymer with molecular weight between M and $M + dM$. Often weight fractions are used instead of weights. Each $W(M)$ is then divided by the area to give $W_N(M)$ and $W_N(M)dM$ is the weight fraction of polymer with molecular weight between M and $M + dM$. The term $W_N(M)$ is called the normalized value of $W(M)$.

Many other properties are also present as distributions. For example, we can consider a distribution of compositions and a distribution of sequence lengths for a copolymer. In this chapter, all such properties are termed *distributed properties*. This is in contrast to the *simple properties* mentioned earlier.

Often we cannot measure the total property distribution of a distributed property, but must be content with an average. For molecular weight distributions the first average is termed number-average molecular weight and is defined by

$$\bar{M}_1 = \bar{M}_n = \int_0^\infty W(M)dM \bigg/ \int_0^\infty \frac{W(M)}{M} dM \tag{4}$$

The second average is termed weight-average molecular weight and is given by

$$\bar{M}_2 = \bar{M}_w = \int_0^\infty MW(M)dM \bigg/ \int_0^\infty W(M)dM \tag{5}$$

The ratio of $\bar{M}_w/\bar{M}_n$ is a measure of the breadth of the molecular weight distribution and is termed the polydispersity.

In general, the qth average is defined by

$$\bar{M}_q = \frac{\int_0^\infty M^{q-1} W(M)dM}{\int_0^\infty M^{q-2} W(M)dM} \tag{6}$$

Averages can be similarly defined for other distributed polymer properties, including copolymer composition. However, an exception to this way of defining averages is what is commonly referred to as "average copolymer composition." It is defined as the weight of monomer A consumed in a copolymerization (equal to the weight of A in the polymer) divided by the weight of polymer present. It can be expressed entirely in terms of simple properties

$$\bar{W}_1 = [w_{10} - (1 - X)w_1]/X \tag{7}$$

where w_1 = weight fraction of monomer 1 in the monomer mixture
w_{10} = initial weight fraction of monomer 1 in the monomer mixture

One property of a polymer that is not really a molecular weight average, but which is often used in the same way, is intrinsic viscosity. Intrinsic viscosity $[\eta]$ is defined in terms of solution viscosities extrapolated to zero concentration (16)

$$\lim_{c \to 0} \frac{1}{c}\left(\frac{\eta - \eta_0}{\eta_0}\right) = [\eta] \tag{8}$$

where c = concentration of polymer in the solution
η = viscosity of the polymer solution
η_0 = viscosity of the pure solvent

For a mixture of n polymers each containing only one molecular weight, the intrinsic viscosity is given by

$$\overline{[\eta]} = \sum_{i=1}^{n} [\eta]_i w_i \tag{9}$$

For each single molecular weight sample, the Mark–Houwink–Sakurada equation can be used to relate intrinsic viscosity to molecular weight

$$[\eta] = KM^a \tag{10}$$

where K and a are the Mark–Houwink constants.

Both K and a depend on temperature, solvent, and polymer type (16).

Combining Eqs. 9 and 10 provides the expression for intrinsic viscosity of a linear homopolymer containing many molecular weights

$$\overline{[\eta]} = K \int_0^{\infty} M^a W_N(M) dM \tag{11}$$

Comparing Eqs. 6 and 11 we can see how intrinsic viscosity can be considered as a type of molecular weight average.

4. DEFINING POLYMER COMPLEXITY

The breadth of the distribution of a *distributed property* determines whether the polymer is homogeneous or heterogeneous with respect to that property.

For example, in the case of molecular weight, if the molecular weight distribution is "broad" then many different molecular weights exist in the polymer and the polymer is considered heterogeneous with respect to molecular weight. A polymer homogeneous with respect to molecular weight has less molecular weight variety and therefore a narrow molecular weight distribution.

A polymer with more than one type of distributed property can be simultaneously heterogeneous in all properties. For example, a linear copolymer may have broad distributions (a wide variety) of molecular weight, composition, and sequence length. If averages of each of these distributions could be determined, then each would have a large value of polydispersity. Note that the terms "heterogeneous," "broad distribution," and "high polydispersity" can be used to indicate a wide variety of values for the property specified. Also, it should be realized that a copolymer is not necessarily heterogeneous in all of its distributed properties. Other alternatives are possible. For example, it may be simultaneously heterogeneous with respect to one property (e.g., molecular weight) and homogeneous with respect to all others (e.g., composition and sequence length).

Using these concepts, we can define monodisperse, polydisperse, and complex polymers in the context of SEC and HPLC analysis.

Monodisperse polymers are homogeneous in all distributed properties. A polystyrene sample that has a narrow molecular weight distribution is the most common example of a monodisperse polymer.

Polydisperse polymers are heterogeneous in one, and only one, distributed property. A polystyrene sample that has a broad molecular weight distribution is a frequently seen example.

Complex polymers are heterogeneous in more than one distributed property. Linear copolymers and branched polymers that have broad distributions in all of their distributed properties are common examples.

Note that *simple properties* do not enter into any of these definitions. It is variety in the *distributed properties* that cause the analytical difficulties and therefore are used to distinguish the different types of polymers. Furthermore, a vast amount of polymer literature recognizes only molecular weight as a property that can be distributed in a polymer. This is partly because we know so little regarding other property distributions. Thus, we find the terms monodisperse and polydisperse usually are employed with only molecular weight in mind. In this chapter the generality of these terms is emphasized. However, molecular weight remains the usual example of a distributed property because it has been the central interest of methods development.

Often we examine only a small portion of the whole polymer. This portion is termed a "fraction" when it originates from our attempt to separate the original polymer into portions based upon the value of a specific property. For example, a polymer may be separated according to molecular weight and so portions containing "one" (or, more realistically, nearly one) molecular weight would be examined. In examining polymer distributed properties we must then specify whether we are dealing with the whole polymer or a single fraction. A polymer may be very homogeneous with respect to molecular weight within a fraction but may be very heterogeneous if the whole polymer is considered. Property values of fractions are termed *local* property values. This is in contrast to those of the total polymer which are termed *whole polymer* values. Thus, when these averages are obtained for fractions they are called *local averages* and, again using molecular weight as the example, the ratio of *local* weight-average molecular weight to *local* number-average molecular weight, is the *local polydispersity*.

The terms monodisperse, polydisperse, and complex can be applied to a whole polymer sample or to a polymer fraction. However, unless otherwise specified, the terms refer to a whole polymer.

5. ORGANIZATION

The remainder of this chapter deals with the problem of selection of SEC–HPLC methods for analysis of complex polymers. Experimental methods and then interpretation methods are described in turn. In each section both *simple property* analysis and *distributed property* analysis are considered.

Experimental methods are divided into "fractionation" and "detection." The objective of fractionation is to separate the polymer according to the values of the property of interest. For example, if molecular weight is of interest then the fractionation is meant to be with respect to molecular weight so that each fraction collected contains a different, unique molecular weight.

Detection is intended to produce a response (e.g., an electrical voltage), which provides the value of the property present in a fraction or in the whole polymer (e.g., the value of the molecular weight) and/or indicates the amount of polymer present with a particular property value (e.g., the concentration of polymer with a specific molecular weight). This chapter deals only with detectors that are directly connected to liquid chromatographs (termed online detectors). The solution present in the detector cell at any time is then a fraction of the whole polymer and the detector is monitoring *local* values. Offline detectors are those not connected to a column and measure whatever is placed in their cells (generally whole polymer). Offline detectors and methods of analysis are examined in other chapters in this book.

Traditionally, the value of a property present in a fraction was determined by reference to an empirical correlation of the value of that property in well-characterized standards versus the retention time of the standards in the chromatograph (calibration curve). Online detectors were only used to establish concentration. Today we find both calibration and online detector technology in use for property value determination.

Interpretation alternatives are classified under the three objectives of the computations:

1. Determination of the quantity of polymer present with a given property value (e.g., the quantity of polymer with a specific molecular weight).
2. Determination of the specific value of the property (e.g., the value of the molecular weight).
3. Assessment and computational enhancement of inadequate fractionation.

6. EXPERIMENTAL METHODS

6.1. Fractionation

6.1.1. Separation of Small Molecules from Polymers

Size exclusion chromatography can readily be used to separate monomers and other small molecules from complex polymers. Figure 2 shows a typical SEC chromatogram. The polymer apppears as the broad peak at early retention times. The small molecules are the sharper peaks seen at later retention times.

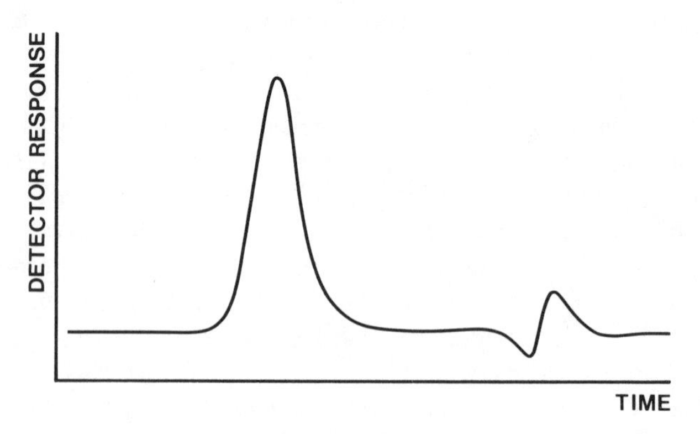

Figure 2. Typical SEC output.

It is much more difficult to separate small molecules from each other by SEC. Separation is often insufficient and solvent impurity peaks frequently interfere with those of other small molecules. Thus, HPLC is frequently used for such analysis (1, 2, 17, 18).

Solvent mixtures used in HPLC often do not dissolve polymers. One approach is to use SEC to separate the polymer from the small molecules and then to inject only collected eluent from the SEC that contains the small molecules into the HPLC for their separation (17). This is shown in Figure 3. The same instrument can be used for both SEC and HPLC but with different columns each time. A typical HPLC output is shown in Figure 4. In this case two fixed wavelength UV detectors were used in series so that two sets of curves (one pointing up [254 nm] and the other pointing down [200 nm] because of the setting of the recorder polarity) were obtained. Note that each peak represents only one chemical species and is therefore much narrower than the usual polymer peak obtained in SEC.

Retention volume, the product of mobile phase flow rate and retention time, is generally used instead of retention time alone because results at different flow rates can be compared. With a few exceptions, this convention is followed in the remainder of this chapter.

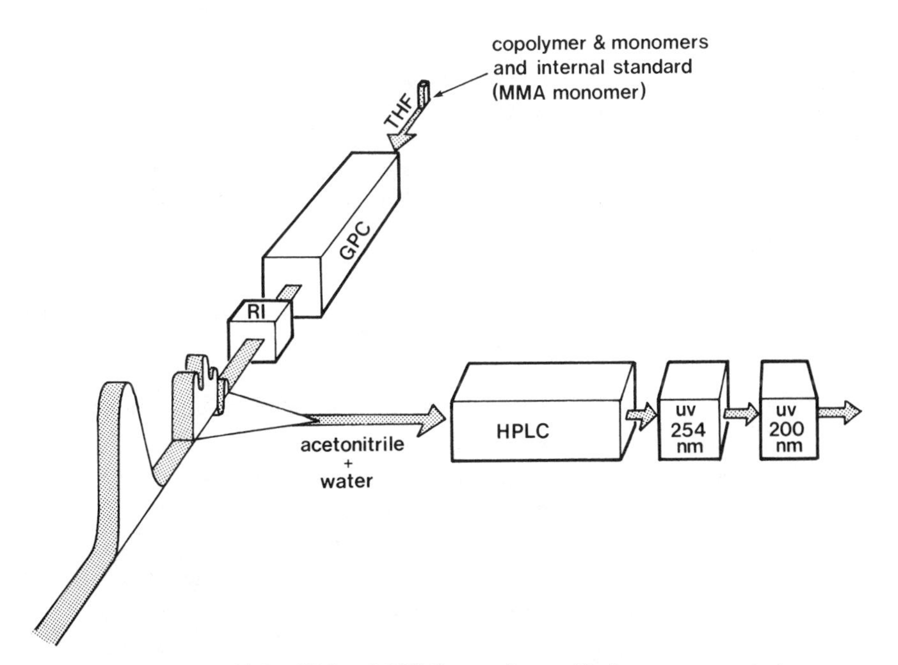

Figure 3. Combining SEC and HPLC to analyze residual monomer content.

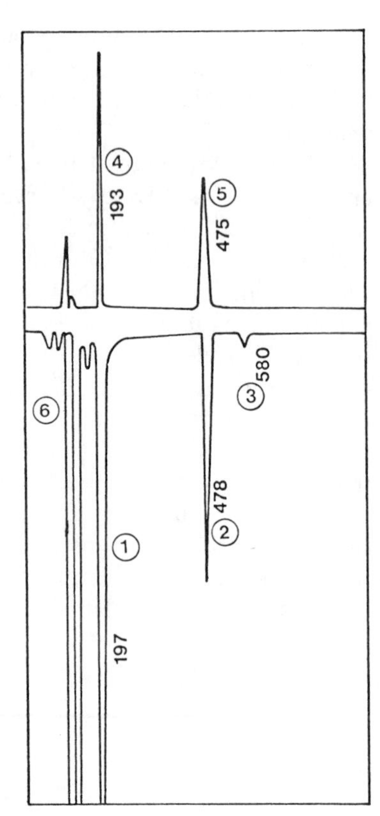

Figure 4. HPLC analysis of monomers showing retention times (s) and identification: 1, 2, and 3 represent methyl methacrylate, styrene, and *n*-butyl methacrylate at 200 nm; 4 and 5 represent methyl methacrylate and styrene at 254 nm; 6 are solvent impurity peaks. Reprinted with permission from S. T. Balke and R. D. Patel, "High-Conversion Polymerization Kinetic Modeling Utilizing Gel Permeation Chromatography," *ACS Symp. Ser.*, **138**, 149 (1980). Copyright © 1980 American Chemical Society.

Sources of error in utilizing SEC and HPLC for separation of monomers from polymer can be summarized as follows:

1. Polymer plugging of columns can be troublesome, particularly for HPLC.
2. Fractionation by SEC may be inadequate because of adsorption of monomer onto the packing and/or because of a low-molecular-weight tail on the polymer molecular weight distribution.
3. Adsorption of monomer onto the polymer can cause fractionation error to vary with the amount of monomer present (19).
4. Procedures involving collection of fractions may be too slow when many samples must be analyzed.

6.1.2. Separation of Microgel and Aggregates from Other Components

Size exclusion chromatography has been used to separate microgel and aggregates from other polymer and from low-molecular-weight species present in a sample (20–22). This application of SEC is similar to those described in Section 6.1.1.

6.1.3. Fractionation of Polymers by Molecular Size

Size exclusion chromatography is most frequently used to separate polymer molecules from each other according to their molecular size in solution. Janca (23) reviewed the wide variety of polymers analyzed. This is a much more difficult task than merely separating polymer from small molecules. The mechanism described above still applies, that is, the larger molecules exit first followed by smaller ones. Also, although sometimes effectiveness of the fractionation is more uncertain for complex polymers than for simple polymers, it is applicable to complex polymers.

The difficulty for all types of polymers analyzed is that now there are many more effects that can significantly influence the results (24).

1. Axial Dispersion. Axial mixing effects in the column mean that, even if the sample contained only one molecular size in solution, it will exit over a range of retention times. It cannot exit as a pulse at one retention time because some of the molecules are displaced from their neighbors by axial mixing. If a sample only contains a few molecular sizes and these sizes are very different from each other, then separate peaks can be seen and the situation is relatively easy to interpret. However, when a sample is injected that contains thousands of consecutive and different molecular sizes, at any retention time there will be a mixture of polymer molecules. These will include those molecules which are expected to exit at that time and many others that are displaced by the mixing effects. Figure 5 shows this effect. The chromatogram is an "envelope" concealing a multitude of unseen, overlapping peaks. This subject is discussed further in Section 7.4.2. However, at this point it should be noted that the symbol $W(v)$ represents the envelope heights when perfect resolution is assumed. That is, each spread curve is assumed to be a pulse. When perfect resolution is not assumed, envelope heights are given by the function $F(v)$ and the individual single molecular size curves by the shape function $G(v)$. A subscript "N" on any of these three symbols means that the curve has been normalized by dividing each height by its area. The area of any normalized curve is unity. With regard to complex polymers, there is some indication that polymer complexity is not a major factor in determining the effect of axial dispersion (25).

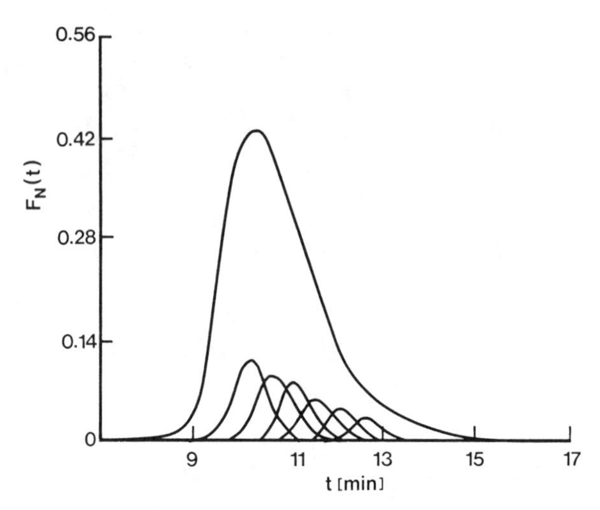

Figure 5. Imperfect resolution in SEC. Reprinted with permission from Ref. 7. Copyright © 1984 Elsevier Science Publishers.

2. *Adsorption and Partition Effects.* Like small molecules, polymers can also be subjected to other separation mechanisms. These effects can vary with all of the molecular properties of the polymer and effectively scramble the desired size fractionation. Furthermore, the greater the variety of polymer molecules present (i.e., as in complex polymers), the higher the probability of such nonexclusion effects and the more difficulty there is in discerning them.

3. *Concentration Effects.* If too many polymer molecules are present then the chromatographic columns become overloaded and molecules bypass pores that they should enter. There is also a more subtle type of concentration effect: a decrease in the hydrodynamic volume (the size) of polymer molecules, because at higher concentrations they are "crowded' by the presence of other macromolecules. The result of this effect for linear homopolymers, for example, is that the retention time of molecules in narrow molecular weight distribution standards can be too high relative to the retention time of polymer molecules of the same molecular weight in a broad molecular weight distribution sample (26).

4. *Shear and Thermal Degradation.* Large polymer molecules may be degraded to smaller ones by the shear stresses experienced as they pass through chromatographic columns (27). When SEC is operated at high temperatures (e.g., 150 °C), so that polymers such as polyethylene can be dissolved, then thermal degradation may result (28).

5. *Inadequate Dissolution.* Sometimes, molecular aggregates are present

in the sample. These aggregates can be solubilized only after special dissolution procedures are employed (e.g., 22, 29).

6. Mobile-Phase Flow Rate Variations. New, low-volume SEC columns demand extremely precise flow rate control. In a recent process control study, Wu et al. (30) found that a variation of only 0.08 mL/min resulted in a 100% deviation in $\bar{M}_w$.

6.1.4. Other Fractionation Approaches

For many years detection has been emphasized over fractionation (14, 30). Copolymers are separated according to molecular size in solution even when a separation according to composition is desired (31). Heavy reliance is placed upon detection to reveal copolymer composition distribution information. The inadequacy of this approach for analysis of complex polymers can be explained with reference to Figure 6. Figure 6*a* shows that when a linear

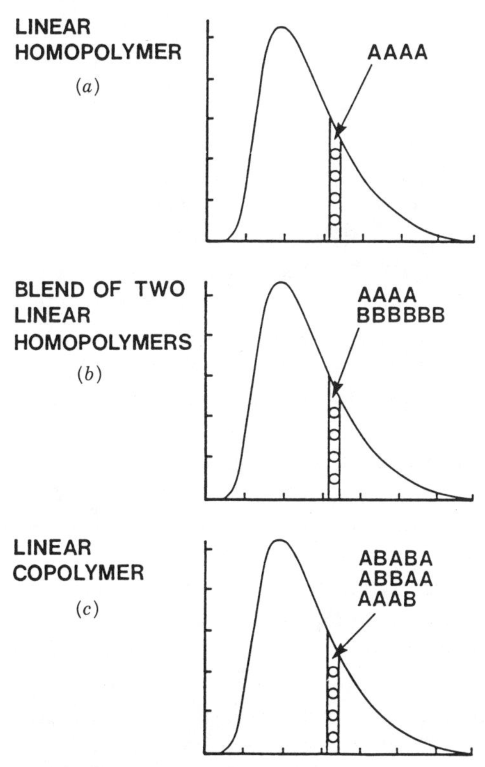

Figure 6. SEC fractionation showing detector contents at a given elution volume. (*a*) linear homopolymer, (*b*) blend of two linear homopolymers, (*c*) linear copolymer.

homopolymer heterogeneous only in molecular weight (i.e., a polydisperse polymer) is fractionated by SEC then only one molecular weight is present at each retention volume (assuming no significant axial dispersion effects or other nonideal fractionation effects). The polymer at each retention volume is monodisperse. Figure 6*b* shows that if a blend of two linear homopolymers is so fractionated then at most two different molecular weights can be present at each retention volume. However, when a linear copolymer (i.e., a complex polymer) is fractionated (see Figure 6*c*), many different combinations of molecular weight, composition, and sequence length can combine to give the same molecular size in solution. The fractionation with respect to molecular size may be completely ineffective in assisting the analysis with respect to composition. A complex polymer can exist at each retention volume.

There has been increasing recognition of the limitations of polymer size fractionation and new emphasis placed upon obtaining fractionation with respect to the property of interest (17). Now HPLC is being used in an attempt to fractionate homopolymers with respect to molecular weight and copolymers with respect to composition (5, 7, 32–43). Size exclusion chromatography instruments are being combined in a method known as "orthogonal chromatography" to obtain a cross-fractionation of copolymers (17, 32, 44–46). These developments will be discussed below.

Methods of using HPLC directly to fractionate polymers with respect to a specific property have shown some notable successes. However, such approaches must deal with interference from size exclusion effects and interference from simultaneous property distributions (for complex polymers). For example, when HPLC is used, it is assumed that the polymer will be separated according to a dominant adsorption–partition mechanism. If the pores in the packing are sufficiently large, however, there will also be a significant size exclusion mechanism participating in the separation. This means that a molecule may elute at a given retention time either because of size exclusion and/or because of adsorption or partitioning. The mechanisms do not necessarily act synergistically. The small molecule may find the surface much less attractive than does a large molecule even if we are dealing with homopolymers. Thus, the fractionation obtained will depend on the molecular weight distribution in an unpredictable manner.

When we consider fractionation of complex polymers by HPLC the situation is even more uncertain. Many different property distributions can be present (e.g., molecular weight, composition, and sequence length for a copolymer). In SEC, even fractionation with respect to molecular size is difficult to interpret in terms of any one of these distributions. Fractionation by HPLC, where different properties appeal differently to different mechanisms, provides even more challenge. Experimental attempts to cause one mechanism to dominate and one property to react to the mechanism have centered about using small pore packings and gradient elution chromatography. Small pore columns can drastically reduce the amount of surface area available for adsorption–partition effects. Gradient operation provides more flexibility, but does not overcome the basic difficulties. It is an accomplishment to show separation of narrow, characterized standards of a polymer. However, it is much more difficult to develop a technique that can reliably be used for polydisperse unknowns, where the individual peaks of each molecular weight or composition cannot be discerned.

Orthogonal chromatography (OC) holds promise for difficult fractionation problems. In OC, one SEC system is connected so that its mobile phase exits into the injection valve of a second SEC system. The first SEC system is operated in a conventional way. The second one, however, utilizes HPLC-type mechanisms by employing a weaker solvent mixture as the mobile phase.

For example, Figure 7 shows the separation of a complex polymer, styrene *n*-butyl methacrylate copolymer by composition. Size exclusion chromatography 1 was run with a good solvent for the polymer (tetrahydrofuran, THF). However, after the sample began to exit from SEC 1, fractionated with respect to molecular size in solution, the flow in SEC 1 was stopped and an injection made into SEC 2. Size exclusion chromatography 2 was run with a mixture of THF and *n*-heptane as the mobile phase. As shown in Figure 8, when the molecules within the injection valve of SEC 2 were in the presence of this solvent mixture, the styrene-rich molecules shrink in size as compared to the *n*-butyl methacrylate rich molecules because of the presence of the heptane (a nonsolvent for polystyrene). Thus, as the sample passed through the columns of SEC 2 the smaller, styrene-rich molecules, permeated more pores than did their *n*-butyl methacrylate rich neighbors and eluted relatively slower. Furthermore, the styrene-rich molecules were exposed to more surface area of the packing than were the *n*-butyl methacrylate rich molecules because of the presence of the heptane and had more opportunity to adsorb. The result was a synergistic combination of size exclusion, adsorption, and partition. The method could be demonstrated for blends of different polymers. That is, OC could readily be used to separate blends according to composition. However, in fractionation of a single copolymer according to composition, the copolymer composition distributions obtained did not agree with theoretically predicted results.

The main limitations of the OC technique are

1. Participation of SEC 1 solvent in the separation obtained in SEC 2: The pulse of solvent injected with the sample actually creates a solvent gradient in the second SEC, which complicates the calibration problem. This may be solved by the use of diode array UV–vis spectrophotometers to provide a calibration curve with each sample injected.

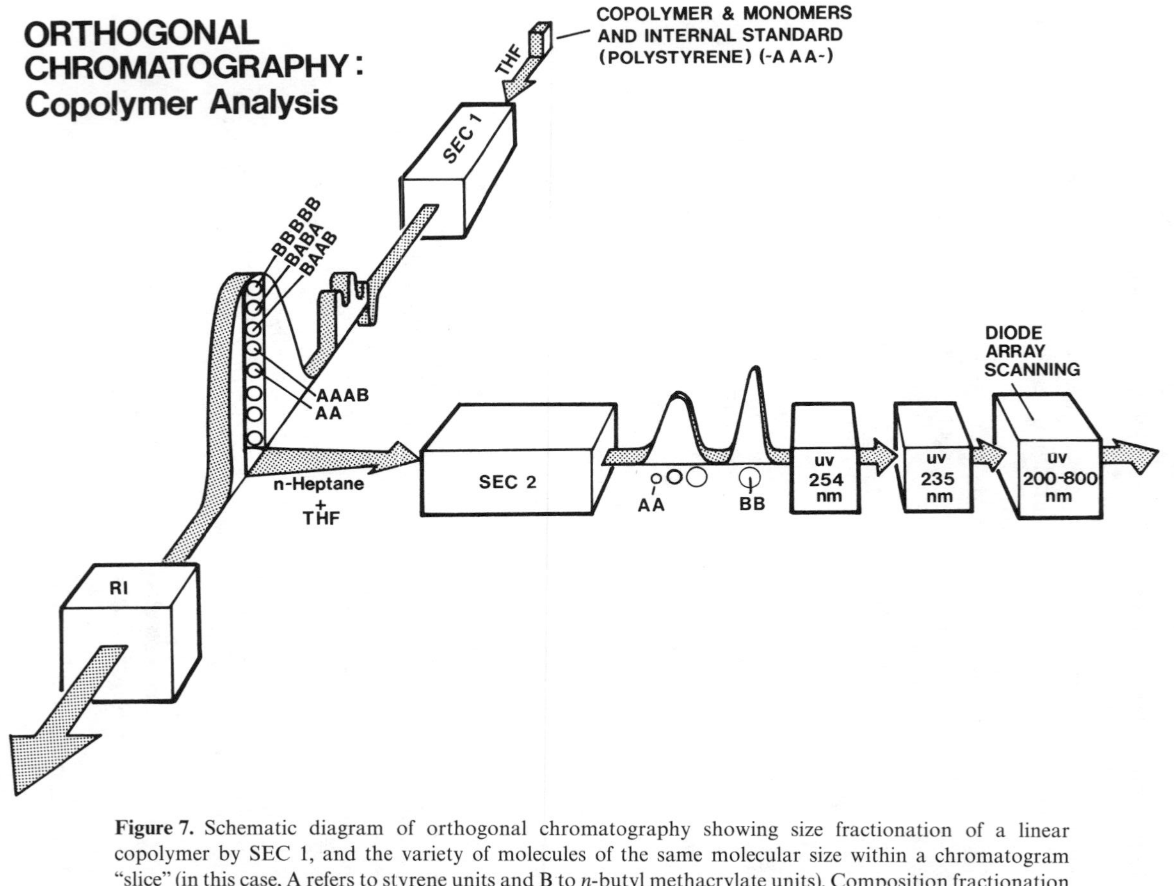

Figure 7. Schematic diagram of orthogonal chromatography showing size fractionation of a linear copolymer by SEC 1, and the variety of molecules of the same molecular size within a chromatogram "slice" (in this case, A refers to styrene units and B to *n*-butyl methacrylate units). Composition fractionation is achieved by SEC 2. Reprinted with permission from S. T. Balke and R. D. Patel, "Orthogonal Chromatography Polymer Cross-Fractionation by Coupled Gel Permeation Chromatography," *Adv. Chem. Ser.*, **203**, 281 (1983). Copyright © 1983 American Chemical Society.

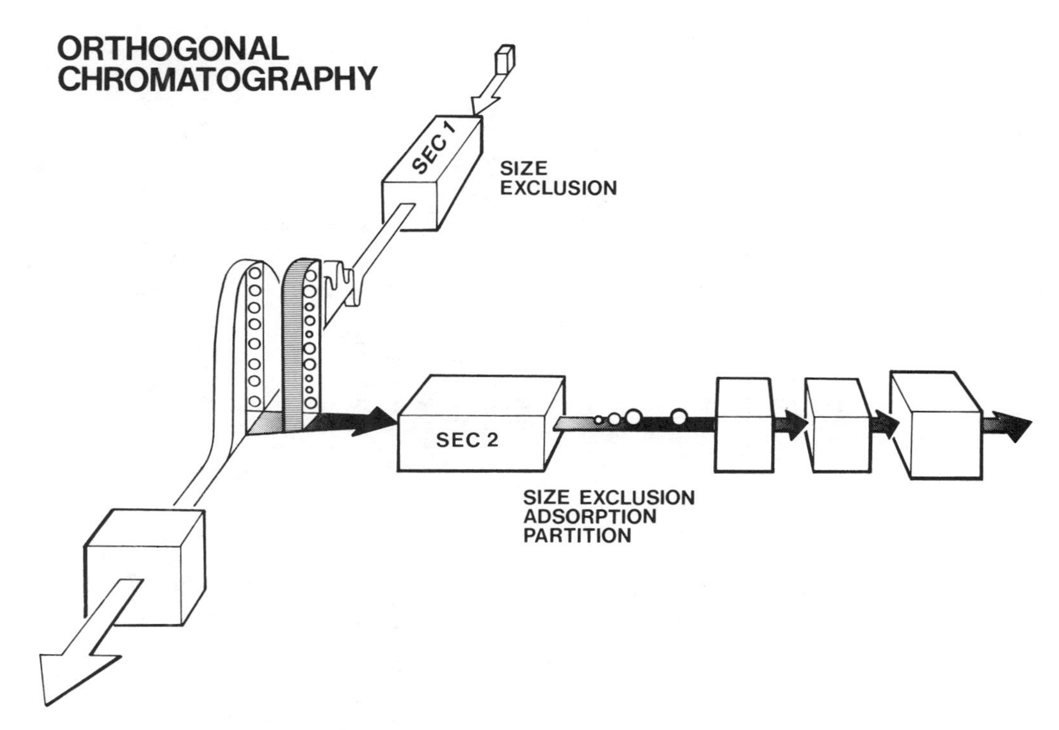

Figure 8. Schematic diagram of orthogonal chromatography showing the separation mechanisms involved. Reprinted with permission from S. T. Balke and R. D. Patel, "Orthogonal Chromatography Polymer Cross-Fractionation by Coupled Gel Permeation Chromatography," *Adv. Chem. Ser.*, **203**, 281 (1983). Copyright © 1983 American Chemical Society.

2. Sequence length effects: Different sequence lengths in the copolymer may affect both fractionation and detection. Detection methods must be developed to examine this aspect.
3. The method requires two SEC systems and generates much data. Automation is a practical necessity.

6.1.5. Recent Developments in Fractionation

Barth (47) reviewed "nonsize" exclusion effects, and reviews utilizing these effects in liquid chromatography have been published (5, 32, 33, 34). Glöckner and co-workers (48–50) published several papers where they achieved impressive separations by coupling an SEC system with an HPLC system operating in a gradient mode and employing packings of small pore size. They show evidence for a polymer precipitation mechanism and try to eliminate any size exclusion effects in the second chromatography by using small-pore packings. With high-resolution packings, they estimated that the surface area of the packings was sufficient to be responsible for the observed separation. In attempts at quantitative analysis, they were concerned with "incomplete" retention of portions of the sample eluting from the second chromatograph and with the effect of sequence length on the UV-detector response (48). Incomplete retention resulted in some of the polymer exiting as an excluded peak or as part of the solvent impurity peak. They suggested the use of a mass detector instead of UV. This has already been done by Mourey and co-workers (51, 52) using a detector, which has been widely advertised as a mass detector: the evaporative light scattering detector. However, as will be summarized below, this detector was found not really to be a mass detector in practice.

Other notable attempts at nonsize exclusion fractionation include the work of Mori and Uno (53), as well as that of Mirabella (54). Mori and co-worker (53) published many papers on the use of gradients to separate copolymers according to composition using gradient liquid–adsorption chromatography. Like other single chromatographic approaches, they assume no molecular weight effect on the fractionation. Although this is true with the polymers that they had examined, it is unlikely to be true in general and is a difficult assumption to prove with complex, polydisperse polymers.

The temperature rising elution fractionation (TREF) method of Mirabella (54) is a method of precipitating polymer by cooling on a column packing according to the degree of crystallization. The polymer fractions are then stripped off by a flowing solvent at different temperatures. This method also does not depend on a molecular weight effect and evidence for this assumption has been shown for polyethylene.

6.2. Detection

6.2.1. *Concentration Detection of Polymers or Small Molecules*

Although many types of detectors (11, 55) may be used to measure concentration of polymers in solution, differential refractometers and UV–vis spectrophotometers are the two main types of detectors employed in both SEC and HPLC.

In the common deflection-type refractometer, light is focused on a single cell containing a reference solvent side and a sample side. The difference in refractive index created by the presence of polymer in the sample side causes the beam to deflect. The change in beam position creates the output signal. The change in refractive index is considered proportional to the polymer concentration

$$\Delta n = kc \tag{12}$$

where Δn = difference in refractive index between the solution in the detector cell and that of the pure solvent
c = concentration of polymer in the solution in the detector cell
k = proportionality constant

The differential refractometer is applicable to many polymers. However, it is generally not as sensitive to concentration as other detectors. Also, it is sensitive to undesirable variables such as temperature and pressure fluctuations.

For complex polymers, the major complication is that the total refractive index change may be caused by many properties of the polymer. For example, for a copolymer, both total concentration of polymer and composition are contributing factors. This can be used to an advantage in determining copolymer composition (see Sections 7.2.1 and 7.3.1). However, other properties may contribute too for example, molecular weight, (56, 57). Sequence length and even polymer conformation are possible influential variables as well.

Ultraviolet–visible spectrophotometers utilize the measurement of UV or visible light to detect concentration. They are the most widely used detectors in HPLC and are rapidly gaining favor over the differential refractometer traditionally used in SEC. The three main types of UV–vis detectors are

1. Single wavelength: utilize filters and detect absorbance of light at one wavelength.
2. Continuously variable wavelength: utilize monochromators so that any

single wavelength can be selected; permit all wavelengths to be examined over a 5–10-min scanning period.

3. Diode-array UV–vis spectrophotometers: multiple wavelengths are incident on the sample and a diode array distinguishes the different wavelengths upon receiving the transmitted light; permit all wavelengths to be examined in as little as 0.01 s (58).

Figure 9 shows the difference in output between two online UV–vis spectrophotometers: (a) a diode-array instrument that provides a complete spectrum at each retention time (or, volume); and (b) a fixed-wavelength instrument that provides only one absorbance value at each retention time.

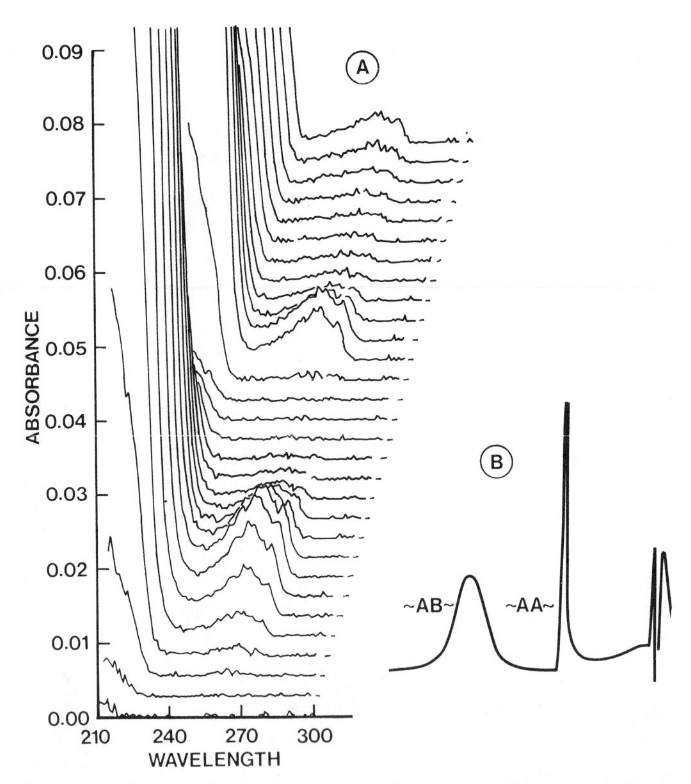

Figure 9. (*a*) Output from a liquid chromatograph using a diode-array UV–vis spectrophotometer. (*b*) Separation as it appears on a fixed wavelength detector. Reprinted with permission from S. T. Balke and R. D. Patel, "Orthogonal Chromatography Polymer Cross-Fractionation by Coupled Gel Permeation Chromatography," *Adv. Chem. Ser.*, **203**, 281 (1983). Copyright © 1983 American Chemical Society.

The absorbance A of UV light by the polymer solution at a specific wavelength is given by the Beer–Lambert law

$$A_\lambda = a_\lambda b\,c = \varepsilon_\lambda c \tag{13}$$

where a_λ = absorptivity (a constant at a specified wavelength)
b = cell path length
c = concentration of absorbing species (e.g., concentration of polymer in the cell)
ε_λ = the combined constant $a_\lambda \cdot b$

UV–vis spectrophotometers have high sensitivity to many polymers and, unlike refractometers, are not as sensitive to temperature and pressure variations. However, one factor that preserves the use of the refractometer in many cases is when UV is used, the polymer must contain chromophoric groups. Like the refractometer, for complex polymers the UV detector can be influenced by many polymer properties. The presence of more than one absorbing species, sequence length, and conformation can all contribute to A (57, 59).

6.2.2. *Weight-Average Molecular Weight*

Weight-average molecular weight can be determined directly by a light scattering detector (60). This type of detector includes a light source whose beam is incident on a sample solution. The intensity of the scattered light at a series of angles can be detected by a photomultiplier. The detector response is then related to $\bar{M}_w$ according to

$$K_o c/\bar{R}_\theta = (1/\bar{M}_w) + 2A_2 c + 3A_3 c^2 \tag{14}$$

where $\bar{R}_\theta$ = the reduced scattering intensity, a value depending on both angle of measurement and light intensity
K_o = the optical constant, a value incorporating wavelength of the light and, dn/dc, the differential refractive index increment of the solution
c = concentration
A_2 = second virial coefficient
A_3 = third virial coefficient

Since macromolecules are sufficiently large to cause interference effects and since these effects diminish as the angle of viewing approaches zero, light scattering measurements have traditionally been used offline and have utilized an extrapolation to both zero angle and zero concentration by using a method termed a "Zimm plot." The result is the whole polymer weight-average molecular weight. With new laser light sources, measurements at very low angles ($\sim 3°$ vs $\lesssim 30°$ in conventional instruments) is possible, and extrapolation to zero angle is unnecessary. The now popular LALLS (low-angle laser light scattering) detector used in SEC analysis is an example of this development (see Chapter 9) (60, 61).

It should be noted that since the differential refractive index increment enters the light scattering equation, this method is very difficult to apply to copolymers since composition variations will cause this term to vary. In contrast, branched homopolymers are well suited to this detector because the refractive index is generally considered unaffected by length or frequency of branching. This area is covered in more detail in Chapter 9.

6.2.3. Solution Viscosity

The Hagen Poiseuille equation forms the basis for measurement of solution viscosity η in liquid chromatography

$$\eta = \frac{\pi \Delta P R^4}{8LQ} \tag{15}$$

where ΔP = the pressure drop across the length of capillary
R = the radius of the capillary
L = the length of the capillary
Q = the volumetric flow rate of liquid

Eluent viscosity from SEC has been measured by two techniques (62, 63): measurement of the time required for a known volume of the eluent to pass through a capillary, and measurement of the pressure drop resulting as the eluent passes through a capillary. Although the first method is suitable for offline determinations, for online measurement it requires so much solution for one data point that significant chromatographic separation must be sacrificed. In contrast, measurement of pressure drop provides a continuous measurement of viscosity. Currently, there are three competing designs of online viscometers (64–68). All utilize measurement of the pressure drop across a capillary. As will be seen in Sections 7.2.3 and 7.3.3, solution viscosity values calculated from Eq. 15 are used along with data from a concentration detector (typically a differential refractometer) to provide intrinsic viscosity values at each retention volume. Reduction of undesired pressure fluctuations, caused mainly by the pumping system, is the primary aim of the various designs. The simplest design employs a single pressure transducer and relies upon system flow dampers and computer smoothing of the data (66). Commercially available instruments are reviewed in Chapter 7.

6.2.4. Other Developments

Detector development is a rapidly advancing area (1, 11, 55, 69). Especially relevant to polymers are densitometry (70), a mass detector based upon "spray drying" (71), nuclear magnetic resonance (72), and a multiangle laser light scattering instrument (73, 74). The use of UV spectrophotometry to measure sequence length in copolymers (57, 59) and interest in quantitative infrared (IR) measurement of polymers (75) are also important trends.

Mourey and Oppenheimer (76, 77) carried out an extensive evaluation of the evaporative light scattering detector. As mentioned in Section 6.1.5, these authors successfully applied it to detection of copolymers separated on the basis of composition by gradient adsorption chromatography. The detector was found to be sensitive to both particle density and refractive index. Thus, the output can vary with composition of the copolymer molecules. Recent work has corroborated and extended these findings (78).

7. QUANTITATIVE INTERPRETATION METHODS

7.1. Concentration Determination

7.1.1. Simple Properties

As mentioned previously, properties that are defined in terms of polymer and individual monomer quantities are all classified as *simple properties* because no property distribution elucidation is required. When chromatography is used to determine the concentration of polymer and/or individual monomers, assuming good chromatographic resolution, separate peaks will result when a differential refractive index detector or single wavelength UV–vis spectrophotometer is used. The total polymer and residual monomer values obtained can then be used to determine conversion (Eq. 1 or Eq. 2), copolymer monomer composition (Eq. 3), and average copolymer composition (Eq. 7).

Considering an online detector and using $G(v)$ to represent the response of the detector at any retention volume (it may be absorbance or differential refractive index) then, since this response is proportional to concentration,

$$G(v) = kc(v) \tag{16}$$

Integrating over all retention volumes to obtain the area A under this peak

$$A = \int_0^\infty G(v)dv = k\int_0^\infty c(v)dv \tag{17}$$

$$A = km_s \tag{18}$$

where m_s is the mass of total polymer or individual monomer injected. Thus, on a retention volume scale, the area under an individual peak is proportional to the mass injected. However, modern instruments provide area under a retention time scale rather than a retention volume scale. Since retention volume is the product of mobile phase flow rate and time, this area is proportional to both flow rate and mass injected. Errors in flow rate then directly affect the computed area: a given percent error in flow rate causes the same percentage error in area (6).

Peak height is sometimes used instead of area. This assumes that peak height is always proportional to peak area and that peak shape is invariant over the conditions used. The primary sources of error in determining these *simple properties* from individual, well-separated chromatographic peaks are

1. The detector may be responding to more than one property. For example, a UV detector can be sensitive to all chromophoric groups present, to sequence length of a copolymer, and to polymer conformation. In each case a different proportionality constant can relate absorbance to concentration. The total detector response is then a weighted sum of the individual concentrations, where the weighting factors and/or the concentrations are unknown. Even if only "pure solutes" are involved, where detector response should be unambiguously related to total concentration, a separated peak may be concealing other solutes in addition to the one expected. The problem of inadequate resolution is examined in Section 7.4.1.
2. The detector may show a nonlinear response to concentration. Either the linear region can be defined and values limited to that region be used, or a nonlinear response can be fit to a nonlinear equation to accommodate the result. The former approach is more easily implemented (6).
3. Erroneous fitting of the response versus concentration curve and imprecise values of the response constants can easily result. Even if only relative concentrations are of interest, the detector response constants must be determined to allow for different response to different solutes. This should be done by regression of response versus injected concentration data and the fit checked by an inspection of a plot of residuals (6, 7, 79). The use of Eq. 16 or 18, particularly the assumption of a zero intercept on the ordinate, must be shown to be valid. The approach of using one data point and the assumption that the response is linear to determine a response factor is a particularly error-prone method of quantitation (7).

4. The use of internal standard methods may sometimes provide worse results (80). Internal standards are reference solutes that are added to the solution injected to provide a "base point" for concentration measurement. However, because modern instruments are equipped to inject very reproducible quantities of solution and the use of internal standards requires that ratios of areas be used [thus enhancing error propagation effects (81, 82)], an internal standard method may be inferior to external calibration.

7.1.2. Distributed Properties

In this case, the concentration of each of a continuous series of different property values must be determined. Using molecular weight as an example, we must first assume that detector response is proportional to concentration, and fractionation is perfect with respect to molecular weight. As we have already seen, for complex polymers these assumptions are not easily satisfied. When they are valid, the relationship between weight of polymer of molecular weight M to $M + dM$ and detector response from retention volume v to $v + dv$ is

$$W(v)dv = -kW(M)dM \tag{19}$$

Or in terms of retention time

$$-W(M)dM = (1/k)W(v)dv \tag{20}$$

The area under the peak A is

$$A = \int_0^\infty W(v)dv = -k\int_\infty^0 W(M)dM = k\int_0^\infty W(M)dM \tag{21}$$

Thus, the normalized values of $W(M)$ and $W(v)$, $W_N(M)$, and $W_N(v)$ are related by

$$W_N(M) = -W_N(v)\,(dv/dM) \tag{22}$$

Now $W_N(M)dM$ is the weight fraction of molecules with molecular weight between M and $M + dM$.

7.1.3. Quantitative Measures of Distributed Properties in the SEC Chromatogram

7.1.3.1. Property Distribution Curves. The ordinate of the property distribution curve for molecular weight is determined by using Eq. 22 to determine

$W(M)$. To obtain the dv/dM for use in this equation and to define the abscissa of the distribution M, an SEC calibration curve is needed. This calibration curve relates the molecular weight of a sample to the retention volume of mobile phase required to elute it from the chromatograph. It assumes that a fractionation with respect to molecular weight has been obtained.

A frequently used calibration curve equation is

$$M(v) = D_1 \exp(-D_2 v) \tag{23}$$

For this "linear" calibration curve

$$-\frac{dv}{dM} = (D_2 M)^{-1} \tag{24}$$

and

$$W_N(M) = \frac{W_N(v)}{D_2 M} \tag{25}$$

In practice, the most sensitive and useful form of the molecular weight distribution is with log M plotted on the abscissa. The ordinate value then must obey the following equation:

$$W_N(v)dv = -W_N(\log M) d \log M \tag{26}$$

where $W_N(\log M) d \log M$ is the weight fraction of polymer with molecules of log M to log $M + d \log M$. The ordinate $W(\log M)$ is then calculated from

$$W_N(\log M) = -W_N(v)\frac{dv}{d \log M} \tag{27}$$

and for the calibration curve of Eq. 23

$$W_N(\log M) = \frac{2.303\, W_N(v)}{D_2} \tag{28}$$

7.1.3.2. Molecular Weight Averages and Intrinsic Viscosity. Using Eq. 19 with the defining equation for $\bar{M}_n$ (Eq. 4) we obtain

$$\bar{M}_n = \left[\int_0^\infty W(v)dv\right] \Big/ \left[\int_0^\infty \frac{W(v)}{M(v)} dv\right] \tag{29}$$

and with Eq. 5 for $\bar{M}_w$

$$\bar{M}_w = \left[\int_0^\infty M(v) W(v) dv \right] \Big/ \left[\int_0^\infty W(v) dv \right] \tag{30}$$

Similarly, considering intrinsic viscosity in terms of a perfectly resolved SEC chromatogram of a polydisperse sample, the weight fraction of each monodisperse polymer fraction is given by $W_N(v)dv$ and

$$[\eta] = \int_0^\infty [\eta](v) W_N(v) dv \tag{31}$$

and, finally, with Eqs. 11 and 19 for the intrinsic viscosity of a polydisperse polymer (e.g., a linear homopolymer)

$$[\eta] = \left[K \int_0^\infty M(v)^a W(v) dv \right] \Big/ \left[\int_0^\infty W(v) dv \right] \tag{32}$$

The chromatogram should always be superimposed on the calibration curve to see the range of elution of the sample compared to the range of separation of the chromatograph. To see what portion of the chromatogram is actually being used to calculate a specific molecular weight average or intrinsic viscosity, moment analysis plots can be used. Moment analysis plots are plots of the normalized value of the integrand for each of the averages versus retention time superimposed on the normalized chromatogram itself (6, 7, 83). Inspection of these plots reveals what primary heights are being used in calculation of each average. Figure 10 shows an example for $\bar{M}_n$, $\bar{M}_w$, and $\bar{M}_z$.

7.2. Determination of Local Polymer Property Values

7.2.1. Copolymer Composition at Each Retention Volume

As mentioned in Section 6.2.1, the output of a differential refractometer can reflect both polymer concentration and composition if a copolymer of different compositions is examined. The assumptions are (a) the total observed detector response is caused only by the sum of the responses from each different type of monomer unit present in the copolymer; (b) the response contribution of each type of monomer unit is proportional to its concentration (i.e., Eq. 12 applies); and (c) the proportionality constant for each type of monomer unit is different.

For example, for a complex polymer, styrene–methyl methacrylate copolymer, the styrene represents one type of monomer unit and the methyl

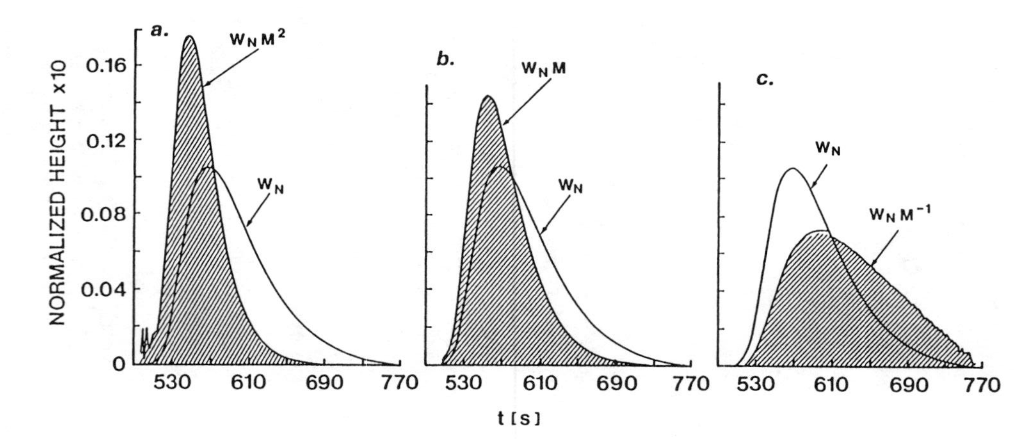

Figure 10. Moment analysis plots for (*a*) $\bar{M}_z$, (*b*) $\bar{M}_w$, and (*c*) $\bar{M}_n$. Reprinted with permission from Ref. 7. Copyright © 1984 Elsevier Scientific Publishers.

methacrylate represents the other. Then using subscripts 1 and 2 to indicate the respective two monomer types present,

$$\Delta n = \Delta n_1 + \Delta n_2 \tag{33}$$

and from Eq. 12

$$\Delta n = K_1 c_1 + K_2 c_2 \tag{34}$$

The same assumptions can be applied to UV–vis detectors via Eq. 13. The total absorbance from a UV–vis detector is then the sum of the contributions from each solute

$$A_\lambda = A_{1\lambda} + A_{2\lambda} \tag{35}$$

Applying Eq. 13,

$$A_\lambda = \varepsilon_{1\lambda} c_1 + \varepsilon_{2\lambda} c_2 \tag{36}$$

If the composition of a copolymer is to be determined then two equations are needed to determine the concentration of each of the two monomer types present in the polymer. Thus, measurement using two detectors in series (e.g., two UV–vis fixed wavelength detectors each set at a different wavelength) is necessary. Figure 11 shows this method. The two concentration detectors are connected sequentially at the exit of a liquid chromatograph. At each retention volume, each detector provides one equation in two unknowns (the concentration of each monomer unit in the copolymer exiting within a small increment of eluent volume encompassing the specified retention volume). This provides the average composition of the polymer at that retention volume.

One practical problem in implementing this method is determining how to find the "corresponding" retention volume in each chromatogram. Accounting for both time delay for transport of polymer between detectors and axial dispersion in detector cells must be considered. A more fundamental difficulty is that determination of the actual distribution of compositions within the polymer depends on the dependency of fractionation with respect to composition. As mentioned in Section 6.1.4, the molecular size fractionation accomplished by SEC is, not in general, synonymous with a fractionation according to composition. It may provide no simplification at all in terms of the variety of molecular compositions and sequence lengths and conformations viewed by the detector. The "composition distribution" determined is the variation of average polymer composition with molecular size and not a

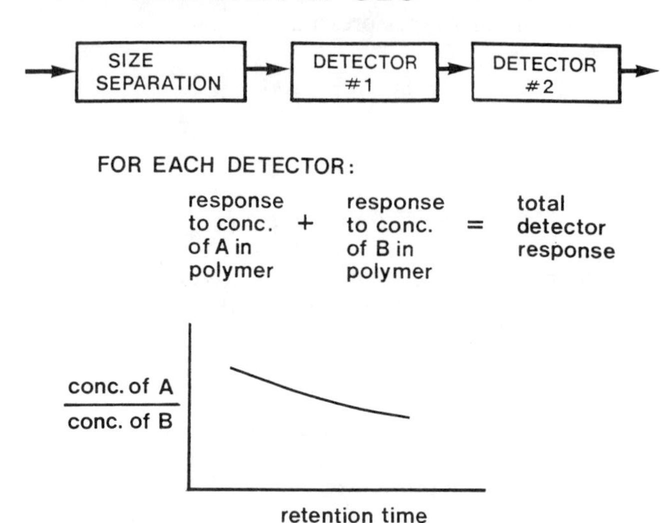

Figure 11. SEC using two detectors to measure copolymer composition. Reprinted with permission from S. T. Balke and R. D. Patel, "Orthogonal Chromatography Polymer Cross-Fractionation by Coupled Gel Permeation Chromatography," *Adv. Chem. Ser.*, **203**, 281 (1983). Copyright © 1983 American Chemical Society.

concentration versus molecular composition plot. When a diode array UV–vis spectrophotometer is used instead of two detectors in series, the problem of time delay between detectors is eliminated and many more absorbance–wavelength pairs are available, but the fundamental problem of inadequate fractionation remains. Fractionation with respect to composition is required. Both the OC and HPLC methods hold possibilities. However, sequence length and polymer conformation effects on both fractionation and detection have yet to be elucidated. Interpretation of UV spectra obtained at each retention time by diode-array instruments holds much potential (6).

García-Rubio and co-workers (84, 85) published on interpretation of copolymer spectra from offline spectrophotometers, Balke (32) emphasized that the method is very sensitive to impurities, such as residual monomer, which can contribute to the copolymer spectra. An interesting result of García-Rubio's work was the significant spectral contribution of end groups of the polymer resulting from the use of benzoyl peroxide as initiator (85).

Absorbance ratioing remains a very common practice in using multiple SEC detectors. Problems with this practice evident in HPLC have now been thoroughly investigated (6, 7, 86, 87) and should discourage the use of ratios in SEC.

In an analysis of propagated error in multiple detector SEC, García-Rubio (88) derived equations for the confidence limits on copolymer composition and total polymer concentration as obtained from a UV and an RI detector, molecular weight from an intrinsic viscosity detector with an RI detector, and $\bar{M}_w$ from LALLS with an RI detector. He showed some data indicating how the confidence limits explained apparent data discrepancies.

7.2.2. Molecular Size at Each Retention Volume: SEC Universal Calibration

For separation of any polymer by SEC, where the size exclusion mechanism is dominating, separation is based on the value of the molecular size parameter J defined by

$$J(v) = \overline{[\eta(v)]\bar{M}_n(v)} \tag{37}$$

For polymers that exhibit only one value of molecular weight M at a given retention volume $M(v) = \bar{M}(v)$, and $J(v)$ may be defined as follows:

$$J(v) = [\eta](v)M(v) \tag{38}$$

The term J is known as the molecular size in solution and as the "hydrodynamic volume." Actually it is a parameter that is proportional to the hydrodynamic volume (16). The important aspect is that if a calibration plot of J versus retention volume (or retention time) is established for one polymer, then that plot is also valid for any other polymer. The only requirement is that the size exclusion mechanism holds. Historically, the universality was demonstrated by Benoit et al. (89) via the use of Eq. 38. Because of contradictory results with branched poly(vinyl acetate), the more general form suitable for complex polymers, Eq. 37, was elegantly derived by Hamielec and Ouano (60, 90).

Assuming adequate resolution, the calibration curve can be applied to a chromatogram of any polymer to change the abscissa from retention time to molecular size. Also, if the detector response is proportional to the total concentration of polymer at a given retention volume, it can then be used to define an ordinate $W_N(J)$, where,

$$W_N(J) = -W_N(v)(dv/dJ) \tag{39}$$

A plot of $W_N(J)$ versus J represents a molecular size distribution. Molecular size averages can be calculated in an analogous manner to those used for molecular weight averages.

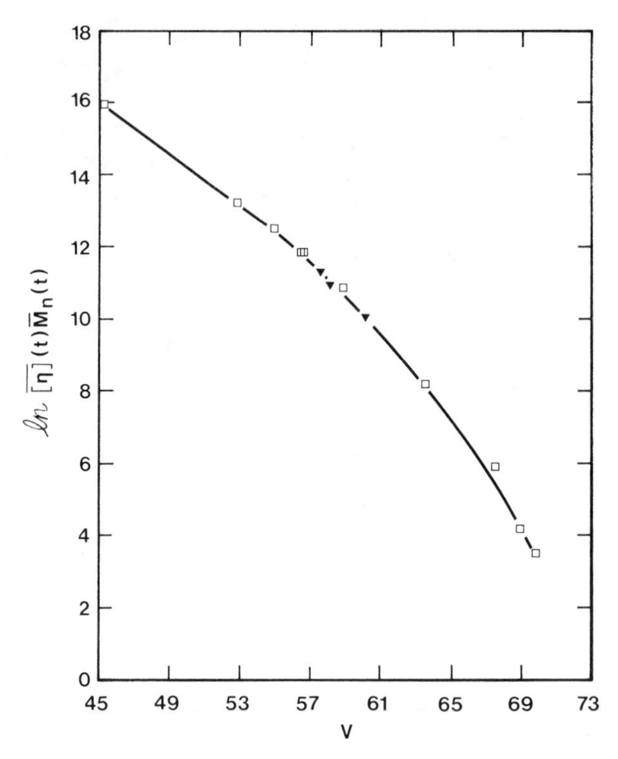

Figure 12. A SEC universal calibration curve: □ polystyrene; ▼ poly(vinyl chloride).

Figure 12 shows an example of a universal calibration curve. Note that both polystyrene samples and poly(vinyl chloride) standards fall on the same curve. If molecular weight is plotted instead of J, two different curves are obtained (Fig. 13). The curve based on molecular size is "universal" because molecular size is the basis of the separation mechanism in SEC. Unlike the curve based upon molecular weight, this universal curve can be determined from narrow fractions of one polymer and used for any other polymer at the same chromatographic conditions. Unfortunately, however, it is the molecular weight versus retention volume curve, which is generally needed. The many methods used to determine it are discussed in Section 7.2.3.

7.2.3. Molecular Weight at Each Retention Volume: SEC Molecular Weight Calibration

Currently, the most widely used method for determining a molecular weight distribution is by interpreting the SEC chromatogram. This requires defining

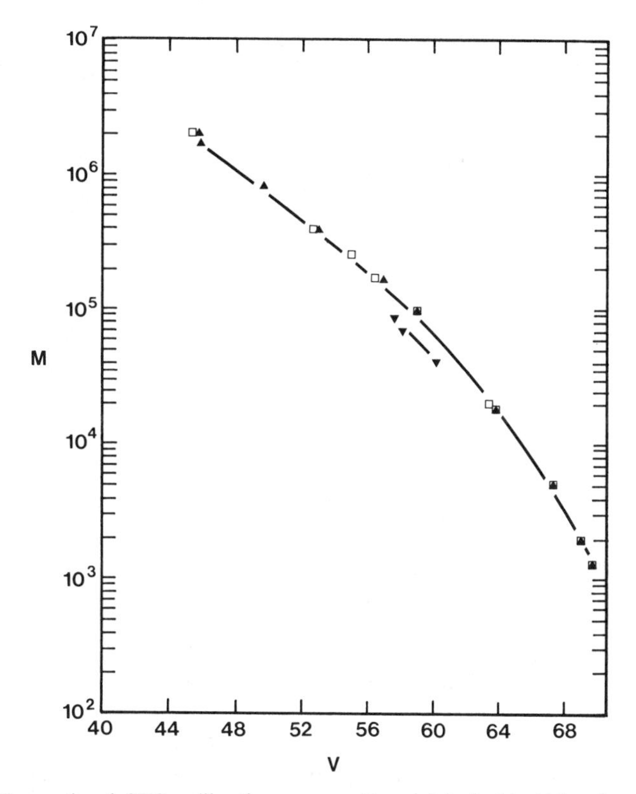

Figure 13. Conventional SEC calibration curves: ▼ poly(vinyl chloride); other points are polystyrene.

the relationship between retention volume in the instrument and molecular weight of the polymer. The mathematical equation used to express this relationship is termed an SEC calibration curve. With it the abscissa of the chromatogram, retention volume of each molecular size present, is converted to molecular weight, and the normalized ordinate, $W_N(v)$, to $W_N(\log M)$ via Eq. 27.

For polymers that have only one distributed property, molecular weight, fractionation by molecular size is synonymous with fractionation by molecular weight. A mathematical relationship between molecular weight and retention volume (e.g., Eq. 23) then can be expected. However, for complex polymers, many different molecular weights can exist at each retention volume. Conceivably, an infinite number of calibration curves can exist—one for each variety of polymer molecule. Equation 23 becomes a family of calibration curves given by (91)

$$\bar{M}_q(v) = D_{1q}\exp(-D_{2q}v) \tag{40}$$

where local molecular weight averages have replaced the local molecular weight. The subscript $q = 1$ for $\bar{M}_n(v)$ and 2 for $\bar{M}_w(v)$. Local intrinsic viscosity can be similarly calibrated and the subscript symbol η then used for q.

There are five primary methods of determining the relationship between molecular weight and retention volume (6, 7, 92).

1. Derivation from a universal calibration curve.
2. Use of monodisperse polymer standards of known molecular weight.
3. Use of polydisperse polymer standards of known molecular weight averages.
4. Use of polydisperse polymer standards of known molecular weight distribution.
5. Use of online detectors that provide a direct measure of molecular weight at each retention volume.

In the following paragraphs, each of these will be described in turn and their applicability to complex polymer commented upon.

7.2.3.1. Derivation from a Universal Calibration Curve. If the polymer can be fractionated so there is only one molecular weight at each retention volume, Eq. 38 can be used to determine M as a function of retention volume. For linear homopolymers, the intrinsic viscosity–molecular weight relationship used is the Mark–Houwink–Sakurada equation (Eq. 10). The calibration curve for the linear homopolymer, denoted by the subscript P, in terms of the universal calibration curve, is

$$M_P(v) = \left[\frac{J(v)}{K_P}\right]^{1/(a_P+1)} \tag{41}$$

For complex polymers, the Mark–Houwink–Sakurada equation is not valid. If some other equation is used or if local intrinsic viscosity is measured directly (Sections 6.2.3 and 7.2.3), then from use of Eq. 37 rather than Eq. 38 a plot of $\bar{M}_n(v)$ versus v is obtained.

7.2.3.2. Calibration from Monodisperse Polymer Standards. If a series of nearly monodisperse polymer standards are available, each with known molecular weight, then the calibration curve for the polymer represented by these standards can be determined by simply determining the peak retention volume of each standard. In general, such standards are not truly monodisperse. The calibration curve is actually a plot of the peak molecular weight calculated assuming a log normal distribution from

$$M(\text{peak}) = (\bar{M}_n \bar{M}_w)^{1/2} \tag{42}$$

versus retention volume of the peak of the chromatogram.

The calibration curve determined is applicable only to polymers of exactly the same type that are polydisperse (or monodisperse) in molecular weight. It is not applicable to complex polymers. However, if the polymer is a linear homopolymer with known Mark–Houwink constants, then the universal calibration curve can be generated from the same data used to obtain the molecular weight calibration.

New methods of correcting for the "overcrowding effect" of injected concentration have been proposed (93, 94). These methods have yet to be thoroughly evaluated. Probably the main obstacle to their use is the need to obtain accurate values of the Huggins constant. Tejero et al. (93) attempted to avoid this by correlating this constant with the intrinsic viscosity at theta conditions. However, they also employed a correlation between two Huggins constants [see Ref. (6) for comment on the practice of using the correlation obtained between two coefficients in the same equation, e.g., between slope and intercept]. The Rudin model (16) likely remains the most practical and well-founded method for concentration correction. However, the situation is complex. In a recent reactive extrusion study (95), the Rudin model concentration correction was found to only slightly improve the accuracy of the distributions. It was found that the polymerization kinetic model developed was unaffected by use of the correction because it focused on the difference between consecutive distributions rather than on their absolute value.

7.2.3.3. Calibration Using Polydisperse Standards of Known Molecular Weight Averages. In this method, a polymer sample of the same type for which a calibration curve is sought is available. The sample has a broad molecular weight distribution with known molecular weight averages, say $\bar{M}_n$ and $\bar{M}_w$ (intrinsic viscosity could be used in place of either). The procedure involves injecting the polymer into the SEC to be calibrated, recording the chromatogram, and using a nonlinear regression method to find the calibration curve that, when applied to that chromatogram, will result in the known molecular weight averages. The regression method actually systematically guesses and evaluates sets of coefficient values for an equation whose form represents the calibration curve. For example, a popular technique has been to assume a "linear" equation for the curve (Eq. 23) and to search for D_1 and D_2. After each guess of D_1 and D_2, the molecular weight averages are calculated and compared to the known true values. The search stops when the calculated and true values correspond.

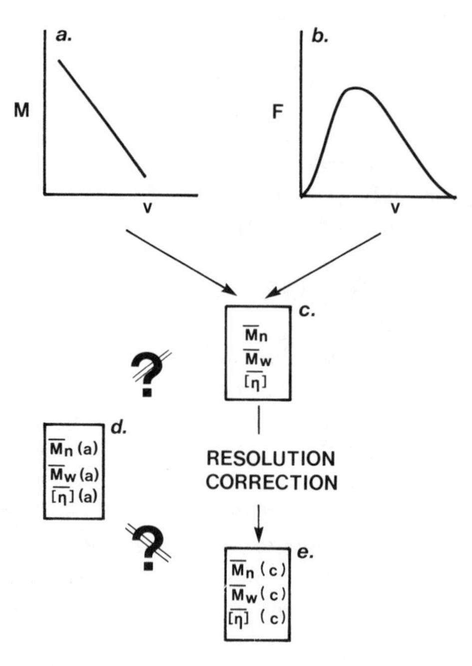

Figure 14. SEC calibration curve search with resolution correction of molecular weight averages. Reprinted with permission from Ref. 7. Copyright © 1984 Elsevier Science Publishers.

There are a variety of such calibration curve search methods (96). Some versions use more complex forms for the calibration equation. Others employ resolution correction methods. Up to this point it has been assumed that resolution is perfect. However, as mentioned earlier, each chromatogram of a polydisperse sample really represents the sum of a series of unseen chromatograms, one for each molecular size present in the sample. Methods of correcting for this effect are discussed in Section 7.4.2. Suffice to say at this point that either the averages calculated from the "imperfectly resolved" chromatogram can be corrected by multiplying each average by a correction factor or the chromatogram itself can be "reshaped" before the calibration curve is applied to calculate averages from it. In Figure 14, route *abcd* represents a search with no resolution correction and *abced* with correction of molecular weight averages. Figure 15 shows the search applied with resolution of the chromatogram before averages are calculated. Intrinsic viscosity has the advantage of using more centrally located heights of the chromatograms and therefore is less influenced by axial dispersion effects (see Section 7.4.2).

Even if axial dispersion effects were negligible, or if resolution corrections were used, use of a conventional calibration curve implies that the polymers to

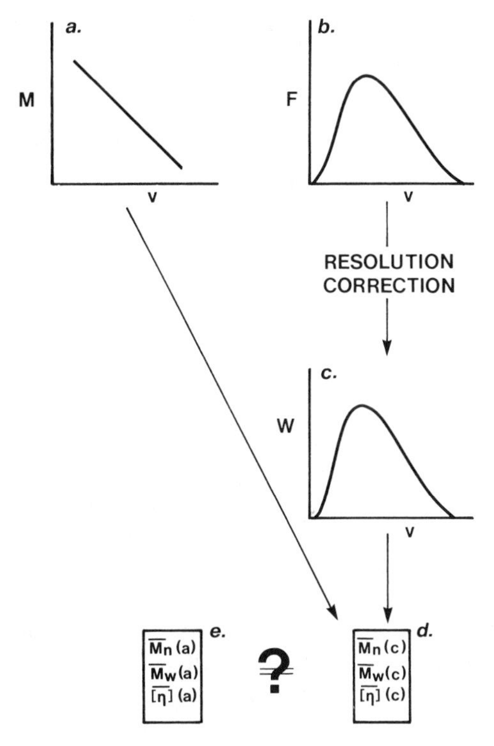

Figure 15. SEC calibration curve search with resolution correction of chromatogram. Reprinted with permission from Ref. 7. Copyright © 1984 Elsevier Science Publishers.

be analyzed in the future using the curve are polydisperse (or monodisperse) in molecular weight and are not complex. If complex polymers are used instead of polydisperse polymers as standards, with the idea of applying the resulting calibration curve to extremely similar complex polymers, it should be recognized that a calibration curve of the form of Eq. 40 for $\bar{M}_w$ must be searched rather than that of Eq. 23. Furthermore, the defining equations for the whole polymer $\bar{M}_n$, $\bar{M}_w$, and $[\eta]$ that must be calculated in the search are not Eqs. 29–31 but rather

$$\bar{M}_n = \frac{\int_0^\infty W(v)dv}{\int_0^\infty \frac{W(v)dv}{\bar{M}_n(v)}} = \frac{\int_0^\infty W(v)dv}{\int_0^\infty \frac{W(v)dv}{D_{11}\exp(-D_{21}v)}} \tag{43}$$

$$\bar{M}_w = \frac{\int_0^\infty \bar{M}_w(v)W(v)dv}{\int_0^\infty W(v)dv} = \frac{\int_0^\infty D_{12}\exp(-D_{22}v)W(v)dv}{\int_0^\infty W(v)dv} \tag{44}$$

$$\overline{[\eta]} = \int_0^\infty \overline{[\eta]}(v)W_N(v)dv = D_{1\eta}\int_0^\infty \exp(-D_{2\eta}v)W_N(v)dv \tag{45}$$

Now, local averages rather than monodisperse molecular weights are under the integration signs. If $\bar{M}_n$ and $\bar{M}_w$ are known for a sample, four unknown parameters ($D_{11}, D_{21}, D_{12}, D_{22}$) instead of only two ($D_1, D_2$) are involved. That is two too many! (See Eq. 40 for the definition of the subscript.)

The problem can be circumvented by utilizing two samples each of known $\bar{M}_w$ and searching for D_{12} and D_{22} only. (Two intrinsic viscosity values or two $\bar{M}_n$ values could be substituted for the two $\bar{M}_w$ values and $D_{1\eta}, D_{2\eta}$, or D_{11}, D_{21} could be used instead.) However, even taking this approach will provide a calibration curve of little practical use for complex polymers, if the shape of the local property distributions vary among the samples to be analyzed.

One improvement on the basic calibration curve search methods described above is use of the universal calibration curve. This improves the methods application to more complex polymers and eliminates the need to assume a form for the molecular weight calibration curve (91, 97). In this method, the parameters in the intrinsic viscosity–molecular weight relationship are guessed by the search, a molecular weight calibration curve is obtained by substituting this relationship into the known universal calibration curve, and then the molecular weight averages are calculated for comparison with the true values. For a linear homopolymer this involves guessing the Mark–Houwink constants, K and a.

Chiantore and Hamielec (98), as well as Yau (99), have pointed out that in using the universal calibration curve, groups of Mark–Houwink constants rather than the individual values can be searched. This means that the molecular weight calibration curve for an unknown can be obtained directly from the molecular weight calibration curve of polystyrene without knowing values of the Mark–Houwink constants of polystyrene. When two broad standards, each of known intrinsic viscosity and $\bar{M}_w$ are used, it is possible to obtain both the molecular weight calibration curve and the Mark–Houwink constants of the unknown. Axial dispersion was judged to be very important to the values of Mark–Houwink constants obtained.

For more complex polymers, the relationship between intrinsic viscosity and $\bar{M}_n(v)$ must be assumed. Figure 16 shows the former situation as path *abcef* and the latter as *adcgf*. The parameters shown for the intrinsic viscosity–

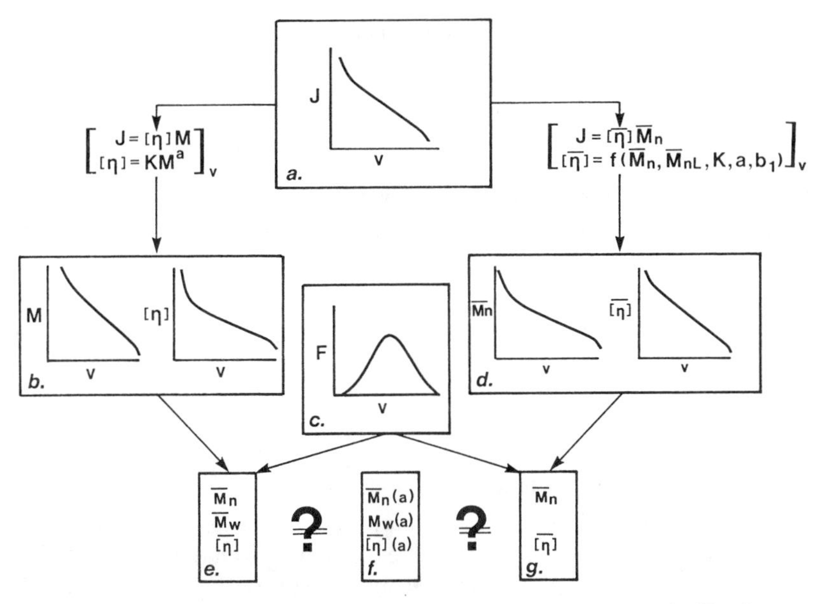

Figure 16. SEC calibration curve search approaches based upon the universal calibration curve. Reprinted with permission from Ref. 7. Copyright © 1984 Elsevier Science Publishers.

$\bar{M}_n$ relationship in this figure were proposed (91, 97) for the analysis of branched polymers and include a lower molecular weight limit for branched molecules, $\bar{M}_{nL}$, and a parameter b_1 as well as the Mark–Houwink constants. At the conclusion of the search both a calibration curve and a quantitative description of the variation of intrinsic viscosity with molecular weight for the branched sample results. Information on degree of branching may thus be deduced.

The same considerations mentioned above for the more conventional search methods and their application to complex polymers apply here. In this case a calibration curve of $\bar{M}_n(v)$ versus v can be obtained for a complex polymer. However, axial dispersion effects influence the result.

7.2.3.4. Calibration Using Standards of Known Molecular Weight Distribution. If the molecular weight distribution of a sample of the polymer of interest is known, from kinetic model predictions or from other chromatographic analyses, then a calibration curve can be derived from the chromatogram of the polymer injected into the system to be calibrated (7, 17, 100–103). The method is based upon a point-by-point comparison of the chromatogram and the molecular weight distribution. The cumulative form of each is used.

For the molecular weight distribution this form is

$$W_{N,\mathrm{cum}}(M)]_{M=M_a} = \int_0^{M_a} W_N(M)dM \tag{46}$$

where M_a is a specific value of molecular weight.

The relationship between the cumulative distribution and the chromatogram of the standard is

$$W_{N,\mathrm{cum}}(M)]_{M=M_a} = 1 - \int_0^{v_a} W_N(v)dv \tag{47}$$

$$= 1 - [W_{N,\mathrm{cum}}(v)dv]_{v=v_a} \tag{48}$$

where v_a is the retention volume corresponding to M_a.

Thus, as shown in Figure 17 for a specified v, the "wt% less than" is read from the ordinate of the cumulative curve calculated from the chromatogram (curve 1). The M corresponding to that same ordinate on $W_{N,\mathrm{cum}}(M)$ versus M (curve 2) provides the needed molecular weight value. This is a powerful method for polymers that are polydisperse in terms of molecular weight because, as will be seen later, chromatogram heights of such polymers are often little affected by imperfect resolution. This is in direct contrast to the consistent sensitivity of calculated molecular weight averages to axial dispersion effects.

The calibration curve obtained is specific to the polymer type used. For complex polymers it would be useful only if the polymers to be analyzed are very similar in all distributed properties other than molecular weight.

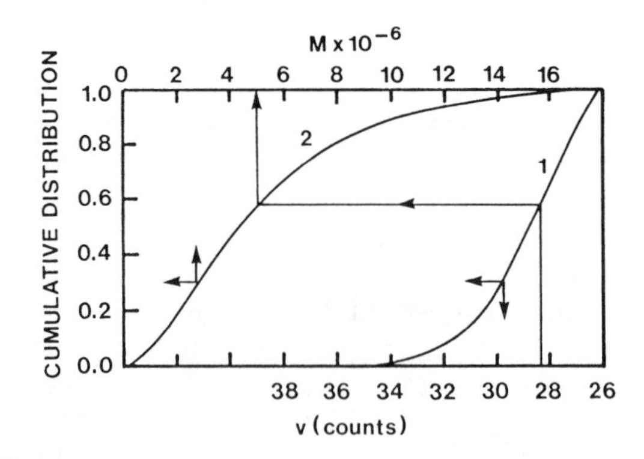

Figure 17. Determination of a SEC calibration curve from a standard of known molecular weight distribution. Reprinted with permission from Ref. 100. Copyright © 1974 John Wiley & Sons, Inc.

7.2.3.5. Use of Online Measurement of Local Weight-Average Molecular Weight. The LALLS detector is used to measure local weight average molecular weight, $\bar{M}_w(v)$. As is evident from Eq. 14, a concentration value at each retention volume is required along with the detector output in order to determine $\bar{M}_w(v)$. Also, concentration is needed to define the ordinate of the molecular weight distribution. This data is provided by a concentration detector, such as a refractometer or UV–vis detector, in series with the LALLS. A typical system is shown in Figure 18. (See Chapter 9 for further discussions.)

As mentioned in Section 6.2.2, the LALLS collects data at such a low angle that it requires only an extrapolation to zero concentration at each retention volume and not to zero concentration at each retention volume and not to zero angle. Even extrapolation to zero concentration may be avoided in many cases. If the concentration is sufficiently low so that the A_2 term is negligible, or if A_2 is known and a constant (not a function of molecular weight), then measurement at a single concentration value will suffice for determination of the local weight average molecular weight from Eq. 14.

Common problems encountered in attempting to interpret the data are

1. The use of a concentration detector in series with the molecular weight detector, means that the chromatograms from each must be superimposed to permit quantitative interpretation. This is a difficult problem because: the

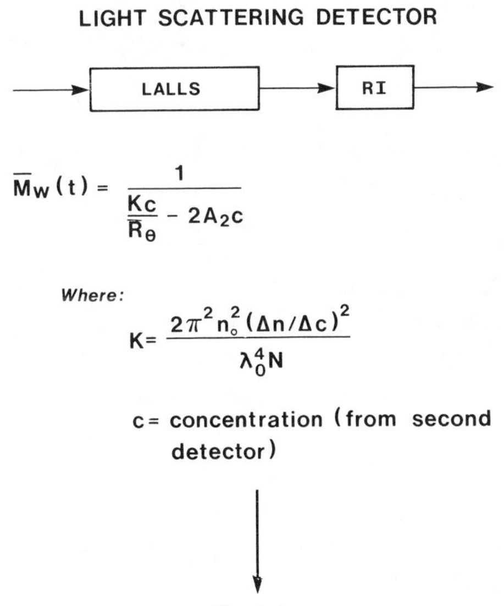

Figure 18. Low-angle laser light scattering (LALLS) detector system for online weight-average molecular weight measurement.

molecular weight detector in each case is more sensitive to the high molecular weights than is the concentration detector (i.e., noise level heights in the latter correspond to significant heights in the former) (90); axial dispersion occurs in each detector cell and causes differences in the shapes of the chromatograms; the time of transport of the polymer molecules from one detector to another can be influenced by concentration and molecular weight (92).

2. For polymers other than linear homopolymers the detector does not provide the information necessary to calculate a molecular weight distribution. Low-angle laser light scattering for branched or linear homopolymers, can provide $\bar{M}_w(v)$ at each retention volume and a final concentration versus $\bar{M}_w(v)$ plot. However, although sometimes useful, for complex polymers (i.e., most branched polymers) this is not the same as a true molecular weight distribution where each point on the abscissa represents only one molecular weight.

3. As mentioned earlier, the output from this detector is very difficult to interpret when used with copolymers, which may contain molecules of different composition at the same molecular size, as well as at different molecular sizes. These factors will influence the refractive index increment and hence light scattering. The optical constant K in Eq. 14 is then not really constant (104).

4. The most fundamental problem associated with the LALLS detector is that it is mismatched to the molecular size parameter, J, forming the basis for SEC separation since J is the product of intrinsic viscosity and $\bar{M}_n(v)$ [not $\bar{M}_w(v)$]. Thus, the LALLS detector cannot readily take advantage of universal calibration (7, 105).

7.2.3.6. Use of Online Measurement of Intrinsic Viscosity. Since, according to its definition (Eq. 8), intrinsic viscosity requires solution viscosities to be extrapolated to zero concentration, intrinsic viscosity is determined by using a concentration detector in series with a viscosity detector. Figure 19 shows a schematic of this system and Figure 20 a typical output from it (see Chapter 7).

In recent years increasing emphasis has been placed upon just how intrinsic viscosity values have been obtained from raw data. Viscosities are traditionally assumed to obey the Huggins equation in the extrapolation to zero concentration

$$\frac{1}{c}\left(\frac{t-t_0}{t_0}\right) = [\eta] + k_H[\eta]^2 c + k'_H[\eta]^3 c^2 \tag{49}$$

where k_H and k'_H are constants.

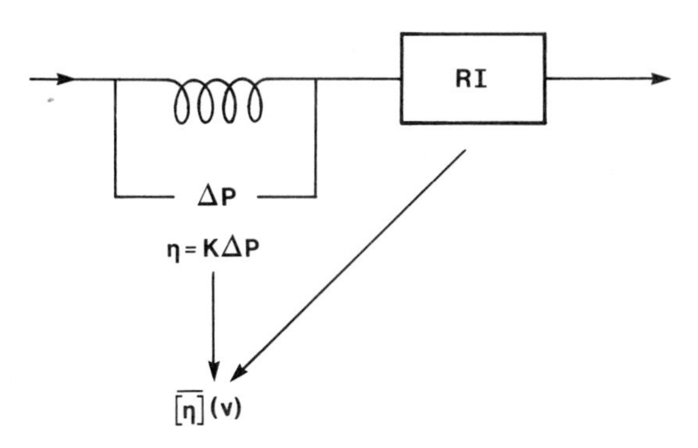

Figure 19. Detector system for online intrinsic viscosity measurement.

Equation 49 is termed a "virial equation" as is Eq. 14 for $\bar{M}_w(v)$.

García-Rubio et al. (106) recently assessed the quantitative interpretation requiring use of such equations, particularly those of exactly the same form as the Huggins equation, with emphasis on the effects of error propagation. The following paragraphs summarize their findings.

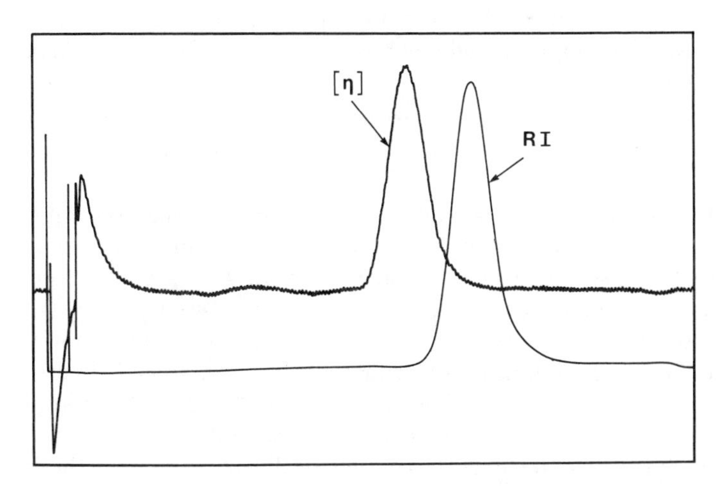

Figure 20. Output from an online differential refractometer coupled to an intrinsic viscosity detector.

For intrinsic viscosity, the equation may be considered to be a polynomial with N terms

$$y = \sum_{i=1}^{N} k_i x^i \tag{50}$$

or, in the "specific" form

$$z = \frac{y}{x} = \sum_{i=1}^{N} k_i x^{i-1} \tag{51}$$

Usually the first two terms of the specific form are used

$$z = \frac{y}{x} = k_1 + k_2 x \tag{52}$$

and k is estimated from a plot of y/x versus x as follows:

$$k = \lim_{x \to 0} z \tag{53}$$

Three major sources of error present are (106)

1. Calculation of y/x can result in large errors caused by error propagation effects.
2. The plot usually used has x in both ordinate and abscissa. This can mislead attempts to evaluate how many terms should be retained in the equation.
3. Both y and x contain experimental error and therefore violate an assumption in the usual linear least-squares procedure.

García-Rubio et al. (106) applied the error propagation equation for random error to examine the equation form. In addition, they showed that the result of instrument bias, synonymous with a nonzero intercept on the y axis in a plot of y versus x, causes a curvature in the usual plot of y/x versus x. Their conclusion was that extrapolation to zero concentration should be avoided. Instead, nonlinear least-squares regression should be used on the polynomial form of the equation with the "objective function" formulated so as to allow for error in both dependent and independent variables. The objective function is the function defining the difference between the fit and the data. In conventional least squares, it is the sum of squares of the deviations of the data from the fitted curve measured along the ordinate.

Error propagation considerations can also influence the actual experimental work of determining the viscosity–concentration data as well. It is interesting to note that by considering error propagation (6, 7, 81, 82), Reilly et al. (107) concluded that only three specifically selected, different concentrations were needed. If more runs were done the result could actually be less precise.

Several "single point" methods have been proposed for determining intrinsic viscosity from a single concentration (108). All of these methods depend on the validity of a proposed model for behavior of the viscosity with concentration.

The result of whatever interpretation method is used is intrinsic viscosity $[\eta](v)$ at each retention volume (i.e., local values of intrinsic viscosity). Thus a concentration versus intrinsic viscosity plot can be determined for each sample. Furthermore, unlike the LALLS detector system, here the data are compatible with the universal calibration curve. The use of this local intrinsic viscosity with the universal calibration curve (Eq. 37) provides $\bar{M}_n(v)$, the local number-average molecular weight at each retention volume, and the final result can then be a concentration versus $\bar{M}_n(v)$ plot. This is sometimes valuable information, but for complex polymers (e.g., branched hompolymers or copolymers) can still be quite different from the desired plot of concentration versus all the molecules present in the sample of a specific molecular weight.

Additional considerations in interpreting the output of this detector system are

1. Unlike the LALLS detector, the viscosity detector provides a measurement that can readily be interpreted as viscosity, even for complex copolymers. The main difficulty for such polymers is associated with interpreting the output of the concentration detector.

2. As in the case of the LALLS detector interpretation, the chromatogram from each detector must be superimposed in order to determine corresponding concentration values. The relative insensitivity of the concentration detector to high molecular weights, axial dispersion in both detector cells, and transport time variabilities can make this task difficult.

3. The viscosity detector is really a pressure detector. Thus, it is influenced by many other variables, notably pump pressure fluctuations (clearly evident as a "sawtooth" variation in Fig. 20). Computer-implemented noise smoothing is feasible. However, considerable care is necessary to avoid distorting or deleting information. The use of a balancing pressure, as in the differential viscometer (64), has provided a practical, commercially available, experimental solution to the problem.

7.3. Determination of Whole Polymer Property Values

7.3.1. Average Copolymer Composition

7.3.1.1. Simple Property Method. Average copolymer composition can be calculated from residual monomer analysis if the initial relative concentration of monomers in the reaction mixture is known (see Eq. 7). This then is essentially a simple property concentration determination problem (Section 7.1.1).

7.3.1.2. Distributed Property Method. Average composition can also be determined from analysis of the copolymer alone. In this case analysis of a distributed property is involved. The measurement is generally accomplished using concentration detectors arranged as shown in Figure 11.

The total area under each detector's chromatogram is proportional to the total absorbance of the sample at the specific wavelength, which in turn is proportional to the mass of polymer (Eq. 18). Thus, no chromatogram superposition is necessary because a point-by-point comparison of the two is not required. Only the area of each independent curve is needed. Equations analogous to Eq. (36) are written with areas replacing total absorbance values.

The primary source of error is the ambiguity often inherent in detector response when complex polymers are analyzed. The response is meant to indicate the concentration of each monomer in the copolymer. However, it is frequently sensitive to sequence length and polymer conformation as well.

7.3.2. Average Molecular Size in Solution

Molecular size averages may be defined in an exactly analogous manner to that of molecular weight averages. Since the separation mechanism of SEC is molecular size, they are much easier to determine for complex polymers than are molecular weights. Unfortunately, they are not often of any practical use.

Goldwasser et al. (109) sought to provide a way of using the molecular size separation parameter as an indicator of molecular weight by defining a new whole-polymer molecular weight average $\bar{M}_x$ in terms of the molecular size distribution and the whole polymer intrinsic viscosity

$$\bar{M}_x = \frac{\int_0^\infty J(v) W_N(v) dv}{\overline{[\eta]}} \tag{54}$$

With the Mark–Houwink–Sakurada equation, $\bar{M}_x$ will lie between $\bar{M}_w$ and $\bar{M}_z$ for a linear homopolymer. This could be a very useful measure for a complex polymer. However, it must be noted that, like molecular size in solution and intrinsic viscosity, the value of $\bar{M}_x$ for a polymer is dependent on the mobile phase used.

7.3.3. Molecular Weight Averages and Intrinsic Viscosity

Molecular weight averages* and intrinsic viscosity can be obtained by calculation from the SEC chromatogram, from online molecular weight-intrinsic viscosity, and from offline molecular weight-intrinsic viscosity detectors. The last mentioned method is discussed in Chapter 7.

The most commonly used method for determining molecular weight averages is calculation from an SEC chromatogram. This method was detailed in Section 7.1.3 and consists of applying Eqs. 29 and 30 to the chromatogram; intrinsic viscosity requires Eq. 31. However, valid application of these equations requires fractionation with respect to molecular weight. The previous section described the use of online LALLS and online intrinsic viscosity detectors with concentration detectors to determine local weight-average molecular weight and intrinsic viscosity values, respectively. Their use to obtain the "whole polymer" molecular weight averages and intrinsic viscosity is based upon the idea that the "local" values are the values of individual fractions of the polymer. Thus, the whole polymer is considered a mixture with values given by Eqs. 44 and 45, respectively.

It should be noted that these equations provide true values, even if perfect resolution is not obtained (6). However, it was found that sometimes the use of the LALLS detector online will result in a different whole polymer weight-average molecular weight than when it is used offline. This was attributed to the high-molecular-weight tail being so diluted in the fractionation process as to be invisible to the detectors (110). It also means that the refractive index detector alone is even less suitable for determination of $\bar{M}_w$ in some simpler cases and represents an additional difficulty in obtaining accurate $\bar{M}_w$ values from SEC.

Lew et al. (111) circumvented this problem, while using only a differential refractometer detector, by using the individual heights of the molecular weight distribution rather than the averages. They analyzed a reactive extrusion process and phrased the molecular weight predictions in terms of these heights. The heights were also less affected by axial dispersion effects than were the averages.

Froment and Revillon (112) examined various aspects of LALLS. From theoretical distributions they concluded that axial dispersion corrections were

*See note added in press at the end of this chapter regarding the calculation of $\bar{M}_n$ for complex polymers using the intrinsic viscosity detector.

important for accurate results. Dumelow et al. (113) applied LALLS to block copolymers and determined the compositional heterogeneity parameter. An important assumption was a narrow molecular weight distribution at each retention volume.

Prochazka and Kratochvil (114) applied the error propagation equation to the LALLS equation and also examined the error in conventional calibration curve fits. Their results reflected the problem of combining a differential refractometer with LALLS when the former is more sensitive at low molecular weights, while the latter is more sensitive a high molecular weights. In mention of the dilution effect on $\bar{M}_w$, they suggested that the ratio of areas from the two detectors be used in order to calculate $\bar{M}_w$ and avoid the effect (115). That requires the assumption that the second virial coefficient term be zero. [It is evident that a similar argument can be made using an analogous equation developed by Lundy and Hester (116) for the intrinsic viscosity detector.] Hunkeler and Hamielec (117) investigated A_2 in some detail for linear homopolymers and developed a new single-point method to obtain $\bar{M}_w$ from offline LALLS.

7.4. Assessment and Correction of Resolution

To this point we have assumed that fractionation was perfect. In chromatography this is synonymous with stating that there is no overlap among peaks. Resolution is the term used to describe degree of peak overlap. Mathematically it is defined for two neighboring peaks by

$$R_s = \frac{t_2 - t_1}{0.5(w_1 + w_2)} \tag{55}$$

where R_s = the "resolution index"
t = peak retention time and $t_2 > t_1$
w = peak width at half-height and subscripts 1 and 2 refer to each respective peak

This simple equation shows that resolution depends on two factors: the distance between the peaks (a function of the calibration curve) and the breadth of the peaks (which depends on axial dispersion effects). Resolution must be quantitatively assessed and, if insufficient, either corrected by changing experimental fractionation conditions or by application of specialized mathematical methods. The resolution required depends on what properties of the peaks are to be used in quantitative interpretation. This in turn depends on whether we are examining chromatograms for simple polymer properties or for distributed polymer properties.

7.4.1. Resolution and Simple Polymer Properties

In this case we desire separate chromatographic peaks for each component present and wish to accurately estimate the area under each peak. When peak overlap occurs, modern liquid chromatographs rely on geometric constructions to resolve the peaks. Figure 21 shows the two most common methods: perpendicular drop and tangent drawing (118). In this figure, the error due to the former method is $V'V''E1$ to AVV' and to the latter is AE_1E_3. Variation in the relative heights of the two peaks from sample to sample will prevent these "corrections" from providing even consistent values of relative area. Other correction methods are available. Curve fitting is a particularly appealing one (6). However, considering that the usual situation involves only a few components, improvement of experimental conditions to affect the desired fractionation is generally the best course of action.

7.4.2. Resolution and Distributed Polymer Properties

Figure 5 shows the "imperfect resolution" situation for distributed polymer properties. The problem becomes very evident when, despite careful calibration, the molecular weight averages calculated from chromatograms of standard linear homopolymers are not the same as the values obtained from other instrumental methods. This situation is described mathematically by the Tung axial dispersion equation (119):

$$F(v) = \int_{-\infty}^{\infty} W(y)G_N(v, y)dy \tag{56}$$

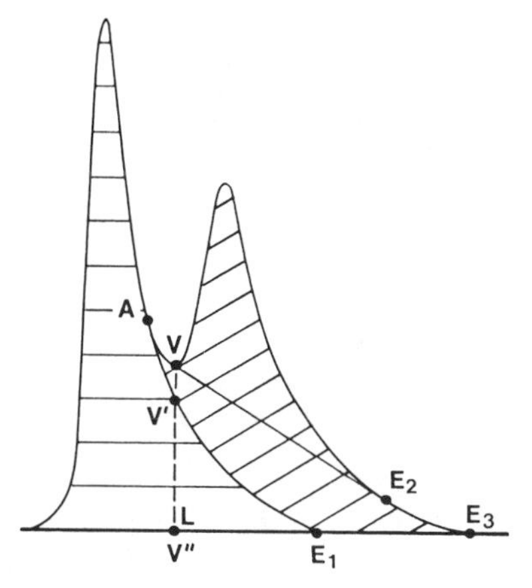

Figure 21. The perpendicular drop method (VV'') and the tangent separation method (VE_2). Reproduced from *Journal of Chromatographic Science* by permission of Preston Publications, Inc. See Ref. 118. Copyright © 1980 Preston Publications, Inc.

where $F(v)$ = the function representing the chromatogram heights at retention volume v

$W(y)$ = the function representing the heights of the chromatogram that would be obtained if resolution were perfect

$G_N(v, y)$ = the normalized function representing the shape of the unseen chromatograms originating from each molecular size present in the sample

If it can be assumed that the unseen individual chromatograms are all of the same shape, we can then write the equation as a convolution integral.

$$F(v) = \int_{-\infty}^{\infty} W(y) G_N(v - y)\, dy \tag{57}$$

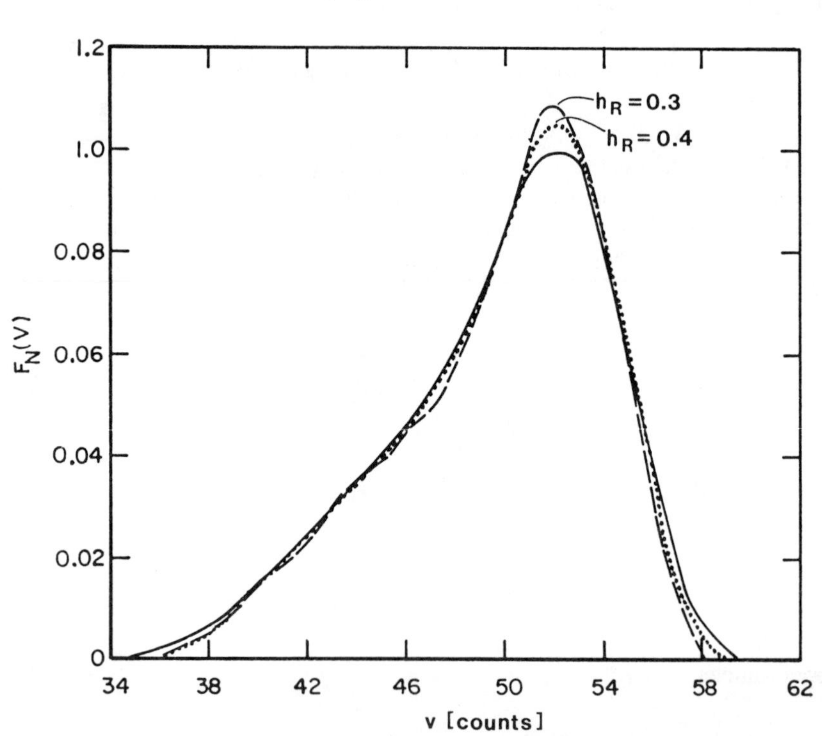

Figure 22. Change in heights of a broad chromatogram with application of resolution correction. Solid line indicates uncorrected experimental chromatogram. Parameter h_R is the resolution factor (the reciprocal of double the variance of the Gaussian shape). Reprinted with permission from S. T. Balke and R. D. Patel, "High-Conversion Polymerization Kinetic Modeling Utilizing Gel Permeation Chromatography," *ACS Symp. Ser.*, **138**, 149 (1980). Copyright © 1980 American Chemical Society.

For many years the primary emphasis was on solving Eq. 57 for $W(y)$ given $F(v)$ and assuming a Gaussian shape function $G_N(v-y)$. Method 2 of Ishige et al. (120) proved to be an extremely effective means of accomplishing this solution, even for non-Gaussian and variable shape functions. The current remaining problem is determination of the correct shape function for the chromatographic conditions and sample injected. Several methods of accomplishing this, including OC, are currently under investigation. However, it is interesting to note that for "broad chromatograms" (i.e., chromatograms of polydisperse samples) the chromatogram heights of the reshaped chromatogram are often not very different from those of the original, uncorrected chromatograms (17, 111). Figure 22 shows two corrected chromatogram superimposed on the uncorrected chromatogram to demonstrate this effect. The main differences are seen at the tails and at the peak of the chromatogram since the corrected version is invariably taller and narrower. This empirical result of the spreading of one molecular size being compensated for by the overlap of its neighboring sizes has been used effectively in polymerization kinetics (17, 111).

An alternative to "reshaping" the experimental chromatogram is the application of "correction factors" to the molecular weight averages calculated from the raw, uncorrected chromatogram of polydisperse polymers (i.e., linear homopolymers with only a broad molecular weight distribution). Unlike chromatogram heights, molecular weight averages are weighted integrals over the whole chromatogram and always exhibit the effects of imperfect resolution. Assuming a linear calibration curve (Eq. 23) and a Gaussian shape function, Hamielec and Ray (122) solved Eq. 57 analytically for the corrected molecular weight averages, $\bar{M}_n(c)$ and $\bar{M}_w(c)$ [rather than for the corrected heights, $W(y)$] and obtained

$$\bar{M}_n(c) = \bar{M}_n(uc)\exp(\sigma^2 D_2^2/2) \tag{58}$$

$$\bar{M}_w(c) = \bar{M}_w(uc)\exp(-\sigma^2 D_2^2/2) \tag{59}$$

where $\bar{M}_n(uc)$ and $\bar{M}_w(uc)$ = the uncorrected averages
D_2 = the "slope" of Eq. 23
σ = the standard deviation of the Gaussian shape function

This development immediately clarified the origin of some empirically derived correction factors and had many consequences. In particular:

1. It led to an analytically based definition of resolution for this complex situation (123). The definition depends on the molecular weight average to be

calculated from the chromatogram. However, for both $\bar{M}_n$ and $\bar{M}_w$ it is the same and is given as the resolution index $R_s(v)$ as follows:

$$R_s(v) = \frac{2}{\sigma^2 D_2^2} \tag{60}$$

2. It led to the use of non-Gaussian shape functions and new analytical solutions (91, 124, 125). Currently, a variety of functions can be used and wide latitudes in variation of these shape functions with retention volume have been introduced.

3. When molecular weight detectors became popular, this solution provided a basis for correcting the local average molecular weights determined by these detectors (91, 124, 125).

As a result of these developments, currently there are three methods of dealing with the "imperfect resolution" problem in SEC. Sometimes combinations of these methods are used in practice. The methods are as follows:

1. Experimental improvement of resolution mostly by using new column technology.
2. Utilization of property measurements, which are least affected by axial dispersion (e.g., "central-side" chromatogram heights of broad chromatograms).
3. Resolution correction using the Tung axial dispersion equation.

Often calibration curve searches now incorporate resolution correction to obtain a "true" calibration curve rather than an "effective" one. Kubin (126) pointed out that where the Mark–Houwink constants are searched, when the ratio of intrinsic viscosities was used in the search objective function, the correction for symmetrical axial dispersion canceled. This coincides with the earlier findings of Dobbin et al. (127) that their use provided the most accurate results. Use of polydispersity in the objective function was least desirable because the ratio was strongly affected by axial dispersion. In a paper by Styring et al. (128) on evaluation of the differential viscometer detector, the resolution-corrected broad standards provided much better results than resolution-corrected narrow standards. The reason for this could be related to the same smoothing effect that causes the heights of broad chromatograms to be little affected by axial dispersion correction. More information on the method of resolution correction used in this detector's software is needed.

A major concern in accomplishing resolution correction with or without simultaneous calibration is defining the shape of the chromatogram of each molecular size present. These shapes are additive and provide the total observed chromatogram. They may vary with both retention time and concentration and are likely non-Gaussian at high molecular weights. Recent attempts to elucidate this function and its variation with retention time (129–132) have yet to be extensively assessed.

8. SELECTION OF SEC AND HPLC METHODS FOR COMPLEX POLYMERS

At this point it can be appreciated that a wide array of options now exist for both experimental methods and quantitative interpretation in the analysis of complex polymers using SEC and HPLC. *Simple properties* are generally much easier to determine than *distributed properties* and should therefore be emphasized wherever possible. For these properties, the best and, thanks particularly to modern column technology, a usually attainable objective, is fractionation that is sufficiently complete to enable the concentration of each fraction to be accurately determined. This means sufficient peak separation to permit individual areas to be measured. Characterization of detector response requires valid application of regression and parameter estimation methods.

Unfortunately, often *distributed properties* must be measured directly. Then a much more complex situation exists. The following guidelines draw on the previously presented material in this chapter and should assist in method selection.

1. Before any experimental work is conducted, the objectives of the analysis should be carefully defined in terms of the ultimate use of the data. The property of interest, the quantitative measures of property to be used (e.g., distribution, whole polymer average, and local average), and acceptable precision and acceptable accuracy, require specification. With regards to accuracy, comparison can be made with offline analysis and property ambiguity should be considered (e.g., effect of copolymer composition on calculated molecular weight).

2. Assessing data obtained from a chromatographic system should include consideration of what is really being measured. Uncertainty in fractionation mechanism combined with broad property distributions dictate against the use of HPLC for complex polymers. Orthogonal chromatography appears potentially attractive but is yet in an early development phase. In contrast, conventional SEC can generally produce a molecular size fractionation no matter what the complexity of the polymer.

Among detectors for complex polymers, the combination of solution viscosity with concentration detection appears the most generally useful. However, much attention must be paid to sources of error mentioned in Sections 7.2.3 and 7.3.3.

3. Assessing computed values should involve examining the actual calculations used along with their assumptions. Points listed in this chapter under concentration determination, property value determination, and resolution assessment and correction should be considered. Moment analysis plots, plots of residuals, objective function formulation, and error propagation should be particularly examined.

4. Iteration should be expected. Communication with the user of the data is important to see if the objectives of the analysis can be somehow modified to better fit the output of the available analytical instrumentation. For complex polymers use of both nonchromatographic and chromatographic methods is highly advisable.

9. CONCLUSIONS

The difficulty associated with analysis of polymers by SEC and HPLC methods depends on several factors: the polymer type (monodisperse, simple, or complex), the property of interest, the type of measure of the property needed (e.g., average, distribution, and local average), the reproducibility and accuracy required, and the instrumentation available. As described in this chapter, modern methods of liquid chromatography, combined with information-rich detectors and molecular weight specific detectors, have the potential of characterizing complex polymers. These approaches, together with quantitative data treatment, have greatly enhanced our ability to study polymeric materials.

Note added in press: This chapter examines literature to about mid-1988. A significant advance in this area since that time was reported (133). Goldwasser showed how true, whole polymer number-average molecular weights could be obtained regardless of the complexity of the polymer by using the intrinsic viscosity detector. With the method, a refractometer detector is not required. However, universal calibration must apply to the polymer and the universal calibration curve needs to be known. Also, operation of the instrument must provide good signal to noise at the low-molecular-weight end of the chromatogram. These latter two requirements generally can be met and do not represent serious obstacles to the application of the method. The equation used to calculate the $\bar{M}_n$ value with absolute accuracy for any complex polymer is

$$\bar{M}_n = \frac{\text{Total Mass Injected}}{\displaystyle\int_0^{\infty} \frac{\eta_{sp}(v)}{J(v)}\, dv}$$

where $\eta_{sp}(v)$ is the specific viscosity at retention volume, v (i.e., the output signal of the differential viscometer at v), and $J(v)$ is the ordinate of the universal calibration curve at v. The total mass injected can be obtained from the product of injected concentration and injected volume.

NOTATION*

a	Mark–Houwink exponent
a_λ	Absorptivity
A	Area
A_λ	Absorbance
[A], [B]	Concentration of free monomers A and B
b	Cell path length
c	Concentration of polymer in solution
D_1, D_2	Calibration curve coefficients, Eq. 23
D_{1q}, D_{2q}	Calibration curve coefficients, Eq. 40
f_1	Mole fraction of monomer A in monomer mixture
$F(v)$	Heights of chromatogram uncorrected for axial dispersion
$G(v)$	Heights of chromatogram of a truly monodisperse sample
$J(v)$	Molecular size parameter
k_i	Coefficients in Eq. 50
k_H, k'	Coefficients in Eq. 49
k	Proportionality constant
K_o	Optical constant
K	Mark–Houwink constant
L	Length of capillary
m_s	Mass of sample injected
M	Molecular weight of monodisperse sample
$\bar{M}_n$	Number-average molecular weight
$\bar{M}_w$	Weight-average molecular weight
$\bar{M}_q$	q^{th}–average molecular weight

*A bar (—) above a symbol indicates that it represents a property averaged over molecules having different values of the property. For example, $\bar{M}_n$ indicates that molecules of different molecular weight are present in the sample. The absence of a bar indicates that the molecules present have the same value of the property (e.g., M_n of a monodisperse polymer). Local property values are distinguished from whole polymer property values by the presence of (v) or (t) after their symbol indicating that they are functions of retention volume or retention time, respectively. For example, $\bar{M}_n(v)$ or $\bar{M}_n(t)$ is the local value of number-average molecular weight. This is in contrast to $\bar{M}_n$, the whole polymer value. In resolution correction it is customary to distinguish corrected from uncorrected values of a property by using (c) and (uc) after its symbol [e.g., $\bar{M}_n(c)$, $\bar{M}_n$(uc)], respectively.

n	Refractive index
P	Pressure
Q	Volumetric flow rate
R	Radius of capillary
$\bar{R}_\theta$	Reduced scattering intensity
t	Time
v	Retention volume
w	Weight fraction
$\bar{W}$	Average weight fraction
W	Monomer weight
$W(M)$	Ordinate of differential molecular weight distribution
$W(v)$	Heights of chromatogram assuming perfect resolution
x	Variable in Eq. 51
X	Conversion of monomer to polymer
y	Variable in Eq. 50
z	Variable in Eq. 51
η	Viscosity
$[\eta]$	Intrinsic viscosity

Subscripts

o	Initial value; solvent
1	Value for monomer 1 (monomer A)
cum	Cumulative rather than differential distribution value
N	Normalized (value divided by area)

REFERENCES

1. H. G. Barth, W. E. Barber, C. H. Lochmuller, R. E. Majors, and F. E. Regnier, "Column Liquid Chromatography," *Anal. Chem.*, **60**, 387R (1988).
2. C. G. Smith, R. A. Nyquist, N. H. Mahle, P. B. Smith, S. J. Martin, and A. J. Pasztor, Jr., "Analysis of Synthetic Polymers," *Anal. Chem.*, **59**, 119R (1987).
3. A. Krishen, "Rubber," *Anal. Chem.*, **59**, 114R (1987).
4. S. D. Brown, T. Q. Barker, R. J. Larivee, S. L. Monfre, and H. R. Wilk, "Chemometrics," *Anal. Chem.*, **60**, 252R (1988).
5. G. Glöckner, *Polymer Characterization by Liquid Chromatography*, Elsevier, Amsterdam, 1987.
6. S. T. Balke, "Chemometrics in Size Exclusion Chromatography," *Am. Chem. Soc. Symp. Ser.*, **352**, 202 (1987).
7. S. T. Balke, *Quantitative Column Liquid Chromatography: A Survey of Chemometric Methods*, Elsevier, Amsterdam, 1984.

8. B. G. Belenkii and L. Z. Vilenchik, *Modern Liquid Chromatography of Macromolecules*, Journal of Chromatography Library, Vol. 25, Elsevier, Amsterdam, 1983.
9. J. Janca (Ed.), *Steric Exclusion Liquid Chromatography of Polymers*, Marcel Dekker, New York, 1984.
10. A. M. Krstulovic and P. R. Brown, *Reversed-Phase High Performance Liquid Chromatography, Theory, Practice, and Biomedical Applications*, Wiley, New York, 1982.
11. T. Vickrey (Ed.), *Liquid Chromatography Detectors*, Marcel Dekker, New York, 1983.
12. W. W. Yau, J. J. Kirkland and D. D. Bly, *Modern Size Exclusion Chromatography, Practice of Gel Permeation and Gel Filtration Chromatography*, Wiley, New York, 1979.
13. S. A. Borman, "Recent Advances in Size Exclusion Chromatography," *Anal. Chem.*, **55**, 384A (1983).
14. L. H. Tung (Ed.), *Fractionation of Synthetic Polymers*, Marcel Dekker, New York, 1977.
15. J. L. White and G. W. Kingry, "Theoretical Analysis and Critique of the Chromatographic Separation of Macromolecules Using Porous Adsorbents," *J. Appl. Polym. Sci.*, **14**, 2723 (1970).
16. A. Rudin, *The Elements of Polymer Science and Engineering: An Introductory Text for Engineers and Chemists*, Academic, New York, 1982.
17. S. T. Balke and R. D. Patel, "High-Conversion Polymerization Kinetic Modeling Utilizing Gel Permeation Chromatography," *Am. Chem. Soc. Symp. Ser.*, **138**, 149 (1980).
18. K. Ito, M. Omi, and T. Ito, "Kinetics of Radical Polymerization with Primary Radical Termination," *Polym. J.*, **14**, 115 (1982).
19. K. L. Petrak and E. Pitts, "Compositional Analysis by GLC of Monomer Feed in Copolymerization—Errors Due to Adsorption of Monomer to Copolymers," *Polymer*, **24**, 729 (1983).
20. F. B. Malihi, C. Y. Kuo, and T. Provder, "Determination of Gel Content of Acrylic Latexes by Size Exclusion Chromatography," *J. Liq. Chromatogr.*, **6**, 667 (1983).
21. M. Y. Hellman, T. Bowmer, and G. N. Taylor, "Determination of Gel Content and Percent Gel in Radiation-Cured Poly(vinyl chloride)-Cross-Linking Monomer Coatings by Combined GPC–LC Techniques," *Macromolecules*, **16**, 34 (1983).
22. A. H. Abdel-Alim and A. E. Hamielec, "Molecular Aggregation in Poly(vinyl chloride)," *J. Appl. Polym. Sci.*, **17**, 3033 (1973).
23. J. Janca, "Polymer Analysis by Size Exclusion Chromatography," *J. Liq. Chromatogr.*, **4**(S-1), 1 (1981).
24. S. Mori, "Effect of Experimental Conditions," in J. Janca (Ed.), *Steric Exclusion Liquid Chromatography of Polymers*, Marcel Dekker, New York., 1984, Chapter 4.

25. L. H. Tung and J. R. Runyon, "Calibration of Instrumental Spreading for GPC," *J. Appl. Polym. Sci.*, **13**, 2397 (1969).
26 P. J. Wang and B. S. Glasbrenner, "Overloading and Degradation Study in GPC Using Low-Angle Laser Light Scattering Detector (LALLS)," *J. Liq. Chromatogr.*, **10**, 3047 (1988).
27. H. G. Barth and F. J. Carlin, Jr., "A Review of Polymer Shear Degradation in Size-Exclusion Chromatography," *J. Liq. Chromatogr.*, **7**, 1717 (1984).
28. L. A. Utracki and M. M. Dumoulin, "Size Exclusion Chromatography of Polyethylenes: Reliability of Data," *Am. Chem. Soc. Symp. Ser.*, **245**, 97 (1984).
29. V. Grinshpun, K. F. O'Driscoll, and A. Rudin, "High-Temperature Size Exclusion Chromatography of Polyethylene," *Am. Chem. Soc. Symp. Ser.*, **245**, 273 (1984).
30. G. Z. A. Wu, L. A. Denton, and R. L. Laurence, "Batch Polymerization of Styrene—Optimal Temperature Histories," *Polym. Eng. Sci.*, **22**, 1 (1982).
31. S. Mori and T. Suzuki, "Problems in Determining Compositional Heterogeneity of Copolymers by Size-Exclusion Chromatography and UV–RI Detection System," *J. Liq. Chromatogr.*, **4**, 1685 (1981).
32. S. T. Balke, "Orthogonal Chromatography and Related Advances in Liquid Chromatography," *Am. Chem. Soc. Symp. Ser.*, **352**, 59 (1987).
33. G. Glöckner, "Analysis of Compositional and Structural Heterogeneities of Polymers by Non-Exclusion HPLC," *Adv. Polym. Sci.*, **79**, 159 (1986).
34. S. Mori, "Determination of Chemical Composition and Molecular Weight Distributions of High-Conversion Styrene–Methyl Methacrylate Copolymers by Liquid Adsorption and Size Exclusion Chromatography," *Anal. Chem.*, **60**, 1125 (1988).
35. S. Mori, "Chromatographic Determination of Copolymer Composition," *Adv. Chromatogr.*, **21**, 187 (1983).
36. D. W. Armstrong and K. H. Bui, "Nonaqueous Reversed-Phase Liquid Chromatographic Fractionation of Polystyrene," *Anal. Chem.*, **54**, 706 (1982).
37. M. Danielewicz and M. Kubin, "High-Performance Column Adsorption Chromatography of Random Copolymers Styrene–Acrylics," *J. Appl. Polym. Sci.*, **26**, 951 (1982).
38. G. Glöckner, J. H. M. van den Berg, N. L. J. Meijerink, T. G. Scholte, and R. Koningsveld, "Size Exclusion and High-Performance Precipitation Liquid Chromatography of Styrene–Acrylonitrile Copolymers," *Macromolecules*, **17**, 962 (1984).
39. J. Klein and G. Leidigkeit, "Adsorption Chromatography of High Molecular Homologous Polystyrenes," *Makromol. Chem.*, **180**, 2753 (1979).
40. J. P. Larmann, J. J. Destefano, A. P. Goldberg, R. W. Stout, L. R. Snyder, and M. A. Stadalius, "Separation of Macromolecules by Reversed-Phase High-Performance Liquid Chromatography. Pore-Size and Surface-Area Effects for Polystyrene Samples of Varying Molecular Weight," *J. Chromatogr.*, **255**, 163 (1983).

41. P. H. Sackett, R. W. Hannah, and W. Slavin, "Polystyrene Fractionation by Liquid Chromatography and Identification of the Compounds by Spectroscopy," *Chromatographia*, **11**, 634 (1978).
42. L. R. Synder, M. A. Stadalius, and M. A. Quarry, "Gradient Elution in Reversed-Phase HPLC Separation of Macromolecules," *Anal. Chem.*, **55**, 1412A (1983).
43. S. Teramachi, A. Hasegawa, Y. Shima, M. Akatsuka, and M. Nakajima, "Separation of Styrene-Methyl Acrylate Copolymer According to Chemical Composition Using High Speed Liquid Chromatography," *Macromolecules*, **12**, 992 (1979).
44. S. T. Balke and R. D. Patel, "Orthogonal Chromatography: Polymer Cross-Fractionation by Coupled Gel Permeation Chromatographs," *Am. Chem. Soc. Adv. Chem. Ser.*, **203**, 281 (1983).
45. S. T. Balke and R. D. Patel, "Coupled GPC/HPLC: Copolymer Composition and Axial Dispersion Characterization," *J. Polym. Sci., Polym. Lett. Ed.*, **18**, 453 (1980).
46. S. T. Balke, "Orthogonal Chromatography: Chromatographic Cross-Fractionation of Polymers," *Sep. Purif. Methods*, **11**, 1 (1982).
47. H. G. Barth, "Nonsize Exclusion Effects in High-Performance Size Exclusion Chromatography," *Am. Chem. Soc. Symp. Ser.*, **352**, 29 (1987).
48. G. Glöckner, "Quantitative Aspects of Gradient HPLC of Copolymers from Styrene and Ethyl Methacrylate," *Chromatographia*, **23**, 517 (1987).
49. G. Glöckner and J. H. M. Van Den Berg, "Copolymer Fractionation by Gradient High-Performance Liquid Chromatography," *J. Chromatogr.*, **384**, 135 (1987).
50. G. Glöckner, "High-Performance Precipitation Liquid Chromatography," *Trends Anal. Chem.*, **4**, 214 (1985).
51. T. H. Mourey, "High-Performance Liquid-Solid Adsorption Chromatography of Spiropyan-End-Labelled Polystyrenes," *J. Chromatogr.*, **357**, 101 (1986).
52. T. H. Mourey, I. Noh, and H. Yu, "High-Performance Liquid–Solid Adsorption Chromatography of Spiropyan-End-Labelled Polystyrenes," *J. Chromatogr.*, **303**, 361 (1984).
53. S. Mori and Y. Uno, "Operational Variables for the Separation of Styrene-Methyl Methacrylate Copolymers According to Chemical Composition by Liquid Adsorption Chromatography," *Anal. Chem.*, **59**, 90 (1987).
54. F. M. Mirabella, "Characterization of Multicomponent Polymer Systems Using Temperature Rising Elution Fractionation," *Proc. Int. Symp. GPC '87*, 180 (1987).
55. R. P. W. Scott, *Liquid Chromatography Detectors*, Journal of Chromatography Library, Vol. 11, Elsevier, Amsterdam, New York, 1977.
56. N. Binboga, D. Kisakurek, and B. M. Baysal, "Effect of Molecular Weight on the Refractive Index Increment of Polystyrene, Poly(ethylene glycol), Poly(propylene glycol), and Poly(dichlorophenylene oxide) in Solution," *J. Polym. Sci., Polym. Phys. Ed.*, **23**, 925 (1985).
57. L. H. García-Rubio, A. E. Hamielec, and J. F. MacGregor, "UV Spectrophotometers as Detectors for Size Exclusion Chromatography of Styrene-Acrylonitrile (SAN) Copolymers," *Am. Chem. Soc. Symp. Ser.*, **197**, 151 (1982).

58. S. A. Borman, "Photodiode Array Detectors for LC," *Anal. Chem.*, **55**, 836A (1983).
59. L. H. García-Rubio, "The Effect of Composition, Sequence Length, and Tacticity on the UV Absorption Analysis of Styrene Copolymers in Solution," *J. Appl. Polym. Sci.*, **27**, 2043 (1982).
60. A. E. Hamielec, A. C. Ouano, and L. L. Nebenzahl, "Characterization of Branched Poly(vinyl acetate) by GPC and Low Angle Laser Light Scattering Photometry," *J. Liq. Chromatogr.*, **1**, 527 (1978).
61. R. B. Green, "Lasers: Practical Detectors for Chromatography?," *Anal. Chem.*, **55**, 20A (1983).
62. F. B. Malihi, C. Kuo, M. E. Koehler, T. Provder, and A. F. Kah, "Development of a Continuous Gel Permeation Chromatography Viscosity Detector for the Characterization of Absolute Molecular Weight Distribution of Polymers," *Am. Chem. Soc. Symp. Ser.*, **245**, 281 (1984).
63. A. C. Quano, "Quantitative Data Interpretation Techniques in Gel Permeation Chromatography," *J. Macromol. Sci., Rev. Macromol. Chem.*, **C9**, 123 (1973).
64. M. A. Haney, "The Differential Viscometer. II. On-Line Viscosity Detector for Size-Exclusion Chromatography," *J. Appl. Polym. Sci.*, **30**, 3037 (1985).
65. D. G. Moldovan and S. C. Polemenakos, "A New Probe into Polymer Characterization Using the Viscotek Differential Viscometer Coupled to a 150C GPC," *Proc. Int. Symp. GPC '87*, 129 (1987).
66. C. Y. Kuo, T. Provder, M. E. Koehler, and A. F. Kah, "Use of a Viscometric Detector for Size Exclusion Chromatography Characterization of Molecular Weight Distribution and Branching in Polymers," *Am. Chem. Soc. Symp. Ser.*, **352**, 130 (1987).
67. J. Lesec, D. Lecacheux, and G. Marot, "The Continuous Viscometric Detection in Size Exclusion Chromatography," *Proc. Int. Symp. GPC '87*, 89 (1987).
68. W. W. Yau, S. D. Abboutt, G. A. Smith, and M. Y. Keating, "A New Stand-Alone Capillary Viscometer Used as a Continuous Size Exclusion Chromatographic Detector," *Am. Chem. Soc. Symp. Ser.*, **352**, 80 (1987).
69. S. R. Abbott and J. Tusa, "Recent Advances in HPLC Optical Detection," *J. Liq. Chromatogr.*, **6**(S1), 77 (1983).
70. W. L. Elsdon, J. M. Goldwasser, and A. Rudin, "Densimeter Detector in GPC of Copolymers," *J. Polym. Sci., Polym. Chem. Ed.*, **20**, 3271 (1982).
71. J. M. Charlesworth, "Evaporative Analyzer as a Mass Detector for Liquid Chromatography," *Anal. Chem.*, **50**, 1414 (1978).
72. H. C. Dorn, "H-NMR: A New Detector for Liquid Chromatography," *Anal. Chem.*, **56**, 747A (1984).
73. P. J. Wyatt, C. Jackson, and G. K. Wyatt, "Part 1: Absolute GPC Determinations of Molecular Weights and Sizes from Light Scattering," *Am. Lab.*, **20**(5), 86 (1988).
74. P. J. Wyatt, D. L. Hicks, C. Jackson, and G. K. Wyatt, "Part 2: Absolute GPC Determinations of Molecular Weights and Sizes," *Am. Lab.*, **20**(6), 108, 1988.

75. K. Holland-Moritz and H. W. Siesler, "Infrared Spectroscopy of Polymers," *Appl. Spectrosc. Rev.*, **11**, 1 (1976).
76. T. H. Mourey and L. E. Oppenheimer, "Principles of Operation of an Evaporative Light-Scattering Detector for Liquid Chromatography," *Anal. Chem.*, **56**, 2427 (1984).
77. L. E. Oppenheimer and T. H. Mourey, "Examination of the Concentration Response of Evaporative Light Scattering Mass Detectors," *J. Chromatogr.*, **323**, 297 (1985).
78. M. Righezza and G. Guiochon, "Effects of the Nature of the Solvent and Solutes on the Response of a Light-Scattering Detector," *J. Liq. Chromatogr.*, **11**, 1967 (1988).
79. N. R. Draper and H. Smith, *Applied Regression Analysis*, 2nd. ed., Wiley, New York, 1981.
80. P. Haefelfinger, "Limits of the Internal Standard Technique in Chromatography," *J. Chromatogr.*, **218**, 73 (1981).
81. P. R. Bevington, *Data Reduction and Error Analysis for the Physical Sciences*, McGraw-Hill, New York, 1969.
82. H. H. Ku, "Statistical Concepts in Metrology," *Handbook of Industrial Metrology*, American Society of Tool and Manufacturing Engineers, Prentice-Hall, New York, 1967, pp. 20–50.
83. K. A. Boni, "Gel Permeation Chromatography," in R. Myers and S. S. Long (Eds.), *Characterization of Coatings: Physical Techniques, Part II*, Marcel Dekker, New York, 1976, pp. 68–121.
84. L. H. García-Rubio and N. Ro, "Detailed Copolymer Characterization Using Ultraviolet Spectroscopy," *Can. J. Chem.*, **63**, 253 (1985).
85. L. H. García-Rubio, N. Ro, and R. D. Patel, "Ultraviolet Analysis of Benzoyl Peroxide Initiated Styrene Polymerizations and Copolymerizations. 1," *Macromolecules*, **17**, 1998 (1984).
86. H. Cheng and R. R. Gadde, "Absorbance Ratio Plots in High Performance Liquid Chromatography: Some Software Problems and Remedies," *J. Chromatogr. Sci.*, **23**, 227 (1985).
87. A. C. J. H. Drouen, H. A. H. Billiet, and L. De Galan, "Dual-Wavelength Absorbance Ratio for Solute Recognition in Liquid Chromatography," *Anal. Chem.*, **56**, 971 (1984).
88. L. H. García-Rubio, "Multiple Detectors in Size Exclusion Chromatography: Signal Analysis," *Am. Chem. Soc. Symp. Ser.*, **352**, 220 (1987).
89. H. Benoit, Z. Grubisic, P. Rempp, D. Decker, and J. G. Zilliox, "Etude par Chromatographie en Phase Liquide de Polystyrenes Lineaires et Ramifies de Structures Connues," *J. Chim. Phys.*, **63**, 1507 (1966).
90. A. E. Hamielec and A. C. Ouano, "Generalized Universal Molecular Weight Calibration Parameter in GPC," *J. Liq. Chromatogr.*, **1**, 111 (1978).
91. A. E. Hamielec, "Characterization of Complex Polymer Systems by SEC—Homopolymers with Long Chain Branching and Copolymers with Composition Drift," *Pure Appl. Chem.*, **54**, 293 (1982).

92. J. V. Dawkins, "Calibration of Separation Systems," in J. Janca (Ed.), *Steric Exclusion Liquid Chromatography of Polymers*, Marcel Dekker, New York, 1984, pp. 53–116.
93. R. Tejero, V. Soria, A. Campos, J. E. Figueruelo, and C. Abad, "Quantitative Prediction of Concentration Effects in Steric Exclusion Chromatography," *J. Liq. Chromatogr.*, **9**, 711 (1986).
94. S. Mingshi and H. Guixian, "Study on the Concentration Effect in GPC. I. A New Model Theory for Concentration Dependence of Hydrodynamic Volumes and Gel Permeation Chromatography Elution Volumes," *J. Liq. Chromatogr.*, **8**, 2543 (1985).
95. R. Lew, P. Cheung, D. Suwanda and S. T. Balke, "Quantitative Size Exclusion Chromatography of Polypropylene II: Analysis Systems," *J. Appl. Polym. Sci.*, **35**, 1065 (1988).
96. M. Kubin, "Calibration of Size Exclusion Chromatography Systems with Polydisperse Standards," *J. Liq. Chromatogr.*, **7**, 41 (1984).
97. A. Ram and J. Miltz, "Role of Polymer Chain Structure in Gel Permeation Chromatography Interpretation," *Polym. Plast. Technol. Eng.*, **4**, 23 (1975).
98. O. Chiantore and A. E. Hamielec, "Molecular Weight Calibration of SEC Using Broad MWD Standards-Application for Poly(*p*-methyl styrene)," *J. Liq. Chromatogr.*, **7**, 1753 (1984).
99. W. W. Yau, "GPC Calibration Methodologies," Conference Proceedings for the Society Plastics Engineers, 44th Annual Technology Conference, **32**, 461 (1986).
100. A. H. Abdel-Alim and A. E. Hamielec, "GPC Calibration for Water-Soluble Polymers," *J. Appl. Polym. Sci.*, **18**, 297 (1974).
101. C. M. Atkinson, R. Dietz, and M. Francis, "Whole Polymer Reference Materials in Analytical GPC," in R. Epton (Ed.), *Chromatography of Synthetic and Biological Polymers: Column Packings, GPC, GF and Gradient Elution*, Vol. 1, Halsted, New York, 1978, pp. 297–303.
102. M. J. R. Cantow, R. S. Porter, and J. F. Johnson, "Method of Calibrating Gel Permeation Chromatography with Whole Polymers," *J. Polym. Sci. Part A1*, **5**, 1391 (1967).
103. A. R. Weiss and E. Cohn-Ginsberg, "Calibration of Gel-Permeation Columns with Unfractionated Polymers," *J. Polym. Sci. Part A2*, **8**, 148 (1970).
104. F. B. Malihi, C. Y. Kuo, and T. Provder, "Determination of the Absolute Molecular Weight of a Styrene-Butyl Acrylate Emulsion Copolymer by Low-Angle Laser Light Scattering (LALLS) and GPC/LALLS," *J. Appl. Polym. Sci.*, **29**, 925 (1984).
105. R. C. Jordan, S. F. Silver, R. D. Sehon, and R. J. Rivard, "Size Exclusion Chromatography with Low-Angle Laser Light-Scattering Detection: Application to Linear and Branched Block Copolymers," *Am. Chem. Soc. Symp. Ser.*, **245**, 295 (1984).
106. L. H. García-Rubio, A. V. Talatinian, and J. F. MacGregor, "Propagation of Errors in Polymer Measurement Equations and the Estimation of Second Order

Effects", in Proc. Symposium on Quantitative Characterization of Plastics and Rubber, J. Vlachopoulos (Ed.), McMaster University, Hamilton, Ontario, June 21–22, 1984.

107. P. M. Reilly, B. M. E. van der Hoff, and M. Zioglas, "Statistical Study of the Application of the Huggins Equation to Measure Intrinsic Viscosity," *J. Appl. Polym. Sci.*, **24**, 2087 (1979).

108. K. K. Chee, "A Critical Evaluation of the Single Point Determination of Intrinsic Viscosity," *J. Appl. Polym. Sci.*, **34**, 891 (1987).

109. J. M. Goldwasser, A. Rudin, and W. L. Elscon, "Characterization of Copolymers and Polymer Mixtures by Gel Permeation Chromatogrophy," *J. Liq. Chromatogr.*, **5**, 2253 (1982).

110. V. Grinshpun, K. F. O'Driscoll, and A. Rudin, "On the Accuracy of SEC Analyses of Molecular Weight Distributions of Polyethylenes," *J. Appl. Polym. Sci.*, **29**, 1071 (1984).

111. R. Lew, D. Suwanda, and S. T. Balke, "Quantitative Size Exclusion Chromatography of Polypropylene I: Method Development," *J. Appl. Polym. Sci.*, **35**, 1049 (1988).

112. P. Froment and A. Revillon, "Some Aspects of Analysis of Polymers by Steric Exclusion Chromatography and On-Line Low Angle Laser Light Scattering (SEC–LALLS)," *J. Liq. Chromatogr.*, **10**, 1383 (1987).

113. T. Dumelow, S. R. Holding, L. J. Maisey, and J. V. Dawkins, "Determination of the Molecular Weight and Compositional Heterogeneity of Block Copolymers Using Combined Gel Permeation Chromatography and Low-Angle Laser Light Scattering," *Polymer*, **27**, 1170 (1986).

114. O. Prochazka and P. Kratochvil, "Analysis of the Accuracy of Determination of Molar Mass Distribution by GPC with On-Line Light-Scattering Detector," *J. Appl. Polym. Sci.*, **31**, 919 (1986).

115. M. Martin, "Polymer Analysis by Fractionation with On-Line Light Scattering Detectors," *Chromatographia*, **15**, 426 (1982).

116. C. E. Lundy and R. D. Hester, "An Eluent Pressure Detector for Aqueous Size Exclusion Chromatography," *J. Liq. Chromatogr.*, **7**, 1911 (1984).

117. D. Hunkeler and A. E. Hamielec, "One-Point Method to Calculate Molecular Weight from Low-Angle Laser Light Scattering Data I. Application to Acrylamide Polymers," *J. Appl. Polym. Sci.*, **35**, 1603 (1988).

118. E. F. G. Woerlee and J. C. Mol, "A Real-Time Gas Chromatographic Data System for Laboratory Applications," *J. Chromatogr. Sci.*, **18**, 258 (1980).

119. N. Friis and A. E. Hamielec, Gel Permeation Chromatography: A Review of Axial Dispersion Phenomena, Their Detection, and Correction," *Adv. Chromatogr.*, **13**, 41 (1975).

120. T. Ishige, S. I. Lee, and A. E. Hamielec, "Solution of Tung's Axial Dispersion Equation by Numerical Techniques," *J. Appl. Polym. Sci.*, **15**, 1607 (1971).

121. S. T. Balke and A. E. Hamielec, "Bulk Polymerization of Methyl Methacrylate," *J. Appl. Polym. Sci.*, **17**, 905 (1973).

122. A. E. Hamielec and W. H. Ray, "An Analytical Solution to Tung's Axial Dispersion Equation," *J. Appl. Polym. Sci.*, **13**, 1319 (1969).
123. A. E. Hamielec, "An Analytical Solution to Tung's Axial Dispersion Equation. Applications in Gel Permeation Chromatography," *J. Appl. Polym. Sci.*, **14**, 1519 (1970).
124. A. E. Hamielec, "Correction for Axial Dispersion," in J. Janca (Ed.), *Steric Exclusion Liquid Chromatography of Polymers*," Marcel Dekker, New York, 1984, pp. 117–160.
125. A. E. Hamielec and H. Meyer, "On-Line Molecular Weight and Long-Chain Branching Measurement Using SEC and Low-Angle Laser Light Scattering," *Dev. Polym. Char.*, **5**, 95 (1986).
126. M. Kubin, "Determination of the Mark–Houwink Constants by Gel Permeation Chromatography with Correction for Longitudinal Spreading," *Collect. Czech. Chem. Commun.*, **51**, 1636 (1986).
127. C. J. B. Dobbin, A. Rudin, and M. F. Tchir, "Evaluation of GPC Methods for Estimation of Mark–Houwink–Sakurada Constants," *J. Appl. Polym. Sci.*, **25**, 2985 (1980).
128. M. G. Styring, J. E. Armonas, and A. E. Hamielec, "An Experimental Evaluation of a New Commercial Viscometric Detector for SEC Using Linear and Branched Polymers," *Am. Chem. Soc. Symp. Ser.*, **352**, 104(1987).
129. Y. H. Chen, M. L. Ye, X. Li, and L. H. Shi, "A New Method of the Peak Spreading Correction in GPC by Means of Wesslau's and Tung's Molecular Weight Distribution Functions," *J. Liq. Chromatogr.*, **9**, 1163 (1986).
130. D. Alba and G. R. Meira, "Calibration for Instrumental Spreading in Size Exclusion Chromatography by a Novel Recycle Technique," *J. Liq. Chromatogr.*, **9**, 1141 (1986).
131. R. S. Cheng, Z. L. Wang, and Y. Zhao, "Dependence of the Spreading Factor on the Retention Volume of Size Exclusion Chromatography," *Am. Chem. Soc. Symp. Ser.*, **352**, 281 (1987).
132. G. Glöckner, "Straightforward Procedure for Estimating the Spreading Factor in SEC," *J. Liq. Chromatogr.*, **7**, 1769 (1984).
133. J. M. Goldwasser, "Determination of Absolute $\bar{M}_n$ Using Gel Permeation Chromatography/Differential Viscometry", Proc. Int. Gel Permeation Chromatography Symposium, Newton, Massachusetts, 1989, pp. 150–157.

CHAPTER

2

DATA REDUCTION IN SIZE EXCLUSION CHROMATOGRAPHY OF POLYMERS

GREGORIO R. MEIRA

INTEC (CONICET and Universidad Nacional del Litoral)
Guemes 3450, Sante Fe, Argentina

Modern Methods of Polymer Characterization, Edited by Howard G. Barth and Jimmy W. Mays
ISBN 0-471-82814-9

1. INTRODUCTION

The molecular structure of a synthetic polymer influences the end-use and processability properties, such as hardness, tensile strength, drawability, elastic modulus, and melt viscosity. The most important characteristics defining the molecular architecture of synthetic macromolecules are

1. The chemical nature of the repeating units.
2. In branched copolymers, the combined distribution of molecular weights, branching frequency, composition, and sequence length.
3. The molecular topology (e.g., helicity and tacticity).

After the pioneering efforts of Moore (1), size exclusion chromatography (SEC) is presently not only the preferred method for fractionating and analyzing the molecular weight distribution (MWD) of macromolecules, but also has the potential for determining the complete molecular architecture. Like most other analytical techniques, SEC requires the polymer sample to be dissolved in a solvent. For this reason, cross-linked hyperstructures cannot be analyzed through SEC.

Several technological advances have contributed to the developments in SEC: (a) the introduction of new low-interaction and high-resolution SEC packings; (b) the development and increased use of online specific and molar-mass-sensitive detectors; and (c) the application of digital techniques for data acquisition and reduction. The field has grown, and during the past decade, the output of journal articles has been in the order of 500 papers published annually. Fortunately a number of monographs have appeared (2–9). Other books specific to SEC that are congress proceedings have been published (10–15).

In this chapter, a review on the use of SEC for the characterization of polymers is given, with emphasis on data treatment associated with instrumental broadening corrections. The discussion is limited to determining the MWD in linear homopolymers, the branching frequency in homopolymers, and the combined MWD–chemical composition distribution in linear copolymers.

2. GENERAL CONSIDERATIONS

2.1. Equipment and Separation Mechanism

Figure 1 schematically represents an SEC, which is basically a special-purpose high-performance liquid chromatograph (HPLC). The “heart” of the instru-

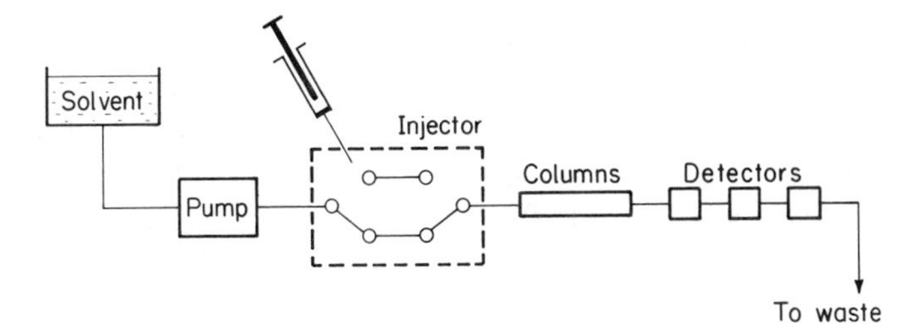

Figure 1. A schematic SEC.

ment is the fractionation columns, which contain porous packings with a porosity distribution in the same range of sizes as the solute polymer molecules.

In ideal SEC, the separation is exclusively by molecular size, and no axial dispersion is present. The separation mechanism (steric exclusion) allows the largest molecules to permeate a small fraction of the pores, while the smallest molecules permeate most of the pore volume. Thus, polymer molecules are separated according to size, with the largest eluting first.

Several detectors have been developed to permit continuous sensing of the emerging species. Detectors may be classified as follows:

1. Total mass detectors (differential refractometer and densimeter).
2. Specific detectors (single or multiple wavelength absorbance spectrophotometers).
3. Molar-mass-sensitive detectors (online viscometer and low-angle laser light scattering photometer, LALLS).

Column packing may be organic (e.g., gels of polystyrene cross-linked with divinylbenzene) or inorganic (e.g., porous glass or silica). The porous material typically consists of nearly spherical microbeads of a diameter of about 10 μm or less, to ensure low axial dispersion in columns. In some cases, polystyrene packings are preferred to inorganic packings, because of their reduced interactions with the solute, which otherwise may lead to secondary fractionation processes. The reasons for such secondary processes are adsorption, partition, ionic inclusion, ionic exclusion, incompatibility, and solvation. In general, adsorption, partition, and ionic inclusion increase retention time t (or retention volume v), while incompatibility and ionic exclusion reduce retention. Solvation will reduce or increase retention time, according to whether it occurs in the matrix or in the solute. As a first approximation, these processes are independent of retention volume. Thus, with homopolymers, they introduce retention time shifts in the chromatograms, without distortion. With copolymers, however, the composition distribution will distort the chromatograms, compared to those produced in the absence of secondary mechanisms.

Apart from secondary mechanisms, two other difficulties further complicate data interpretation from SEC measurements: instrumental broadening and the presence of complex polymers.

2.2. Instrumental Spreading

Except perhaps for low-molecular-weight oligomers, the resolution in SEC is insufficient to separate each of the individual species. Instead, a distribution of molecular sizes is normally present in the detector cell. There are many reasons for imperfect resolution, the most important being: (a) axial dispersion in the columns; (b) parabolic flow profiles in capillaries; (c) finite injection volumes; (d) finite detection cell volumes; (e) inhomogeneities in the column packing; and (f) column-end effects. While a–c introduce symmetrical spreading, d–f induce skewed broadening. The instrumental broadening is complicated because both the spreading breadth and its skewness are a function of the mean retention volume of the individual molecular species.

In practice, all methods for correcting instrumental spreading are based on the integral equation introduced by Tung (16)

$$F(t) = \int_{-\infty}^{\infty} g(t, \tau) F^c(\tau) d\tau \tag{1}$$

where t, τ = retention time for both terms

$F(t)$ = the measured chromatogram, from a mass detector

$g(t, \tau)$ = the spreading function, or the normalized set of noncausal impulse responses of hypothetical monodisperse polymers applied at different mean retention times τ

$F^c(t)$ = the chromatogram corrected for instrumental broadening

The following points should be noted concerning Eq. 1:

1. The model of Eq. 1 involves continuous functions because it implies a very large number of chemical species. In the particular case of oligomers, the following discrete function is preferable

$$F(t) = \sum_{i=1}^{n} b_i g_i(t) \tag{2}$$

where n = the number of chemical species (e.g., $\leqslant 10$)

b_i = the area under the peak (proportional to the mass) corresponding to the i-th species
$g_i(t)$ = the broadening due to the i-th species

2. If the chromatogram is narrow, then the spreading may be assumed constant (or time invariant). In this case, Eq. 1 transforms into a simple convolution integral with $g(t, \tau)$ replaced by $g(t - \tau)$.

3. The two basic problems associated to Eq. 1 are (a) the spreading calibration or the estimation of $g(t, \tau)$; and (b) the deconvolution or the calculation of $F^c(t)$ given $F(t)$ and $g(t, \tau)$. These problems will be considered further below.

4. Equation 1 with the same spreading function $g(t, \tau)$ is also applicable to molecular weight detectors. Assume that a strictly monodisperse sample of molecular weight M and retention time τ is injected into a chromatograph equipped with a molecular weight detector, whose signal is proportional to the product of the instantaneous mass and the molecular weight. Since M is constant at all retention times t, then the measured chromatogram is proportional to the spreading $g(t)$ of average retention time τ. This is an interesting concept that needs to be exploited further.

2.3. Main Calibration

When no secondary mechanisms are present, theory predicts that the elution behavior of all polymers can be characterized by a universal size parameter. Studies of the dilute solution viscosity of synthetic polymers indicate that the polymer molecule can be represented by an equivalent sphere of hydrodynamic volume V_h.

Assume that the theory of separation was sufficiently precise with regards to the basic exclusion mechanism, secondary fractionation, and instrumental spreading, and that the distribution of pores could be accurately measured. In this case, retention time or volume (v or t) could be directly transformed into hydrodynamic volume V_h without the need of intermediate calibrations. Unfortunately, this is not yet the case, but even if it were, V_h depends on the solvent that is used and on the temperature, and therefore V_h is not a fundamental property inherent to the polymer alone, such as molecular weight M. The unsolvable difficulty in SEC stems from the fact that the technique fractionates according to V_h and not M.

Assuming infinite resolution, in the special case of linear homopolymers or linear copolymers with constant composition, then the instantaneously emerging species would be monodisperse. Such polymers are classified as *simple* from the point of view of SEC, because at least in theory, a one-to-one relationship exists between V_h and the molecular weight. Mixtures of linear homopolymers, heterogeneous copolymers, polymers with long-chain branching, and so on, are considered *complex* because even under conditions of perfect resolution, a variety of species of different molcular weights will coexist in the detector cell. The reasons for this are

1. For a given molecular weight, the hydrodynamic volume of a branched homopolymer is smaller than that of its linear counterpart.
2. In a copolymer, the partial molar volumes M_i/V_{h_i} for each of its repeating units, are in general different from each other.

In standard SEC, the chromatograph is equipped with a mass detector, and a calibration is required to transform retention time into molecular weight. If an online viscometer or a LALLS detector were also available, however, then the technique may be considered "absolute," in the sense that calibration is no longer required. Typically, such detectors: (a) produce noisy signals, (b) are insensitive at low molecular weights (in the case of LALLS), and (c) require the simultaneous measurement of instantaneous concentration. Consequently, large errors in the molecular weights result at the chromatogram tails, and particularly at the low-molecular-weight end of the distribution (17).

Because of the inaccuracies and difficulties of instantaneous molecular weight detection, one normally uses *absolute* molecular weight detectors for establishing an appropriate calibration curve, and then employ such calibration for routine measurements. Thus, even with absolute detectors, the main calibration (i.e., the transformation of t or v into M) is of vital importance for an appropriate interpretation of the raw data in SEC.

The calibration may be expressed by the function

$$v = f(M) \tag{3}$$

However, it is more common to consider v as the independent variable, and choose the logarithmic relationship (Fig. 2)

$$\ln M = f_1(v) \tag{4}$$

In a certain range, $\ln M$ is linearly related to $f(v)$, and we can write

$$\ln M = C_1 - C_2 v \tag{5}$$

where C_1 and C_2 are positive constants. In Eqs. 3–5, M represents different quantities according to whether a simple or a complex polymer is analyzed. In

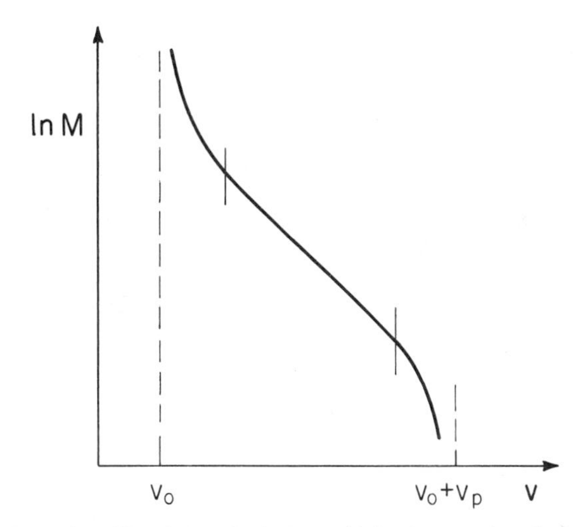

Figure 2. The main calibration, v_0 is the interstitial volume and v_p the pore volume.

the first case, M is the molecular weight of a hypothetical monodisperse species of retention volume v. In the second case, it may be associated to some molecular weight average ($\bar{M}_n$, $\bar{M}_v$, $\bar{M}_w$, $\bar{M}_z$, etc.) of all species with retention volume v. Thus, in theory, there are infinite calibration curves for a complex polymer.

The main calibration may be established: (a) through absolute molecular weight detectors or (b) by using calibration standards. Such methods are discussed below.

2.4. Calculation of the Molecular Weight Distribution and Average Molecular Weights

Let $F(v)$ be the continuous, base line corrected chromatogram, obtained from a mass detector. For quantitative work, one requires to sample the continuous curve at regular intervals Δv_i. We shall call $F_i(v_i)$ the corresponding continuous step-wise function. After correction for instrumental broadening, $F_i(v_i)$ provides $F_i^c(v_i)$.

With G representing the weight fraction, let $G(M)$ be the "continuous" weight MWD, in the sense that a smooth curve and a linear M axis are represented. If one determines $G(M)$, then all of the molecular weight averages may be readily calculated by sampling such a continuous curve at regular discrete intervals of M. However, it is numerically preferable to calculate such averages from $F_i^c(v_i)$ directly, as shown below. The total solute mass may be represented by

$$\sum_i G_i(M_i)\Delta M_i = -\sum_i F_i^c(v_i)\Delta v_i \tag{6}$$

where $G_i(M_i)$ is the continuous step-wise weight MWD corresponding to $F_i(v_i)$, and $-\Delta M_i$ represent the variable increments in M corresponding to the different Δv_i values. Assume the calibration to be represented by Eq. 3. Thus,

$$\frac{dv_i}{dM_i} = f'(M_i) \cong \frac{\Delta v_i}{\Delta M_i} \tag{7}$$

From Eqs. 6 and 7

$$\sum_i G_i(M_i)\frac{\Delta v_i}{f'(M_i)} = -\sum_i F_i^c(v_i)\Delta v_i \tag{8}$$

Therefore,

$$G_i(M_i) = F_i^c(v_i) f'(M_i) = -F_i^c(v_i)\left(\frac{\Delta v_i}{\Delta M_i}\right) \tag{9}$$

Because $f'(M_i)$ changes exponentially with M_i, Eq. 9 illustrates the deformation that must be introduced in the chromatogram to obtain the weight MWD. Multiplying both sides of Eq. 9 by $(M_i \Delta M_i)$ and summing, one obtains

$$\sum M_i G_i(M_i)\Delta M_i = -\sum M_i F_i^c(v_i)\Delta v_i \tag{10}$$

Dividing Eq. 10 by Eq. 6, and remembering that Δv_i is constant,

$$\frac{\sum M_i G_i(M_i)\Delta M_i}{\sum G_i(M_i)\Delta M_i} \equiv \bar{M}_w = \frac{\sum M_i F_i^c(v_i)}{\sum F_i^c(v_i)} \tag{11}$$

Analogously, it is easy to show that

$$\bar{M}_n = \frac{\sum M_i F_i^c(v_i)}{\sum [F_i^c(v_i)/M_i]} \tag{12}$$

$$\bar{M}_z = \frac{\sum M_i F_i^c(v_i) M_i^2}{\sum F_i^c(v_i) M_i} \tag{13}$$

$$\bar{M}_v = \left[\frac{\sum M_i F_i^c(v_i) M_i^a}{\sum F_i^c(v_i)} \right]^{1/a} \quad (14)$$

Equations 11–14 indicate that the pairs of values of $F_i^c(v_i)$ can be directly used to calculate the molecular weight averages, with mathematically identical expressions to those for $G(M)$. This result has been used for a long time, but the exact derivation is fairly recent (18). As opposed to the classical approach based on smooth curves, the basic assumption of the given derivation is that the polymer sample is built up of hypothetical species of molecular weight M_i and mass $G_i(M_i)\Delta M_i$.

With an appropriate interpolation, Eq. 9 may be used to obtain $G(M)$. Alternatively, one can write

$$G(M) = -F^c(v) \left[\frac{dv}{d(\ln M)} \right] \left[\frac{d(\ln M)}{dM} \right] \quad (15)$$

$$G(M) = -\left[\frac{F^c(v)}{M} \right] \left[\frac{dv}{d(\ln M)} \right] \quad (16)$$

and in the case of a linear calibration, $dv/d(\ln M)$ is a constant. If needed, the differential number MWD can be obtained by dividing $G(M)$ by M

$$N(M) = \frac{G(M)}{M} \quad (17)$$

Note that the divisions by M in Eqs. 16 and 17 introduce errors as $M \to 0$ (i.e., when the chromatogram contains low-molecular-weight tails).

In summary, the suggested calculation procedure for obtaining $G(M)$ from $F(v)$ involves: (a) an inverse filtering operation to obtain $F^c(v)$; (b) the transformation of the horizontal axis into M, through the calibration curve; and (c) the nonlinear transformation of Eq. 9 or Eq. 16 to find $G(M)$. Alternatively, methods based on the analytical solution of Eq. 1 have been developed, which allow the direct calculation of the average molecular weights without requiring the explicit deconvolution of $F(v)$. Such methods will be briefly discussed below.

3. SIMPLE POLYMERS

A simple polymer can be characterized by its MWD. For its determination, only a mass detector and possibly an absolute molecular weight detector are required.

3.1. Detector Equations

3.1.1. Differential Refractometer

The differential refractometer is the most "universal" detector in SEC, because practically all polymer solutions exhibit a refractive index that differs from that of the pure solvent.

The response of a differential refractometer Δn is proportional to the polymer mass concentration c in the detector cell as follows:

$$\Delta n = \nu c \qquad \nu = \partial n/\partial c \tag{18}$$

where ν is the specific refractive index increment. The difficulty is that ν exhibits a slight dependence with molecular weight, which can be approximated by

$$\nu = C_3 - \frac{C_4}{M} \tag{19}$$

where C_3 and C_4 are positive constants that may be determined from the values of $\partial n/\partial c$ for the high polymer (ν_p) and for the monomer (ν_m) of molecular weight M_m, as follows:

$$\text{For } M \cong \infty \qquad C_3 \cong \nu_p \tag{20}$$

$$\text{For } M = M_m \qquad C_4 = M_m(\nu_p - \nu_m) \tag{21}$$

Equations 18 and 19 indicate that the detector tends to underestimate the low-molecular-weight tail of the chromatogram. To correct for this effect, the following expression can be used (19, 20):

$$\tilde{F}_i(v_i) = F_i(v_i)(\nu_h/\nu_i) \tag{22}$$

where $\tilde{F}_i(v_i)$ = a representation of the corrected chromatogram
ν_i, ν_h = the specific refractive index increments for a generic ith species and for the highest molecular weight species in the sample, respectively

3.1.2. UV–Vis Spectrophotometer

In general, absorbance spectrophotometers are preferable to differential refractometers, because of their reduced sensitivity to temperature fluctuations. For this type of detector, Beer's law is applied,

$$A(\lambda) = \varepsilon(\lambda) l c \tag{23}$$

where $A(\lambda)$ = the absorbance response function
$\varepsilon(\lambda)$ = the extinction coefficient function (assumed known)
l = the light path length in the cell

For simple polymers, detection at a single wavelength is sufficient. If a multidiode array spectrophotometer were used, then the instantaneous detected mass is proportional to $\sum [A(\lambda)/\varepsilon(\lambda)]$.

3.1.3. Online Viscometer

The use of continuous online viscometers for HPLC was first suggested by Ouano et al. (21), and since then, relatively simple and highly sensitive detectors were produced. Briefly, the pressure drop of the eluting sample Δp_i is measured across a capillary and compared to the pressure drop caused by the solvent Δp_0. The technique is based on the determination of the instantaneous intrinsic viscosity $[\eta]_i$ through

$$[\eta]_i = \frac{1}{c_i}\log\left(\frac{\eta_i}{\eta_0}\right) \cong \frac{1}{c_i}\log\left(\frac{\Delta p_i}{\Delta p_0}\right) = \left(\frac{1}{c_i}\right)\left(\frac{\Delta p_i - \Delta p_0}{\Delta p_0}\right) \tag{24}$$

where c_i is the polymer concentration.

The instantaneous molecular weight M_i may be determined through the Mark–Houwink expression

$$[\eta]_i = K M_i^a \tag{25}$$

where the parameters K and a may be considered constant for a given solvent and temperature, and M_i is strictly speaking, the viscosity–average molecular weight of the detector contents. If the Mark–Houwink constants are not known, then the universal calibration concept may be used (see Section 3.3.2).

3.1.4. Low-Angle Laser Light Scattering Photometry

In LALLS experiments (22), the measured intensity $I(\theta)$ of the light scattered by a polymer solution at a low angle θ (typically 4°–6°) is generally expressed through the Rayleigh ratio

$$R_\theta = r^2\left(\frac{I_\theta}{I_0 V}\right) \tag{26}$$

where r is the distance of observation and $I_\theta/I_0 V$ is the light scattering intensity I_θ per unit volume V normalized with respect to the intensity of the incident beam I_0. The excess Rayleigh factor

$$\bar{R}_\theta = R_\theta|\,\text{solution} - R_\theta^0|\,\text{solvent} \tag{27}$$

is related to the scattering by the solute molecules through

$$\frac{K_\theta c}{\bar{R}_\theta} = \frac{1}{M} + 2A_2 c \tag{28}$$

where A_2 is the second virial coefficient, and K_θ is is an optical constant for a given polymer

$$K_\theta = \frac{2\pi^2 n_0^2}{\lambda_0^4 N_A}\left(\frac{dn}{dc}\right)^2 \tag{29}$$

where N_A is Avogadro's number, λ_0 is the *in vacuo* wavelength of light, n_0 is the refractive index of the solvent, and dn/dc is the specific refractive index increment of the polymer solution.

In SEC, the instantaneous concentration c may be independently determined through a mass detector. Then, Eqs. 26–29 allow the calculation of the corresponding values of M, and the pairs (c_i, M_i) constitute the MWD, obtained in an absolute fashion. Strictly speaking, M is the weight-average molecular weight of the detector cell contents, and A_2 depends on molecular weight according to

$$A_2 = C_5 M^{-\gamma} \tag{30}$$

with $0.15 \leqslant \gamma \leqslant 0.35$ for many polymers in good solvents (23). In practice, owing to the low values of γ, it is usually assumed that A_2 is independent of M, and a mean value of A_2 is determined from LALLS offline measurements. Furthermore, because of the low concentration, the $A_2 c$ term in Eq. 28 is usually taken as zero.

3.2. Instrumental Broadening

Consider the case of a simple polymer with a mass detector. The instrumental broadening process is directly related to the diffusion coefficients of the polymer solutes. If one assumes that such coefficients are dependent on hydrodynamic volume only, then polymer standards of a chemical nature different from that of the analyzed product could be used to estimate peak broadening. This conclusion is known as the "principle of universal peak broadening calibration" (4, 24).

3.2.1. Spreading Calibration for Mass Detectors

The broadening calibration for mass detectors [i.e., obtaining the function $g(t, \tau)$] would be simple if strictly monodisperse polymers were available. However, only proteins are monodisperse, and all synthetic polymers exhibit a distribution of molecular weights. Even if monodisperse polymers were available, owing to instrumental spreading there still remains the uncertainty regarding the assignment of τ. Assuming uniform spreading and polydisperse standards, two uncertainties arise: (a) the determination of the time origin in the calculated function $g(t)$, and (b) the specification of τ. In the first case, the normal procedure is to assign the origin of $g(t)$ at some measure of central tendency of that curve. In the second, it has been suggested (25) to assign τ at some measure of central tendency of the chromatogram corrected for axial spreading. It is easily shown that for uniform spreading, and if the origin of $g(t)$ is assigned at the mean value of that curve, then the mean of $F(t)$ coincides with the mean of the corrected chromatogram $F^c(t)$.

For determining uniform spreading, both narrow and broad-distributed standards can be used. The simplest approximation is to define $g(t)$ as the normalized chromatogram of the monomer or the pure solvent.

The proposed techniques for the determination of $g(t)$ may be classified as follows:

1. Methods that assume such standards to be monodisperse (26). (This assumption is clearly very crude, yielding overcorrected chromatograms.)
2. Methods that employ standards of known MWDs (e.g., Ref. 27).
3. Reverse flow techniques (e.g., Refs. 24 and 28).
4. Recycle techniques (e.g., Refs. 25, 29, and 30). (Reference 25 is possibly the only one that enables the calculation of the spreading without requiring any *a priori* assumption with regard to its shape.)

For the determination of nonuniform spreading, all existing methods employ narrow standards, and adopt the basic assumption that the spreading is uniform in the range of elution times covered by the corresponding peaks. After determination of $g(t)$ at several values of τ, interpolations are required for finding $g(t, \tau)$.

3.2.2. Chromatogram Correction

According to the employed methodology, the approaches that have been proposed for the correction of chromatograms for instrumental spreading may be classified as analytical and phenomenological.

The *analytical* approach is based on studying the broadening at the detector cell level, simultaneously considering the instrumental broadening and the main calibration. For example, in Hamielec et al. (31), a correction equation for dispersion in the case of nonuniform Gaussian instrumental spreading with nonlinear calibration is analytically derived. Assume a calibration defined by

$$\ln M = \ln D_1(v) - D_2(v)v \tag{31}$$

or

$$M(v) = D_1(v)e^{-D_2(v)v} \tag{32}$$

Also, assume that the nonuniform spreading is defined by the set (Y, σ), where σ is the standard deviation of a Gaussian distribution with average retention volume Y. Then, it can be shown (31) that the corrected chromatogram may be obtained from

$$F^c(v) = F(v)\frac{\sigma(v)}{\bar{\sigma}(v)}\exp\left\{-\frac{[v - \bar{Y}(v)]^2}{2\,\bar{\sigma}(v)^2}\right\} \tag{33}$$

with

$$\bar{Y}(v) = v + \frac{1}{D_2(v)}\ln\left\{\frac{F[v + D_2(v)\,\sigma(v)^2]}{\sqrt{F[v - D_2(v)\,\sigma(v)^2]\,F[v + D_2(v)\,\sigma(v)^2]}}\right\} \tag{34}$$

$$\bar{\sigma}(v) = \sigma(v)^2 + \frac{1}{D_2(v)^2}\ln\left\{\frac{F[v - D_2(v)\,\sigma(v)^2]\,F[v + D_2(v)\,\sigma(v)^2]}{F(v)^2}\right\} \tag{35}$$

where $\bar{Y}(v)$ and $\bar{\sigma}(v)^2$ are the mean retention volume and variance of the distribution of the different-sized species in the detector cell.

In the analytical approach (4), expressions used to obtain the molecular weight averages of the whole polymer from the measured chromatogram without calculating $F^c(v)$ have been derived. Such expressions may be generalized to include molecular weight detectors and complex polymers.

The *phenomenological* approach to the deconvolution problem considers the system as a "black box" and does not attempt to investigate the broadening process at the detector cell level. Numerous techniques have been proposed to solve the deconvolution problem. Most of them require the spreading function to be uniform, for example (16, 32–37). Relatively few investigations have attempted the direct solution of the discrete version of Eq. 1, with no assumptions on $g(t, \tau)$, for example, (38–42). A comparison between several existing techniques was given by Gugliotta et al. (43) based on a synthetic example originally proposed by Chang and Huang (44), but also attempted

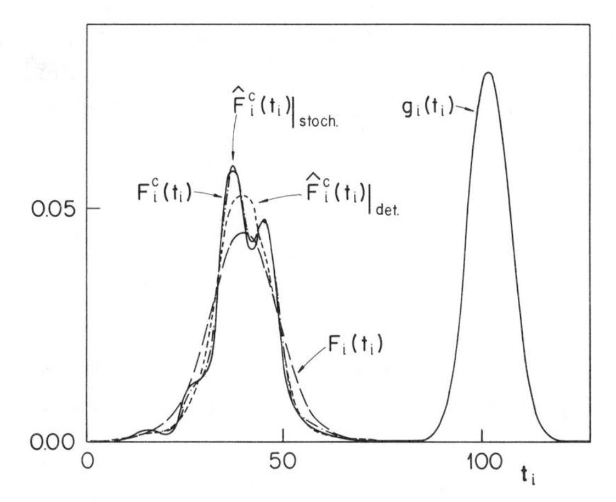

Figure 3. Comparison of deconvolution techniques (———): exact result; (- - - - -): estimation of a deterministic technique; (—·—·): estimation of a stochastic technique. See text for details.

by others (39, 40, 43). The problem is illustrated by Figure 3, which represents the following: $F^c(t_i)$ is considered the "real" solution to the deconvolution problem; $g_i(t_i)$ is the uniform spreading function, $F_i(t_i)$ is the result of convoluting $F^c(t_i)$ with $g_i(t_i)$; then $\hat{F}^c(t_i)|_{\text{det}}$ is the estimated solution via a deterministic method presented in Ishige et al. (39); and $\hat{F}^c(t_i)|_{\text{stoch}}$ is the estimated solution via a stochastic approach presented by Alba and Meira (40). The function $\hat{F}^c(t_i)|_{\text{det}}$ in Figure 3, is very similar to the estimated solution via other deterministic techniques (44, 37); while $\hat{F}^c(t_i)|_{\text{stoch}}$ is close to the solution provided by other stochastic methods (41, 43). Clearly, stochastic methods provide the best numerical results. The reason for this is that they allow better use of all prior information on the measurement noise and on the shape of the expected solution.

Synthetic examples are ideal for comparing the ability of the different techniques to solve the inversion problem. This is so because from a "true" solution and its corresponding estimated curve, the associated mean-square error may be calculated. The best technique may be sought as that which minimizes this parameter. In a real problem, however, one must resort to other checks to validate different possible estimates $F_c(t_i)$. Obvious checks, are (a) the solution must be nonnegative; (b) by processing $F_c(t)$ through the system spreading function, the noise-free measured chromatogram should be recuperated; and (c) the area under the corrected chromatogram should be equal to that of the measured curve. It should be strongly emphasized that item b is only a necessary (but not a sufficient) condition for good results; the reason being the algorithmic singularity of Eq. 1, or its discrete counterpart. This implies that there are, in principle, infinite possible numerical solutions that can recover the noise-free chromatogram. Another useful check is based on minimizing the variance of the "innovations" or "residuals"; that is the difference between the measured chromatogram and the expected noise-free chromatogram, calculated from $F_c(t_i)$.

3.3. Main Calibration

Assume that a molecular weight detector is not available. For finding the molecular weights at each elution time, one may use: (a) standards of the same chemical nature to the analyzed polymer (direct calibration); or (b) standards of a different polymer, usually polystyrene, PS (indirect or universal calibration).

3.3.1. Direct Calibration

The direct calibration is, in principle, more accurate than the indirect calibration. The limitations are that standards for relatively few polymers are available. Calibrations can be made with both narrow or broad-distributed polymers, for which the values of $\bar{M}_n$ and $\bar{M}_w$ have been determined by some independent absolute method.

If a *narrow standard* is employed, a normal procedure is to assign the geometric mean of $\bar{M}_n$ and $\bar{M}_w$ at the elution volume of the chromatogram peak maximum. Narrow polymer standards of polystyrene, poly(α-methyl styrene), polyisoprene, polybutadiene, poly(methyl methacrylate), and so on, are commercially available. Once the pairs (M, v) have been determined, then a curve, such as that of Figure 2, can be presented. The upper and lower limits of the sigmoidal curve correspond, respectively, to the cases of total exclusion from the pores (v_0) and of total inclusion $(v_0 + v_p)$, where v_p is the pore volume. The linear segment of the curve may be represented by Eq. 5. Alternatively, the following expression may be used (45)

$$\log M = C_6 + C_7 v + C_8 v^2 + C_9 v^3 + \cdots \tag{36}$$

where C_6–C_9 are the coefficients of the polynomial.

Consider the calibration with *broad standards*. Balke et al. (46) proposed a method involving the linear calibration

$$M = D_1 e^{-D_2 v} \tag{37}$$

Substituting Eq. 37 into Eqs. 11 and 12, the parameters D_1 and D_2 may be obtained from a single standard covering the molecular weight range of the sample, through

$$\bar{M}_w = \sum F(v) D_1 e^{-D_2 v} \tag{38}$$

$$\bar{M}_n = \frac{D_1}{\sum F(v) e^{D_2 v}} \tag{39}$$

The method by Balke et al. (46) does not include a correction for chromatogram broadening. Such correction could be performed before the calculation of D_1 and D_2. Alternatively, Yau et al. (47) proposed a reformulation of the technique, which included a correction for variable Gaussian broadening. The relevant equations are

$$\bar{M}_w = \exp\left[\frac{-(D_2\sigma)^2}{2}\right] \sum F(v) D_1 \exp(-D_2 v) \tag{40}$$

$$\bar{M}_n = \frac{D_1 \exp[(D_2\sigma)^2/2]}{\sum F(v) \exp(D_2 v)} \tag{41}$$

where σ is the peak standard deviation caused by column dispersion. The searching procedure to find the constants D_1 and D_2 in Eqs. 40 and 41 may be simplified by deriving an equation for the polydispersity (48)

$$\frac{\bar{M}_w}{\bar{M}_n} = \exp[-(D_2\sigma)^2] [\sum F(v) \exp(-D_2 v)] \sum F(v) \exp(D_2 v)] \tag{42}$$

Through Eq. 42, D_2 may be calculated by a single-variable search routine, and then D_1 may be found with either Eq. 40 or Eq. 41. The value of σ may be evaluated in a separate experiment, or found from two broad reference standards, after determination of D_1 and D_2 (48).

There are commercially available broad standards of polypropylene, poly(vinyl chloride), linear poly(vinyl acetate), poly(vinyl alcohol), and so on.

3.3.2. Universal Calibration

According to the Einstein viscosity relation, the intrinsic viscosity $[\eta]$ is defined by

$$[\eta] M = 0.025\, N_A\, V_h \tag{43}$$

where V_h is the hydrodynamic volume of an equivalent sphere, and N_A is Avogadro's number. If hydrodynamic volume or a related parameter controls the separation, a plot of $[\eta]M$ versus v will be the same for all polymers. Experimental evidence of $[\eta]M$ for universal calibration was presented by Grubisic et al. (49), who studied homopolymers and copolymers having various chemical and geometrical structures. In spite of the many deviations that have been reported since its publication, the implications of that work are very important, and for this reason, it is possibly the most important publication in SEC so far. When in addition to size exclusion, adsorption and partition are also present, then according to Dawkins and Hemming (50), an even more general calibration may be obtained by representing $\ln([\eta]M)$ versus $(v - v_0)/K_p$, where v_0 is the interstitial volume indicated in Figure 2, and K_p is the distribution coefficient due to adsorption or partition. ($K_p = 1$ in the case of pure exclusion, and $K_p > 1$ when secondary mechanisms are present.) The main difficulty is the estimation of K_p (dependent on the temperature and on the polymer–solvent–packing interactions).

For quantitative work, and for a given temperature and solvent, the universal curve $\log([\eta]M)$ versus v needs not be utilized as such, but rather its concept. In effect, at any given retention volume, Eq. 43 implies that

$$[\eta]_X(v)M_X(v) = [\eta]_{PS}(v)M_{PS}(v) = J_{PS}(v) \tag{44}$$

where subscript X indicates the analyzed polymer and subscript. PS the polystyrene standard. Two ways of determining M_X from Eq. 44 are possible. One is to measure $[\eta]_X$ through an online viscometer. In this case, at each v

$$M_X = \frac{J_{PS}(v)}{[\eta]_X(v)} \tag{45}$$

Alternatively, if the Mark–Houwink constants K and a for the analyzed polymer and for the standard are known, then, substituting Eq. 25 into Eq. 44, one obtains

$$M_X = \left(\frac{K_{PS}M_{PS}^{a_{PS}+1}}{K_X}\right)^{1/(a_X+1)} \tag{46}$$

4. COMPLEX POLYMERS

Complex polymers require further specifications apart from the MWD. For such characterization, additional detectors are necessary, and in general, at least one detector signal per property is required. In what follows, the general problems related to the determination of the MWD are first considered. Then, the specific analyses of long-chain branching frequency in homopolymers, and of composition distribution in linear copolymers are discussed.

4.1. Calibration and Instrumental Broadening

A complex polymer has an infinite number of calibration curves, one for each molecular weight average. Under conditions of perfect resolution, the detector cell contains polymer species having the same hydrodynamic volume, but possibly very different copolymer compositions and molecular weights.

The universal molecular weight calibration (49), involving the product $[\eta](v)M(v)$, was generalized for complex polymers by Hamielec and Ouano (51), and the derivation follows. Assume that for a given retention volume, several molecular species $X_i (i = 1,2,3,\cdots)$ of the analyzed polymer X have the same hydrodynamic volume. Then, with no instrumental spreading, and in the absence of secondary mechanisms, Eq. 44 provides

$$[\eta]_{X_1}(v)M_{X_1}(v) = [\eta]_{X_2}(v)M_{X_2}(v) = \cdots = [\eta]_{PS}(v)M_{PS}(v) \tag{47}$$

$$J_1(v) = J_2(v) = \cdots = J_{PS}(v) \tag{48}$$

Defining G_i as the weight fraction associated with X_i, then the intrinsic viscosity of the detector cell contents may be calculated from

$$[\eta]_X(v) = G_1 [\eta]_{X_1}(v) + G_2[\eta]_{X_2}(v) + \cdots \tag{49}$$

$$= G_1 \frac{J_1(v)}{M_{X_1}(v)} + G_2 \frac{J_2(v)}{M_{X_2}(v)} + \cdots \tag{50}$$

$$= J_{PS}(v)\left(\frac{G_1}{M_{X_1}(v)} + \frac{G_2}{M_{X_2}(v)} + \cdots\right) \tag{51}$$

But,

$$\bar{M}_n(v) = \left(\frac{G_1}{M_1} + \frac{G_2}{M_2} + \cdots\right)^{-1} \tag{52}$$

Hence,

$$J_{PS}(v) = [\eta]_{PS}(v)\, M_{PS}(v) = [\eta]_X(v)\, \bar{M}_n(v) \tag{53}$$

Thus, under ideal chromatographic conditions, and assuming that: (a) the universal calibration curve $\{[\eta]_{PS}(v)\, M_{PS}(v)\}$ versus v for a linear polystyrene standard is available; and (b) a continuous measurement of the instantaneous intrinsic viscosity $[\eta]_X(v)$ is feasible, then the number-average molecular weight of the complex polymer $\bar{M}_n(v)$ may be obtained through Eq. 53. Considering Eq. 53, it is rather unfortunate that continuous $\bar{M}_n$ detectors are not yet commercially available; but promising prototypes based on the shrinkage of soft gel particles by osmotic pressure are presently being developed (52). Also note that prior to the calculation of $[\eta]_X(v)$ through Eq. 24, the viscosity measurement Δp_i versus v, ought to be corrected for instrumental broadening.

4.2. Long-Chain Branching in Homopolymers

The characterization of branched polymers introduces additional parameters such as: (a) the functionality of branch points, (b) the number of branches per macromolecule, and (c) the length and distribution (star, comb, or random) of branches (see Chapters 3 and 7). Besides the functionality of branch points (that can be estimated from the polymerization mechanism), only two parameters will be of interest here: the number of branches per polymer chain m and the chain branching frequency λ_b (i.e., the number of branches per molecular weight unit):

$$\lambda_b = \frac{m}{M} \tag{54}$$

Short-chain and long-chain branching do not affect the same polymer properties; the first greatly alters solid-state properties, whereas the latter introduces large changes in viscometric properties.

The basic concept regarding the use of SEC is that the occurrence of long branches in a molecule reduces its hydrodynamic volume with respect to a linear chain of the same chemical nature and molecular weight.

To calculate the extent of long-chain branching, Stockmayer and Fixman (53) defined a branching parameter g as the ratio of the average values of the square radius of gyration of a branched molecule $\langle S^2 \rangle_b$, to the square of the radius of gyration of a linear chain $\langle S^2 \rangle_l$ of the *same* molecular weight

$$g = \frac{\langle S^2 \rangle_b}{\langle S^2 \rangle_l} \qquad M_b = M_l \tag{55}$$

The parameter g can be classically determined by light scattering if a linear sample with the same molecular weight is available, or if the relationship $\langle S^2 \rangle = f(M)$ is known for the linear structure. Also, g can be theoretically related to the number-average value of the number of branches per molecule

$\bar{m}_n$. The most commonly used relationships, valid for monodisperse chains in theta solvents are (54)

Trifunctional branch points

$$g = \left[\left(1 + \frac{\bar{m}_n}{7}\right)^{1/2} + \frac{4\bar{m}_n}{9\pi} \right]^{-1/2} \tag{56}$$

Tetrafunctional branch points

$$g = \left[\left(1 + \frac{\bar{m}_n}{6}\right)^{1/2} + \frac{4\bar{m}_n}{3\pi} \right]^{-1/2} \tag{57}$$

The evaluation of g is complicated in practice, because whole polymer samples are polydisperse. This is why SEC fractionation, prior to any measurement, simplifies the problem of branching evaluation. Furthermore, because intrinsic viscosity measurements are much easier to carry out than measurements radius of gyration, a second branching parameter g' was introduced (55), as the ratio of intrinsic viscosities of homologous branched and linear samples

$$g' = \frac{[\eta]_b}{[\eta]_l} < 1 \qquad M_b = M_l \tag{58}$$

The evaluation of $\bar{m}_n$ from g' is not feasible, since no theoretical expressions for g' have been published until now. However, empirical relationships between g and g' of the type

$$g' = g^x \tag{59}$$

are available, and in the case of low-density polyethylene, $x = 1.2 \pm 0.2$ (56).

Size exclusion chromatography studies of long-chain branching requires a universal calibration from well-characterized fractions of the corresponding linear polymer. For each chromatographic fraction of a branched sample

$$[\eta]_{i,b} M_{i,b} = [\eta]^*_{i,l} M^*_{i,l} \equiv K M^{*(a+1)}_{i,l} \tag{60}$$

where $[\eta]^*_{i,l}$, $M^*_{i,l}$ correspond to the linear polymer fraction with the same retention volume. From Eq. 60

$$[\eta]_{i,b} = \frac{K M^{*(a+1)}_{i,l}}{M_{i,b}} \tag{61}$$

Also,

$$[\eta]_{i,l} = K(M_{i,l})^a = K(M_{i,b})^a \tag{62}$$

Substituting Eqs. 61 and 62 into Eq. 58 provides

$$g' = \left[\frac{M_{i,l}^*}{M_{i,b}}\right]^{a+1} \tag{63}$$

or,

$$g' = \frac{([\eta]_{i,b} M_{i,b})/K}{\{([\eta]_{i,l}^* M_{i,l}^*)/[\eta]_{i,b}\}^{a+1}} \tag{64}$$

$$g' = \left(\frac{[\eta]_{i,b}}{[\eta]_{i,l}^*}\right)^{a+1} \tag{65}$$

Equation 65 indicates that g' can be determined by SEC if the viscosity law of a linear polymer is known, and if each $[\eta]_{i,b}$ is measured.

Continuous online LALLS detection has been used for the determination of g and g' (57). Online continuous viscometry has been utilized by Lecacheaux et al. (58, 59) for the rapid characterization of chain branching in polyethylene and complex polyolefins. Taken from Ref. (59), Figure 4 illustrates a complete analysis by a refractometer–viscometer detection system. In Figure 4*a*, the experimental viscosity law deduced from the viscometric signal and Eq. 60 is compared to the Mark–Houwink relationship for linear polyethylene. Also, g' versus $\log M$ may be obtained through Eq. 58, Figure 4*b*. If $x = 1.2$ in Eq. 65, and instead of Eq. 56 we utilize the simple relationship (54)

$$g = \frac{3}{2}\left(\frac{\pi}{\lambda_b M}\right)^{1/2} - \frac{5}{2\lambda_b M} \tag{66}$$

where $\bar{m}_n$ has been replaced by $\lambda_b M$, then,

$$g' = \left[\frac{3}{2}\left(\frac{\pi}{\lambda_{i,b} M_{i,b}}\right)^{1/2} - \frac{5}{2\lambda_{i,b} M_{i,b}}\right]^{1.2} \tag{67}$$

Equation 67 allows the determination of $\lambda_{i,b}$, which is represented in Figure 4*d*, while the MWD is shown in Figure 4*c*. Finally, it is important to emphasize that axial dispersion correction is indispensable if rational values for g' are to be obtained (60).

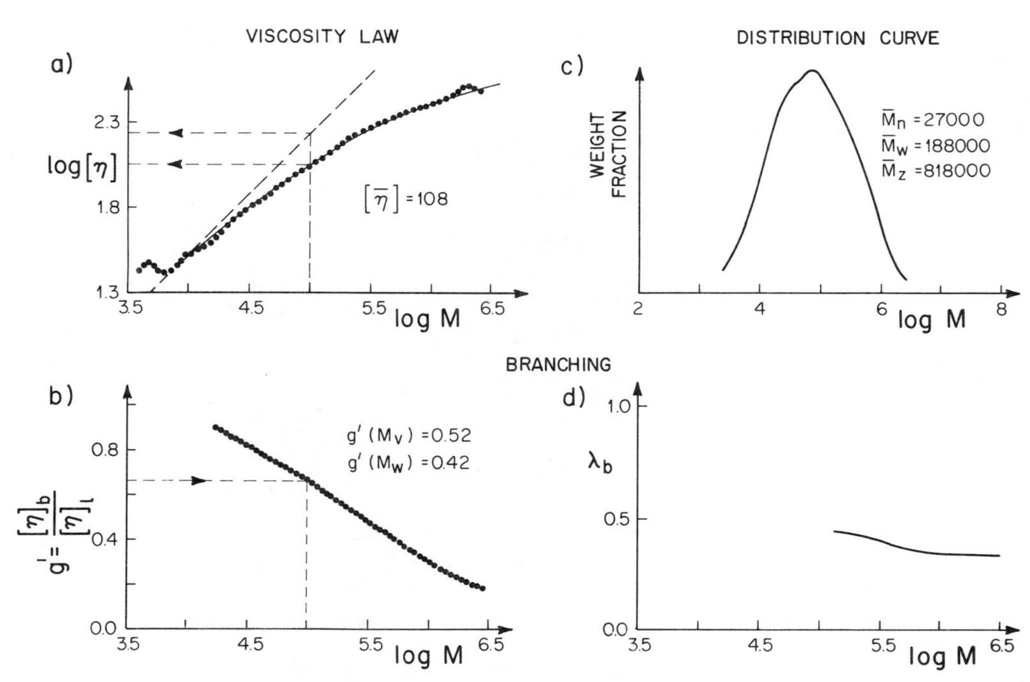

Figure 4. A long-chain branching study [after Lecacheaux et al. (58)]. See text for details.

4.3. Molecular Weight and Chemical Composition Distributions in Linear Copolymers

Let us restrict ourselves to the simplest case of linear copolymers with repeating units A and B, where the sequence length distribution and the

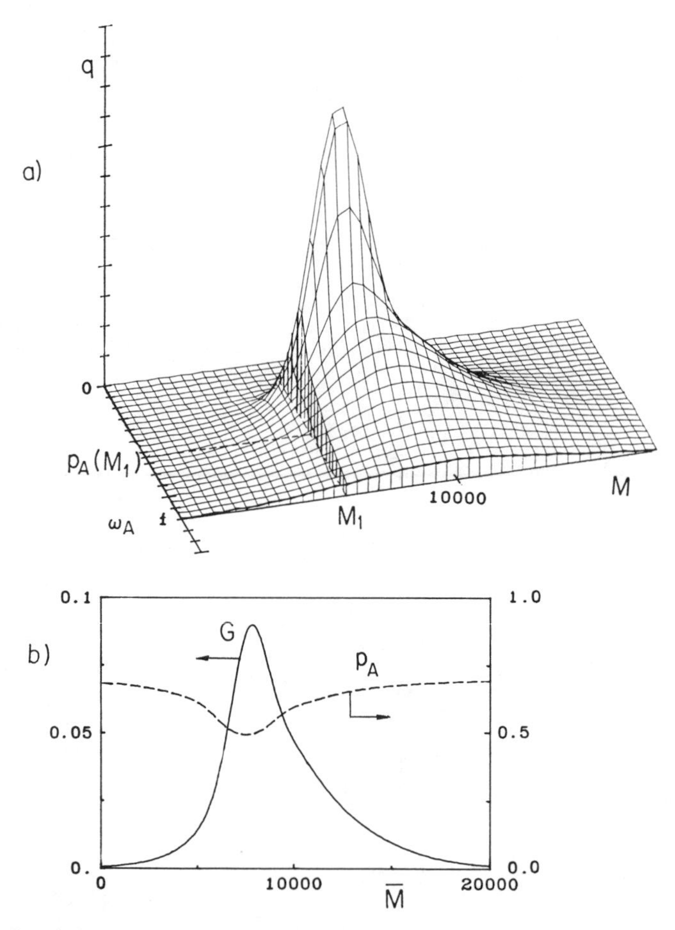

Figure 5. Copolymer representations. (*a*) The combined molecular weight–chemical composition distribution. Every type of molecular species is characterized by its molecular weight M, the weight fraction of comonomer A in the species w_A, and the mass associated with the molecular type q. For any given M_1, the average weight fraction of A is $p_A(M_1)$. (*b*) The average molecular weight–average composition distribution. It consists of two functions of the same independent variable $\bar{M}$ (i.e., any generic molecular weight average of the detector cell contents). G represent the total mass for a given M.

stereoregularity distribution will not be considered. In what follows, the problems associated with the measurement of the combined distribution of molecular weights and chemical composition will be discussed on the basis of the approach by Meira and García–Rubio (61).

Possibly, the most "natural" representation of the combined molecular weight–chemical composition distribution (MWD–CCD) of a linear copolymer is through the function $q(n_A, n_B)$, where q is the mass (in grams) of a molecular species with n_A repeating units of comonomer A and n_B repeating units of comonomer B. However, no direct way of measuring such a function is available. An alternative representation, that lends itself more to SEC measurements, is given by $q(M, w_A)$, where M is the molar mass (in grams per mole) of each molecular species, and w_A is the weight fraction of A corresponding to the polymer fraction with a common molar mass M.

At each M, and except for the cases of constant comonomer composition, a distribution in w_A is to be expected. Strictly speaking, all of the mentioned distributions are discrete. However, in the case of high-molecular-weight polymers, continuous surfaces such as that of Figure 5*a* may be utilized to represent such distributions. Discrete distributions are made continuous with the criterion that the volume determined by the continuous surface and any "base" $\Delta w_A \times \Delta M$ on the $w_A - M$ plane is proportional to the total mass of the molecular species contained within such a base.

Standard SEC equipment is presently unable to determine a distribution of compositions at each hydrodynamic volume. To solve this problem, the use of two SEC instruments, each running with a different mobile phase has been proposed (62, 63). This technique, called "cross" or "orthogonal" chromatography is very promising, but so far does not provide acceptable quantitative information (see Chapter 1).

Let us define $G(M)$ as the total mass associated with a given M, and $p_A(M)$ the average weight fraction of comonomer A, of a given M, that is:

$$G(M) = \sum_{w_A} q(M, w_A) \tag{68}$$

$$p_A(M) = \left[\sum_{w_A} q(M, w_A) w_A(M) \right] \Big/ \left[\sum_{w_A} q(M, w_A) \right] \tag{69}$$

Every retention volume in SEC corresponds to an average value of M, indicated by $\bar{M}$. The pair of functions $G(\bar{M})$ and $p_A(\bar{M})$ are called the average molecular weight–average composition distribution (AMWACD). For example, the combined distribution of Figure 5*a* reduces to the representation of Figure 5*b*. The problems associated with the measurement of the AMWACD are considered in the following discussions.

The refractometer–UV dual detection system has been mostly investigated because of the variety of copolymers based on styrene (or aromatic monomers), which are easily detected by UV spectroscopy (absorption at 254 nm). Typical studies are, for example, given in Refs. 64–66. Consider now a way of calculating the AMWACD from the signals of a refractometer Δn and of a spectrophotometer A. At each retention volume, note that

$$p_A G = \text{mass of comonomer A in detector cell} \tag{70}$$

Therefore, assuming that the spectrophotometer responds to comonomer A only, the following may be written

$$\Delta n = [\nu_A p_A + \nu_B (1 - p_A)] G \tag{71}$$

$$\mathrm{A} = k_A p_A G \tag{72}$$

From the last two equations

$$p_A = \frac{\nu_B (\mathrm{A}/\Delta n)}{k_A + (\nu_B - \nu_A)(\mathrm{A}/\Delta n)} \tag{73}$$

$$G = \frac{\Delta n}{\nu_B} + \frac{(\nu_B - \nu_A)\mathrm{A}}{k_A \nu_B} \tag{74}$$

From Eqs. 73 and 76 and after calibration, $G(M)$ and $p_A(M)$ may be readily derived. In all cases, the dead volume between the two detector cells should be taken into account. In the determination of p_A through Eq. 73, a particular problem is created at the extreme limits of the MWD curve, due to the propagation of errors (17), and to possible variations of the response factor (especially the UV response) with the concentration (65, 67). A review of data interpretation of SEC detectors for copolymer analysis is presented in Ref. 68. Calculation of the AMWACD requires a calibration and corrections for secondary fractionations and for instrumental broadening.

The average molecular weight $\bar{M}$ of the copolymer species appearing at a given retention volume v can be estimated roughly by using the calibration curves (molecular weight vs retention volume) of each of the homopolymers composing the copolymer (64)

$$\log \bar{M} = p_A \log M_A + p_B \log M_B \tag{75}$$

where M_A and M_B are the molecular weights of homopolymers A and B at a given retention volume. Alternatively, the following expression has been proposed (65)

$$\bar{M} = x_A M_A + x_B M_B \tag{76}$$

where x_A and x_B are the average mole fractions for each of the comonomers. It should be noted that an accurate calculation of the instantaneous $\bar{M}_w$ with a LALLS detector is in principle impossible, unless the complete combined MWD–CCD is measured. This occurs because the refractive index and the specific refractive index increment are, in general, different for each of the comonomers.

Consider the correction for secondary fractionations. With SEC only an average composition is measured at each retention volume. At every hydrodynamic volume, the broadening due to adsorption, partition and so on is directly related to the complete composition distribution. Therefore, without prior information about that distribution (except for its mean), only an approximate correction may be performed that is based on the observed shifts for A and B homopolymers (61). In fact, even if the combined distribution of molecular weights and composition could be measured by orthogonal chromatography, corrections for secondary fractionation distortion are always difficult to implement (61).

Finally, let us examine the instrumental broadening correction. Two solution paths seem possible: (1) to correct the raw measurements $\Delta' n(t)$ and $A'(t)$ from spreading through

$$\Delta' n(t) = \int_{-\infty}^{\infty} g_{\Delta n}(t, \tau)\, \Delta n_i(\tau)\, d\tau \tag{77}$$

$$A'(t) = \int_{-\infty}^{\infty} g_A(t, \tau)\, A(\tau)\, d\tau \tag{78}$$

prior to the calculation of $p_A(t)$ and $G(t)$; or (2) to calculate $G'(t)$ and $p'_A(t)$ from the raw measurements, and correct those derived variables for instrumental broadening with

$$G'(t) = \int_{-\infty}^{\infty} g_G(t,\tau)\, G(\tau)\, d\tau \tag{79}$$

$$p'_A(t) = \int_{-\infty}^{\infty} g_{P_A}(t, \tau) p_A(\tau)\, d\tau \tag{80}$$

A recent publication (69) indicates that the path involving Eqs. 82 and 83 is preferable. Consider now the way of estimating the broadening functions $g_{\Delta n}$, $g_{\rm A}$, $g_{\rm G}$ and $g_{p_{\rm A}}$. Assume that strictly monodisperse copolymers in both hydrodynamic volume and molecular weight were available. In this case, $p_{\rm A}$ is some arbitrary but constant value, and from Eqs. 73 and 74, it follows that Δn and A are both proportional to the mass G. For this reason, and calling $g_{\rm PS}(t, \tau)$ the spreading calibration obtained through normal techniques with a mass detector and narrow polystyrene standards (25, 29, 30), one can write

$$g_{\Delta n}(t, \tau) = g_{\rm A}(t, \tau) = g_{\rm G}(t, \tau) = g_{p_{\rm A}}(t, \tau) = g_{\rm PS}(t, \tau) \tag{81}$$

Thus, correction for instrumental broadening with dual detection may involve two independent deconvolutions (Eqs. 79 and 80), with the spreading functions obtained in standard fashion.

5. CONCLUSIONS

There are many problems associated with detector signal analysis in SEC, particularly when accurate molecular characteristics of polymers are to be inferred. The most important limitations are

1. Even with an ideal chromatographic system, separation is produced according to molecular size and not to molecular weight. For this reason, in the case of complex polymers, a distribution of molecular weights will be always present in the detector cell, and at most, average values of such distributions can be estimated.
2. The instrumental spreading is unavoidable, and the resolution capacity is important at low molecular weights only. The instrumental broadening problem is relatively well solved only in the case of mass detectors and linear homopolymers. However, when online molecular weight detectors are utilized, or when complex polymers are analyzed, only relatively crude solutions are possible.
3. In the case of homopolymers or strictly alternate copolymers, presence of secondary fractionation mechanisms introduce uniform time shifts in the chromatogram, but no peak distortion. In the case of statistical or block copolymers, however, secondary fractionations may severely distort the chromatogram shape, and with the current state of technology, it is virtually impossible to recuperate the true curve.

In spite of all its limitations, SEC is the most important technique for the characterization of synthetic polymers. The measurement of MWDs in linear homopolymers is the best solved situation. For long-chain branching studies in homopolymers, the solution is not so elegant, because it is based on the rather indirect concept of hydrodynamic volume shrinkage as the number of linear chains is increased (chromatographic concentration effect). For linear copolymers of two repeating unit types, the difficulties are related to the calibration with absolute detectors, to the errors introduced in the calculation of the average composition at the low and high-molecule-weight limits, and to the unsurmountable problems introduced by secondary interactions with the packing.

In the near future, the following advances are to be expected. Points 3 and 4 are essential for the characterization of branched copolymers.

1. The more generalized use of SEC in online process monitoring and control.
2. The more generalized use of multiple detection, in particular molar mass detectors and diode array spectrophotometers.
3. New theoretical developments to enable the measurement of long-chain branching from a more direct or fundamental basis.
4. The development of quantitative methods for the fractionation of copolymers according to composition.

ACKNOWLEDGMENTS

The assistance of L. H. García-Rubio (University of South Florida) for help in revising the manuscript is gratefully appreciated. The financial support of CONICET, Universidad Nacional del Litoral and the National Science Foundation (NSF) is acknowledged.

NOTATION

a	Mark–Houwink exponent (Eq. 25)
A, $A(\lambda)$	Undistorted UV–vis spectrophotometer response (Eqs. 23 and 72)
A'	UV response, distorted for instrumental broadening (Eq. 78)
A_2	Second virial coefficient (Eqs. 28 and 30)
b_i	Mass of ith oligomer (Eq. 2)
c, c_i	Polymer concentration ($g/100\,cm^3$)
$C_i (i = 1, \ldots, 9)$	Constant coefficient
D_1, D_2	Coefficients of linear SEC calibration (Eq. 37)

$D_1(v)$, $D_2(v)$	Coefficients of nonlinear SEC calibration (Eq. 31)
f'	Slope of calibration curve (Eqs. 3 and 8)
$F(t)$, $F(v)$	Measured chromatogram
$F^c(t)$	Chromatogram, corrected for instrumental broadening
$\hat{F}^c(t)$	Estimation of the "true" undistorted chromatogram
$\tilde{F}(t)$	Chromatogram corrected for refractive index dependence with M (Eq. 22)
g	Original branching parameter (Eq. 55)
g'	Second branching parameter (Eq. 58)
$g(t, \tau)$	Nonuniform instrumental spreading function
$g(t - \tau)$	Uniform instrumental spreading function
$G(M)$	Undistorted weight MWD, with G representing mass or weight fraction
$G'(M)$	Weight MWD, distorted by instrumental spreading
I_θ, I_o	Scattering light intensity at angle θ and of incident beam, respectively (Eq. 26)
$J_{PS}(v)$	"Universal" hydrodynamic volume, obtained with polystyrene standards (Eq. 44)
K	Mark–Houwink preexponential parameter (Eq. 25)
K_θ	Optical constant in LALLS (Eqs. 28 and 29)
K_p	Distribution coefficient in Ref. 50
k_A	Calibration constant in UV spectrophotometry (Eq. 72)
l	Cell path length in UV–vis spectrophotometry (Eq. 23)
m	Number of branches per polymer molecule
$\bar{m}_n$	Number-average value of m
M	Molecular weight, g/mol
$M^*_{i,l}$	Molecular weight of a linear polymer with retention time i (Eq. 60)
$\bar{M}_n$, $\bar{M}_v$, $\bar{M}_w$, $\bar{M}_z$	Number-, viscosity-, weight-, and z-average molecular weights, respectively
$\bar{M}$	Generic average molecular weight
M_m	Monomer molecular weight
n, n_o	Refractive index of polymer solution and pure solvent, respectively
$N(M)$	Number MWD
N_A	Avogadro's number
$p_A(M)$	Undistorted composition distribution of comonomer A, with p_A representing the average weight fraction
$p'_A(M)$	Composition distribution, distorted by instrumental spreading (Eq. 80)
$q(M, w_A)$	Combined molecular weight–chemical composition distribution, with q representing mass

r	Observation distance in LALLS (Eq. 26)
R_θ, $\bar{R}_\theta$	Rayleigh ratio and excess Rayleigh factor, respectively (Eqs. 26 and 27)
$\langle S^2 \rangle$	Mean–square radius of gyration
t	Retention time
v	Retention volume
v_o, v_p	Interstitial and pore volumes, respectively
V	Light scattering volume (Eq. 26)
V_h	Hydrodynamic volume
w_A	Weight fraction of comonomer A for the type of molecular species considered
x	Exponent relating g with g' (Eq. 59)
x_A	Average mole fraction of comonomer A (Eq. 76)
Y, $\bar{Y}$	Average retention volume of nonuniform Gaussian instrumental spreading and instantaneous average retention volume in detector cell contents, respectively (Eqs. 33 and 34)
γ	Exponent in second virial coefficient relationship (Eq. 30)
Δn	Undistorted differential refractometer signal
$\Delta' n$	Differential refractometer signal, distorted by instrumental spreading (Eq. 77).
$\varepsilon(\lambda)$	UV–vis extinction coefficient (Eq. 23)
$[\eta]$	Intrinsic viscosity
$[\eta]^*_{i,l}$	Intrinsic viscosity of a linear polymer of retention time i (Eq. 60)
θ	Scattering angle
λ	Wavelength of light
λ_b	Number of branches per molecular weight unit (Eq. 54)
ν	Specific refractive index increment (Eq. 18)
σ, $\bar{\sigma}$	Standard deviation in nonuniform Gaussian instrumental spreading and instantaneous average standard deviation in detector cell contents, respectively (Eqs. 33 and 35)
τ	Retention time (dummy variable)

REFERENCES

1. J. C. Moore, "GPC. I. New Method for MWD of High Polymers," *J. Polym. Sci. A2*, 835 (1964).
2. W. W. Yau, J. J. Kirkland, and D. D. Bly, *Modern Size Exclusion Liquid Chromatography*, Wiley-Interscience, New York, 1979.

3. B. G. Belenkii and L. Z. Vilenchik, *Modern Liquid Chromatography of Macromolecules*, Elsevier, Amsterdam, 1983.
4. J. Janča (Ed.), *Steric Exclusion Liquid Chromatography of Polymers*, Marcel Dekker, New York, 1984.
5. H. Determann, *Gel Chromatography*, Springer-Verlag, New York, 1967.
6. T. Kremmer and L. Boross, *Gel Chromatography*, Wiley-Interscience, New York, 1979.
7. L. Fisher, *Gel Chromatography*, Elsevier, Amsterdam, 1980.
8. G. Glöckner, *Polymer Characterization by Liquid Chromatography*, Elsevier, Amsterdam, 1986.
9. P. L. Dubin (Ed.), *Aqueous Size-Exclusion Chromatography*, Elsevier, Amsterdam, 1988.
10. K. H. Altgelt and L. Segal (Eds.), *Gel Permeation Chromatography*, Marcel Dekker, New York, 1971.
11. J. Cazes and X. Delamare (Eds.), *Liquid Chromatography of Polymers and Related Materials*, Chromatographic Science Series, Vol. 8, Marcel Dekker, New York, 1980.
12. J. Cazes and X. Delamare (Eds.), *Liquid Chromatography of Polymers and Related Materials II*, Chromatographic Science Series, Vol. 13, Marcel Dekker, New York, 1980.
13. T. Provder (Ed.), *Size Exclusion Chromatography (GPC)*, *Am. Chem. Soc. Symp. Ser.*, **138**, (1980).
14. T. Provder (Ed.), *Size Exclusion Chromatography*, *Am. Chem. Soc. Symp. Ser.*, **245**, (1984).
15. T. Provder (Ed.), *Detection and Data Analysis in Size Exclusion Chromatography*, *Am. Chem. Soc. Symp. Ser.*, **352**, (1987).
16. L. H. Tung, "Method of Calculating MWD Function from Gel Permeation Chromatograms," *J. Appl. Polym. Sci.*, **10**, 375 (1966).
17. L. H. García-Rubio, "Multiple Detectors in SEC. I. Signal Analysis," *Am. Chem. Soc. Symp. Ser.*, **352**, 220 (1987).
18. J. Lesec, "Problems Encountered in the Determination of Average Molecular Weights by GPC Viscometry," in J. Cazes and X. Delamare (Eds.), *Liquid Chromatography of Polymers and Related Materials II*, Chromatographic Science Series, Vol. 13, Marcel Dekker, New York, 1980, pp. 1–17.
19. J. M. Evans, Rapra Bulletin (Nov.), 334 (1972).
20. G. R. Meira and J. F. Johnson, "Gel Permeation Chromatography: Automatic Data Acquisition and Reduction in Real-Time with a Process Computer," *Polym. Eng. Sci.*, **21**, 57 (1981).
21. A. C. Ouano, D. L. Horne, and A. R. Gregges, "GPC. IX. Instrumental Design and Computer Interfacing of a GPC Molecular Weight Detector System," *J. Polym. Sci., Polym. Phys. Ed.*, **12**, 307 (1974).

22. A. C. Ouano, "Recent Advances in GPC," *Rubber Chem. Technol.*, **54**, 535 (1981).
23. C. Quivoron, in G. Champetier (Ed.), *Chimie Macromoleculaire* II, Hermann, Paris, 1972.
24. L. H. Tung and F. R. Runyon, "Calibration of Instrumental Spreading for GPC," *J. Appl. Polym. Sci.*, **13**, 2397 (1969).
25. D. Alba and G. R. Meira, "Calibration of Instrumental Spreading in SEC by a Novel Recycle Technique," *J. Liq. Chromatogr.*, **9**, 1141 (1986).
26. M. Hess and R. F. Kratz, "Axial Dispersion of Polymer Molecules in GPC," *J. Polym. Sci.*, Part A-2, **10**, 375 (1966).
27. A. Husain, A. E. Hamielec, and J. Vlachopoulos, "A New Method for Identifying and Estimating the Parameters of the Instrumental Spreading in SEC. Application to Particle Analysis," *J. Liq. Chromatogr.*, **4**, 459 (1981).
28. L. H. Tung, J. C. Moore, and G. W. Knight, "Calculating MWD Function [of Polymer] from Gel Permeation Chromatograms. II. Evaluation of the Method by Experiments," *J. Appl. Polym. Sci.*, **10**, 1261, (1966).
29. J. L. Waters, "Determination of the Width of a Narrow MWD by GPC," *J. Polym. Sci. Part A2*, **8**, 411 (1970).
30. F. L. McCrackin and H. L. Wagner, "Measurement of Polydispersity of Narrow Fractions and Column Spreading Parameters by Recycle Liquid SEC," *Macromolecules*, **13**, 685 (1980).
31. A. E. Hamielec, H. J. Ederer, and K. H. Ebert, "SEC of Complex Polymers. Generalized Analytical Corrections for Imperfect Resolution," *J. Liq. Chromatogr.*, **4**, 1697 (1981).
32. L. H. Tung, "Method of Calculating MWD from Gel Permeation Chromatograms. III. Application of the Method," *J. Appl. Polym. Sci.*, **10**, 1271 (1966).
33. L. H. Tung, "Correction of Instrumental Spreading in GPC," *J. Appl. Polym. Sci.*, **13**, 775 (1969).
34. S. T. Balke and A. E. Hamielec, "Polymer Reactors and MWD. A Method of Interpreting Skewed GPC Chromatograms," *J. Appl. Polym. Sci.*, **13**, 1381 (1969).
35. P. E. Pierce and J. E. Armonas, "A Method of Solving the Tung Axial Diffusion Equation for GPC," *J. Polym. Sci. Part C*, **21**, 23 (1968).
36. R. V. Figini, "Relationship between Apparent and True Molecular Weight in GPC. Part 1. A New Analytical Solution of the Tung's Axial Dispersion Integral," *Polym. Bull.*, **1**, 619 (1979).
37. D. Alba and G. R. Meira, "Instrumental Broadening Correction in SEC through Fast Fourier-Transform Techniques," *J. Liq. Chromatogr.*, **6**, 2411 (1983).
38. K. S. Chang and Y. M. Huang, "A New Method for Calculating and Correcting Molecular Weight Distribution from GPC," *J. Appl. Polym. Sci.*, **13**, 1459 (1969).
39. T. Ishige, S. I. Lee, and A. E. Hamielec, "Solution of Tung's Axial Dispersion Equation by Numerical Techniques," *J. Appl. Polym. Sci.*, **15**, 1607 (1971).
40. D. Alba and G. R. Meira, "Inverse Optimal Filtering Method for the Instrumental Spreading Correction in SEC," *J. Liq. Chromatogr.*, **7**, 2833 (1984).

41. L. M. Gugliotta, D. Alba, and G. R. Meira, "Correction for Instrumental Broadening in SEC through a Stochastic Matrix Approach Based on Wiener Filtering Theory," *Am. Chem. Soc. Symp. Ser.*, **352**, 287 (1988).
42. E. M. Rosen and T. Provder, "Instrument Spreading Correction in GPC. III. General Shape Function using Singular Value Decomposition with a Nonlinear Calibration Curve," *J. Appl. Polym. Sci.*, **15**, 1687 (1971).
43. L. M. Gugliotta, J. R. Vepa, and G. R. Meira, "Instrumental Broadening Correction in SEC, Comparison of Several Deconvolution Techniques", *J. Liq. Chromatogr.*, **13**, 1671 (1990).
44. K. S. Chang and Y. M. Huang, "Generalized Method for Correcting Instrumental Spreading in GPC," *J. Appl. Polym. Sci.*, **16**, 329 (1972).
45. S. Mori and T. Suzuki, "Optimized Mathematical Approximation of Calibration Curve for SEC," *J. Liq. Chromatogr.*, **3**, 343 (1980).
46. S. T. Balke, A. E. Hamielec, B. P. Le Clair, and S. L. Pearce, "GPC Calibration Curve from Polydisperse Standards," *Ind. Eng. Chem., Prod. Res. Dev.*, **8**, 54 (1969).
47. W. W. Yau, H. J. Stoklosa, and D. D. Bly, "Calibration and Molecular Weight Calculations in GPC using a New Practical Method for Dispersion Correction," *J. Appl. Polym. Sci.*, **21**, 1911 (1977).
48. A. E. Hamielec and S. N. E. Omoridon, "Molecular Weight and Peak Broadening Calibration in SEC. Use of Multiple Broad MWD Standards for Linear Polymers," *Am. Chem. Soc. Symp. Ser.*, **138**, 183 (1980).
49. Z. Grubisic, P. Rempp, and H. Benoit, "A Universal Calibration for GPC," *J. Polym. Sci. Part B*, **5**, 753 (1967).
50. J. V. Dawkins and M. Hemming, "GPC with Crosslinked Polystyrene Gels and Poor and Theta Solvents for Polystyrene. 2. Separation Mechanism," *Makromol. Chem.*, **176**, 1795 (1975).
51. A. E. Hamielec and A. C. Ouano, "Generalized Universal Molecular Weight Calibration Parameter in GPC," *J. Liq. Chromatogr.*, **1**, 111 (1978).
52. W. W. Yau, "An Online Osmometer for Size Exclusion Chromatography", *J. Appl. Polym. Sci., Appl. Polym. Symp.*, (1991), in press.
53. W. H. Stockmayer and M. Fixman, "Dilute Solutions of Branched Polymers," *Ann. N. Y. Acad. Sci.*, **57**, 334 (1953).
54. B. H. Zimm and W. H. Stockmayer, "The Dimensions of Chain Molecules Containg Branches and Rings," *J. Chem. Phys.*, **17**, 1301, (1949).
55. B. H. Zimm and R. W. Kilb, "Dynamics of Branched Polymer Molecules in Dilute Solution," *J. Polym. Sci.*, **37**, 19 (1959).
56. R. Casper, V. Biskup, H. Lange, and U. Phol, "Determination of Long-Chain Branching in High Pressure Polyethylene by Light-Scattering and Viscosity Measurements," *Makromol. Chem.*, **117**, 1111 (1976).
57. D. E. Axelson and W. C. Knapp, "SEC and LALLS. Application to the Study of Long-Chain Branched Polyethylene," *J. Appl. Polym. Sci.*, **25**, 119 (1980).
58. D. Lecacheaux, J. Lesec, and C. Quivoron, "Characterization of Branched

Polyolefins by Coupling High-Speed GPC with a High-Temperature Continuous Viscometer," *Am. Chem. Soc. Polym. Div. Prepr.*, **23**(2), 126 (1982).

59. D. Lecacheaux, J. Lesec, and C. Quivoron, "High-Temperature Coupling of High-Speed GPC with Continuous Viscometry. I. Long-Chain Branching in Polyethylene," *J. Appl. Polym. Sci.*, **27**, 4867 (1982).
60. L. Marais, Z. Gallot, and H. Benoit, "Recycle GPC: Its Use for Determination of Separation Efficiency and Polydispersity of Narrow Standards," *Analusis*, **10**, 443 (1976).
61. G. R. Meira and L. H. García-Rubio, "Correction for Instrumental and Secondary Broadening in the Chromatographic Analysis of Linear Copolymers," *J. Liq. Chromatogr*, **12**, 997, (1989).
62. S. T. Balke and R. D. Patel, "Coupled GPC/HPLC: Copolymer Composition and Axial Dispersion Characterization," *J. Polym. Sci., Polym. Lett. Ed.*, **18**, 453 (1980)
63. S. T. Balke and R. D. Patel, "Orthogonal Chromatography: Polymer Cross-Fractionation by Coupled GPC," *Am. Chem. Soc. Polym. Div. Prep.*, **21**(1), 290 (1981).
64. J. R. Runyon, D. E. Barnes, J. F. Rudd, and L. H. Tung, "Multiple Detectors for Molecular Weight and Composition Analysis of Copolymers by GPC," *J. Appl. Polym. Sci.*, **13**, 2359 (1969).
65. S. Mori and T. Suzuki, "Problems in Determining Compositional Heterogeneity of Copolymers by SEC and UV–RI Detection System," *J. Liq. Chromatogr.*, **4**, 1685 (1981).
66. A. De Chirico, S. Arrighetti, and M. Bruzzone, "GPC and Viscometric Investigation on Grafting of Styrene-*Co*-Acrylonitrile Polymer to Ethylene Propylene Elastomer," *Polymer*, **22**, 529 (1981).
67. S. Mori, "Chromatographic Determination of Copolymer Composition," *Adv. Chromatogr.*, **21**, 187, Marcel Dekker, New York, 1983.
68. L. H. García-Rubio, J. F. MacGregor and A. E. Hamielec, "Size Exclusion Chromatography of Copolymers," *Am. Chem. Soc. Symp. Ser.*, **203**, 311, 1983.
69. R. O. Bielsa and G. R. Meira, "Linear Copolymer Analysis through Dual-Detection SEC", Proc. of the 2nd. Latin American Polymer Symp. SLAP '90, Mexico, Sept. 1990.

CHAPTER

3

MEASUREMENT OF LONG-CHAIN BRANCH FREQUENCY IN SYNTHETIC POLYMERS

ALFRED RUDIN

Guelph-Waterloo Centre for Graduate Work in Chemistry
Department of Chemistry
University of Waterloo
Waterloo, Ontario, Canada

1. INTRODUCTION

This chapter summarizes the current state of size exclusion chromatography (SEC) techniques for measuring the variation of long-chain branching with molecular weight in synthetic polymers. The analysis of molecular weight distribution of branched polymers is not treated explicitly, although this information is necessary for the measurement of long-chain branching.

It has been widely recognized that long branching may have profound effects on the properties of polymers in which this feature may occur. Although many attempts have been made to quantify long-branch concentrations, these efforts have been hampered by analytical problems and the estimates that have been reported have perforce rested on unverifiable assumptions of relations between molecular weight and long-branch frequency. The more modern equipment and techniques of recent years have improved the analysis for long-branch concentrations and permitted more objective testing of the assumptions on which such estimates are based. Not surprisingly, SEC plays a central role in such advances.

Full characterization of a polymer that contains long branches requires at least the measurement of the concentration of such branches as a function of

Modern Methods of Polymer Characterization, Edited by Howard G. Barth and Jimmy W. Mays
ISBN 0-471-82814-9

polymer molecular weight. In a polydisperse material, the more extended molecules are more likely to participate in entanglements than more compact species. Melt elasticity and flow behavior are thought to be influenced particularly by the rheological characteristics of the larger macromolecules in the sample. The low-molecular-weight tail of the molecular weight distribution, however, may have a very significant effect on mechanical properties. In general, a branched molecule will be more compact and less entangled with its neighbors than a linear molecule with the same molecular weight. Therefore, we may expect that the processability of a thermoplastic may be relatively unaffected by long branching in the low-molecular-weight region, while such isomerism in high-molecular-weight polymer molecules may be extremely important in this regard.

Size exclusion chromatography is the only current analytical technique that may provide information on the variation of long branching with molecular weight, without fractionation of the sample. Information about details of branch length and structure are highly desirable, but these parameters are beyond our reach at present, at least for long branches.

2. MEASUREMENT OF LONG-CHAIN BRANCHING

Universal calibration based on hydrodynamic volumes of linear polymers is not valid for isomers with branches that are longer than a certain minimum. This is because such long-branched species are more compact than linear macromolecules with the same molecular weight. More is said later on the minimum branch length for long branching.

At equal SEC elution volume and infinite dilution, the molecular weights of linear and branched species with the same constitution are related by (1–3)

$$[\eta]_b M_b = [\eta]^* M^* \tag{1}$$

where the subscript b and superscript * refer to branched and linear macromolecules, respectively, that have the same SEC retention time. This implies that branched species with molecular weight M_b have the same solvodynamic volume in the SEC solvent as linear polymers with molecular weight M^*.

In general $M_b \geqslant M^*$, with the equality prevailing only for short branches. (The lengths of short and long branches are defined operationally below.) Presumably, also, the inequality would not apply even for short branches if these were very frequent.

Now, consider a branched and a linear polymer, both with the same molecular weight. In that case the intrinsic viscosity $[\eta]_l$ of the linear species will be greater than that of the branched polymer $[\eta]_b$ in the SEC solvent. The ratio of intrinsic viscosities may be expressed as follows:

$$g' = \frac{[\eta]_b}{[\eta]_l} \tag{2}$$

Note that the corresponding molecular weights considered to this point are related by $M_b = M_l \geqslant M^*$.

For monodisperse versions of the linear polymer in the SEC solvent

$$[\eta] = KM^a \tag{3}$$

where K and a are the Mark–Houwink constants. Then,

$$[\eta]_b = g'[\eta]_l = g'KM_l^a = g'KM_b^a \tag{4}$$

in which $M_b = M_l$ and a is the value from the linear polymer. From Eqs. 1 and 3

$$[\eta]^*M^* = K(M^*)^{a+1}$$

and Eq. (1) is therefore equivalent to

$$g'KM_b^{a+1} = K(M^*)^{a+1} \tag{5}$$

So that

$$g' = \left(\frac{M^*}{M_b}\right)^{a+1} \tag{6}$$

In order to relate g' to actual molecular size it is necessary to consider the ratio g of the mean-square radii of gyration $\langle S^2 \rangle$ of the branched and linear polymers with respective molecular weights equal to $M_l = M_b$. That is

$$g = \frac{\langle S^2 \rangle_b}{\langle S^2 \rangle_l} < 1 \tag{7}$$

Various relations have been proposed of the form (4, 5):

$$g' = g^k \tag{8}$$

the value of k will be considered later in this chapter. With Eqs. 6 and 8

$$g' = \left(\frac{M^*}{M_b}\right)^{a+1} = g^k \tag{9}$$

Recall that the exponent a is the Mark–Houwink constant for the linear polymer in the SEC solvent.

Equation 9 can serve as the basis for a technique to measure the relation between long-chain branching and molecular weight using SEC with a detector that measures molecular weight directly. At any given elution volume M_b is measured directly, while M^* is calculated from the universal calibration curve for linear species. At present, there are two commercially available detectors that permit the direct measurement of the molecular weights of eluting polymers. One is the low-angle laser light scattering detector (LALLS), which measures $\bar{M}_w$ of the macromolecules (6) (see Chapter 9).

An alternative detector is an online viscometer that can be used to measure intrinsic viscosities of the polymers in the eluant stream. Several designs have been reported (7, 8), based on monitoring pressure drops across capillaries. Room temperature operation of a continuous viscometer detector is less difficult than high temperature use, where the output may be affected by thermal as well as mechanical and electrical noise. Analysis of the most widely used branched polymer, that is, low-density polyethylene (LDPE), requires experiments at temperatures near 145 °C. A continuous viscometer has been described for operation under these conditions (9, 10) (also see Chapter 7).

The viscometer detector is the more generally appropriate device for studying branched polymers. The LALLS detector gives correct results only if all species viewed by the detector are uniform with respect to long-chain branching. If the polymers in the detector cell consist of species with different extents or types of long branching then M_b in Eq. (1) is actually equal to $\bar{M}_n$ (11). Since the LALLS detector measures $\bar{M}_w$, it is not suitable for measurements of molecular weights or long-chain branch distribution of polymers in which various species may have the same solvodynamic volumes and different molecular weights.

Such a situation will arise when the branched polymers have higher molecular weights than less branched or linear analogs. One expects such a mixture to be produced in the isothermal free-radical polymerization of vinyl monomers, where long branches result from radical chain transfer to dead polymer (12, 13). Larger molecules are more likely targets for such chain transfer and are therefore more likely to contain long branches from purely statistical reasoning.

Poly(vinyl chloride) is possibly a polymer of the type mentioned. Universal calibration and direct detection of molecular weights by LALLS may be invalid if the sample contains an appreciable concentration of long branches.

The LALLS technique appears to provide correct results, however, for high-pressure LDPE (14). This is because radical chain transfer to polymer, which produces long branches, occurs most frequently at higher temperatures in this nonisothermal polymerization. These are also the conditions that yield lower molecular weight polyethylene (15).

Equation 9 has been used recently to measure long-chain branch frequency in polyethylene (15, 16). At any given elution volume, M_b is measured directly with a LALLS detector, while M^* is calculated from the universal calibration curve for linear polymers. The long-chain branch frequency is expected to be reflected in the value of the parameter g'. Several assumptions must be invoked in order to proceed further and to estimate the actual number of long branches per molecule. In particular, one must assume a branch structure for the branched macromolecules and a value for the exponent k in Eq. 9.

Most such calculations use the Zimm–Stockmayer (17) relation for a randomly branched polymer with trifunctional branch points. This is,

$$g = \frac{6}{n_w} \cdot \frac{1}{2} \left[\frac{2 + n_w}{n_w} \right]^{1/2} \ln \left[\frac{(2 + n_w)^{1/2} + n_w^{1/2}}{(2 + n_w)^{1/2} - n_w^{1/2}} - 1 \right] \tag{10}$$

where n_w is the number of long branches per weight average molecule.

An alternative Zimm–Stockmayer relation may be used to estimate the number of trifunctional long branches per number average molecule n_n

$$g = \left[\left(1 + \frac{n_n}{7} \right)^{1/2} + \frac{4n_n}{9\pi} \right]^{-1/2} \tag{11}$$

If one assumes that the branch points are tetrafunctional then the appropriate equation is:

$$g = \left[\left(1 + \frac{n_n}{6} \right)^{1/2} + \frac{4n_n}{3\pi} \right]^{-1/2} \tag{12}$$

Analytical solutions to Eqs. 10–12 are not available. An iterative computer program is used to calculate n_w or n_n at each value of M_b, from Eqs. 9 and 10, 11 or 12, with a given value of k (15).

There are evidently some serious assumptions involved in proceeding from the experimental values of M^*/M_b to the estimated value of n_w. One may question, for example, whether the polyethylenes are randomly branched. Theoretically, short branches in polyethylene made by free radical initiation are expected to be clustered and to contain quaternary as well as tertiary carbon branch points (12, 18). Evidence on long-branch occurrence is lacking.

Furthermore, the Zimm–Stockmayer model was derived for molecular sizes in theta solvents, whereas the SEC analyses should be made in good solvents. It has been reported, however, that the solvent–temperature system does not affect the measurement of long-chain branches in polyethylene (19, 20).

The absolute value of branch frequency may be uncertain for these reasons, and because the exponent k may have been assumed incorrectly in the calculations. The data should be valid at least in a relative sense, however, if the same value of k in Eq. 9 can be assumed for species of all molecular weights.

Various values of k have been suggested between 0.5 and 1.5 (4, 21, 22). Some experimental information on this point has become available recently. Copolymers of ethylene with 1-olefins were analyzed by SEC and it was shown that k varied from 0.68 to 0.88 as the number of long branches increased from 10/1000 carbon atoms to 49/1000 carbon atoms (22). Clearly, k is not the same for samples with different long-branch concentrations, at least in this concentration range.

Studies with copolymers of 1-olefins and ethylene have shown that the C_6 branches are not registered as long branches in SEC–LALLS analyses, whereas C_{12} branches are measured as long branches. The minimum branch length for "long branching" is therefore between C_6 and C_{12} (15, 23), at least for such copolymers in which the 1-olefin concentration is less than 7 mol%.

Nuclear magnetic resonance (NMR) analyses can be used with SEC characterizations to estimate k for different polymers if k is assumed to be uniform through the molecular weight distribution. Present-day ^{13}C NMR analyses measure polyethylene branches of six carbon atoms or longer. This is approximately the minimum length that is apparently recognized by SEC. NMR results will produce a mean long-chain branch frequency, in terms of branches per 1000 carbon atoms, for the whole polymer sample. The same value can be estimated from SEC. In this case, it is appropriate to use Eq. (11) with Eq. (9) and to take a weighted sum across the molar size distribution. Presumably, then, k can be iterated in Eq. (9) to produce a value of g that yields the same number average of n_n as that determined by ^{13}C NMR. A variation of this approach has been reported by Bugada and Rudin (24). The method suggested here is straightforward in principle. It will, however, be subject to the uncertainty in SEC and ^{13}C NMR measurements, which are not insignificant. Also, while Eq. (11) is correct for comparisons with the results of NMR analyses, this relation has been reported to provide unreasonably high estimates of long-branch frequency as compared to those from application of Eq. (10) (25–27).

Hama et al. (20) characterized low-density polyethylenes fractions and reported in this case that $g' = g^{1.0 \pm 0.3}$. In studies with ethylene copolymers with 1-olefins k was found to vary with branch frequency and to lie between 0.68 and 0.81 (23). These figures are compatible with those derived by Hert and Strazielle (28) from measurements on polyethylene fractions. Other workers have studied different polymers and reported or assumed different values of k (29–32). It seems likely that this exponent may depend on branch point functionality and branch frequency, at least. Assumption of any particular value is likely to produce estimates of long-branch frequency that are relatively, but not necessarily absolutely accurate.

The present rather cloudy state of the art can probably be summarized best by comparing reports on the long-chain branch character of low-density polyethylene, which is the most widely studied polymer.

Constantin (33) reported a study of long branching in polyethylene, using continuous detection of the intrinsic viscosity of the SEC eluant. In this case g' (Eq. 2) decreased with increasing molecular weight, indicating that long branching increased at higher molecular weights, but the author notes that the absolute values of long-branch frequency estimated at high molecular weights are not credible.

More recently, Hert and Strazielle (28) studied fractions of polyethylene obtained from large scale SEC. These fractions were characterized by light scattering and intrinsic viscosity. It was shown that the parameter k varied with molecular weight and even with the source of the LDPE. This is consistent with later SEC–LALLS conclusions (23). Also, long-branch frequency decreased with increasing molecular weight.

Lecacheux et al. (34) characterized branched polyethylene and ethylene–vinyl acetate copolymers by coupling the SEC apparatus to a high temperature continuous viscometer of their design. Long-chain branching was found not to vary with molecular weight, for the range of samples studied by these workers.

The most recent analyses (15, 27) agree that long-branch frequency decreases with increasing molecular weight in low-density polyethylenes. This is consistent with the results of Hama et al. (20) and also with theoretical considerations of the polymerization process (35).

3. CONCLUSIONS

The conclusions of different laboratories regarding long-branch frequency–molecular weight relations in polyethylene appear to vary widely. One possible explanation is that these discrepancies reflect differences in the polymers studied. It is more likely, however, that there are fundamental differences in analytical techniques or interpretations that have not been identified. The analysis is hampered at present by uncertainties in the tails of the SEC chromatograms because of differing sensitivities of the concentration and molecular weight detectors (36, 37).

Long branches are usually defined as being approximately as long as the main polymer chain. The branches that are detected as long branches in SEC analyses may be much shorter than this length. The connection has yet to be made between SEC measurement of long-branch concentrations and effects on polymer melt rheology or mechanical properties.

REFERENCES

1. Z. Grubisic, P. Rempp, and H. Benoit, "A Universal Calibration for Gel Permeation Chromatography," *J. Polym. Sci. Part B*, **5**, 753 (1967).
2. A. Rudin and R. A. Wagner, "Solvent and Concentration Dependence of Hydrodynamic Volumes and GPC Elution Volumes," *J. Appl. Polym. Sci.*, **20**, 1483 (1976).
3. H. K. Mahabadi and A. Rudin, "Effect of Solvent on Concentration Dependence of Hydrodynamic Volumes and GPC Elution Volumes," *Polymer J.*, **11**, 123 (1979).
4. B. H Zimm and R. W. Kilb, "Dynamics of Branched Polymer Molecules in Dilute Solution," *J. Polym. Sci.*, **37**, 19 (1959).
5. G. C. Berry, "Thermodynamic and Conformational Properties of Polystyrene. III. Dilute Solution Studies on Branched Polymers," *J. Polym. Sci. Part A2*, **9**, 687 (1971).
6. A. C. Ouano and W. Kaye, "GPC. A Molecular Weight Detection by Low Angle Laser Light Scattering," *J. Polym. Sci., Polym. Chem. Ed.*, **12**, 1151 (1974).
7. A. C. Ouano, "Quantitative Data Interpretation Techniques in GPC," *J. Macromol Sci., Rev. Macromol Chem.*, **C9**, 123 (1973).
8. F. B. Malihi, C. Kuo, M. E. Kohler, T. Provder, and A. F. Kah, "Application of an In-line Continuous GPC Viscosity Detector for the Characterization of Molecular Weight and Branching in Polymers," *Org. Coat. Appl. Polym. Sci. Proc.*, **48**, 760 (1983).
9. M. Haney, "The Differential Viscometer. I. A New Approach to the Measurement of Specific Viscosities of Polymer Solutions," *J. Appl. Polym. Sci.*, **30**, 3023 (1985).
10. M. Haney, "The Differential Viscometer. II. On-line Viscosity Detector for SEC," *J. Appl. Polym. Sci.*, **30**, 3037 (1985).
11. A. E. Hamielec and A. C. Ouano, "Generalized Universal Molecular Weight Calibration Parameter in GPC," *J. Liq. Chromatogr.*, **1**, 111 (1978).
12. M. J. Roedel, "The Molecular Structure of Polyethylene. I. Chain Branching in Polyethylene during Polymerization," *J. Am. Chem. Soc.*, **75**, 6110 (1953).
13. J. K. Beasley, "The Molecular Structure of Polyethylene. IV. Kinetic Calculations of the Effect of Branching on Molecular Weight Distribution," *J. Am. Chem. Soc.*, **75**, 6123 (1953).

14. V. Grinshpun, A. Rudin, and D. Potter, "Comments on the Measurement of Long-Chain Branching by SEC," *Polym. Bull.*, **13**, 71 (1985).
15. A. Rudin, V. Grinshpun, and K. F. O'Driscoll, "Long-Chain Branching in Polyethylene," *J. Liq. Chromatogr.*, **7**, 1809 (1984).
16. D. E. Axelson and W. C. Knapp, "SEC and Low-Angle Laser Light Scattering," *J. Appl. Polym. Sci.*, **25**, 119 (1980).
17. B. H. Zimm and W. H. Stockmayer, "The Dimensions of Chain Molecules Containing Branches and Rings," *J. Chem. Phys.*, **17**, 1301 (1949).
18. A. H. Willbourn, "Polymethylene and the Structure of Polyethylene: Study of Short-Chain Branching, Its Nature and Effects," *J. Polym. Sci.*, **34**, 569 (1959).
19. G. Williamson and A. Cervenka, "Characterization of Low-Density Polyethylene by GPC. II. The Drott Method Using Fractions," *Eur. Polym. J.*, **10**, 295 (1974).
20. T. Hama, K. Yamaguchi, and T. Suzuki, "Long-Chain Branching and Solution Properties of Low-Density Polyethylene," *Makromol Chem.*, **155**, 283 (1972).
21. G. C. Berry and E. F. Casassa, "Thermodynamic and Hydrodynamic Behavior of Dilute Polymer Solutions," *J. Polym. Sci. D, Macromol. Rev.*, **4**, 1 (1970).
22. M. Kurata, H. Okamoto, M. Iwama, M. Abe, and T. Homma, "Randomly Branched Polymers. II. Computer Analysis of the Gel-Permeation Chromatogram," *Polym. J.*, **3**, 739 (1972).
23. V. Grinshpun, A. Rudin, K. E. Russel, and M. V. Scammell, "Long-Chain Branching Indexes from SEC of Polyethylenes," *J. Polym. Sci., Polym. Lett. Ed.*, **24**, 1171 (1986).
24. D. C. Bugada and A. Rudin, "Sizes of Long Branches in Low-Density Polyethylenes," *J. Appl. Polym. Sci.*, **33**, 87 (1987).
25. E. E. Drott and R. A. Mendelson, "Determination of Polymer Branching with GPC. I. Theory," *J. Polym. Sci. Part A2*, **8**,1361, 1373 (1970).
26. L. Westermann and J. C. Clark, "Low-Density Polyethylene: Variation of Branching Frequency with Molecular Weight and Its Influence on Molecular Weights as Determined by GPC," *J. Polym. Sci., Phys. Ed.*, **11**, 559 (1973).
27. F. H. Mirabella and L. Wild, "Determination of Long-Chain Branching Distributions of Polyethylenes by Combined Viscometry SEC Technique," *Am. Chem. Soc. Polym. Mater. Sci. Eng. Div. Prepr.*, **59**, 7 (1988).
28. M. Hert and C. Strazielle, "Study of the Branching Structure of Low-Density Polyethylene by Viscosity and Light Scattering Measurements on Fractions from GPC," *Makromol. Chem.*, **184**, 135 (1983).
29. R. Dietz and M. A. Francis, "Characterization of Long-Chain Branching in Poly (vinyl acetate): A Comparison of Two Methods," *Polymer*, **20**, 450 (1979).
30. H. Voelker and F. J. Luig, "Long-Chain Branching in Polyethylenes," *Angew. Makromol. Chem.*, **12**, 43 (1970).
31. K. Yamamoto, "An Analytical Method for Long-Chain Branching in Free-Radical Polymerization of Ethylene," *J. Macromol. Sci. Chem.*, **A17**, 415, (1982).
32. S. Shiga, "Modern Characterization of Long-Chain Branching," *Polym. Plast Technol. Eng.*, **28**, 17 (1989).

33. D. Constantin, "Coupling of GPC and Automatic Viscometry. Application to the Study of Long-Chain Branched Polyethylene," *Eur. Polym. J.*, **13**, 907 (1977).

34. D. Lecacheux, J. Lesec, and C. Quivoron, "Characterization of Branched Polyolefins by Coupling High-Speed GPC with a High Temperature Viscometer," *Am. Chem. Soc. Polym. Div. Prepr.*, **23**(2), 126 (1982).

35. A. Rudin, "Molecular Weight Distributions in Free Radical Polymerizations," *Makromol. Chem., Macromol. Symp.*, **10/11**, 273 (1987).

36. V. Grinshpun, K. F. O'Driscoll, and A. Rudin, "On the Accuracy of SEC Analysis of Molecular Weight Distributions of Polyethylene," *J. Appl. Polym. Sci.*, **29**, 1071 (1984).

37. O. Prochazka and P. Kratochvil, "Analysis of the Accuracy of Determination of Molar Mass Distribution by GPC with On-line Light-Scattering Detector," *J. Appl. Polym. Sci.*, **31**, 919 (1986).

CHAPTER

4

POLYMER ANALYSIS BY FIELD-FLOW FRACTIONATION

KARIN D. CALDWELL

Departments of Chemistry and Bioengineering
University of Utah
Salt Lake City, Utah

1. INTRODUCTION

Recent years have seen an explosive growth in the design and production of new polymeric materials whose various properties are intimately linked to their molecular weight, polydispersity, and level of branching. The importance of these characteristics has led to the development of numerous techniques, which permit their rapid and accurate determination. By far the most widely

Modern Methods of Polymer Characterization, Edited by Howard G. Barth and Jimmy W. Mays
ISBN 0-471-82814-9

used characterization techniques are those based on size exclusion chromatography (SEC), which essentially yield information on the molecular size and size distribution for a sample in the solvent chosen for the analysis (1).

Despite the widespread application of these techniques they suffer from some inherent drawbacks, such as a strict limitation in the range of molecular sizes resolvable by a given SEC column, and the lack of a rigorous theory for retention and column dispersion. Furthermore, the high shear stresses to which samples are exposed during passage through modern high-resolution SEC columns result in a certain amount of shear degradation of long-chain polymer molecules (2, 3).

Among alternative tools for polymer analysis, the field-flow fractionation (FFF) class of separation methods presents some attractive features, which make it a useful complement to the very popular SEC techniques (4). Here, the separation takes place in a thin, open channel through which samples are transported by a laminar flow of liquid. The action of a field, applied perpendicularly to the channel flow, causes a partitioning of the various components into regions of different flow velocities. The downstream flow will therefore result in differential migration, and ultimately separation, of these species. In choosing the field, one is largely governed by the nature of the sample. At present, centrifugal fields are routinely used for the characterization of colloidal materials (5–13), and electrical fields have been used for protein separation (14, 15). Aside from reports on the purification and molecular weight determination of DNA samples (16, 17), and a study of the size distribution of cartilage proteoglycans (18), both by means of sedimentation FFF, the analysis of natural and synthetic polymers, in turn, has mainly involved the use of thermal (19–23) and hydraulic gradients (24–30). It is therefore appropriate to limit the present discussion to the two subtechniques termed thermal FFF (TFFF) and flow FFF (FFFF).

Common to all FFF subtechniques is the thin, ribbonlike channel or cylindrical fiber (30) through which the carrier liquid is flowing in a manner that lends itself to exact mathematical description. The strength of the externally applied field is likewise subject to exact control, and the coupled effects of field and flow of the sample can therefore be described in exact mathematical terms. As a result, an observed retention in the channel relates directly to that property of the sample which made it susceptible to the field; frequently, this property is some simple function of the sample's molecular weight.

In contrast to SEC, the FFF analysis takes place in an open channel of well-defined dimensions. The straightforward geometry together with the simple retention mechanism have permitted the formulation of rigorous theories for both retention and column dispersion. As a result of this mathematical tractability, the FFF process leads not only to the separation of complex samples, but it also provides accurate analytical information about their constituents without necessitating elaborate calibration procedures. In addition, this theoretical framework provides guidelines for the selection of experimental conditions for optimal sample resolution in a given time. In the following discussion we will examine the usefulness of the FFF approach to polymer characterization. Where appropriate, this will involve a comparison with the universally accepted SEC analysis. Rather than strongly advocating one approach over the other, it is the intent to outline the relative merits of each of these characterization methods.

2. RETENTION MECHANISM

2.1. General FFF

As a sample enters the FFF channel and becomes exposed to the field, it is forced to move in order to minimize its potential energy and will soon acquire a steady state velocity U, proportional to the strength of its interaction with the field. This migration is hindered by the channel wall, where the sample concentrates to form an exponential layer whose thickness is determined by the magnitude of U on the one hand, and the sample's diffusion coefficient D, on the other

$$c(x) = c(0)e^{-xU/D} = c(0)e^{-x/l} \tag{1}$$

Here, $c(x)$ represents the concentration at distance x from the accumulation wall where $x = 0$, and $c(0)$ is the concentration at the wall. The right-hand formulation in Eq. 1 expresses the concentration profile in terms of the parameter l, which has significance as a measure of the thickness of the distribution

$$l = D/U \tag{2}$$

Under the influence of an applied field of strength ϕ a molecule, whose susceptibility to the field is symbolized by parameter s, experiences a force F given by

$$F = \phi s \tag{3}$$

The molecule's steady state migration in the field will therefore occur at a drift velocity U, specified by the magnitude of F and the molecular friction coefficient f

$$U = F/f \tag{4}$$

From a combination of Eqs. 2–4 it is clear that an increase in field strength ϕ will result in a more compact sample layer, characterized by a smaller value for parameter l. The second quantity with bearing on l is the sample's diffusion coefficient D which, like the drift velocity, is a function of f according to Stokes' law

$$D = kT/f \tag{5}$$

Here, k is the Boltzmann constant and T the temperature. By inserting Eqs. 3–5 into Eq. 2 we can now obtain a general expression for layer thickness l (31)

$$l = kT/\phi s \tag{6}$$

A sample is normally injected into the FFF channel in the absence of axial flow, to allow for the relaxation of its components into exponential equilibrium distributions of the type given in Eq. 1. In the thin FFF channel migration distances are short, and equilibrium is often established within a few minutes (32). The subsequent onset of axial flow will result in the downstream migration of the various sample components according to the compactness of their respective layers, as seen in Figure 1. The thin channel ensures laminar flow in all practical velocity regimes, and the carrier will consequently move less compact zones (large l:s) at rates exceeding those whose l values are small.

Chromatographic retention is often discussed in terms of the dimensionless retention ratio R, defined as the ratio of zonal migration velocity V to the average linear velocity $\langle v \rangle$ of the carrier (33). Although channels of cylindrical geometry have been used (30, 34), most FFF analyses performed to date have involved channels with rectangular cross section. These thin FFF channels are designed to minimize the effects of the side walls and the velocity distribution is to a good approximation that of flow between infinite parallel plates (35). As a result, the zonal velocity is found by averaging the product of concentration and velocity for all values of coordinate x

$$R = \frac{V}{\langle v \rangle} = \frac{\langle c(x) \cdot v(x) \rangle}{\langle c(x) \rangle} \bigg/ \langle v \rangle \tag{7}$$

Here, the angular brackets symbolize cross-sectional averages. For most applications the velocity profile is parabolic with its axis of symmetry

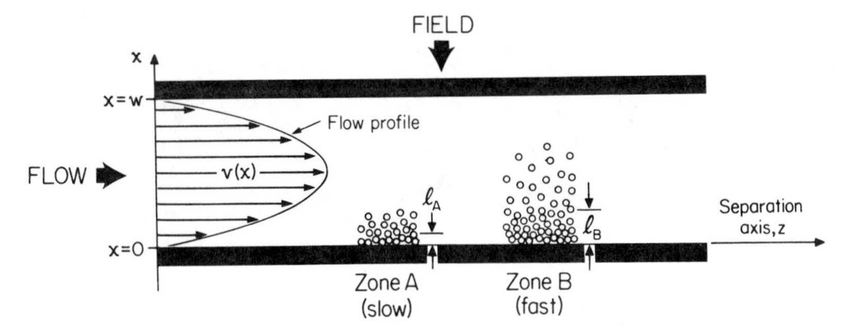

Figure 1. Side view of the FFF channel with equilibrated zones of different thickness in motion along the separation coordinate z. The field is applied in the perpendicular direction (i.e., along the x coordinate).

located at the center of the channel (36)

$$v(x) = 6\langle v \rangle \left(\frac{x}{w} - \frac{x^2}{w^2} \right) \tag{8}$$

In this case, a combination of Eqs. 1, 7, and 8 gives the following expression for R (33)

$$R = 6l/w[\coth(w/2l) - (2l/w)] \tag{9}$$

The retention is therefore exclusively controlled by the ratio of l to channel thickness w. For convenience, this dimensionless ratio is given the symbol λ. The bracketed function in Eq. 9 rapidly approaches unity as l is reduced, so that for many applications the retention process is adequately described by the simple relationship

$$R = 6\lambda \tag{10}$$

The assumption of a parabolic velocity profile falls short in situations where the carrier viscosity varies with position in the channel. Since thermal gradients are often used to accomplish polymer fractionation in the FFF mode, it is necessary to let R reflect those departures from the parabolic velocity profile of Eq. 8, which result from the temperature dependence of the carrier viscosity. These corrections make the retention expression somewhat cumbersome (37). However, their effect is minor for small temperature differentials ($\Delta T < 20\,°C$), and will be ommitted from this discussion in the interest of clarity. Figure 2 gives a comparison of the two retention expressions

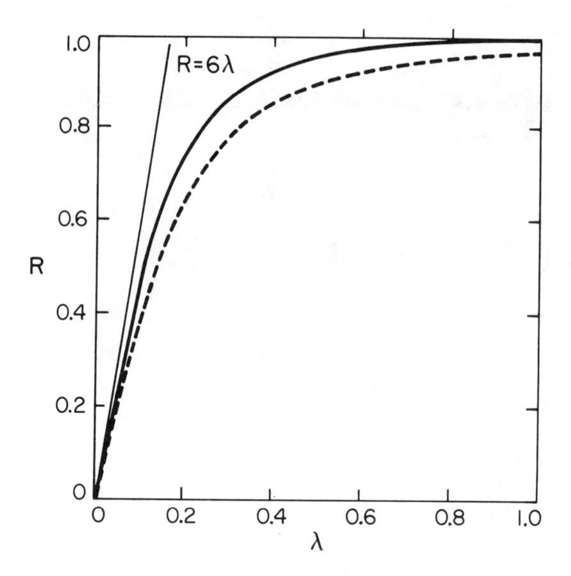

Figure 2. The relationship between retention ratio $R(=V^0/V_r)$ and reduced layer thickness λ. ——— symbolizes isoviscous conditions, whereas ---- represents the carrier ethylbenzene under a ΔT of 100 °C for a cold wall temperature of 10 °C (37).

(Eqs. 9 and 10) applicable to parabolic flow for different values of parameter λ. For the purpose of generating an order-of-magnitude appreciation for the effects that a departure from parabolic flow will have on retention, the figure also includes a graph relating R to λ for the case of thermal FFF under the rather extreme temperature drop of 100 °C. The carrier liquid in this case is ethyl benzene, whose viscosity shows a relatively strong variation with temperature (38).

Experimentally, R is determined from measurements of retention volumes V_r, or retention times t_r. Due to the inverse relationship between zone velocities and elution times, or elution volumes if the flow rate remains constant, one can easily rewrite Eq. 7 in a form that more directly relates to these measured experimental quantities

$$R = \frac{t^0}{t_r} = \frac{V^0}{V_r} \tag{11}$$

Parameters V^0 (void volume) and t^0 (void time) are either measured from the elution of nonretained components, or calculated from the known dimensions of the channel. The retention volume, or retention time, associated with a given component can therefore be directly translated into a value for λ through use of Eqs. 9 or 10, or their viscosity corrected analog, as appropriate.

The relationship between an experimentally determined λ and the physical properties of a sample molecule or particle varies with the choice of FFF subtechnique. In the following sections we will examine the specific expressions for λ, which are associated with the two subtechniques of thermal and flow FFF. These subtechniques have been used most extensively for polymer characterization.

2.2. Thermal FFF

Macromolecules in solution have long been known to migrate towards lower temperatures when exposed to a thermal gradient (39–41). Although the transport mechanism is not yet fully understood, we can express the field induced velocity U of Eq. 1 in phenomenological terms (42)

$$U = D_T\left(\frac{dT}{dx}\right) \tag{12}$$

where D_T is the coefficient of thermal diffusion, and dT/dx is the applied gradient in temperature T. For most liquids the thermal conductivity varies with T, so that the gradient shows some departure from linearity (38). The average drift velocity of molecules within a thermal FFF zone is therefore determined by $(dT/dx)_{x=l}$, that is the magnitude of the gradient at average layer thickness l; since l is of the order of only a few micrometers (43) for retention of any significance, the gradient is commonly evaluated at the cold wall. In the case of small temperature drops ($\Delta T < 40\,°\mathrm{C}$), the gradient remains virtually constant across the entire channel, so that dT/dx is well approximated by $\Delta T/w$. In either event, the reduced layer thickness λ_T, which applies to TFFF can be obtained from Eq. 2 by inserting the expression for drift velocity U, given in Eq. 12

$$\lambda_T = \frac{D}{D_T(dT/dx)w} \simeq \frac{D}{D_T\Delta T} \tag{13}$$

Equation 13 clearly demonstrates the two factors responsible for retention, that is, the strength of the applied field (proportional to ΔT) and the solute specific quantity D/D_T.

Of the two diffusion coefficients, D_T has been found to vary insignificantly, if at all, with molecular weight M (21, 44) for a given polymer type. Its value does, however, vary both with the type of polymer and the nature of the solvent (45). By contrast, D shows a strong molecular weight dependence, which is often expressed in the general form

$$D = A \cdot M^{-b} \tag{14}$$

where coefficients A and b depend on the nature of the polymer–solvent pair.

To explore the factors that influence b, we recall the Stokes–Einstein equation (46), which predicts an inverse relationship between D and the hydrodynamic radius r_h of the molecule. Since the universal calibration used in SEC is based on the relationship between M and the molecular volume (proportional to r_h^3) (47), there is an obvious link between D and M, that is, between the exponents b in Eq. 14 and a of the Mark–Houwink equation (48)

$$b = \frac{a+1}{3} \tag{15}$$

While a in general has values of 0.8 or below for random-coil molecular conformations (49), these values increase significantly for charged polymers dissolved in media of high dielectric constant, as a result of intramolecular repulsion (50). The values for exponent b in Eq. 14 can therefore be expected to

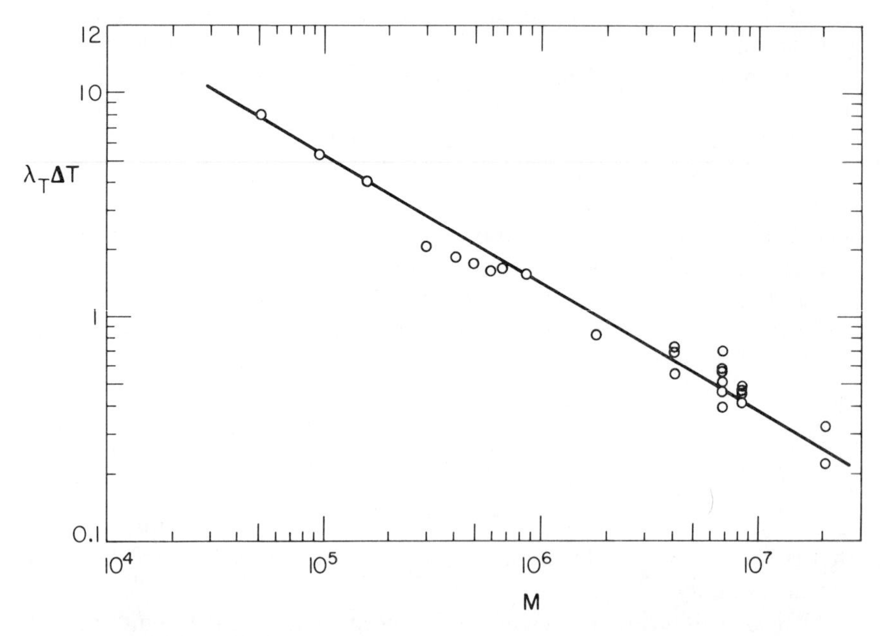

Figure 3. Double-logarithmic plot of $\lambda_T \Delta T$ versus sample molecular weight M. The carriers are ethylbenzene and THF, both good solvents for the sample polystyrene (51). The slope of the line ($-b'$ in Eqs. 15–17) has the value -0.53. The cold wall temperature was 15 °C, and ΔT ranged from 8 to 81 °C. Reprinted with permission from Y. S. Gao, K. D. Caldwell, M. N. Myers, and J. C. Giddings, "Extension of Thermal Field-Flow Fractionation to Ultra-High Molecular Weight Polystyrenes," *Macromolecules*, **18**, 1272 (1985). Copyright © 1985 American Chemical Society.

range from 0.33 for a nonpenetrating sphere (whose radius is proportional to $M^{1/3}$) to somewhere above 0.7 for extended polymer chains. Single chain polymers in θ solvents are characterized by b values of about 0.5 (49).

With virtually all molecular weight dependence residing in the diffusion coefficient, one may write the ratio D/D_T in a form similar to Eq. 14

$$D/D_T = B \cdot M^{-b'} \tag{16}$$

By combining Eqs. 13 and 16 one can therefore formulate relationships between sample molecular weight M and the observed retention parameter λ_T

$$\lambda_T = \left[B \Big/ \left(\frac{dT}{dx} \right) w \right] M^{-b'} \simeq \left(\frac{B}{\Delta T} \right) M^{-b'} \tag{17}$$

or

$$\log \lambda_T \Delta T = \log B - b' \log M \tag{18}$$

Figure 3 is a double logarithmic representation of retention parameters measured for a large set of linear polystyrene standards in the (good) solvents ethylbenzene and tetrahydrofuran (THF). This data set, which was assembled under a variety of field strengths (51), supports the linear relationship predicted by Eq. 18.

Despite the wide variations in field strength used to generate Figure 3, great care was taken to maintain the cold wall at a constant temperature of 15 °C. This arrangement had proven necessary as the result of a series of studies (52, 53), which had indicated that retention is strongly dependent not only on ΔT but on the cold wall temperature as well. Under otherwise identical experimental conditions, retention was shown to decrease significantly with higher temperatures of the cold wall. This effect, which is illustrated in Figure 4, suggests operation at the lowest practical temperatures, in order to maximize retention at a given field strength.

Since polymer solubility is enhanced by higher temperatures, one is, however, often forced to work at elevated cold wall temperatures in order to maintain the sample in solution. An extreme demonstration of this need was made in a TFFF analysis of a high-molecular-weight polyethylene sample, which was sparingly soluble in the carrier tetrachloroethylene, even at its boiling point of 121 °C and atmospheric pressure (54). For the purpose of generating the necessary temperature drop across the channel, while maintaining the carrier in the liquid phase, the unit was pressurized to 15 atm. This allowed the application of a ΔT of 86 °C to a system whose cold wall was held at 107 °C. However, temperature-enhanced solubility and the associated increased sample detectability were obtained at the expense of decreased resolution.

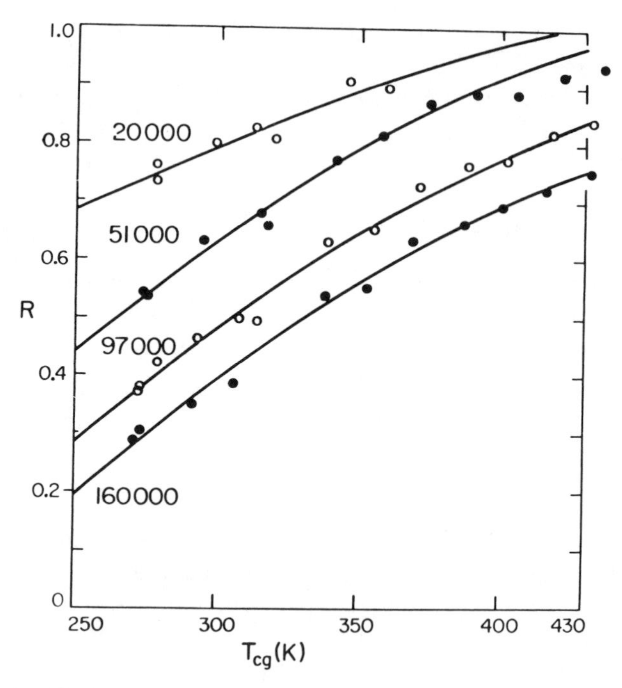

Figure 4. Relationship between R and T_{cg}, the temperature of the zone's center of gravity (approximately equal to the cold wall temperature T_c) for an applied temperature differential ΔT of $40 \pm 1\,^{\circ}C$. The carrier is ethylbenzene, and the samples are linear polystyrene standards of the indicated molecular weights (53). Reproduced with permission from John Wiley & Sons (53). Copyright © 1985.

Just as the ratio of diffusion coefficients D/D_T is sensitive to the goodness of the solvent used as carrier for a given polymer type, it is also sensitive to the level of branching in polymeric samples of a given composition and molecular weight. Since both affect the hydrodynamic radius of the molecule, one expects to see the largest b' values for linear species, and progressively lower values for the more compact molecular structures. This tendency is clearly demonstrated in Figure 5, which compiles retention data for linear- and star-branched polystyrenes collected under identical experimental conditions. The slopes of the double logarithmic plots range from -0.67 for the linear to -0.58 for the 12-armed star polymer (22).

2.3. Flow FFF

The FFFF subtechnique is unique in that the applied field, or cross-flow, imparts the same drift velocity U to all species in the channel. This cross-flow is

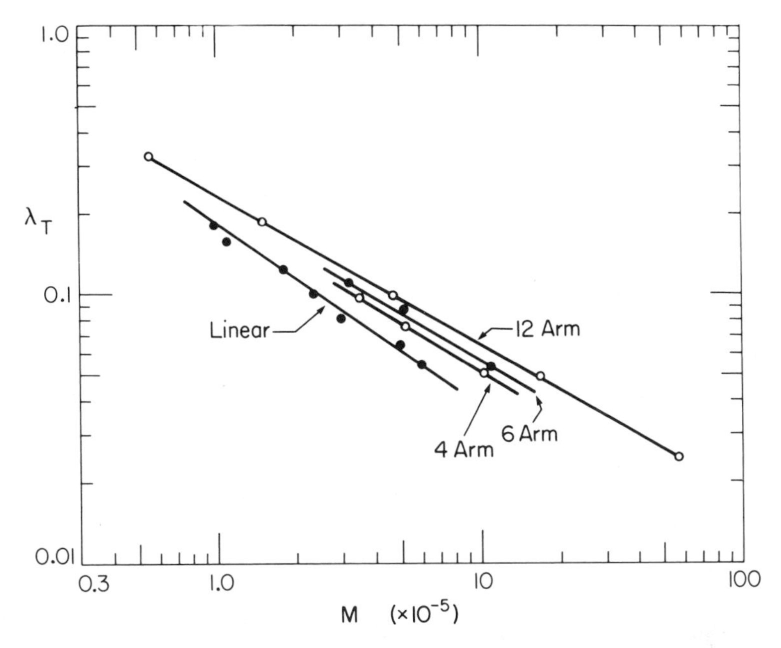

Figure 5. Relationship between λ_T and molecular weight M for polystyrene standards of different conformation. The data derive from TFFF with an applied ΔT of 47 °C and a cold wall temperature of 25 °C, using THF as the carrier. The slope for the linear polystyrene was -0.67, whereas slopes for the 4-, 6-, and 12-arm star shaped polystyrenes were -0.63, -0.60, and -0.58, respectively (22).

made possible by constructing the channel walls out of semipermeable material that permits passage of carrier liquid while retaining macromolecules and particles. A sample injected into the channel is swept along towards one of the semipermeable walls with a velocity equal to that of the cross-flow. Unable to permeate the wall, the sample will accumulate and establish a concentration gradient, which in turn will generate a diffusive flux in the normal manner. At equilibrium the sample distribution is described by (24)

$$\lambda_F = \frac{D}{Uw} = \frac{DV^0}{\dot{V}_c w^2} \tag{19}$$

The right-hand expression in Eq. 19 is obtained by replacing U with the volumetric cross-flow $\dot{V}_c$ divided by the area of the accumulation wall (V^0/w). The predicted inverse relationship between retention derived λ_F values and the magnitude of the applied cross-flows is shown in Figure 6 for a dextran sample with a molecular weight of 98,000. The linearity of the plot is an indication that

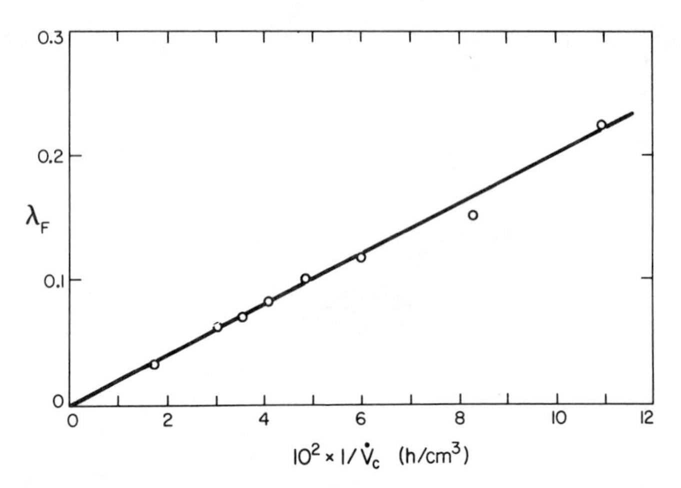

Figure 6. Relationship between retention-derived values for parameter λ_F and the inverse of the applied cross-flow $\dot{V}_c$. The linearity of the plot, and the fact that the line passes through the origin, both indicate that the sample (linear dextran $\bar{M}_w = 98{,}000$) behaves ideally over a wide range of experimental conditions.

the sample is behaving ideally (D constant), even at the high concentrations present in the most compact zones.

Unlike its thermal analog, FFFF retention reflects one single sample characteristic, namely, the diffusivity as seen in Eq. 19. This quantity, in turn, is related to sample molecular weight through Eq. 14, where constants A and b depend on the polymer–solvent system, as noted earlier. For any given polymer type, Eq. 14 predicts the logarithm of retention derived diffusion coefficients (from Eqs. 9, 11, and 19) to form a straight line with a slope equal to $-b$, when plotted against the logarithm of sample molecular weight. That this is indeed the case is demonstrated by Figure 7, where experimental D values are related to values for M for a series of sulfonated polystyrene standards analyzed in an aqueous buffer under slightly alkaline conditions and moderate ionic strength. The goodness of this solvent, in combination with some degree of conformational extension due to charge repulsion, is indicated by a b value of 0.65 (25). Similar relationships have been recorded for linear polystyrenes dissolved in ethylbenzene (51), although the b value determined for this system was somewhat lower (0.51–0.56).

3. FRACTIONATION RANGE

Figures 3 and 7 bear evidence of the large range of molecular weights accessible for analysis by either thermal or flow FFF. Indeed, the lower

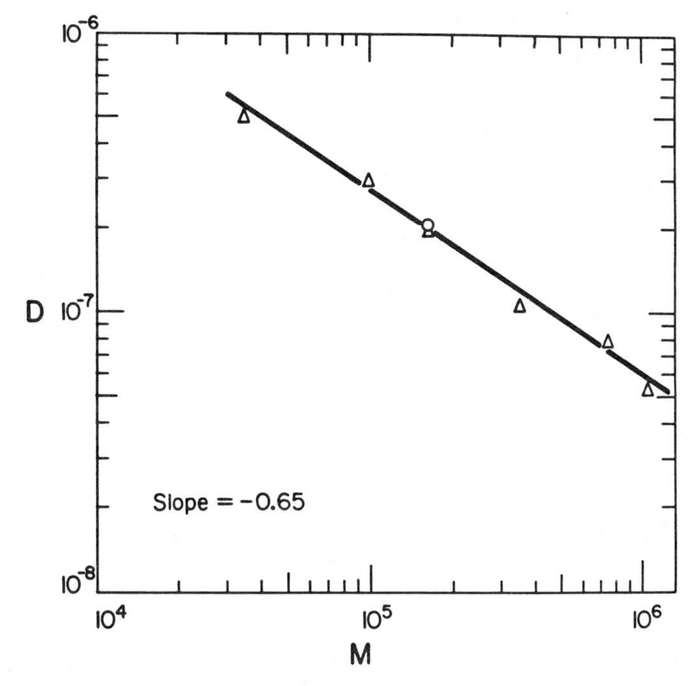

Figure 7. Diffusion coefficients D, derived from retention in FFFF, plotted as a function of sample molecular weight M for sulfonated polystyrenes in a Tris–HNO_3 buffer of 0.025 M ionic strength and a pH of 8.6 (25). The slope ($-b$ in Eq. 14) was determined as -0.65. Reprinted with permission from K.-G. Wahlund, H. S. Winegarner, K. D. Caldwell, and J. C. Giddings, "Improved Flow Field-Flow Fractionation System Applied to Water-Soluble Polymers: Programming, Outlet Stream Splitting, and Flow Optimization," *Anal. Chem.* **58**, 573 (1986). Copyright © 1986 American Chemical Society.

molecular weight limit for sample retention in TFFF is set by the temperature interval in which the carrier remains in the liquid state. For most solvents at atmospheric pressure operation is therefore restricted to a ΔT of 100–150 °C (38). Pressurizing the system will increase the boiling point of the carrier which, as a result, will be able to accommodate larger temperature drops and retain smaller molecules. Under a channel pressure of 8 atm the boiling point for ethylbenzene was raised by 100 °C, which permitted the application of a 158 °C temperature drop across the channel. This was sufficient to significantly ($R = 0.9$) retain a 600 g/mol polystyrene standard sample (55).

In FFFF the lower limit of resolvable molecular weights is set by the permeability of the membrane walls. In selecting the appropriate membrane, one must balance the demands on smaller pore size for retention of low-molecular-weight solutes, against the increased cross-flows, which are needed to offset the higher diffusivities of these smaller molecules. The finer the pore size the higher is the pressure needed to maintain a given cross-flow $\dot{V}_c$, and as a matter of convenience we are currently employing commercial (Millipore PTGC or Amicon YM 50) membranes whose molecular weight cutoff lies around 10^4 g/mol for globular macromolecules.

All FFF techniques have an upper limit to the size of molecules or particles, which may be analyzed using the theory summarized in Eqs. 8, 13, and 19. This limit, which represents the onset of steric effects (56), is only reached for particles whose size is comparable to the thickness l of the sample layer under the chosen experimental conditions. Typically, these effects are negligible for particle sizes less than 0.7 μm (corresponding to molecular weights of about 10^{10} g/mol).

The velocity gradients generated in the open FFF channels are significantly less (by a couple of orders of magnitude) than those present in high-performance SEC columns, where shear degradation appears to become a problem for sample molecular weights above 10^6 g/mol (2, 3). In order to test whether the weaker shear stresses present in the open FFF column would leave even ultrahigh-molecular-weight samples intact during the fractionation, we performed a series of TFFF separations of a polydisperse, linear polystyrene sample with an assigned molecular weight of 2×10^7 g/mol. (51). Cuts were taken at different points in the fractogram and refractionated under experimental conditions comparable to those used in the original run. Even cuts containing the highest molecular weights eluted with identical retention ratios the first and second time through the system, and we could therefore conclude that the FFF process is relatively nondestructive.

4. RESOLVING POWER

The goodness of a separation of two components is often characterized by the resolution index R_s (57), defined as the distance Δz (along the separation coordinate) between component peaks at the time of elution of the lead component, divided by the sum of the half-widths of the two distributions. This sum is generally approximated by $4\bar{\sigma}$, where $\bar{\sigma}$ is the average standard deviation for the two zones along the z coordinate

$$R_s = \Delta z / 4\bar{\sigma} \tag{20}$$

A resolution of unity or better implies complete separation of the two components.

The term Δz is a measure of the difference in retention between the two components, and is therefore a reflection of the selectivity of the process, whereas σ represents the zone spreading indicative of the process efficiency. These two concepts will be examined below.

4.1. Selectivity

The FFF techniques, as well as other elution methods used to separate and characterize the molecular weight of polymers, will generally perform a sorting of the various components of a polydisperse sample, so that the elution pattern becomes a molecular weight spectrum of the composite, where each elution volume V_r is uniquely associated with a particular molecular weight M

$$V_r = V_r(M) \tag{21}$$

An elution-based fractionation method is obviously more selective the further apart in the elution spectrum one will find two components with a given difference ΔM in molecular weight. The selectivity of the process is frequently defined as (57)

$$S = \left| \frac{d \ln V_r}{d \ln M} \right| \tag{22}$$

where the absolute value of the differential is used to make the definition equally valid whether V_r increases (as in FFF) or decreases (as in SEC) with molecular weight. Equations 9 and 11 can now be used in conjunction with the appropriate relationship between λ and M to develop expressions for the selectivity offered by a particular FFF subtechnique. As a first step it is convenient to rewrite Eq. 14 as the product of two differentials of which one ($d \ln V_r/d \ln \lambda$) is common to all FFF techniques with parabolic flow in parallel plate channels, and the other ($d \ln \lambda/d \ln M$) is specific for a given subtechnique

$$S = \left| \left(\frac{d \ln V_r}{d \ln \lambda} \right) \left(\frac{d \ln \lambda}{d \ln M} \right) \right| \tag{23}$$

Taking the logarithm of both sides in Eq. 11 and differentiating gives

$$-\frac{d \ln V_r}{d \ln \lambda} = \frac{d \ln R(\lambda)}{d \ln \lambda} = \left(\frac{\lambda}{R} \right) \left(\frac{dR}{d\lambda} \right) \tag{24}$$

where $dR/d\lambda$ is found through differentiation of Eq. 9

$$\frac{dR}{d\lambda} = \frac{3}{\lambda} \left(\frac{R^2}{36\,\lambda^2} + R - 1 \right) \tag{25}$$

By inserting Eqs. 24 and 25 into Eq. 23 one obtains an expression demonstrating how the system's selectivity varies with level of retention

$$S = \left| 3\left(\frac{R}{36\,\lambda^2} + 1 - \frac{1}{R}\right) \right| \cdot \left| \frac{d \ln \lambda}{d \ln M} \right| \tag{26}$$

A graphical illustration of Eq. 26 is given in Figure 8. As expected, R values close to unity are seen to offer very low selectivities. However, S increases rapidly as R decreases, and approaches asymptotically its maximum value $S_{\max}$ at high retention (low R)

$$S_{\max} = \left| \frac{d \ln \lambda}{d \ln M} \right| \tag{27}$$

Operation at maximum selectivity occurs when Eq. 10 becomes a valid approximation to the relationship between R and λ.

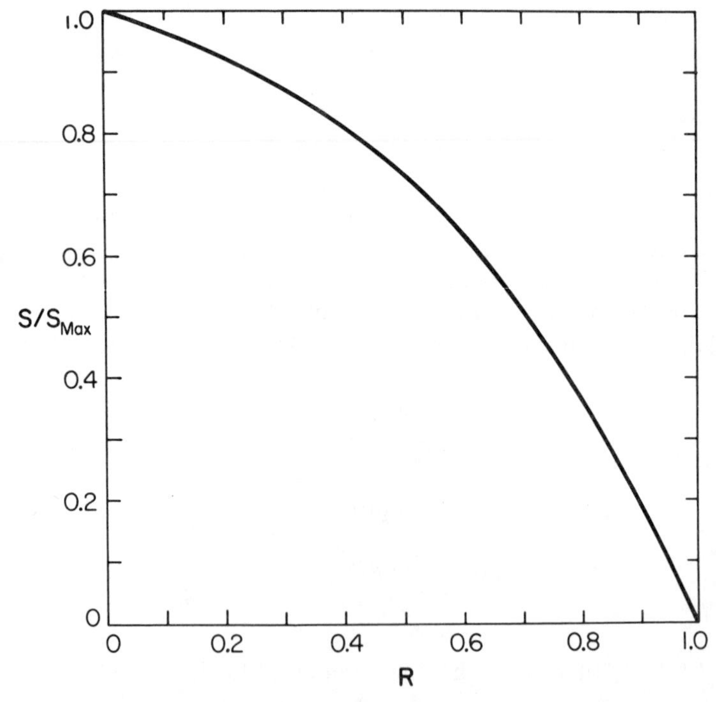

Figure 8. Relationship between retention R and selectivity S in FFF (see Eq. 26). The diagram depicts the normalized selectivity $S/S_{\max}$, where $S_{\max}$ is the selectivity gained by a given technique in the limit of high retention.

The value for S_{max} differs from subtechnique to subtechnique; for polymeric samples fractionated by either thermal or flow FFF it will likewise vary with the nature of both sample and solvent. Differentiation of the relationship between λ and M given by Eq. 16 shows S_{max} for TFFF to equal

$$S_{max,\,T} = b' \tag{28}$$

Similarly, the maximum selectivity for FFFF is found from a combination of Eqs. 14 and 19 to equal

$$S_{max,\,F} = b \tag{29}$$

Values for b will vary between 0.33 and about 0.7 depending on the sample's molecular conformation and the goodness of the solvent, as discussed above. This range of values is in general characteristic also for $S_{max,\,T}$. However, notable exceptions to the correlation between solvent strength and the value for b' are demonstrated by many aqueous systems (23, 52), where highly soluble polymers such as dextrans, nucleic acids, and proteins, show little or no TFFF retention regardless of molecular weight, and where the maximum selectivity therefore is close to zero. In general, the maximum selectivities associated with the two separation techniques of thermal and flow FFF compare favorably with the values for S_{max} of 0.2 or less, which are characteristic for SEC (57).

4.2. Efficiency

The efficiency of a separation is intimately linked to the level of dispersion inflicted on various sample components during the process. The zone broadening associated with chromatographic or FFF-type separations is conveniently discussed in terms of the plate height, H (58). For uniform columns H is the variance (along the separation coordinate) of the eluting peak divided by the length of the column L

$$H = \frac{\sigma^2}{L} \tag{30}$$

A variety of broadening effects contribute to the dispersion of a zone, so that HL represents the sum of variances of different origin (58). The general expression for FFF plate height is given by (59)

$$H = \chi(\lambda)\frac{w^2}{D}\langle v \rangle + H_p + \sum H_i \tag{31}$$

where the first (velocity dependent) term on the right represents the nonequilibrium contribution caused by the finite times needed for sample molecules to diffuse in and out of flow lines of different velocity. Mass transfer terms of this type are common to all forms of chromatography (57, 60) and generally remain constant or increase with increased retention. By contrast, FFF is unique in that this term decreases rapidly with a decrease in R, due to decreased diffusion distances characteristic of the more compact sample layers. This decrease is reflected in the functional form of the nonequilibrium coefficient χ, which varies with the third power of λ. The exact relationship between the two is rather complex but can be closely approximated by (61)

$$\chi(\lambda) \simeq 24\,\lambda^3(1 - 8\,\lambda + 12\,\lambda^2) \tag{32}$$

χ values calculated from Eq. 32 are in error of less than 2% for $R < 0.4$.

Equation 31 also includes a number of apparatus induced zone broadening contributions, collectively labeled $\sum H_i$. Among such nonideal effects are those caused by finite injection times and volumes, and zone distortions caused by drag at the side walls of the channel. Extra column effects introduced in the connecting tubing and detector are also part of this composite term, which must be minimized through good experimental procedures. In general, we find this term to be negligible in comparison with other plate height contributions, except at very high retention of monodisperse ($H_p = 0$) samples where the first and second terms of Eq. 31 are infinitesimal or zero. For low to moderate retention, the FFF plate heights are larger by about an order of magnitude than those observed in SEC (62).

Since Δz in Eq. 20 is linearly related to system selectivity S, while σ (through Eq. 30) relates to the square root of plate height H, the order-of-magnitude difference in plate height is often compensated for by the roughly threefold higher selectivity offered by TFFF as compared to SEC (see also Eq. 43 below).

As selectivity in SEC is based on molecular size, and therefore relates to sample diffusivity D, while selectivity in TFFF is based on the ratio D/D_T, of which the thermal diffusivity is size independent but reflects composition, the two techniques may well complement one another. This was shown in Ref. (21), where a mixture of 240,000 g/mol of PMMA and 200,000 g/mol of PS was analyzed both by SEC and TFFF. Due to similarities in hydrodynamic radius the two were not resolved by SEC, whereas apparent differences in D_T resulted in base line resolution by TFFF. One can easily envision pairs of polymers with different composition and molecular weights for which SEC would cause resolution and TFFF would not.

Polymeric samples are seldom strictly monodisperse, but are instead distributions characterized by an average molecular weight. The weight average $\bar{M}_w$ and the number average $\bar{M}_n$ are used most frequently; the discrepancy between the two numbers is an indication of the width of the distribution, and their ratio is commonly used as a polydispersity index ($\mu = \bar{M}_w/\bar{M}_n$). Any fractionation process sensitive to sample molecular weight will tend to sort the various components of a polydisperse material in the order of increasing (as in FFF) or decreasing (as in SEC) M, as specified by the separation mechanism. The sample will therefore leave the fractionator as a zone whose width is influenced by its polydispersity, in combination with the selectivity of the process. For samples of low to moderate polydispersity, this contribution to FFF plate height (included as H_p in Eq. 31) can be shown to equal (59)

$$H_p = LS^2\left(1 - \frac{1}{\mu}\right) \tag{33}$$

where S is obtained either from the average level of retention through use of Eq. 26, or from the slope of a calibration curve (relating $\ln V_r$ to $\ln M$) evaluated at $\bar{M}_w$ in accordance with the definition of selectivity given in Eq. 22.

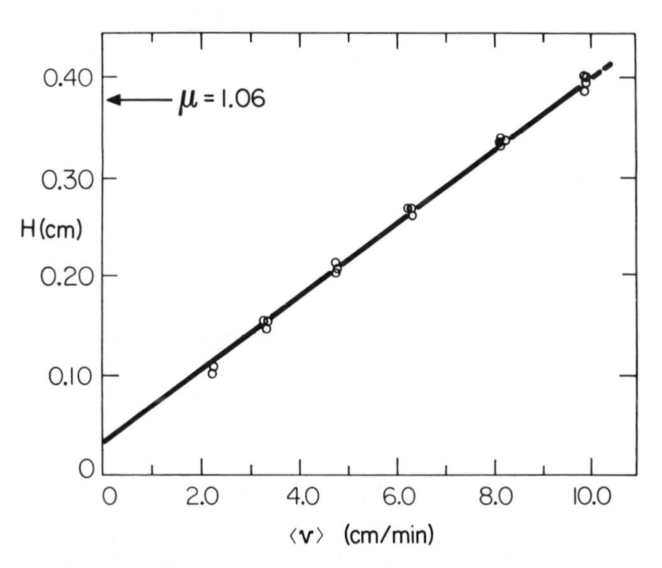

Figure 9. Plot of a plate height H versus linear carrier velocity $\langle v \rangle$. Extrapolation to zero velocity eliminates the nonequilibrium contribution to H, leaving as the intercept the sum of polydispersity derived H_p and instrumentally induced ΣH_i plate heights. Neglecting the latter, one calculates a polydispersity μ of 1.003 for this linear polystyrene sample of $M = 200{,}000$. The manufacturer's value of $\mu \leqslant 1.06$ would result in an intercept of $H_p \leqslant 0.38$ (indicated by an arrow in the figure) (63).

Among all the plate height contributions listed in Eq. 31, the nonequilibrium term alone shows a significant dependence on carrier velocity. By observing the zone broadening H as a function of average velocity $\langle v \rangle$ of the carrier, and performing a linear extrapolation to zero velocity, one should find a value for the intercept that equals $H_p + \sum H_i$. If the nonideal zone broadening, $\sum H_i$ is either negligible or can be determined using a probe of well-characterized polydispersity, the value for H_p associated with the actual sample is easily obtained and translated into the corresponding polydispersity index μ through use of Eq. 33. This process is illustrated in Figure 9 for a relatively narrow fraction of a linear polystyrene with a supplier-furnished weight-average molecular weight of 200,000 g/mol (63). Since the eluting peaks, which form the basis for this plot, were very nearly Gaussian, experimental H values were determined from the peak width at half-height (60). Occasionally, plate heights were rigorously determined from the second moments of the peaks (60), but a comparison between the two indicated the simple graphical method to yield values, which were in error by less than 2% for this sample. The data set in Figure 9 appeared to be affected by a negligible instrument induced zone broadening, and the entire intercept at zero velocity was therefore assumed to result from polydispersity effects. Through use of Eq. 33 the experimentally determined H_p could then be translated into a value for the polydispersity μ of 1.003, to be compared with the value of "< 1.06" specified by the manufacturer. The error incurred by neglecting nonideal zone broadening terms must therefore be considered minor.

5. CHARACTERIZATION OF POLYDISPERSE MATERIALS

Samples of significant polydispersity are in general inadequately described by the two parameters $\bar{M}_w$ (or $\bar{M}_n$) and μ. Instead, characterization of such samples requires determination of their complete molecular weight distribution. In these cases it is often permissible to neglect the dispersion contributions to the peak width and consider the fractogram as a nearly accurate representation of the molecular weight distribution in the sample, as demonstrated in Figure 10. The shaded area in the figure represents the nonequilibrium zone broadening associated with the molecular weight at the peak maximum. In order to estimate this broadening one must first identify the λ value characteristic of the maximum (given, e.g., by Fig. 2), and the resulting nonequilibrium coefficient $\chi(\lambda)$, which is calculated from Eq. 31. Inserting values for χ at the peak maximum, and diffusion coefficient D for the molecular weight at the maximum (obtained, e.g., from Eq. 14), together with the experimental parameters w (channel thickness) and $\langle v \rangle$ (average linear carrier velocity), into the first term to the right in Eq. 31, yields the required value for

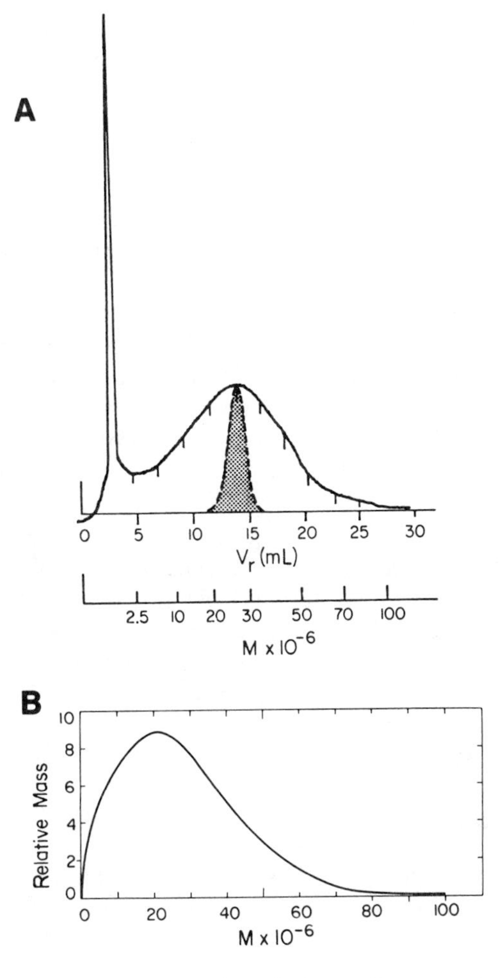

Figure 10. (*a*) Thermal FFF fractogram of a polydisperse linear polystyrene sample with a manufacturer assigned molecular weight of 20×10^6 g/mol. Experimental conditions were $\Delta T = 8\,°C$, $T_c = 13\,°C$; the carrier was THF (51). The figure shows a cross-hatched area symbolizing the peak shape corresponding to a sample whose molecular weight is that associated with peak maximum ($M = 25 \times 10^6$). Instrumental band broadening is assumed negligible. The peak width is seen to largely derive from sample polydispersity (estimated as $\mu = 1.52$). (*b*) Unnormalized molecular weight distribution derived from the fractogram in Figure 10*a* by division with the scale correction function dV_r/dM (Eq. 39). Reprinted with permission from Y. S. Gao, K. D. Caldwell, M. N. Myers, and J. C. Giddings, "Extension of Thermal Field-Flow Fractionation to Ultra-High Molecular Weight Polyethylene," *Macromolecules*, **18**, 1272 (1985). Copyright © 1985 American Chemical Society.

the nonequilibrium (neq) plate height H_{neq}. From this value one obtains the appropriate variance (with respect to the fractogram) $\sigma_{f'}^2$

$$\sigma_{f'}^2 = \frac{f'^2}{L} \cdot H_{\text{neq}} \tag{34}$$

Here, f' is the distance on the chart from start to peak maximum, and L is the length of the column. The shaded area in Figure 10*a* is bounded by a Gaussian curve, which is centered at the peak maximum, and has a variance equal to $\sigma_{f'}^2$. It is evident that nonequilibrium dispersion makes only a minor contribution to the width of the peak in Figure 10*a*, whose shape therefore is essentially a profile of the sample's molecular weight distribution. In order to transform this profile into a distribution curve, one must correct for the differences in resolution associated with different levels of retention. From Figure 8 it is clear that the selectivity is insignificant at low retention (large R) and increases rapidly with increased retention. A given volume element in the early, low retention, part of the fractogram will therefore contain a broader range of molecular weights than an element of comparable size, which emerges with high retention and superior resolution. As a result, the detector signal $c(V_r)$ will not accurately reflect the mass of components of a given molecular weight, $m(M)$. By application of a scale correction, this problem is removed and the fractogram transformed into a molecular weight distribution curve (51). This correction is best understood by considering the following relationship between the fractogram, represented by $c(V_r)$, and the true distribution, $m(M)$

$$m(M) = c(V_r)(dV_r/dM) \tag{35}$$

The scale correction factor dV_r/dM is obtained either from a calibration curve or from theoretical relationships between retention ratio and molecular weight M

$$\frac{dV_r}{dM} = \left(\frac{dV_r}{dR}\right)\left(\frac{dR}{d\lambda}\right)\left(\frac{d\lambda}{dM}\right) \tag{36}$$

By combining Eqs. 9 and 11 and differentiating one obtains an expression for the first differential to the right in Eq. 36

$$\frac{dV_r}{dR} = -\frac{V_r}{R} \tag{37}$$

The second differential is given in Eq. 25, and the third is easily obtained from relationships between λ and M for the desired subtechnique (Eq. 16 for TFFF or a combination of Eqs. 14 and 19 for FFFF). In the case of TFFF, it has the following form

$$\frac{d\lambda}{dM} = -b'\lambda M^{-1} \tag{38}$$

The corresponding expression for flow FFF contains the factor b instead of b'.

A combination of Eqs. 25 and 36–38 will therefore give an expression for the scale correction in terms of the retention related parameters R, λ, and V_r, and the corresponding value for M

$$\frac{dV_r}{dM} = \frac{-3V_r b'}{M}\left(\frac{R}{36\lambda^2} + 1 - \frac{1}{R}\right) \tag{39}$$

Deconvolution of the fractogram in Figure 10*a* with this scale correction function gives the unnormalized molecular weight distribution curve shown in Figure 10*b*.

6. OPTIMIZING SPEED AND RESOLUTION

The existence of exact mathematical relationships describing FFF retention and zone broadening enables the analyst to select the field strength and axial flow rate, which will accomplish a desired separation in the shortest possible time. This optimization process is best initiated by an examination of the resolution expression given in Eq. 20.

At the point of elution of the first of two components, which differ in molecular weight by ΔM, the two are positioned a distance Δz apart along the separation axis. If this elution occurs at time t_r, Δz is given by

$$\Delta z = R_1\langle v\rangle t_r - R_2\langle v\rangle t_r = t_r\langle v\rangle \Delta R \tag{40}$$

where $R_1\langle v\rangle$ and $R_2\langle v\rangle$ are the zonal velocities of the two components and ΔR represents $(R_1 - R_2)$. Time t_r is the time needed by component 1 to pass the full column length L, and may therefore be replaced in Eq. 40 by the ratio $L/R_1\langle v\rangle$ in order to express the resolution in terms of ΔR

$$R_s = \left(\frac{L}{4\bar{\sigma}}\right)\left(\frac{\Delta R}{R_1}\right) \simeq \left(\frac{L}{4\bar{\sigma}}\right)\left(\frac{\Delta R}{\bar{R}}\right) \tag{41}$$

Equation 41 is often generalized by replacing R_1 with the average retention for the pair, $\bar{R}$ (64).

The average zone width along the separation coordinate is characterized by an average standard deviation $\bar{\sigma}$, which relates to the plate height through Eq. 30. If $\bar{\sigma}$ in Eq. 41 is replaced by $\sqrt{HL}$ one obtains an expression for the resolution in terms of $\bar{R}$ and $\bar{H}$, both of which can be evaluated for a given set of experimental conditions (64)

$$R_s = \frac{1}{4}\sqrt{\frac{L}{\bar{H}}}\cdot\frac{\Delta R}{\bar{R}} \tag{42}$$

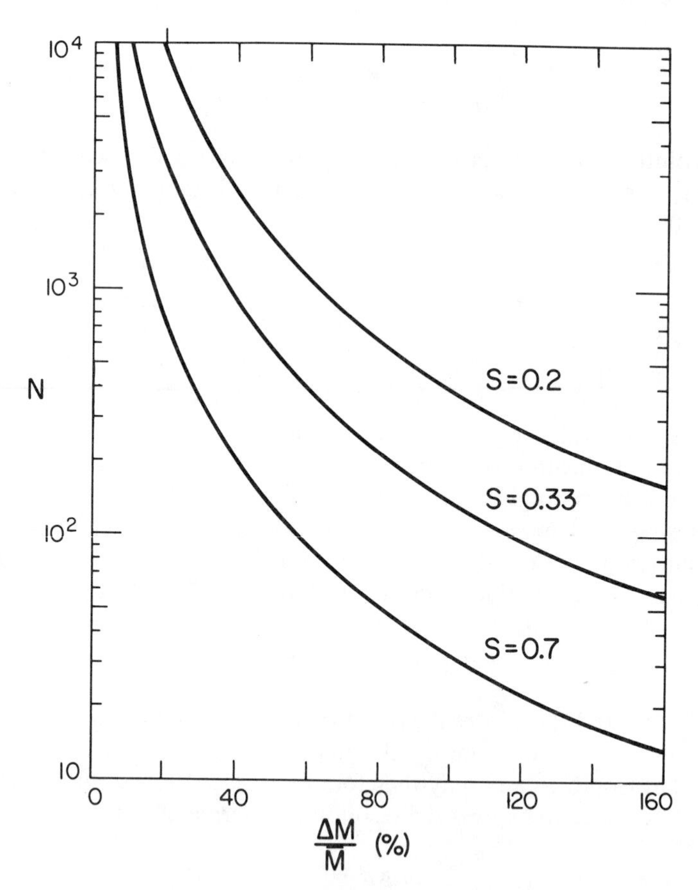

Figure 11. Graphs representing the number of plates $N(=L/H)$ required for unit resolution of samples differing by $\Delta M/\bar{M}$ in relative molecular weight at different levels of system selectivity. The S values correspond to maximum values recorded in thermal and flow FFF for randomly coiled linear polymers ($S = 0.7$) and nondraining spheres ($S = 0.33$). SEC generally displays selectivities below $S = 0.2$.

For FFF and SFC alike, the ratio $\Delta R/\bar{R}$ can be shown to equal the product of average column selectivity $\bar{S}$ (Eq. 22) and the relative molecular weight difference for the pair $\Delta M/\bar{M}$ (57)

$$R_s = \frac{1}{4}\sqrt{\frac{L}{\bar{H}}}\,\bar{S}\left(\frac{\Delta M}{\bar{M}}\right) \tag{43}$$

Figure 11 summarizes the efficiency requirements for unit resolution of pairs of compounds as a function of their relative molecular weight difference. To avoid specifying column length the diagram is based on the number of plates $N(=L/H)$. The requirements are obviously different for different levels of selectivity. The curves in Figure 11 are all computed assuming that the system operates at maximum selectivity, which for FFF is specified by Eqs. 27–29. The figure also contains a curve representing the $S_{\max}$ value of 0.2, which can be considered characteristic for SEC (57, 62). The plate requirements are obviously more stringent the less selective the system; the slightly more than threefold difference in selectivity between SEC and thermal or flow FFF in good solvents can only be compensated for by an order of magnitude difference in plate height.

The selection of operating conditions to minimize analysis time must begin by a determination of the average plate height $\bar{H}$ needed for resolution of a pair of components with specified relative molecular weight difference, using a technique of known selectivity. The determination of optimal conditions is based on the assumption that both components are monodisperse, so that H_p (in Eq. 31) equals zero. Furthermore, the apparatus-induced zone broadening is assumed negligible, making the nonequilibrium term the focus of attention in this context. Once one has determined the $\bar{H}$ required for unit resolution, for example, from Eq. 43 or Figure 11, the optimization analysis proceeds by examining the effects of carrier velocity $\langle v \rangle$ and field strength ϕ on the elution time under conditions of unit resolution. It becomes convenient to replace λ in Eq. 31 by the ratio of average zonal layer thickness $\bar{l}$ (Eq. 6) and channel thickness w

$$\bar{H} = 24\left(\frac{\bar{l}}{w}\right)^3\left(\frac{w^2\langle v \rangle}{\bar{D}}\right) = 24\left(\frac{kT}{\phi\bar{S}}\right)^3\left(\frac{\langle v \rangle}{w\bar{D}}\right) \tag{44}$$

The average retention time t_r can be expressed in terms of void time t^0 and $\bar{l}$, using Eqs. 10 and 11

$$t_r = \frac{t^0}{6\lambda} = \frac{(L/\langle v \rangle)w}{6\bar{l}} \tag{45}$$

In the right-hand expression, t^0 has been replaced by the time required to travel column length L at average velocity $\langle v \rangle$. Substituting Eq. 45 into Eq. 44 and regrouping terms gives

$$\frac{\bar{H}\bar{S}^2\bar{D}}{4\,L(kT)^2} = \text{const} = \frac{1}{\bar{t}_r \phi^2} \tag{46}$$

As a consequence of Eq. 46, the time t_r required for unit resolution of a pair of components is inversely related to the square root of field strength ϕ. Therefore, even modest increases in ϕ lead to significant reductions in t_r.

7. OPTIMIZING THE CHANNEL GEOMETRY

Field-flow fractionation is similar to other elution-based separation techniques in affording better resolution of sample mixtures with longer separation columns. Indeed, the general resolution equation (Eq. 42) shows R_s to increase with $\sqrt{L}$ for a given pair of components. In practice, our thermal and flow FFF columns range in length from 40 to 60 cm. Detailed descriptions of their basic design are given elsewhere (65). Since publication of Ref. 65, TFFF has undergone no significant changes, although modern equipment universally uses thinner channels (0.0127 cm or less). The walls are still manufactured from copper blocks, although they are now chrome plated and polished to a mirror finish. The arsenal of flow systems has been expanded to include asymmetric channels, which show great potential for high resolution (29). In these new channels, the upper semipermeable wall is replaced by a wall of solid glass. This construction has increased the dimensional consistency of the apparatus and has also permitted focusing of the sample during relaxation.

Difficulties in keeping the field uniform along the column increase markedly with an increase in L. This is particularly true in the case of TFFF, where the channel is sandwiched between two copper blocks, of which the upper one is heated by high-powered heating rods and the lower one is cooled by running tap water. Even if a given temperature differential, ΔT, can be maintained between these blocks, the temperature of the cold wall will tend to increase with distance from the intake of coolant and thereby weaken the retention, as was shown in Figure 4. Only through an arrangement with multiple inlets and outlets for the coolant can this problem be eliminated. Thus, for practical purposes, a lengthening of the column must not unduly increase the dimensions of the heating and cooling blocks. An increase in L has been achieved by construction of a hairpin channel (66), as demonstrated in Figure 12. In order to avoid "race track" zone broadening as the sample moves around the bends in the serpentine column, the channel breadth is

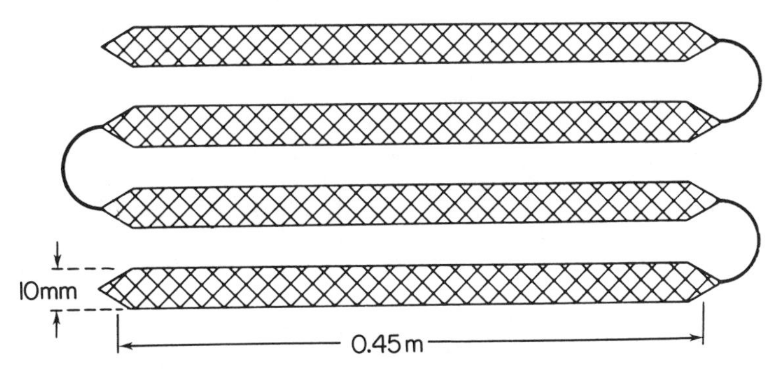

Figure 12. Principle of lengthening the FFF channel without increasing the end-to-end dimensions of the heating and cooling blocks (66).

reduced to minimal dimensions in these bands (~ 0.15 cm), while the main portion of the channel consists of straight, 1-cm wide parallel segments with a cumulative length of 180 cm.

The breadth b of an FFF channel must remain significantly larger than its thickness w, in order for the system to be treated as having "infinite parallel plate" geometry. Assumption of this geometry forms the basis for developments of both FFF retention theory (i.e., establishment of Eq. 9 from Eq. 8), as well as the theory for zone broadening (Eq. 31), which neglects any effects of the side walls. The infinite parallel plate model has shown to hold for breadth–thickness ratios exceeding 20:1 (67).

In the previous discussion of optimization, summarized by Eq. 46, there is no apparent influence of channel thickness w. This stems from the assumption that field strength ϕ can be varied at will. In reality, however, one is constrained to work within some range of practically feasible ϕ:s. In thermal FFF, for instance, the heat flux across the narrow gap is limited by the power of the heaters and the rate of heat transfer between copper block and coolant. At a given field strength, Eq. 43 predicts maximum resolution to occur for minimal values of $\bar{H}$, which implies operation at the lowest practical levels of the ratio of linear velocity $\langle v \rangle$ to channel thickness w. Expressed in terms of volumetric flow rate $\dot{V}$, which has to be delivered uniformly by the pump, this ratio becomes

$$\frac{\langle v \rangle}{w} = \frac{\dot{V}}{bw^2} \tag{47}$$

The narrower the channel, therefore, the slower must be the delivery of carrier.

Thin channels, characterized by small values of w, are generally desirable

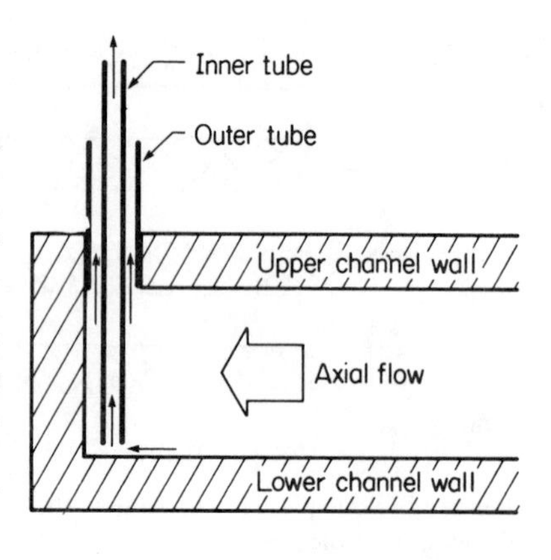

Figure 13. The principle of stream splitting in FFF. A minor fraction of the effluent is collected near the sample accumulation wall letting the bulk of the carrier exist (free of sample) from a separate port (25). Represented with permission from K.-G. Wahlund, H. S. Winegarner, K. D. Caldwell, and J. C. Giddings, "Improved Flow Field-Flow Fractionation System Applied to Water-Soluble Polymers: Programming, Outlet Stream Splitting, and Flow Optimization," *Anal. Chem.*, **58**, 573 (1986). Copyright © 1986 American Chemical Society.

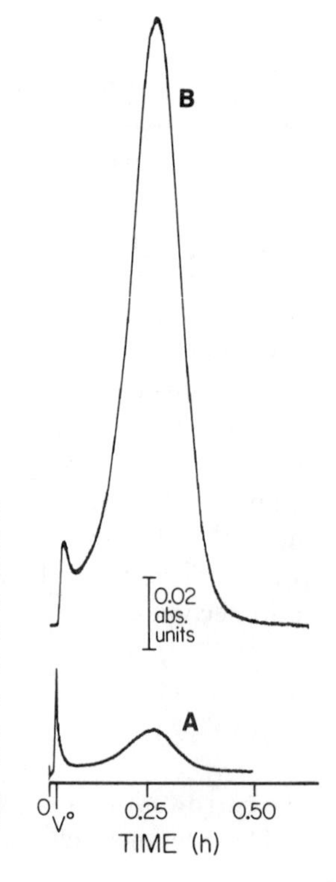

Figure 14. Illustration of the signal enhancement obtained through stream splitting (25). By collecting 1:40 of the effluent near the accumulation wall it was possible to obtain a 20-fold (less than the ideal 40-fold) increase in the detector's response to a fixed amount of sulfonated polystyrene (*b*), as compared to the response recorded when all sample was forced to exit through one port (*a*). Represented with permission from K.-G. Wahlund, H. S. Winegarner, K. D. Caldwell, and J. C. Giddings, "Improved Flow Field-Flow Fractionation System Applied to Water-Soluble Polymers: Programming, Outlet Stream Splitting, and Flow Optimization," *Anal. Chem.*, **58**, 573 (1986). Copyright © 1986 American Chemical Society.

from the standpoint of causing less dilution of the zone as it exits the channel. This fact is easily understood if one expresses the zone width in terms of volume units, rather than in the units of distance along the separation coordinate, which were used in Eq. 30 to define plate height

$$\sigma_v = \sigma_L bw/R \tag{48}$$

Here, σ_v and σ_L are the standard deviations for the zone width in volume and length units, respectively, with y and w having the customary meaning of channel breadth and thickness. A given separation efficiency, represented by a given value for H and hence for σ_L, will therefore result in larger values for σ_v (more dilute zones), the larger the value of w. The detrimental dilution associated with thicker channels has recently been circumvented by splitting the effluent stream, as shown in Figure 13 (25). Since the thickness of the FFF zone, represented by parameter l, is only a few percent of w even at modest levels of retention, it is profitable to collect only a minor fraction of the effluent from the region near the accumulation wall, while letting the major portion exit through a different port. The significant enhancement in detectability afforded by this approach is illustrated in Figure 14 (25), where a 20-fold increase in the detector signal is seen to result from a 40:1 split of the effluent stream in a flow FFF analysis of sulfonated polystyrenes. Although the experiment clearly illustrates the positive effects of stream splitting, it falls short of the theoretically predicted 40-fold increase in detector response.

8. PROGRAMMING

Optimal resolution of multicomponent samples in a given time period may require a variation of the experimental conditions during the course of the analysis (68–70). Both field strength ϕ (71) and axial flow rate $\langle v \rangle$ (72) lend themselves to programming in today's microprocessor-controlled FFF systems. The basic equation describing chromatographic elution, whether in FFF or SEC is formulated as follows (68)

$$L = \int_0^{t_r} R\langle v \rangle dt \tag{49}$$

where L is the length of the column and t_r is the elution time, that is, the time required for a compound to cover the distance L. In principle, either or both of retention ratio R and average carrier velocity $\langle v \rangle$ may vary in time. To date, most programmed polymer analyses involving FFF have been based on a decaying field (25, 69–74), while $\langle v \rangle$ has been maintained at a constant level.

In this approach, Eq. 49 can be simplified to give

$$t^0 = \int_0^{t_r} R(t)dt \tag{50}$$

Here t^0, the time for elution of an inert compound, replaces $L/\langle v \rangle$.

The selection of program type is dependent on the composition of the sample; most often the field has been allowed to decay exponentially (25, 69) or with a parabolic time course (71) following a lag period during which the initial field strength is held at a constant level. This type of experimental arrangement is illustrated by the thermal FFF resolution of a nine component mixture of polystyrenes, shown in Figure 15. Appropriate equations relating elution time to sample molecular weight under conditions of programmed field of either the flow or the thermal type, have been derived elsewhere (25, 71).

The time required for completion of the run in Figure 15 was 7 h; had the elution been forced to proceed at a constant field, the run would have lasted approximately twice as long, and resulted in peaks that were barely, if at all, detectable. Clearly, field reductions are beneficial in shortening elution times and enhancing detectability. By a judicious choice of program, this can be accomplished without undue loss in resolution, as seen in Figure 15.

A word of caution: The high fields needed to resolve components of

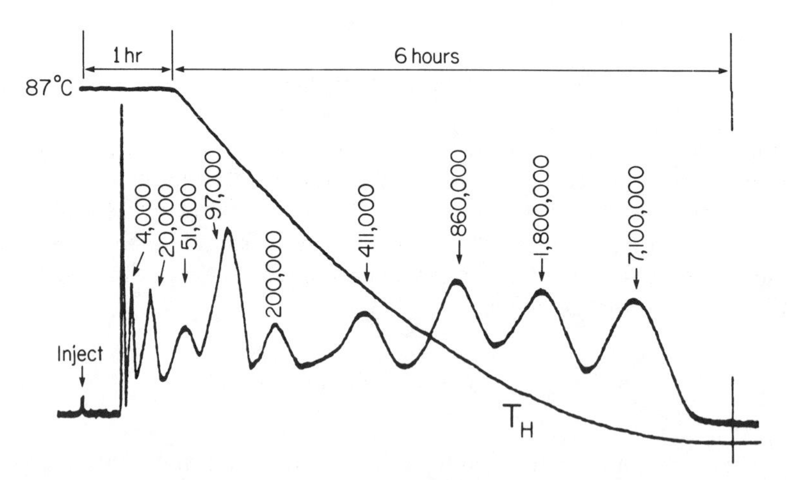

Figure 15. Thermal FFF of a nine-component mixture of linear polystyrenes using field programming. The samples, spanning a large range of molecular weights, eluted close to base line resolution in a 7 h period, during which the field was held constant at $\Delta T = 70\,°C$ for 1 h and forced to decay to zero in 6 h using a parabolic temperature program. The carrier was ethylbenzene (71).

relatively low molecular weight will create significant compression (small values of λ) of zones containing higher molecular weight species (75). Because of the exponential distribution of solute within the zone, the concentration $c(0)$ at the accumulation wall will reach progressively higher levels the higher the sample retention. Through integration of Eq. 1 one obtains the following relationship between the cross-sectional concentration average $\langle c \rangle$, the wall concentration $c(0)$, and retention parameter λ

$$c(0) = \frac{\langle c \rangle}{\lambda(1 - e^{-1/\lambda})} \simeq \frac{\langle c \rangle}{\lambda} \tag{51}$$

It is evident that the small λ values associated with high retention will lead to large values for $c(0)$. Since high polymer concentrations often lead to nonideal behavior, such as molecular entanglement and the formation of microgels (76, 77), it is essential to verify that a given fractionation pattern remains invariant with concentration of injected material. Likewise, whether or not the fractionation involves gradient elution, it is prudent to verify retention-derived molecular weight values under different conditions of field strength (i.e., varying λ) and channel flow (i.e., varying H and thereby the dilution of the zone) to ensure accurate characterization of the sample.

9. CONCLUSION

The need for rapid and accurate determination of polymer molecular weights and polydispersities has led to the development of sophisticated techniques for polymer fractionation, and for the extraction of analytical information from the fractogram. To this end, methods based on SEC are most commonly used. This chapter described the FFF methods and their application to polymer analysis. The text makes occasional comparisons between SEC and thermal and flow FFF; in order to add perspective to the presentation these comparisons are summarized in Table 1.

The well-defined geometry of the FFF channel, and the external control of the applied field, are both factors that enable the operator to control retention and to optimize the resolution of a given sample. In this respect there is a significant difference between SEC and FFF, with the former being less adaptable to different sample requirements.

The one-phase nature of the FFF process makes it suitable for processing of samples over an exceptionally wide range of molecular weights. Although "polymer fractionation" generally entails the separation of soluble macromolecules, the FFF methods work equally well with colloidal particles, and their fractionation range therefore spans about eight orders of magnitude in

TABLE 1. Brief Comparison of SEC and Thermal or Flow FFF

	SEC	FFF
Retention mechanism known from first principles	No	Yes
Fractionation range in decades of molecular weight (for single pore-size SEC packing material)	~2	~8
Maximum selectivity of separation process	⩽ about 0.2	⩽ about 0.7
Capability for programmed retention	No	Yes
Column dispersion known from first principles	No	Yes
Efficiency (at modest FFF retention, $R = 0.1$)	High	Moderate
Relationship between efficiency and retention	About constant	Increasing
Commercial availability of instrumentation	Yes	Yes[a]

[a] Thermal FFF: FFFractionation, Inc., P. O. Box 8718, Salt Lake City, UT 84108.

molecular weight. By selecting an SEC packing of a given porosity, one is in general limited to a resolution range of about two orders of magnitude.

A direct comparison of the two techniques in terms of resolving power is difficult to make, since the efficiency of SEC remains virtually constant with retention, whereas that of FFF increases sharply with increased retention. At low retention levels the FFF efficiency is low and its selectivity moderate, yielding weaker resolving power than that offered by SEC. At modest retention, the selectivity of FFF surpasses that of SEC by a factor of three or better. Since SEC efficiencies in this case are about an order of magnitude better than those offered by FFF, the two techniques show comparable resolution. High retentions, in turn, give superior resolving power to the FFF techniques, as a result of increases in efficiency.

ACKNOWLEDGMENT

The preparation of this article has been made with support from the National Science Foundation, through grant CHE-8218503.

NOTATION

a	Mark–Houwink exponent (intrinsic viscosity $[\eta] = \text{const} \cdot M^a$)
A	Constant in empirical relationship between D and M
B	Constant in empirical relationship between D/D_T and M
b	Exponent in relationship between D and M
b'	Exponent in relationship between D/D_T and M
b	Column breadth

c	Solute concentration
D_T	Thermal diffusion coefficient
D	Ordinary translational diffusion coefficient
F	Field induced force acting on molecule
f	Molecular friction coefficient
f'	Distance on the fractogram from start of flow to peak maximum
H	Plate height
H_{neq}	Nonequilibrium plate height contribution, $\chi w^2 \langle v \rangle / D$
H_p	Polydispersity
H_i	Instrument–operation induced plate height contribution
k	Boltzmann's constant
L	Column length
l	Characteristic height of solute layer ($l = D/U$)
M	Molecular weight
$\bar{M}_n$	Number-average molecular weight
$\bar{M}_w$	Weight-average molecular weight
$m(M)$	Mass of polymer of molecular weight M
N	Number of theoretical plates ($N = L/H$)
r_h	Hydrodynamic radius
R	Retention ratio $V/\langle v \rangle$
R_s	Resolution index
s	Field interaction parameter
S	Selectivity ($d\ln V_r / d\ln M$)
S_{max}	Maximum selectivity
$S_{max,F}$	Maximum selectivity in FFFF ($S_{max,F} = b$)
$S_{max,T}$	Maximum selectivity in TFFF ($S_{max,T} = b'$)
T	Temperature
T_c	Cold wall temperature
ΔT	Temperature differential across channel
t	Time
t^0	Void time
t_r	Retention time
U	Field-induced drift velocity
V_r	Retention volume
V^0	Void volume
$\dot{V}$	Volumetric channel flow rate
$\dot{V}_c$	Volumetric cross-flow rate
v	Local velocity
$\langle v \rangle$	Average carrier velocity
V	Average velocity of a zone
w	Channel thickness
x	Altitude above accumulation wall

z	Flow axis coordinate (separation coordinate)
λ	Retention parameter ($= l/w$)
λ_F	Retention parameter in FFFF
λ_T	Retention parameter in TFFF
μ	Polydispersity index, $\bar{M}_w/\bar{M}_n$
σ	Standard deviation in zone width
σ_L	Standard deviation in zone width in units of distance in the channel
σ_v	Standard deviation in zone width in units of elution volume
$\sigma_{f'}$	Standard deviation of zone width in units of distance on chart representing the fractogram
ϕ	General field strength parameter
χ	Nonequilibrium coefficient, $H_{\text{neq}} = \chi w^2 \langle v \rangle / D$

REFERENCES

1. W. W. Yau, J. J. Kirkland, and D. D. Bly, *Modern Size Exclusion Chromatography*, Wiley, New York, 1979, Chapter 1.
2. H. G. Barth and F. J. Carlin, Jr., "A Review of Polymer Shear Degradation in Size-Exclusion Chromatography," *J. Liq. Chromatogr.*, **7**, 1717 (1984).
3. W. W. Yau, J. J. Kirkland, and D. D. Bly, *Modern Size Exclusion Chromatography*, Wiley, New York, 1979, Chapter 7.
4. J. C. Giddings, "Field-Flow Fractionation. A Versatile Method for the Characterization of Macromolecular and Particulate Materials," *Anal. Chem.*, **53**, 1170A (1981).
5. G. Karaiskakis, M. N. Myers, K. D. Caldwell, and J. C. Giddings, "Verification of Retention and Zone Spreading Equations in Sedimentation Field-Flow Fractionation," *Anal. Chem.*, **53**, 1314 (1981).
6. J. C. Giddings, K. D. Caldwell, and H. K. Jones, "Measuring Particle Size Distribution of Simple and Complex Colloids Using Sedimentation Field-Flow Fractionation," *Am. Chem. Soc. Symp. Ser.*, **332**, 215 (1987)
7. J. C. Giddings, B. N. Barman, and H. Li, "Colloid Characterization by Sedimentation Field-Flow Fractionation," *J. Colloid Interface Sci.*, **132**, 554 (1989).
8. J. J. Kirkland and W. W. Yau, "Simultaneous Determination of Particle Size and Density by Sedimentation Field-Flow Fractionation," *Anal. Chem.*, **55**, 2165 (1983).
9. K. D. Caldwell and J. Li, "Emulsion Characterization by the Combined Sedimentation Field-Flow Fractionation-Photon Correlation Spectroscopy Methods," *J. Colloid Interface Sci.*, **132**, 256 (1989).
10. J. Janca, D. Pribylova, K. Bouchal, V. Tyrackova, and E. Zurkova, "Characterization of Polystyrene and Poly(styrene-glycidyl methacrylate) Copolymer Latexes by Sedimentation Field-Flow Fractionation," *J. Liq. Chromatogr.*, **9**, 2059 (1986).

11. C. R. Yonker, H. K. Jones, and D. M. Robertson, "Nonaqueous Sedimentation Field-Flow Fractionation," *Anal. Chem.*, **59**, 2574 (1987).
12. R. Beckett, "The Application of Field-Flow Fractionation Techniques to the Characterization of Complex Environmental Samples," *Environ. Tech. Lett.*, **8**, 339 (1987).
13. L. E. Oppenheimer and G. A. Smith, "Sedimentation Field-Flow Fractionation of Colloidal Metal Hydrosols," *J. Chromatogr.*, **461**, 103 (1989).
14. K. D. Caldwell, L. F. Kesner, M. N. Myers, and J. C. Giddings, "Electrical Field-Flow Fractionation of Proteins," *Science*, **176**, 296 (1972).
15. J. C. Giddings, G. C. Lin, and M. N. Myers, "Electrical Field-Flow Fractionation in a Rigid Membrane Channel," *Sep. Sci.*, **11**, 553 (1976).
16. L. E. Schallinger, W. W. Yau, and J. J. Kirkland, "Sedimentation Field-Flow Fractionation of DNA's," *Science*, **225**, 434 (1984).
17. L. E. Schallinger, J. E. Gray, L. W. Wagner, S. Knowlton, and J. J. Kirkland, "Preparative Isolation of Plasmid DNA with Sedimentation Field Flow Fractionation," *J. Chromatogr.*, **342**, 67 (1985).
18. L. E. Schallinger, E. C. Arner, and J. J. Kirkland, "Size Distribution of Cartilage Proteoglycans Determined by Sedimentation Field-Flow Fractionation," *Biochem. Biophys. Acta*, **966**, 231 (1988).
19. G. H. Thompson, M. N. Myers, and J. C. Giddings, "Thermal Field Flow Fractionation of Polystyrene Samples," *Anal. Chem.*, **41**, 1219 (1969).
20. J. C. Giddings, M. N. Myers, and J. Janca, "Retention Characteristics of Various Polymers in Thermal Field-Flow Fractionation," *J. Chromatogr.*, **186**, 37 (1979).
21. J. J. Gunderson and J. C. Giddings, "Chemical Composition and Molecular-size Factors in Polymer Analysis by Thermal Field-Flow Fractionation and Size Exclusion Chromatography," *Macromolecules*, **19**, 2618 (1986).
22. M. E. Schimpf and J. C. Giddings, "Characterization of Thermal Diffusion in Polymer Solutions by Thermal Field-Flow Fractionation: Effects of Molecular Weight and Branching," *Macromolecules*, **20**, 1561 (1987).
23. J. J. Kirkland and W. W. Yau, "Thermal Field-Flow Fractionation of Water-Soluble Macromolecules," *J. Chromatogr.*, **353**, 95 (1986).
24. J. C. Giddings, G. C. Lin, and M. N. Myers, "Fractionation and Size-Distribution of Water-Soluble Polymers by Flow Field-Flow Fractionation," *J. Liq. Chromatogr.*, **1**, 1 (1978).
25. K.-G. Wahlund, H. S. Winegarner, K. D. Caldwell, and J. C. Giddings, "Improved Flow Field-Flow Fractionation System Applied to Water-Soluble Polymers: Programming, Outlet Stream Splitting, and Flow Optimization," *Anal. Chem.*, **58**, 573 (1986).
26. K.-G. Wahlund and J. C. Giddings, "Properties of an Asymmetrical Flow Field-Flow Fractionation Channel Having One Permeable Wall," *Anal. Chem.*, **59**, 1332 (1987).
27. R. Beckett, J. C. Bigelow, J. Zhang, and J. C. Giddings, "Analysis of Humic Substances Using Flow FFF," *Am. Chem. Soc. Adv. Chem. Ser.*, **219**, 65 (1989).

28. K.-G. Wahlund and A. Litzen, "Application of an Asymmetrical Flow Field-Flow Fractionation Channel to the Separation and Characterization of Proteins, Plasmids, Plasmid Fragments, Polysaccharides and Unicellular Algae," *J. Chromatogr.*, **461**, 73 (1989).
29. A. Litzen and K.-G. Wahlund, "Improved Separation Speed and Efficiency for Proteins, Nucleic Acids and Viruses in Asymmetrical Flow Field-Flow Fractionation," *J. Chromatogr.*, **476**, 413 (1989).
30. J. A. Jönsson and A. Carlshaf, "Flow Field-Flow Fractionation in Hollow Cylindrical Fibers," *Anal. Chem.*, **61**, 11 (1989).
31. J. C. Giddings, S. R. Fisher, and M. N. Myers, "Field-Flow Fractionation. One-Phase Chromatography for Macromolecules and Particles," *Am. Lab.*, **10** (5), 15 (1978).
32. F. J. Yang, M. N. Myers, and J. C. Giddings, "Peak Shifts and Distortion Due to Solute Relaxation in Flow Field-Flow Fractionation," *Anal. Chem.*, **49**, 659 (1977).
33. M. E. Hovingh, G. H. Thompson, and J. C. Giddings, "Column Parameters in Thermal Field-Flow Fractionation," *Anal. Chem.*, **42**, 195 (1970).
34. H.-L. Lee, J. F. G. Reis, J. Dohner, and E. N. Lightfoot, "Single-Phase Chromatography. Solute Retardation by Ultrafiltration and Electrophoresis," *AIChE J.*, **20**, 776 (1984).
35. J. Happel and H. Brener, *Low Reynolds Number Hydrodynamics*, Prentice-Hall, Englewood Cliffs, NJ, 1965, p. 34.
36. J. C. Giddings, "Conceptual Basis of Field-Flow Fractionation," *J. Chem. Ed.*, **50**, 667 (1973).
37. J. J. Gunderson, K. D. Caldwell, and J. C. Giddings, "Influence of Temperature Gradients on Velocity Profiles and Separation Parameters in Thermal Field-Flow Fractionation," *Sep. Sci. Technol.*, **19**, 667 (1984).
38. R. C. Reid, J. M. Prausnitz, and T. K. Sherwood, *The Properties of Gases and Liquids*, 3rd ed, McGraw-Hill, New York, 1977.
39. A. H. Emery and H. G. Drickamer, "Thermal Diffusion in Polymer Solutions," *J. Chem. Phys.*, **23**, 2252 (1955).
40. F. J. Bonner, "Thermal Diffusion Measurements on Polystyrene in Toluene," *Ark. Kemi*, **27**, 115 (1967).
41. K. E. Grew, "Thermal Diffusion," in H. J. M. Hanley (Ed.), *Transport Phenomena in Fluids*, Dekker, New York, 1969, Chapter 10.
42. S. R. DeGroot, *Thermodynamics of Irreversible Processes*, North-Holland, Amsterdam, 1951, Chapter 7.
43. J. C. Giddings, "Field-Flow Fractionation. Extending the Molecular Weight-Range of Liquid Chromatography to One-Trillion," *J. Chromatogr.*, **125**, 3 (1976).
44. J. C. Giddings, K. D. Caldwell, and M. N. Myers, "Thermal Diffusion of Polystyrene in Eight Solvents by an Improved Thermal Field-Flow Fractionation Methodology," *Macromolecules*, **9**, 106 (1976).

45. M. E. Schimpf and J. C. Giddings, "Characterization of Thermal Diffusion in Polymer Solutions by Thermal Field-Flow Fractionation: Dependence on Polymer and Solvent," *J. Polym. Sci., Polym. Phys. Ed.*, **27**, 1317 (1989).
46. C. Tanford, *Physical Chemistry of Macromolecules*, Wiley, New York, 1961, Chapter 6.
47. W. W. Yau, J. J. Kirkland, and D. D. Bly, *Modern Size Exclusion Chromatography*, Wiley-Interscience, New York, 1979, Chapter 10.
48. N. C. Ford, Jr., R. Gabler, and F. E. Karasz, "Self-Beat Spectroscopy and Molecular Weight," *Am. Chem. Soc. Adv. Chem. Ser.*, **125**, 25 (1973)
49. P. J. Flory, *Principles of Polymer Chemistry*, Cornell University, Ithaca, NY, 1953, Chapter 14.
50. J. Marra, H. A. van der Schee, G. J. Fleer, and J. Lyklema, "Polyelectrolyte Adsorption from Saline Solutions," in R. H. Ottewill, C. H. Rochester, and A. L. Smith (Eds.), *Adsorption from Solution*, Academic, London, 1983, p. 245.
51. Y. S. Gao, K. D. Caldwell, M. N. Myers, and J. C. Giddings, "Extension of Thermal Field-Flow Fractionation to Ultra-High Molecular Weight Polystyrenes," *Macromolecules*, **18**, 1272 (1985).
52. M. N. Myers, K. D. Caldwell, and J. C. Giddings, "Retention in Thermal Field-Flow Fractionation," *Sep. Sci.*, **9**, 47 (1974).
53. S. L. Brimhall, M. N. Myers, K. D. Caldwell, and J. C. Giddings, "Study of Temperature Dependence of Thermal Diffusion in Polystyrene/Ethylbenzene by Thermal Field-Flow Fractionation," *J. Polym. Sci., Polym. Phys. Ed.*, **23**, 2443 (1985).
54. S. L. Brimhall, M. N. Myers, K. D. Caldwell, and J. C. Giddings, "High Temperature Thermal Field-Flow Fractionation for the Characterization of Polyethylene", *Sep. Sci. Technol.*, **16**, 671 (1981).
55. J. C. Giddings, L. K. Smith and M. N. Myers, "Thermal Field Flow Fractionation: Extension to Lower Molecular Weight Separations by Increasing the Liquid Temperature Range Using a Pressurized System," *Anal. Chem.*, **47**, 2389 (1975).
56. J. C. Giddings, "Displacement and Dispersion of Particles of Finite Size in Flow Channels with Lateral Forces. Field-Flow Fractionation and Hydrodynamic Chromatography," *Sep. Sci. Technol.*, **13**, 251 (1978).
57. J. C. Giddings, "Field-Flow Fractionation of Polymers: One-Phase Chromatography," *Pure Appl. Chem.*, **51**, 1459 (1979).
58. J. C. Giddings, *Dynamics in Chromatography*, Dekker, New York, 1955, Chapter 2.
59. L. K. Smith, M. N. Myers, and J. C. Giddings, "Peak Broadening Factors in Thermal Field-Flow Fractionation," *Anal. Chem.*, **49**, 1750 (1977).
60. W. W. Yau, J. J. Kirkland, and D. D. Bly, *Modern Size Exclusion Chromatography*, Wiley-Interscience, New York, 1979, Chapter 3.
61. J. C. Giddings, Y. H. Yoon, K. D. Caldwell, M. N. Myers, and M. E. Hovingh, "Nonequilibrium Plate Height for Field-Flow Fractionation in Ideal Parallel Plate Columns," *Sep. Sci.*, **10**, 447 (1975).

62. J. C. Gunderson and J. C. Giddings, "Comparison of Polymer Resolution in Thermal Field-Flow Fractionation and Size-Exclusion Chromatography," *Anal. Chim. Acta*, **189**, 1 (1986).

63. M. E. Schimpf, M. N. Myers, and J. C. Giddings, "Measurement of Polydispersity of Ultra-Narrow Polymer Fractions by Thermal Field-Flow Fractionation," *J. Appl. Polym. Sci.*, **33**, 117 (1987).

64. J. C. Giddings, M. Martin, and M. N. Myers, "High-Speed Polymer Separations by Thermal Field-Flow Fractionation," *J. Chromatogr.*, **158**, 419 (1978).

65. J. C. Giddings, M. N. Myers, K. D. Caldwell, and S. R. Fisher, "Analysis of Biological Macromolecules and Particles by Field-Flow Fractionation," in D. Glick (Ed.), *Methods of Biochemical Analysis*, Vol. 26, Wiley, New York, 1980, p. 79.

66. J. C. Giddings, M. Martin, and M. N. Myers, "High-Resolution Polymer Separations in a Four-Pass Hairpin Thermal Field-Flow Fractionation Column," *J. Polym. Sci., Polym. Phys. Ed.*, **19**, 815 (1981).

67. J. C. Giddings and M. R. Schure, "Theoretical Analysis of Edge Effects in Field-Flow Fractionation," *Chem. Eng. Sci.*, **42**, 1471 (1987).

68. J. C. Giddings and K. D. Caldwell, "Field-Flow Fractionation: Choices in Programmed and Nonprogrammed Operation," *Anal. Chem.*, **56**, 2093 (1984).

69. J. J. Kirkland, S. W. Rementer, and W. W. Yau, "Molecular-weight Distributions of Polymers by Thermal Field Flow Fractionation with Exponential Temperature Programming," *Anal. Chem.*, **60**, 610 (1988).

70. J. C. Giddings, V. Kumar, P. S. Williams, and M. N. Myers, in C. D. Craver and T. Provder (Eds.), *Polymer Characterization by Interdisciplinary Methods, Am. Chem. Soc. Symp. Ser.* (in press).

71. J. C. Giddings, L. K. Smith, and M. N. Myers, "Programmed Thermal Field-Flow Fractionation," *Anal. Chem.*, **48**, 1587 (1976).

72. J. C. Giddings, K. D. Caldwell, J. F. Moellmer, T. H. Dickinson, M. N. Myers and M. Martin, "Flow Programmed Field-Flow Fractionation," *Anal. Chem.*, **51**, 30 (1979).

73. J. C. Giddings, M. N. Myers, G. C. Lin, and M. Martin, "Polymer Analysis and Characterization by Field-Flow Fractionation (one-Phase Chromatography)," *J. Chromatogr.*, **142**, 23 (1977).

74. J. J. Kirkland and W. W. Yau, "Thermal Field-Flow Fractionation of Polymers with Exponential Temperature Programming," *Macromolecules*, **18**, 2305 (1985).

75. K. D. Caldwell, S. L. Brimhall, Y. Gao, and J. C. Giddings, "Sample Overloading Effects in Polymer Characterization by Field-Flow Fractionation," *J. Appl. Polym. Sci.*, **36**, 703 (1988).

76. P. G. DeGennes, "Dynamics of Entangled Polymer Solutions. I. Rouse Model," *Macromolecules*, **9**, 587 (1976).

77. P. T. Callaghan and D. N. Pinder, "Dynamics of Entangled Polystyrene Solutions Studied by Pulsed Field Gradient Nuclear Magnetic Resonance," *Macromolecules*, **13**, 1085 (1980).

CHAPTER

5

POLYMER CHARACTERIZATION USING INVERSE GAS CHROMATOGRAPHY

PETR MUNK

Department of Chemistry and Biochemistry and Center for Polymer Research
University of Texas at Austin
Austin, Texas

1. INTRODUCTION

Smidsrod and Guillet (1) showed in 1969 that gas chromatography (GC) using a polymeric stationary phase can contribute extensively to the knowledge of polymers. The method is called inverse gas chromatography (IGC) because

Modern Methods of Polymer Characterization, Edited by Howard G. Barth and Jimmy W. Mays
ISBN 0-471-82814-9

the stationary phase is the phase of interest in contrast to traditional GC in which the separation of components in the mobile phase is the objective. In the early 1970s, IGC was found to be useful for studying the phase and transport properties of polymers. The study of crystallinity was based on the fact that probe molecules have much larger solubility in amorphous polymers than in crystalline ones. The rate of penetration of small probe molecules into a polymer varies widely. Polymers below the glass transition temperature are virtually not penetrated at all under the conditions of an IGC experiment. Hence, the value of IGC had been for the study of the glass transition. As the temperature increases the probe enters the polymer more freely. The resistance of the polymer against probe diffusion leads to peak broadening. The diffusion coefficients of probes in bulk polymers may be found from the width of the chromatographic peaks. When the resistance against diffusion is small enough, a phase equilibrium exists between the mobile and stationary phases. The retention volume of the probe on the column becomes a sensitive measure of the distribution coefficient. Thus, the measurement of retention volumes leads to a wealth of information about the polymer–probe interaction coefficients and their dependence on the nature of the probe, temperature, and so on. If a mixture of two polymers (a polymer blend) is used as a stationary phase, the retention data may yield the polymer–polymer interaction coefficient.

In this chapter we plan to demonstrate all these capabilities of the IGC method. However, we want also to warn the reader about the pitfalls of the method and to stress all the precautions to be taken when reliable results are desired.

In Section 2 we will review the basic concepts of GC with special emphasis on those aspects of the method that follow from the polymeric nature of the stationary phase. In Section 3 we will describe some instrumental modifications, which are necessary for IGC work. The emphasis will not be on the separation and identification of components, as in ordinary GC, but on the precision of the retention values, half-widths of the peaks, and other experimental parameters. Section 4 is devoted to analysis of experimental data with special emphasis on necessary corrections. Finally, in Section 5 we present a short review of the results that were obtained by the IGC method. General references and reviews on IGC are listed at the end of this chapter.

2. THEORY

The physical and theoretical principles underlying IGC are the same as the ones related to traditional GC, only the emphasis is different. In this section, we will present a short review of the theory of GC; and will discuss in more detail the quantitative aspects related to retention volumes and to peak spreading.

2.1. Basic Concepts in Gas Chromatography

A stream of inert *carrier gas* flows at a constant flow rate through a chromatographic column. The *injection port* in front of the column and the *detector* at the end of the column serve as reference points for the measurement of chromatographic quantities. The *void* or *dead volume* V_o is the volume of the *gas phase* between the injection point and the detector. It is composed of two parts: the void volume of the column proper V_c and the apparatus dead or void volume V_d, that is the volume of the injection port, detector, and the connecting plumbing

$$V_o = V_c + V_d \tag{1}$$

In well-designed experiments, V_d is much smaller than V_c and is about equally distributed before and after the column. Flow rate is measured on a gas stream that is either diverted to the measuring device before reaching the detector (in the case of flame ionization and similar detectors) or after leaving the detector (in the case of a thermal conductivity detector). Usually the measuring device is a bubble-flow meter and the flow is measured in volume–time units at the temperature of the thermostated bubble meter T_{th} and atmospheric pressure P_o.

The retention time for a given compound (probe) is the time interval t_p between probe injection (the injection is virtually instantaneous) and probe detection. However, during the passage through the chromatograph, the probe band spreads and the retention time should be mass averaged for the eluting probe. When the elution peak is symmetrical, the mass averaged time is equal to the elution time for the peak maximum. Hence, t_p is defined as the elution time of the peak maximum. The probe, which does not interact with the column at all, is called a marker and its elution time is t_m.

The chromatographic column exhibits a hydrodynamic resistance against the gas flow and this resistance must be overcome by a pressure gradient. It may be shown that the pressure P varies along the column as

$$P = P_{\text{i}}[1 - (x/L)(1 - P_0^2/P_{\text{i}}^2)]^{1/2} \tag{2}$$

where P_{i} and P_{o} are the column inlet and outlet pressures, respectively, L is the column length, and x is the length coordinate. Because of the compressibility of the carrier gas, both the volume flow rate F and linear flow velocity u vary along the column as

$$PF = P_o F_o \tag{3a}$$

$$Pu = P_o u_o \tag{3b}$$

where the subscript o refers to the column outlet. The linear flow velocity is related to F_o as

$$u_o = LF_o/V_c \tag{4}$$

Other routinely used relations are strictly valid if the dead volume V_d is negligible. For finite values of V_d, they remain good approximations provided that (1) V_d is much smaller than V_c; (2) V_d is about equally distributed in front and after the column; and (3) the pressure drops within injector and detector are negligible. If these assumptions are valid, it can be shown that

$$V_o = t_m F_o \frac{3}{2} \frac{(P_i/P_o)^2 - 1}{(P_i/P_o)^3 - 1} \equiv t_m F_o j \tag{5}$$

where j is defined by the identity in Eq. 5. Equation 5 is useful when we are trying to minimize the experimental error. The quantity V_o for a given column must be a constant. Once this constant is established by a series of careful experiments, the least precise quantity of Eq. 5, F_o, can be calculated from easily measured values of t_m, P_i and P_o.

Coefficient of hydrodynamic resistance r_h is another quantity useful for checking the consistency of experimental data. The coefficient r_h relates the pressure drop dP on the chromatographic column to the linear gas velocity u and the element of column length dx as

$$dP = -r_h u\,dx \tag{6}$$

A simple calculation shows that r_h is related to experimental values as follows:

$$r_h = (P_i^2 - P_o^2)/2LP_o u_o = (V_c/L)(\Delta P/F_o)(1 + \Delta P/2P_o) \tag{7}$$

where $\Delta P \equiv P_i - P_o$; V_c/L is a constant for a given column. The resistance r_h depends on the column packing. For a given column, it is proportional to the viscosity of the carrier gas. The viscosity of ideal gases (all carrier gases are virtually ideal) is independent of pressure, consequently, r_h does not change along the column and should be independent of flow rate at a given temperature. It depends on temperature only through the changing viscosity of the gas.

When the probe interacts with the column, it progresses along the column slower. The portion of the probe that is in the gas phase moves ahead with the same velocity as the carrier gas, while the interacting portion does not move at all. Under most chromatographic conditions, the fraction of the probe in gas phase R_f does not change as the probe travels along the column. Thus, the average probe velocity u_p is given as

$$u_p = R_f u \tag{8}$$

The fraction R_f is usually called the *retardation factor.*

Retention volume V_R is another frequently used quantity. It is the volume of carrier gas (volume measured *inside* the chromatograph), which is needed to elute the probe. The retention volume of marker is equal to V_o. The retention volume V_R and the net retention volume V_N are easily calculated as

$$V_R = V_o t_p / t_m \tag{9}$$

$$V_N \equiv V_R - V_o = (t_p - t_m) V_o / t_m \tag{10}$$

Substitution of Eq. 5 yields the well-known relation

$$V_N = (t_p - t_m) F_o j \tag{11}$$

However, once V_o has been measured, Eq. 10 may be more convenient than Eq. 11.

The relation between the retention of the marker and of the probe on the column itself is described either by the retardation factor R_f or by the *partition ratio k*, which is the ratio of the weight of the probe in the stationary phase to its weight in the mobile phase. It follows from the definition of these two quantities that they are related as

$$R_f = 1/(1 + k) \tag{12}$$

The calculation of R_f is based on Eq. 8 which applies only to the column itself and not to the plumbing. Inside the plumbing, both marker and probe move with the same velocity as the carrier gas. Recognizing that the elution times are inversely proportional to the linear velocities and correcting for the time t_d needed for the passage of the probe through the dead volume V_d, we may rewrite Eq. 8 as

$$R_f = (t_m - t_d)/(t_p - t_d) \tag{13}$$

With good precision, t_d may be approximated as

$$t_d = t_m V_d / V_o \tag{14}$$

Obviously, V_d must be known from an auxiliary experiment. Combination of Eqs. 1 and 12–14 yields for k

$$k = (t_p - t_m) V_o / t_m V_c \tag{15}$$

2.2. Retention Mechanisms

The retention of a probe by a chromatographic column should be studied not only from the thermodynamic (equilibrium) viewpoint but also from kinetic considerations.

At equilibrium, the probe distributes itself throughout the system in such a way that its chemical potential is the same everywhere. This includes not only the bulk phases (gas and stationary phase) but also all types of surfaces present: the polymer–gas interface, the uncoated surface of the inert support, even the walls of the column and plumbing connections. The retention by the bulk stationary phase at quasiequilibrium conditions is governed by the theory of solutions. Its analysis can yield a wealth of information about the thermodynamics of the probe–polymer interaction. At low overall concentrations of the probe, the ratio of concentrations in gas and liquid phases is independent of concentration, which simplifies the thermodynamic analysis considerably. However, the adsorption capacity of some surfaces is limited and the adsorption isotherm has a pronounced Langmuir character. In this situation, the proportionality between the probe concentrations in the gas phase and on the surface holds only at extremely small (often experimentally unattainable) concentrations. This effect is most conspicuous in the interactions of polar probes with the "inert" support, which is usually a silicate-like material.

The interaction of probes with the polymer surface is a controversial subject. It was reported many times for polymers at temperatures below the glass–transition temperature and was postulated to exist also at temperatures above the glass–transition temperature. It may be, however, possible to explain kinetically all the phenomena attributed to the probe–surface interaction (adsorption).

In an ideal chromatographic experiment, the equilibrium between the mobile and stationary phases is established instantaneously and no concentration gradient exists in the direction perpendicular to the gas stream in either mobile or stationary phase. In such a situation, the retention data provide thermodynamic information without any controversy.

In a real experiment, the transport of the probe across the phase boundary creates concentration gradients. Let us consider the situation closer. When the front side of the probe zone interacts with the stationary phase, the probe enters the polymer. This causes a depletion of the probe in the gas stream in the close vicinity of the surface; the top-most layer in the polymer is enriched in the probe. The two surface layers are essentially in equilibrium and the transport stops even if the bulk of both phases is still far from equilibrium. Meanwhile, the carrier gas transports the probe along the column, and the probe is not retained as much as expected. The faster the gas velocity, the larger is the amount of "missing" retention. However, the diffusion of the probe into the polymer lowers the surface concentration. The diffusion of the probe in the gas phase brings more probe to the boundary and the transport across the boundary continues. Obviously, the amount of "missing" retention depends on the ratio of the flow velocity u and the diffusion constants of the probe. There are two diffusion terms: one related to the diffusion in the gas phase and the other to the diffusion in the stationary phase.

When the tail of the peak passes the same part of the column, the situation is reversed. The diffusion in the polymer delivers the probe toward the surface too slowly and the retention is excessively high. Thus, the diffusion leads mainly to peak spreading but the total retention time is not changed much.

The situation was studied in detail by Gray and Guillet (2). They recognized two limiting situations. When the diffusion is reasonably fast, the average concentration within the polymer layer does not differ much from the concentration near the surface (the former is lower when the front of the peak is passing, and it is higher during the passage of the tail). In this limit, the retention volume is virtually not changed from its ideal value.

In the other limit, the diffusion is so slow that the whole probe peak passes along the polymer before any significant amount of the probe can penetrate inside, and the retention volume decreases dramatically. When the amount of probe is not small enough, the surface layer of the polymer is saturated along the column and the rest of the probe passes through without any retention and emerges from the column together with the marker. The small amount that was absorbed desorbs very slowly producing a long tail. In this case, the elution behavior is very similar to the case of a surface adsorption exhibiting saturation (Langmuir isotherm). The present author believes that many experimental phenomena ascribed to the probe–polymer surface interaction are actually manifestations of the very slow diffusion. Especially, this may be the case when the polymer is in a glassy state with very small diffusion constant for the probe.

Historically, IGC was used first for studying the changes accompanying a sudden change in diffusion behavior during the glass transition. In the present study, we will proceed from theoretically simpler pseudoequilibrium thermodynamic phenomena to the more complicated kinetic aspects.

2.3. Thermodynamic Analysis

When the only retention mechanism is the bulk absorption by the liquid stationary phase, the partition ratio k may be written as

$$k = m_1^l/m_1^g = V_l c_1^l / V_c c_1^g = m_2 v_2 c_1^l / V_c c_1^g \tag{16}$$

where m_1 and c_1 are the mass and concentration of the probe, respectively. The term V_l is the volume of the liquid phase; m_2 and v_2 are the mass and specific volume, respectively, of the polymer forming the liquid phase. The superscript l and g refer to the liquid and gas phases. Combination of Eqs. 10, 15, and 16 yields

$$V_g \equiv V_N/m_2 = (c_1^l/c_1^g)v_2 \tag{17}$$

where the *specific retention volume* V_g is defined by the identity in Eq. 17.

Frequently, V_g is corrected to the standard temperature $T_o = 273.15$ K. The corrected specific retention volume V_g^0 is defined as

$$V_g^0 \equiv V_g(T_o/T) \tag{18a}$$

where T is the temperature of the column in kelvins. The term V_g^0 is calculated from direct experimental data using relations

$$V_g^0 = (t_p - t_m)(T_o/T)F_o j/m_2 = (t_p - t_m)(T_o/T)V_o/m_2 t_m \tag{18b}$$

According to Eq. 17, V_g is related to the distribution coefficient (c_1^l/c_1^g). This relation may be employed for the calculation of a number of thermodynamic quantities. In polymeric systems where the molecular weight is frequently unknown, molar fraction is an impractical quantity, and the mass/volume concentration c, mass fraction w, or volume fraction ϕ are used instead. Thus, Henry's law for ideally dilute solutions may be written for the probe as

$$P_1 = k_H c_1^l \tag{19}$$

where P_1 is the partial pressure of the probe and k_H is Henry's constant. Combination of Eqs. 17–19 with the ideal gas law (always applicable in conjunction with ideally dilute solutions) yields for k_H

$$k_H = RT_o v_2 / M_1 V_g^0 \tag{20}$$

where R is the gas constant and M_1 is the molecular weight of the probe.

Some authors introduce a quantity called the partial molar free energy of sorption $\Delta\bar{G}_{1,\text{sorp}}$ defined as

$$\Delta\bar{G}_{1,\text{sorp}} \equiv -RT\ln(M_1 V_g / RT_o) \tag{21}$$

The partial molar enthalpy of sorption $\Delta\bar{H}_{1,\text{sorp}}$ may be derived from IGC data as follows. For a pure solvent, the liquid–vapor equilibrium is described by the Clapeyron equation

$$dP_1/dT = \Delta\bar{H}_{1,\text{vap}}/T(\bar{V}^g_{1,0} - \bar{V}^l_{1,0}) \tag{22}$$

Here $\bar{V}_1$ is the partial molar volume of the solvent, the second subscript zero refers to pure compound. An analogous relation is easily derived for sorption equilibrium:

$$(\partial P_1/\partial T)_{c_1^l} = \Delta\bar{H}_{1,\text{sorp}}/T(\bar{V}_1^g - \bar{V}_1^l) \tag{23}$$

Neglecting as usual $\bar{V}_1^l$ as compared to $\bar{V}_1^g$, substituting from the ideal gas equation for P_1 (this is always legitimate in the IGC limit of ideally dilute solutions), realizing that $\bar{V}_1^g c_1^g = M_1$, and employing Eqs. 17 and 18a, Eq. 23 may be transformed into

$$d\ln V_g^0 d(1/T) = -\frac{\Delta\bar{H}_{1,\text{sorp}}}{R} \tag{24}$$

The calculation of k_H and $\Delta\bar{H}_{1,\text{sorp}}$ is meaningful for all probes without regard to whether their critical temperature is above or below the column temperature. For probes, which are under their critical temperature, the IGC data are frequently used for determination of the polymer–solvent miscibility. In this case, pure probe at the column temperature and at the pressure of its saturated vapor P_1^0 is selected as the thermodynamic reference state for both liquid and gas phases. Obviously, the reference chemical potentials $\mu_{1,0}$ of both phases are equal to each other

$$\mu_{1,0}^l = \mu_{1,0}^g \tag{25}$$

Realizing that the gaseous probe at the conditions of IGC is an ideal gas while the saturated vapor may be a nonideal gas, we may write for the chemical potential of the probe in the gas phase μ_1^g

$$\mu_1^g = \mu_{1,0}^g + RT\ln(P_1/P_1^0\gamma_1^0) \tag{26}$$

where γ_1^0 is the fugacity coefficient of the probe in its standard state.

The thermodynamics of nonideal gases yield

$$\gamma_1^0 = \exp\left[\int_0^{P_1^0}\left(\frac{Z-1}{P_1}\right)dP_1\right] \tag{27}$$

where $Z = P_1V_1/RT$ is the compressibility factor of the gas. For gases at low pressures, Z is frequently represented by a virial expansion as

$$Z = 1 + (B_{11}/V_1) + \cdots \tag{28}$$

Here, B_{11} is the second virial coefficient. Substitution of Eq. 28 into Eq. 27 yields approximately

$$\gamma_1^0 = \exp(B_{11}P_1^0/RT) \tag{29}$$

Combination with Eq. 26 yields finally

$$\mu_1^g - \mu_{1,0}^g = RT\ln\left(\frac{P_1}{P_1^0}\right) - B_{11}P_1^0 = RT\ln\left(\frac{RTc_1^g}{M_1P_1^0}\right) - B_{11}P_1^0 \tag{30}$$

where the last equality in Eq. 30 was obtained employing the ideal gas law. Equation 30 is routinely used in GC. However, in many IGC experiments, temperatures above 100 °C and often up to 200 °C are used; frequently, alkanes as low as pentane are used at these temperatures. Under these circumstances, P_1^0 may be very high (and the significance of the last term in Eq. 30 is also high) and the truncation of the virial expansion in Eq. 28 may not be warranted. In fact, if enough care is not exercised, the critical temperature of the probe may be exceeded, invalidating the whole thermodynamic analysis. In any case, when P_1^0 is very high, the nonideal term should be treated with caution.

When calculating the chemical potential of the probe in the liquid phase two contributions must be considered: change of chemical potential caused by the transition from P_1^0 to P_1 and the change caused by conversion of concentration from pure liquid to concentration c_1^l. Consequently,

$$\mu_1^l(P_1, c_1^l) = \mu_{1,0}^l(P_1^0) - V_{1,0}^l P_1^0 + RT\ln a_1^l \tag{31}$$

$$a_1^l = \Omega_1 w_1^l \tag{32}$$

where a_1 is the activity of the probe and w_1 is its weight fraction. The weight fraction activity coefficient is $\Omega_1 \equiv a_1^l/w_1^l$. At equilibrium, μ_1^l and μ_1^g must be equal, and at high dilution $w_1^l = c_1^l v_p$. Consequently, combination of Eqs. 30 to 32 with Eqs. 17 and 18a yields

$$\Omega_1^\infty = (RT_0/V_g^0 P_1^0 M_1)\exp[(V_{1,0}^l - B_{11})/RT] \tag{33}$$

The superscript infinity refers to infinite dilution of the probe in the polymer. The calculation of the activity coefficient Ω_1^∞ is quite general and does not refer to any model of polymer solutions.

When the thermodynamic behavior of the system is described by some model, the IGC data may be used for calculating the parameters of the model. The Flory–Huggins model gives the change of Gibbs function in mixing ΔG_{mix} as

$$\Delta G_{mix} = RT[n_1 \ln \phi_1 + n_2 \ln \phi_2 + n_1 \phi_2 g] \tag{34a}$$

where n_i represents the number of moles, ϕ_i is the volume fraction, and indexes 1 and 2 refer to the probe and polymer respectively. The interaction function g is related to the Flory–Huggins parameter χ as

$$\chi = g + \phi_1(\partial g/\partial \phi_1)_{P,T} \tag{34b}$$

When g is independent of concentration, parameters g and χ are equal. They are also equal for all systems in the limit of $\phi_1 \to 0$ (ideally dilute solutions as in IGC experiments). Realizing that the chemical potential at constant P and T is related to ΔG_{mix} as

$$\mu_1^l(P) = \mu_{1,0}^l(P) + (\partial \Delta G_{mix}/\partial n_1)_{n_i=1,P,T} \tag{35}$$

we find that for the Flory–Huggins model Eq. 31 reads

$$\mu_1^l(P_1, c_1^l) = \mu_{1,0}^l(P_1^0) - \bar{V}_{1,0}^l P_1^0 + RT[\ln \phi_1 + \phi_2(1 - \bar{V}_1^l/\bar{V}_2^l) + \phi_2^2 \chi] \tag{36}$$

where $\bar{V}_2^l$ is the molar volume of the polymer. Within the Flory–Huggins theory $\phi_1 = c_1^l v_1$ and $\bar{V}_i^l = v_i M_i$, where v_i is specific volume of ith component. Using these relations, we may combine Eqs. 30 and 36 with Eqs. 17 and 18a to obtain in the limit of vanishing c_1^l

$$\chi = \ln(RT_o v_2/V_g^0 \bar{V}_1^l P_1^0) - P_1^0(B_{11} - \bar{V}_1^l)/RT - 1 + \bar{V}_1^l/M_2 v_2 \tag{37}$$

The last term in Eq. 37 is negligible for polymers with moderate or high molecular weights.

The values of parameter χ are often measured as a function of the temperature and nature of the probe and the dependencies are further analyzed either within the Flory–Huggins theory or by other theories, for example, by the equation-of-state theory. This detailed analysis is beyond the scope of the present chapter.

It is, however, necessary to include the analysis of the systems where the stationary phase is a blend of two polymers. As we will show, the IGC method may yield in this case the polymer–polymer interaction coefficient that is measurable otherwise only rarely and with great difficulties.

When describing an interaction of a polymer blend with a probe (a three-component system) we will use again subscript 1 for the probe; the two polymers will be identified by subscripts 2 and 3. With this notation all our relations through Eq. 33 will remain valid; except m_p (in Eq. 16) should be replaced by $(m_2 + m_3)$ and v_p (in Eqs. 17 and 20) should be replaced by $(m_2 v_2 + m_3 v_3)/(m_2 + m_3) \equiv w_2 v_2 + w_3 v_3$, where w_i is the weight fraction of polymer i.

Many thermodynamic studies of polymer blends model the change of Gibbs energy in mixing by a simple expression, which is a generalization of the original Flory–Huggins expression for three components

$$\Delta G_{\text{mix}} = RT[n_1 \ln \phi_1 + n_2 \ln \phi_2 + n_3 \ln \phi_3 + n_1 \phi_2 \chi_{12} + n_1 \phi_3 \chi_{13} + n_2 \phi_3 \chi_{23}] \tag{38}$$

Here the binary interaction coefficients χ_{ij} are considered to be parameters independent of the composition of the mixture. The derivative of Eq. 38 reads

$$\begin{aligned}&(\partial \Delta G_{\text{mix}}/\partial n_1)_{n_2,n_3,P,T}\\ &= RT[\ln \phi_1 + 1 - (V_1/V_2)\phi_2 - (V_1/V_3)\phi_3 + \phi_2 \chi_{12} + \phi_3 \chi_{13} - (V_1/V_2)\phi_2 \phi_3 \chi_{23}]\end{aligned} \tag{39}$$

where the superscript l and the bar were omitted from the V_i terms. Substituting Eq. 39 into Eq. 35 we obtain after some manipulations an equivalent of Eq. 37

$$\begin{aligned}&\phi_2(\chi_{12} - V_1/M_2 v_2) + \phi_3(\chi_{13} - V_1/M_3 v_3) - (V_1/V_2)\phi_2 \phi_3 \chi_{23}\\ &= \ln[RT_o(w_2 v_2 + w_3 v_3)/V_g^0 V_1 P_1^0] - P_1^0(B_{11} - V_1)/RT - 1\end{aligned} \tag{40}$$

Usually a new parameter χ'_{23} is introduced as

$$\chi'_{23} \equiv (V_1/V_2)\chi_{23} \tag{41}$$

We will also introduce a parameter $\chi'_{23,\,\mathrm{app}}$ as a value evaluated from Eq. 40 as the unknown. The parameters χ'_{23} and $\chi'_{23,\,\mathrm{app}}$ are equal to each other when Eq. 38 is a satisfactory description of ΔG_{mix}. Thus, if the parameters χ_{12} and χ_{13} are known (e.g., from IGC experiments on appropriate homopolymers), χ'_{23} may be calculated from Eq. 40 as the only unknown.

Comparison of Eqs. 37 and 40 suggests that the IGC study of blends is performed in the following way: three columns are prepared—two from homopolymers and a third from a blend using the same samples of homopolymers to make the blend. The three columns should be studied under conditions as identical as possible. Auxiliary parameters (P_1^0, $T, M_2 M_3$, V_1, v_2, v_3, B_{11}) will be identical for the three experiments and combination of Eqs. 37 (taken twice for two homopolymers) and 40 for the blend will yield

$$\chi'_{23} = \chi'_{23,\mathrm{app}} \equiv \{\ln[V^0_{g,\,\mathrm{blend}}/(w_2 v_2 + w_3 v_3)] - \phi_2 \ln(V^0_{g,2}/v_2) - \phi_3 \ln(V^0_{g,3}/v_3)\}/\phi_2\phi_3 \tag{42}$$

Here the second subscript of V_g identifies the nature of the column. From Eq. 42, χ'_{23} may be calculated even for probes for which the parameters P_1^0, B_{11}, and V_1 are not known or are known with insufficient accuracy.

The theory embodied in Eq. 38 predicts that $\chi'_{23}/V_1 \equiv \chi_{23}/V_2$ is independent of the nature of the probe and depends only on the nature of the two polymers. However, a number of experimental studies produced values that depended on the nature of the probe. This suggests that a more general relation should be used for ΔG_{mix}. The phenomenological relation of Pouchlý, et al. (3) (Eq. 43), is completely general with the binary parameters g_{ij} defined by the experimental behavior of appropriate binary systems and the ternary parameter g_T by the behavior of the ternary system

$$\Delta G_{\mathrm{mix}} = RT(n_1 \ln \phi_1 + n_2 \ln \phi_2 + n_3 \ln \phi_3 + n_1 \phi_2 g_{12} + n_1 \phi_3 g_{13} + n_2 \phi_3 g_{23} + n_1 \phi_2 \phi_3 g_T) \tag{43}$$

If Eq. 43 is used in the IGC calculations, Eq. 44 is obtained for the value of $\chi'_{23,\mathrm{app}}$ as defined by Eq. 42

$$\chi'_{23,\mathrm{app}} = g'_{23} + \phi_2\, dg'_{23}/d\phi_2 - g_T \equiv \chi'_{23} - g_T \tag{44}$$

where the composition dependent value χ'_{23} defined by the second relation of Eq. 44 is related to g'_{23} in the usual way. Here, all the symbols with primes imply the multiplication by V_1/V_2. Equation 44 predicts that the measured interaction parameters should change with the composition of the blend, and with the nature of the probe due to the probe dependent values of g_T. (The g_T

values also depend on the composition of the mixture.) In other words, the IGC method does not measure directly the elusive parameter χ'_{23} but only its combination with another probe dependent parameter.

2.4. Crystallinity and Glass Transition

The study of crystallinity and glass transition by IGC is based on reduced interaction of the probe with crystalline and glassy polymers as compared to rubberlike polymers. The two phenomena are rather similar, but they are based on quite different physical principles. The solubility of all probes in crystals is much less than in liquids, melts, or rubbers (made from the same molecules). Thus, the reduced retention in crystalline polymers is a result of a thermodynamic phenomenon: low solubility. However, the solubility of most probes in glassy polymers is comparable to their solubility in corresponding liquids. The diffusion constants are, however, orders of magnitude lower in glassy polymers than in their rubbery counterparts. Under the conditions of an IGC experiment, the probe does not have enough time to penetrate the polymer beyond the surface layer, and its retention is drastically reduced. The reduction of retention is a kinetic phenomenon in this case.

The study of both phenomena is based on an observation that for rubbery polymers the dependence of $\ln V_g^0$ on inverse temperature $(1/T)$ (the retention diagram) is almost linear in a broad range of temperatures. However, at lower temperatures, when the polymer crystallizes, a rather sharp break results on the retention diagram (4) (Fig. 1).

Crystalline polymers have crystalline regions (often spherulitic) that are imbedded in an amorphous matrix. However, this matrix may comprise as little as 5–10% of the total material. When the amorphous portion of the polymer is above its glass-transition temperature, the probe may penetrate easily around the crystallites and through the whole composite system; however, it dissolves only in the amorphous portion of the polymer.

The evaluation of the polymer crystallinity is in principle very simple. The fraction of the crystalline polymer x_{cr} is computed as

$$x_{cr} = 1 - V^0_{g,\text{ sample}}/V^0_{g,\text{ amorphous}} \tag{45}$$

where $V^0_{g,\text{ sample}}$ is the actual measured value and $V^0_{g,\text{ amorphous}}$ is the value, which would be exhibited by the sample if it were completely amorphous. Obviously, both values have to be corrected for other sources of retention: adsorption on the inert support, surface adsorption, and so on. The difficult part is to estimate $V^0_{g,\text{ amorphous}}$. Originally, the linear portion of the retention diagram above the melting point of the polymer was extrapolated into the crystalline region. Unfortunately, the plot of $\ln V_g^0$ versus $1/T$ is not very

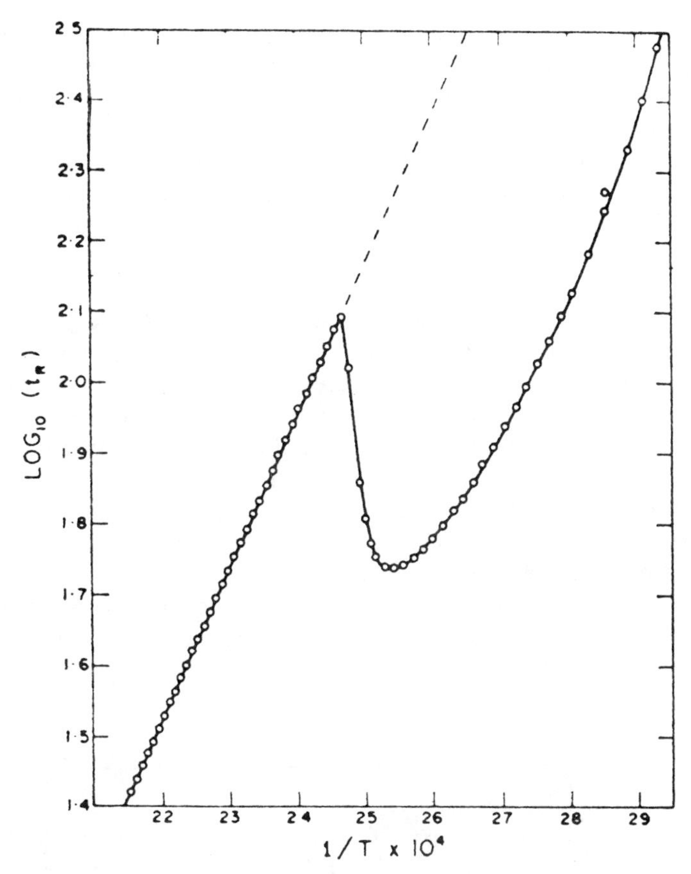

Figure 1. Retention diagram for decane on high-density polyethylene. Reprinted with permission from D. G. Gray and J. E. Guillet, "The Application of the Molecular Probe Technique to a Study of Polymer Crystallization Rates," *Macromolecules*, **4**, 129 (1971). Copyright © 1971 American Chemical Society.

sensitive. Even slight nonlinearity, when neglected, leads to rather large errors in the computed crystallinity. In a more sophisticated procedure, a more sensitive parameter (e.g., the activity coefficient Ω_1^∞ or parameter χ) is computed as a function of temperature above the melting point and then extrapolated into the crystalline region; the corresponding value of $V^0_{g,\,\mathrm{amorphous}}$ is calculated from it.

When evaluating the crystallinity, two special effects should be kept in mind. Whenever the sample is heated to its melting point and above it, it melts. On subsequent cooling it crystallizes again, but the crystallinity and morphology of the sample could be changed. The other possible effect is related to the coating method and to the thinness of the polymer layer; the crystallinity of a very thin layer may not be typical for the polymer. Sometimes it is recommended that the analysis is performed for several loadings and extrapolated to infinite thickness of the layer.

The retention diagrams of polymers undergoing glass transition display similarly two (more or less) linear regions above and below the glass-transition temperature T_g (Fig. 2) (5). However, the transition does not lead to such a sharp break on the retention diagram as was characteristic for polymer melting, it is much more gradual. This behavior is usually described as a result of probe adsorption on the polymer surface. Below T_g the probe cannot penetrate the polymer at all, the only interaction is with the polymer surface. Thus, the surface adsorption is considered to be of a different origin than the bulk absorption. As a consequence, when the bulk data are of interest, they are corrected for surface effects. The correction is done in one of two ways. Either the surface adsorption is extrapolated from the below T_g region toward higher temperatures and subtracted from the bulk data, or the total retention is extrapolated to infinite thickness of the polymer making the surface contribution negligible.

In our opinion, the two sorption mechanisms are not necessarily different. Deshpande and Tyagi (6) studied in detail the T_g transition using poly(vinyl acetate) as the stationary phase. They found that the transition is shifted

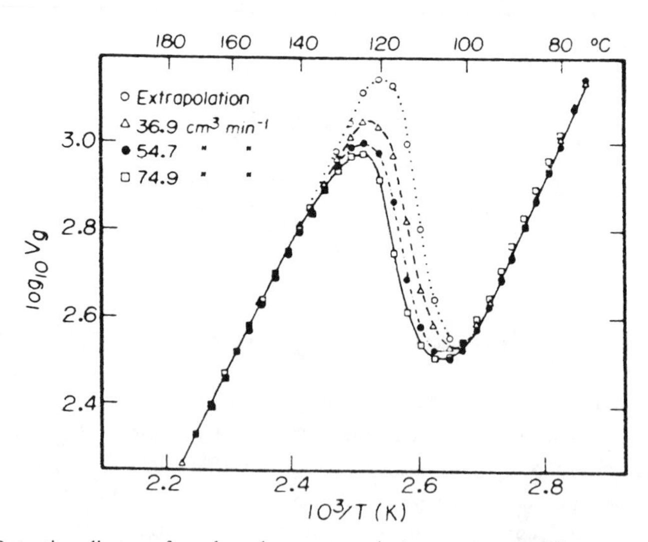

Figure 2. Retention diagram for *n*-hexadecane on polystyrene at several flow rates. Reprinted with permission from J.-M. Braun and J. E. Guillet, "Studies of Polystyrene in the Region of the Glass Transition Temperature by Inverse Gas Chromatography," *Macromolecules*, **8**, 882 (1975). Copyright © 1975 American Chemical Society.

toward lower temperatures for lower flow rates and/or thinner polymer layers (see also Fig. 2). This behavior shows that the IGC data in the region of the glass transition are governed by kinetic (diffusion) phenomena. As the temperature is lowered the probe cannot reach the polymer except for the uppermost layer on the polymer surface. In this sense, the interaction is a surface adsorption, but the mechanism of the adsorption and, especially, the thermodynamic factors governing the interaction remain essentially the same as above T_g. If this model is correct, then no correction for the adsorption on the polymer surface is needed. The extrapolation toward infinite loading may be actually harmful: The diffusion effects, which possibly lower retention in thick layers, are magnified by such extrapolation. However, another surface adsorption must be corrected for: adsorption on the surface of the inert chromatographic support. We will treat this correction in more detail in Section 4.2.

2.5. Diffusion

As we have already mentioned, the diffusion processes within the IGC column spread the chromatographic peaks. It opens the possibility that peak broadening may be used for a measurement of diffusion constants.

In traditional GC, peak spreading is directly related to the capability of the GC columns to separate different compounds. As such, it received extensive theoretical interest. In this study we will present only a simplified theory, and will stress those aspects of the phenomenon, which may lead to the measurement of diffusion constants.

The computational tool for the analysis of peak spreading is the similarity of the chromatographic process with countercurrent extraction using a large number of discrete phase equilibration steps in volume elements of finite size, these elements are called plates. The distribution of the solute in these elements is the same as if the procedure were performed on a chromatographic column. The theory of countercurrent extraction (the theory is actually a rather simple statistical exercise) shows that the width of the solute band (measured as a fraction of the total number of plates) becomes smaller with increasing total number of plates. A chromatographic column with the same separation power is said to have the same number of theoretical plates as the countercurrent apparatus. Obviously, the number of theoretical plates N is proportional to the physical length of the column L. The separating power of a given stationary phase is then described by the ratio $H \equiv L/N$ called the height equivalent to one theoretical plate (HETP). From the above, it should be clear that H should depend on all factors influencing the diffusion processes in the column.

A large number of theories has been offered. However, most of them are variations of the van Deemter theory, which separates the factors influencing peak spreading into three groups: factors independent of flow rate, factors proportional to time, and factors proportional to the flow rate (η). Recognizing that time spent on the column is inversely proportional to the flow rate u, we may write the van Deemter equation as (7)

$$H = A + B/u + Cu \equiv A + B/u + (C_g + C_l)u \tag{46}$$

The flow rate independent factors are described by the constant A. These factors are mainly related to differences in time spent on the column by different portions (streamlines) of the carrier gas. With packed columns, A is called the eddy diffusion term and is of the same order of magnitude as the size of the support particle. It is often negligible with respect to other factors. However, it may become significant in capillary columns or in those parts of the chromatograph where the gas path broadens, that is, in the plumbing.

The time dependent factors are described by the constant B. There is only one significant contribution to B, namely, the longitudinal diffusion of the probe in the stream of the carrier gas. Accordingly, B may be expressed as

$$B = 2\gamma D_g \tag{47}$$

where D_g is the mutual diffusion coefficient of the probe and the carrier gas. γ is the *tortuosity* or *structure* factor and varies from column to column. It has usually a value between 0.5 and 0.7 for diatomaceous type supports.

The factors proportional to flow rate are described by the constant C, referred to as the resistance to mass transfer term. They are mainly related to the radial diffusion, that is, to the diffusion toward and away from the polymer–gas boundary. This diffusion is a result of retention of the probe. When the probe is not retained (e.g., if it is a marker) the third term of Eq. 46 vanishes. Similarly, this term becomes very small when the retention becomes very large, that is, when R_f approaches zero and the partition ratio k approaches infinity. In this case, the diffusion processes have sufficient time to relax concentration differences and C vanishes. Radial diffusion consists of two terms: diffusion in the liquid phase characterized by the constant C_l and diffusion in the gas phase—constant C_g. For columns with uniform liquid phase thickness d_l, the constant C_l is written according to van Deemter as (7)

$$C_l = (8\,d_l^2/\pi^2)R_f(1 - R_f)/D_l \tag{48}$$

where D_l is the diffusion constant of the probe in the liquid phase.

However, in a real column, the thickness of the liquid is not uniform and d_l^2 should be replaced by the average of the square of the film thickness. This average is unfortunately not easily accessible experimentally. It is therefore convenient to replace the expression in parentheses of Eq. 48 by another structural factor γ_l. Equation 48 then reads

$$C_l = \gamma_l R_f(1 - R_f)/D_l \tag{49}$$

Theoretical expressions for C_g are numerous and are complicated and difficult to use. We find it convenient to postulate it similarly to Eq. 49 as

$$C_g = \gamma_g R_f(1 - R_f)/D_g \tag{50}$$

where γ_g is another structural factor.

Substituting Eqs. 47, 49, 50 into Eq. 46 we obtain

$$H = A + 2\gamma D_g/u + R_f(1 - R_f)u(\gamma_l/D_l + \gamma_g/D_g) \tag{51}$$

Thus, HETP of the column depends on the probe-related parameters D_g and D_l and on the column dependent parameters A, γ, γ_l, and γ_g. The latter parameters, which are the same for all probes, vary from column to column. However, for a set of columns prepared in a similar way, we would expect that the column parameters will have similar values. Specifically, for a set of columns differing only in the polymer loading, we would expect that the parameters A, γ, and γ_g that depend only on the column packing will be similar for all columns. However, the parameter γ_l will be strongly dependent on the polymer loading.

The above relations for H were derived assuming constant flow rate of the gas. However, in actual practice the gas velocity changes along the column. Several effects have to be taken into account. In both terms related to gas phase diffusion, the diffusion constant and gas velocity is in the form of a radio D_g/u. For ideal gases, the diffusion constant is inversely proportional to pressure and so is also the gas velocity in the column (Eq. 3b). Thus, the ratio D_g/u does not change along the column and may be replaced by its value at the column outlet D_g^0/u_0, where D_g^0 is the diffusion constant at the outlet pressure P_0. Furthermore, because of the expansion of the gas along the column, the peak in the gas phase broadens. This effect is taken into account by multiplying the gas related terms in H by a factor f defined as

$$f = [(P_i/P_o)^2 + 1]j^2/2 \tag{52}$$

where j was defined by Eq. 5. Under most circumstances, f is within a few percent of unity. Finally, for the term related to the diffusion in the liquid phase, the average gas velocity $u \equiv u_o j$ must be used in place of u_0. Thus, the expression for the average value $\bar{H}$ applicable for the whole column reads

$$\bar{H} = f(A + 2\gamma D_g^0/u_0) + R_f(1 - R_f)u_o(j\gamma_l/D_l + f\gamma_g/D_g^0) \tag{53}$$

From a judiciously chosen series of experiments (using few auxiliary data), extensive information about diffusion constants D_g^0 and D_l may be obtained. The evaluation procedures involved are described in Section 4.3.

3. EXPERIMENTAL TECHNIQUES

The basic experimental techniques in IGC are essentially the same as in traditional GC. Both flame ionization detectors (FID) and thermal conductivity detectors (TCD) are used. Capillary columns are used only rarely: The thickness of the polymer layer would make the diffusion phenomena too prominent. Most researchers prefer larger columns, columns 5-ft long with 1/4-in o.d. are used most often. Isothermal elution is used almost exclusively, otherwise, the operation of the GC is quite routine and we will not include any more details. However, for acceptably precise measurement of V_g^0, several experimental variables must be measured with much better precision than in other GC experiments; namely, the amount of polymer on the column and the flow rate. The shape of the peak is important when analyzing the diffusion phenomena. It might be distorted if the detector operates in a nonlinear region.

In the following sections we will present several techniques which were developed by us that, in our opinion, improve the accuracy of the method considerably.

3.1. Coating the Polymer onto the Support

Similar to other GC methods, in IGC it is usually desirable to maximize the contact area between the gas and liquid phases. This is achieved by coating the polymer onto a porous support. (Chromosorb W is the support used by most researchers.) To minimize the interaction of the support itself with the probes, the support is usually acid washed and treated with dimethyldichlorosilane to derivatize the highly polar groups on the surface of the support, which is essentially a silicate. We will see later that this treatment is not fully successful.

The amount of polymer on the column enters into the calculation of V_g^0 and must be therefore known with high precision. The traditional method of coating the polymer onto the support consists of dissolving the polymer in a volatile solvent, mixing the solution with the support, and evaporating it slowly in a rotary-vacuum evaporator. However, during this procedure the polymer coats not only the support but also the walls of the evaporating vessel and the quantitative relationship between the masses of the support and polymer is lost. The coated support is then analyzed for polymer content by either extraction or calcination; however, both methods are subject to much larger errors than acceptable for thermodynamic purposes (even 1% error in a *single* experimental parameter is not acceptable).

In the newly designed soaking method the weighted amount of polymer is transferred quantitatively to the column by using the following procedure (8). The polymer is dissolved in a solvent as usual. The support is piled on a watch glass and a small amount of the solution is added to the top of the support pile. The pile must be wet as much as possible without letting the solution touch the glass surface either under or around the pile. After the solvent evaporates, the pile is thoroughly mixed. Then the procedure is repeated again and again until all the solution is used. The solution flask is rinsed and the rinsings applied to the pile. When the procedure is done correctly, no polymer is left on the surface of the dish. The support is then dried in an oven and transferred quantitatively into the column using standard precautions. The method is fast (10–20 applications of the solution during a few hours), the amount of the polymer is precisely known, and analysis of polymer on the packing material is avoided.

The homogeneity of the coating is as good as with the traditional method. If the solution is applied 20 times and each time 80% of the support is soaked, then each support particle is coated on average 16 times and the probability that a particle is not coated at all is only about 1×10^{-14}.

3.2. Flow Rate Measurements

Flow rate is among the quantities needed for calculation of V_g^0. Therefore, it must be known with a precision much better than 1%. It is necessary that the fluctuation of the flow rate during the day is kept at a minimum. The value measured should be precise and dependable.

The flow of the gas is regulated by several reduction valves, which provide the hydrodynamic resistance needed for the necessary pressure reduction. The hydrodynamic resistance is proportional to the viscosity of the fluid (carrier gas) and, through it, depends on the temperature. Daily fluctuation of the ambient temperature leads to an unacceptably high fluctuation of flow rate. This problem was easily solved by thermostating the valves by enclosing them in a styrofoam box, thermostated by a copper coil through which a thermostated liquid was circulated. This simple device lowered the fluctuation of the flow rate to about 0.2% during the day.

Notwithstanding the advances in modern technology, the soap-bubble flow meter is still the most sensitive instrument for precise measurement of small flows. However, its design needed a number of improvements.

When FID is used as a detector, the flow of gas must be diverted to the flow meter before it reaches the detector. This is easily accomplished by a three-way valve between the column outlet and the detector. However, when the hot gas enters the alternate outlet, it starts warming the tubing and the accompanying volume changes influence the flow reading. To suppress this effect as much as possible, we led the gas through a copper coil immersed in thermostated liquid. The coil was positioned immediately after the exit of the gas from the body of the chromatograph. From the coil, the gas was led to the bubble meter. It was, of course, necessary to thermostate the bubble meter as well. The meter was manufactured from a 50-mL buret with a thermostated mantle. A single circulating bath may thermostat all the elements, which should be attached in the sequence: bubble meter, valves, and exchange coil.

Another artifact may seriously distort the measurement of the flow rate when helium is used as a carrier gas (9). The helium gas under the bubble diffuses through the bubble and then raises through the atmosphere and is replaced by heavier air. Similarly, air diffuses through the bubble in the opposite direction and immediately falls to the bottom of the meter. Thus, a constant composition gradient is maintained around the bubble. The diffusion

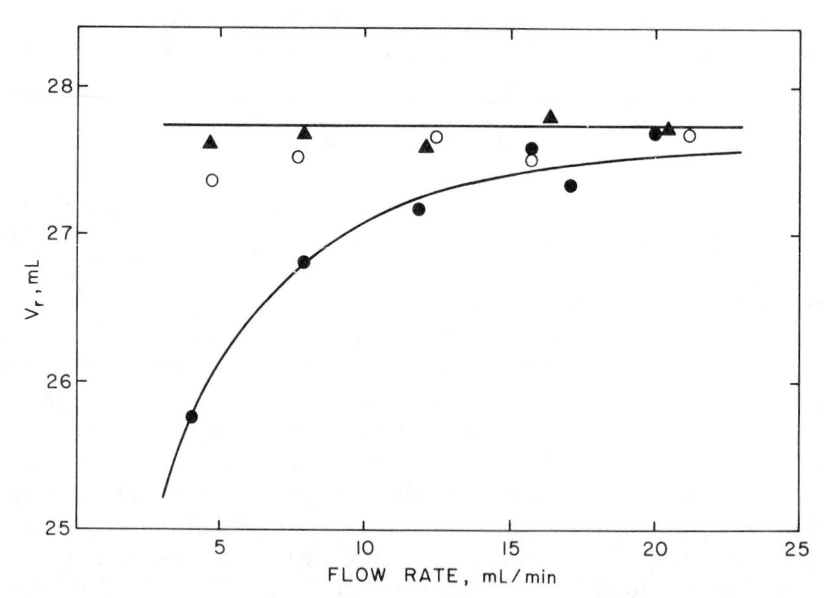

Figure 3. The effect of flow rate on the apparent retention volume of marker. ● helium, old design of flow meter; ○ helium, new design; ▲ nitrogen. Reprinted with permission from T. W. Card, Z. Y. Al-Saigh, and P. Munk, "Diffusion in the Bubble Flow Meter in Inverse Gas Chromatography Experiments," *J. Chromatogr.* **301**, 261 (1984). Copyright © 1984 Elsevier Science Publishers.

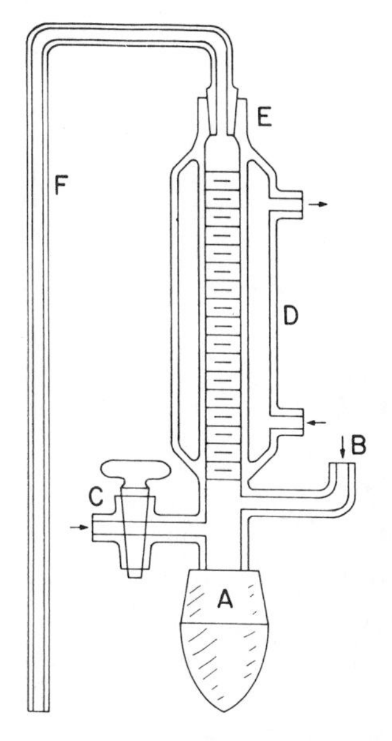

Figure 4. New design of soap-bubble flow meter: A, soap solution reservoir; B, incoming helium gas from the column; C, inlet valve for fast flushing of the bubble meter by helium gas; D, thermostated bubble meter; E, ground joint; and F, inverted U-tube. Reprinted with permission from T. W. Card, Z. Y. Al-Saigh, and P. Munk, "Diffusion in the Bubble Flow Meter in Inverse Gas Chromatography Experiments," *J. Chromatogr.*, **301**, 261 (1984). Copyright © 1984 Elsevier Science Publishers.

constant of helium through the bubble is higher than that of air. There is a constant net outflow from the bubble meter and the measured flow rates are underestimated by about 0.3 mL/min. This effect is detected when the column volume V_0 is measured at several flow velocities using Eq. 5: a constant error in the value F_0 results in an apparent dependence of V_0 on the flow rate (Fig. 3). The effect is eliminated when an inverted U-tube is attached to the outlet of the bubble flow meter (Fig. 4). A second inlet at the bottom of the flow meter is used for flushing the meter with helium before starting the flow measurements. Thus the helium cannot escape from the top of the flow meter; it contacts the bubble from both sides and no net diffusion takes place. It should also be mentioned that no such effect is observed with nitrogen as a carrier gas. The diffusion constants of nitrogen and air across the bubble are virtually identical.

3.3. Detector Linearity

The apparent (recorded) shape of chromatographic peaks at larger injection volumes is strongly influenced by the nonlinear response and saturation of the detector. In our experiments using FID, the dependence of apparent peak height on the injected amounts starts to deviate significantly from linearity when the peak height reaches about 20–30% of the saturated value (10). In experiments with an uncoated column and pentane as a probe (eluting

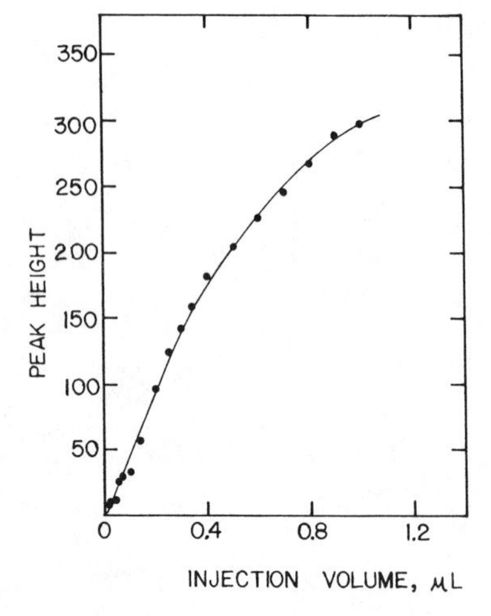

Figure 5. Dependence of peak height on the amount of pentane injected. Uncoated column, 50 °C, 10 mL of He/min, FID. Reprinted with permission from T. W. Card, Z. Y. Al-Saigh, and P. Munk, "Inverse Gas Chromatography 2. The Role of 'Inert' Support," *Macromolecules*, **18**, 1030 (1985). Copyright © 1985 American Chemical Society.

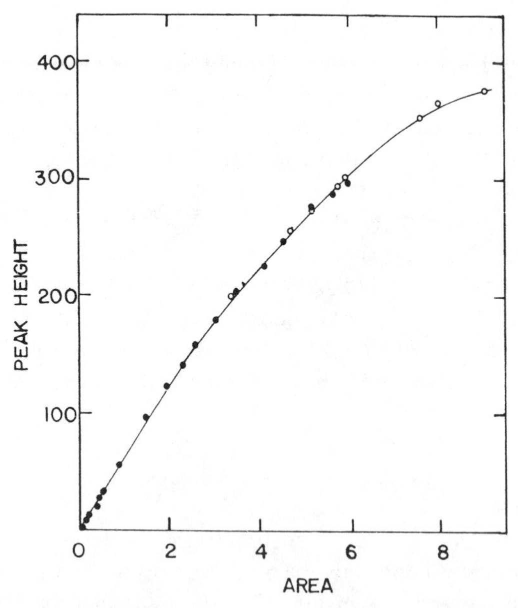

Figure 6. Dependence of peak height on peak area of pentane. The same experiment as in Figure 5. Reprinted with permission from T. W. Card, Z. Y. Al-Saigh, and P. Munk, "Inverse Gas Chromatography 2. The Role of 'Inert' Support," *Macromolecules*, **18**, 1030 (1985). Copyright © 1985 American Chemical Society.

together with the marker) at 60 °C, the deviation point was reached with injections of about 0.1–0.2 μL at a flow rate of about 10 mL/min (Fig. 5). The higher the flow rate, the greater the number of probe molecules reaching the detector per unit time, and the earlier the deviation point is reached for the same concentration of the probe in the carrier gas.

As the apparent height of the peak is reduced by detector saturation, so is the apparent peak area. However, the fractional reduction is less because only part of the peak is above the deviation point. Consequently, the dependence of the peak height on the area is curved also (Fig. 6). This dependence is not influenced by the uncertainty of the injected amount and serves as the most useful criterion of detector nonlinearity.

When the peak height exceeds the deviation point only moderately, the apparent shape of the peak is not much different from the undistorted peaks. The position of the peak maximum is still recorded correctly. However, the detector responses at the full height of the peak and at the half-height are reduced to different extents. The half-height point cannot be located and the width at the half-height cannot be measured.

When the injected amount is high, the detector may become saturated and the top of the peak may be almost flat. This phenomenon is different from the true flat-top peaks. The latter are observed when a probe is injected that has extremely low vapor pressure. The probe may condense inside the column and the carrier gas may be saturated with it for some period of time with a resulting constant signal. The effect is observable when the probe retention is small despite its low volatility, for example, when an uncoated column is used.

4. DATA ANALYSIS

Even the best IGC data are not useful when they are not interpreted correctly. In the case of thermodynamic data, all factors related to nonequilibrium effects must be eliminated as well as effects due to experimental arrangement (support effects). In the case of diffusion data, extra-column peak spreading must be accounted for and the several contributions to the HETP must be carefully separated. Some of the concepts needed for data evaluation will be presented in this section.

4.1. Adsorption on the Support

In GC, the stationary phase is deposited either on a support or is coated onto the inner wall of a chromatographic capillary. The latter method frequently leads to better separation of the mixture being chromatographed. However, in IGC the thickness of the polymer layer inside the capillary leads to extensive diffusion effects. Also, the miniscule amount of polymer within the capillary is difficult to measure exactly. Consequently, most researchers are depositing their polymers onto solid support. A solid support should have a large surface–volume ratio, should be coated uniformly with polymer, and should not interact with probes. Most researchers are using either Chromosorb W, or tetrafluoroethylene powders, or a similar spongelike granular fluorocarbon resin, Fluoropak 80. The fluorinated supports exhibit very small interactions with polar probes; however, they are rather difficult to handle and display significant interactions with hydrocarbon probes. Thus, Chromosorb W is the most common material for IGC. Chromosorbs are materials based on diatoms whose large surface area are partially reduced by thermal treatment. Chromosorbs are essentially silicates, as such they display significant surface interaction with polar probes. To reduce this interaction, they are often treated with dichlorodimethylsilane: The dimethylsilane group may bridge two surface hydroxyls and form a nonpolar surface. The reaction may also lead to formation of short dimethylsiloxane chains between two anchoring hydroxyl groups. Indeed, elemental analysis of Chromosorb W treated with dichlorodimethyl silane revealed the presence of about 0.2% of hydrocarbon residue (presumably methyl groups).

We have recently measured probe retention on Chromosorb W that was acid washed, treated with dimethyldichlorosilane, and uncoated (10). All probes tested were retained significantly but the type of chromatographic behavior varied with the nature of the probe; its dependence on the amount of

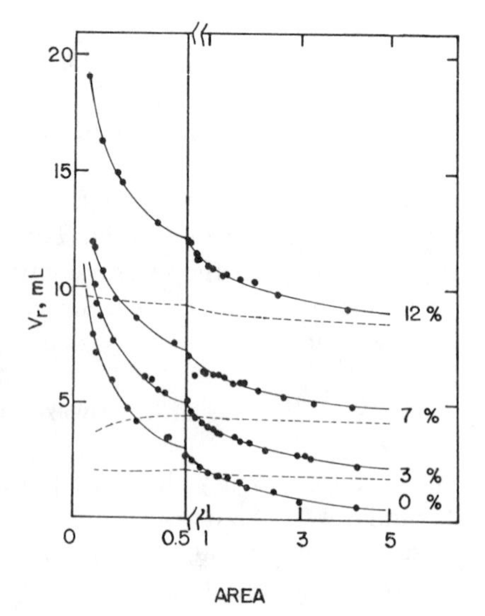

Figure 7. Dependence of the net retention volume of ethanol on Chromosorb W, AW–DMCS treated columns loaded with 0, 3, 7, and 12% polyisobutylene at 60 °C and 10 mL/min flow rate. Full lines—experimental data before correction; broken lines—data after subtracting the dependence for the unloaded column. Note the change of scale on the injected amount axis. Reprinted with permission from T. W. Card, Z. Y. Al-Saigh, and P. Munk, "Inverse Gas Chromatography 2. The Role of 'Inert' Support," *Macromolecules*, **18**, 1030 (1985). Copyright © 1985 American Chemical Society.

injected probe was most conspicuous. For moderately polar ethyl acetate at very small injection volumes, we observed very asymmetric peaks with a sharp onset and pronounced tailing. Retention volumes were very high. At higher injection volumes, peaks became more Gaussian and retention volumes leveled off at a lower value. The strongly polar ethanol exhibited this behavior to an even higher degree (Fig. 7, bottom line). We interpret this behavior as a result of the presence of a very small number of hydroxyl groups on the support that had not reacted with dimethyldichlorosilane. These groups exhibit strong retention but are saturated with miniscule amounts of polar probes. In experiments with untreated Chromosorb, the adsorption effects were so strong as to prevent meaningful measurement altogether. Attempts to remove the interacting groups by further treatment with trimethylchlorosilane were not successful: The adsorption behavior was not changed.

Nonpolar probes exhibited only negligible dependence of retention time on the amount of injected probe but the retention volumes were surprisingly large. For example, hexane at 60 °C had a net retention volume of 2 mL on a column with about 8 g of the uncoated support. This behavior is similar to retention on columns coated with polymer. We believe this polymer to be polydimethylsiloxane that was deposited on and attached to the support during its dimethyldichlorosilane treatment.

In the next series of experiments, three supports were coated using 3, 7, and 12% polyisobutylene, respectively, and used for the same measurement. The dependences of the retention volumes on the amount of the probe injected exhibited the same characteristics for the coated columns as for the uncoated one for all probes tested (Fig. 7).

The parallel curves in Figure 7 suggest that the support and the polymer retain the probe more or less independently. Let us discuss in some detail the theoretical and practical consequences of this somewhat surprising result. Theoretically, the fact that the polar adsorbing moieties on the support are not screened off by the polymer may be explained in two ways: (a) Either the probe can penetrate freely through the polymer and interact with the support unhindered by the polymer; or (b) the polymer is not distributed over all (not even over a significant portion of) the support surface but forms mainly pools in some pores of the support.

Both these concepts seem to defy the accepted models of gas–liquid chromatography but actually both are quite plausible. (a) When the probe is assumed to be fully equilibrated with the polymer, it may penetrate to the support and the stronger polar interaction may play a role. (b) The expectation that the stationary phase would first form a monomolecular layer on the surface of the support and then thicken more or less uniformly is based on the model of adsorption of gaseous or volatile materials on surfaces. The mutual attraction among the adsorbed molecules is only of secondary significance for such materials. However, when a polymer is coated onto another polymer (remember poly(dimethyl siloxane) originating from dimethyldichlorosilane treatment!) the surface forces play the decisive role similar to the case of oil spreading (or nonspreading) on a water surface. We might expect that two incompatible polymers will choose not to maximize their contact area and not form a uniform layer.

At the present time, it is not clear which explanation is correct. The question might be solved if the above experiment is repeated below the glass-transition temperature: A glassy polymer may screen effectively the polar groups hidden under it but would not affect the groups on an uncoated surface.

From a practical viewpoint, the retention of the probe by the support is an unwanted effect and the data should be corrected for it. In our experience, the subtraction of the dependence of retention volume on the injected volume for the uncoated column from the same dependence for the polymer coated column produces retention volumes virtually independent of the injection volumes even for probes as polar as ethanol (broken lines in Fig. 7). Moreover, when the corrected retention volumes are used for calculation of specific retention volumes V_g, the results are essentially independent of column loading—at least for polyisobutylene at 60 °C. The uncorrected values for the polar probes seem to be useless in comparison (Table 1). For hexane on polyisobutylene the correction seems to be minor. However, it becomes again significant when a (nonpolar) probe interacts strongly with polydimethylsiloxane (part of support) but only weakly with the (polar) polymer being studied.

Table 1. Specific Retention Volume (V_g) at 60 °C of Three Probes at Different Loadings of Polyisobutylene, Before and After Correction for Retention by the Uncoated Support[a]

Probe	Loading (%)	V_g Uncorrected (mL/g) 0.01 μL[b]	V_g Uncorrected (mL/g) 0.1 μL[b]	V_g Corrected (mL/g)
Hexane	3	74.5	74.4	68.7
Hexane	7	72.7	72.5	70.0
Hexane	12	69.5	69.3	68.1
Ethyl acetate	3	72.2	61.7	50.4
Ethyl acetate	7	59.0	54.6	48.8
Ethyl acetate	12	53.9	51.3	48.6
Ethanol	3	40.6	21.8	8.5
Ethanol	7	22.2	14.3	8.1
Ethanol	12	16.8	12.3	9.4

[a]Reproduced from reference 10 with permission.
[b]Interpolated for injection volume as noted.

4.2. Thermodynamic Data

The thermodynamic analysis of IGC data as developed in Section 2.3 assumes that the retention is caused only by the bulk polymer and that there is an instantaneous equilibrium between the bulk polymer and the mobile phase. In actual experiments, the probe may be retained also by the support, by the inner walls of the chromatograph (in case of very high-boiling probes), and by the surface of the polymer. Slow equilibration between the stationary and mobile phases may cause not only spreading of the peaks but also reduced retention if the equilibration is too slow. All these effects must be taken into account when thermodynamic analysis is contemplated. A consensus has developed among researchers in the field that the following steps are satisfactory for obtaining meaningful values of V_g.

1. The data should be either extrapolated to vanishing amounts of probe injected or measured using injections arbitrarily defined as vanishingly small. This procedure was designed for elimination of the problems connected with large injections; specifically, with the concentration dependence of the partition coefficients. The danger of this procedure is in potential magnification of the problems related to the presence of residual polar sites on the support. At the smallest injections (in the nanogram range) the retention is governed by these sites when polar probes are employed.

2. The data should be extrapolated to vanishing flow rate. This procedure supposedly provides sufficient time for the probe to equilibrate fully between the phases. There are two pitfalls: (a) In the vicinity of glass transition even the slowest experimentally feasible flow rate is too fast for equilibration (diffusion constants are too small). (b) The problems related to imprecise measurement of flow rate are magnified. In fact, if the diffusion of helium in the bubble-flow meter is neglected, the extrapolation procedure may completely ruin the data.

3. The data should be extrapolated to infinite loading (i.e., to vanishing inverse loading). This procedure makes the contribution of the surface retention negligible as compared to the bulk retention by an infinite amount of bulk polymer. This extrapolation is the most controversial since it exaggerates the effect of slow diffusion and negates the extrapolation to vanishing flow rate.

In our opinion, a routine evaluation procedure cannot be designed and each experiment should be judged on its own merits. Nevertheless, we would suggest several advisable approaches.

1. Retention volume of the marker V_0 should be carefully evaluated first, preferably at several flow rates, to guarantee the elimination of flow meter

errors. Then the V_0 value should be employed for calculation of retention volumes through Eqs. 9 and 10 and flow rate through Eq. 5, and so on.

2. The retention of the probe should be measured on an uncoated column as a function of injected volume. If the dependence is significant, the measurement on the polymer-coated column should be also done as a function of injected volume. Otherwise (nonpolar probes) few measurements at moderate volumes are sufficient. In any case, the two dependences should be subtracted to yield the value corrected for adsorption on the support and on the chromatograph walls. In many cases, it will be found that the effect of adsorption on the polymer surface becomes undetectable too, after this correction.

3. It is not clear whether the dependence of corrected V_g on the column loading should be extrapolated to vanishing loading (that would eliminate the effect of slow diffusion) or to infinite loading (that eliminates the effect of adsorption on polymer surface, if any). Further experiments are needed if the dependence is significant; often it is negligible.

4. The dependence on the flow rate should be extrapolated toward vanishing rate; however, the dependence is frequently insignificant.

5. Too small probe injections should be avoided even for nonpolar probes. It is safer to use a moderate injection and to estimate the correction needed for the dependence of the partition coefficient on concentration.

We have calculated (10) the correction assuming that the distribution coefficient (c_1^l/c_1^g) could be calculated from the Flory–Huggins relation (Eqs. 34a and 34b). We have further assumed that the peak moves as if the distribution coefficient had everywhere within the peak a value corresponding to the mass-weighted average of probe concentration (this is the average applicable to most transport processes). The average probe concentration within the peak decreases as the peak spreads while traveling along the column. Accordingly, the time average of the mass-weighted average was assumed to be the representative value of probe concentration. The final expression for net retention volume of the probe V_N at finite injected volume of probe V_{inj} reads

$$V_N \doteq V_N^* + \left[\frac{4\sqrt{2(\ln 2)/\pi}(1+\chi)}{v_2(1+V_{h/2}^d/V_{h/2}}\right]\left(\frac{V_g V_N^*}{V_{h/2}}\right)V_{\mathrm{inj}} \tag{54}$$

where V_N^* is the net retention volume corresponding to vanishing injection; the parameter χ may be obtained from Eq. 37. $V_{h/2}$ is the width of the chromatographic peak at the half-height measured in units of (corrected) elution volume (same units as for V_0), $V_{h/2}^d$ is the peak width caused by the peak spreading within the dead volume. (It is measurable in a separate experiment as presented in Section 4.3.) All the quantities in Eq. 54 are either easily measurable from a single IGC experiment or can be confidently estimated. Typical dependence is shown in Figure 8.

The main significance of Eq. 54 is in estimating an injection V_{inj} that is small enough for making the correction term in Eq. 54 negligibly small (i.e., smaller than 0.2% of the leading term). If, for any reason (e.g., when the peaks are very broad), it is necessary to use a larger injection volume, Eq. 54 provides means for correcting for thermodynamic nonideality.

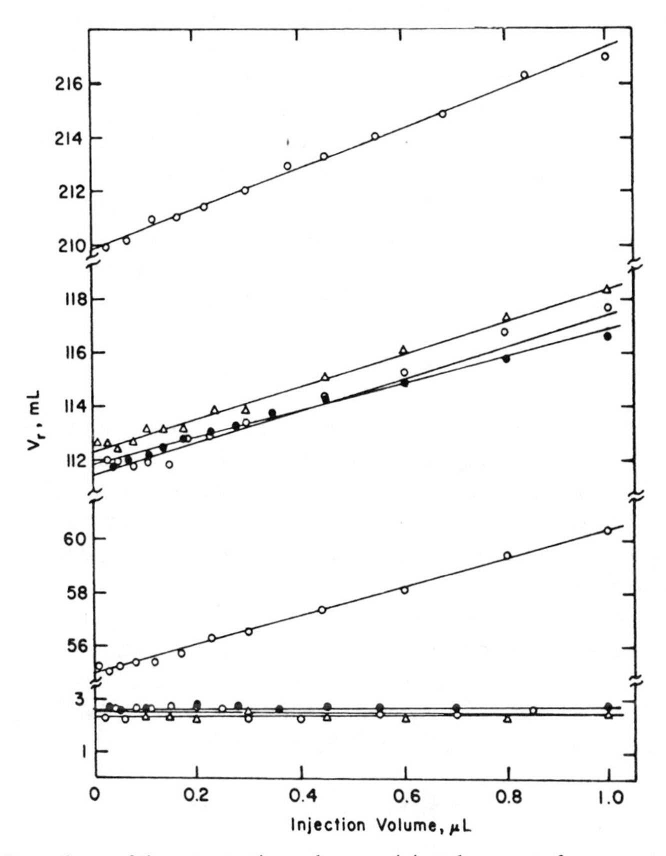

Figure 8. Dependence of the net retention volume on injected amount of nonane at 100 °C on columns loaded with 12% (top line), 7% (three middle lines), 3%, and 0% (three bottom lines) of polyisobutylene. Flow rate: ● 10 mL/min; ○ 20 mL/min; △ 28 mL/min. Reprinted with permission from P. Munk, Z. Y. Al-Saigh, and T. W. Card, "Inverse Gas Chromatography 3. Dependence of Retention Volume on the Amount of Probe Injected," *Macromolecules*, **18**, 2196 (1985). Copyright © 1985 American Chemical Society.

4.3. Evaluation of Diffusion Coefficients

Equation 53 from Section 2.5 offers a way of calculating the polymer–probe diffusion coefficient D_l from the peak-width related parameter $\bar{H}$ (average height equivalent to one theoretical plate). Indeed, Gray and Guillet (12) measured the dependence of $\bar{H}$ on the flow rate up to high values of u_0, where the first two terms of Eq. 53 are negligible with respect to the last term. From the asymptotic slope they calculated D_l. A capillary column was used, because in this column, the structural factor γ_l could be reasonably estimated.

In this section, we will describe a procedure of detailed analysis of the diffusion behavior in a packed column. The analysis will be based on the fact that the parameters A, γ, γ_l, and γ_g depend only on the column and packing and are the same for all probes. A cursory look at Eq. 53 shows that there is no easy way of separating the diffusion constants from their respective structure factors: Only relative values for diffusion constants for different probes may be found. However, when any of the diffusion constants is known from an auxiliary experiment, the remaining ones are calculated easily.

The first step of the analysis consists of calculation of the number of theoretical plates on the column N. This quantity is usually calculated assuming that the elution curve is Gaussian using

$$N = 5.54(t_p/t_{h/2})^2 = 5.54(V_R/V_{h/2})^2 \tag{55}$$

where $t_{h/2}$ is the width of the peak at the half-height in time units. Equation 55 implies that peak spreading occurs only in the column itself and that the initial profile of the peak immediately after the injection was a delta function. Actually, the peak spreads also within the apparatus dead volume. Since we are interested only in the number of theoretical plates in the column itself, we must correct the elution volume for the dead volume and the peak width for the peak spreading within the apparatus dead volume. The latter correction is based on the well-known behavior of Gaussian peaks. If a peak is broadened subsequently by several processes, then the square of the final peak width is the sum of the squares of peak widths of all the contributing processes. Accordingly, the expression for N should be written as

$$N = 5.54(V_R - V_d)^2/(V_{h/2}^2 - V_{h/2}^{d\,2}) \tag{56}$$

Here, V_R and $V_{h/2}$ are calculated most conveniently as $V_R = V_0 t_p/t_m$ and $V_{h/2} = V_0 t_{h/2}/t_m$. The extra-column volume V_d and $V_{h/2}^d$ are measured as follows.

The chromatographic column is replaced by a short capillary with a small volume directly connecting the injection port with the plumbing at the column outlet. The marker and all probes of interest are then injected and their retention analyzed at several flow rates. The retention volume of the marker is the dead volume V_d and the peak width is the dead volume peak width $V_{h/2}^d$. It should be noted that $V_{h/2}^d$ has significant convection terms and significant gas-probe diffusion terms. Thus, it varies with the flow rate and also from probe to probe. The recommended procedure is to measure flow rate dependence for each probe separately, and then to interpolate for the flow rates used in the actual experiments with full columns. The recorder speed for these measurements needs to be rather high. The whole curve is usually much shorter than 1 min and the peak widths correspond to a few seconds.

Once N is known, $\bar{H}$ is calculated as $\bar{H} = L/N$. The linear gas velocity at the column outlet u_0 is calculated from a modification of Eq. 4

$$u_0 = LF_0/V_c = LV_0/jt_m(V_0 - V_d) \tag{57}$$

After $\bar{H}$ and u_0 values have been calculated for the marker and a number of probes at several flow rates, we may proceed with the analysis of the factors contributing to HETP. First, we analyze the data for the marker. For the marker $R_f = 1$, the last term in Eq. 53 vanishes. After rearranging, we arrive at

$$\bar{H}u_0/f = 2\gamma D_g^0 + Au_0 \tag{58}$$

Thus, for the marker, the dependence of $\bar{H}u_0/f$ on u_0 should be linear; its slope is the column parameter A and the intercept is $2\gamma D_g^0$, where D_g^0 is the carrier gas-marker mutual diffusion coefficient.

After A has been found, Eq. 53 is rearranged as

$$(\bar{H}/f - A)u_0 = 2\gamma D_g^0 + [\gamma_l/D_l + f\gamma_g/D_g^0 j]R_f(1 - R_f)u_0^2 j/f \tag{59}$$

Let us inspect the expression in brackets in Eq. 59. For most chromatographic columns, the second term is much smaller than the first one; also, for low-pressure operation the fraction f/j is only a slowly varying function of flow rate. It follows that the plot of $(\bar{H}/f\text{-}A)u_0$ versus $R_f(1 - R_f)u_0^2 j/f$ should be almost linear for all probes. (An example of such a plot is given in Fig. 9 for several columns loaded with different amounts of the same polymer.) Its intercept is $2\gamma D_g^0$, where D_g^0 is the carrier-gas mutual diffusion coefficient, and the slope is $(\gamma_l/D_l + f\gamma_g/D_g^0 j)$, where the ratio f/j is some average value for the appropriate range of u_0.

In the first approximation and, especially for higher loads of polymer, which lead to higher values of γ_l, the second term in the brackets of Eq. 59 may be neglected altogether. For a series of probes, the values of γD_g^0 are then

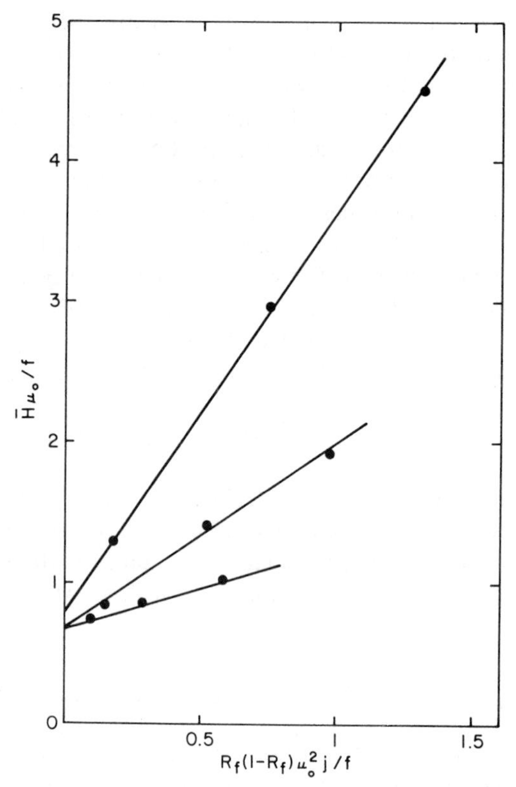

Figure 9. Dependence of $(\bar{H}/f\text{-}A)u_0$ on $R_f(1 - R_f)u_0^2j/f$ for butane at 38.4 °C on columns loaded with 0, 3, 7, and 12% polyisobutylene (from bottom to top). Helium was the carrier gas.

obtained from the intercepts and the values γ_l/D_l from the slopes. If the D_g^0 is known from auxiliary data for any probe or marker, the D_g^0 values for the remaining probes are calculated easily. Similarly, when D_l is known for the given polymer and a probe at any temperature in the range of temperatures measured, all other values are easily calculated from γ_l/D_l because the structural parameter γ_l is independent of temperature (as is also γ).

In the second approximation, the following strategem is employed, which is based on the assumption that the structural parameters γ and γ_g do not vary too much among columns prepared in a nominally identical manner using the same amounts and type of support but loaded to a different degree. (The structural parameter γ_l varies, of course, extensively for such set of columns.) A probe is selected that is retained on one of these columns (i.e., $R_f < 1$) even if its structural parameter γ_l is negligibly small on this column. These circumstances occur for an uncoated column and for a very high-boiling nonpolar probe (e.g., decane at 60 °C). The probe is retained because of the dimethylsiloxane groups on the surface of the support but there is no penetration inside the polymer, hence no C_l term. For such an experiment, the plot of $(\bar{H}/f - A)u_0$ versus $R_f(1 - R_f)u_0^2$ gives an intercept $I = 2\gamma D_g^0$ and a slope $S = \gamma_g/D_g^0$. Hence,

$$IS = 2\gamma\gamma_g \tag{60}$$

and γ_g can be calculated if γ is known from the previous analysis. Once γ_g is found (which is independent of the type of probe and similar for all columns), the appropriate term in Eq. 59 may be evaluated and subtracted. The γ_l/D_l values are then calculated with better confidence.

5. EXPERIMENTAL RESULTS

A rather extensive review of the IGC literature was presented by Lipson and Guillet (13), which covers the literature through about the beginning of 1980. We refer the readers to this article and will review only some of the more recent work not included in it.

While in the 1970s most IGC studies were concerned with phase transitions, effects caused by glass transition, and the surface adsorption phenomena, in the 1980s the emphasis shifted to the more quantitative aspects of IGC, mainly to thermodynamic analysis of polymer–solvent systems and polymer blends.

According to the Hildebrand–Scatchard theory, applicable to polymer–solvent systems without specific interactions, the Flory–Huggins parameter χ (measurable by IGC using Eq. 37) is related to the solubility parameters $\delta_i (i = 1$ for solvent, $i = 2$ for polymer) as

$$\chi = (V_1/RT)(\delta_1 - \delta_2)^2 \tag{61}$$

While δ_1 for solvents is easily measured, δ_2 is usually obtained indirectly. The term δ_2 may be calculated from Eq. 61 but is obtained with insufficient accuracy. DiPaola–Baranyi and Guillet (14) showed that Eq. 61 rearranged to the form

$$(\delta_1^2/RT - \chi/V_1) = (2\,\delta_2/RT)\delta_1 - \delta_2^2/RT \tag{62}$$

allows a combination of data for all probes measured by plotting the experimental left-hand side versus δ_1. The plot is a straight line and δ_2 is found with good precision from its intercept. Later, Eqs. 61 and 62 were replaced by

$$\chi = (V_1/RT)(\delta_1 - \delta_2)^2 + \chi_s \tag{63}$$

$$(\delta_1^2/RT - \chi/V_1) = (2\,\delta_2/RT)\delta_1 - (\delta_2^2/RT + \chi_s/V_1) \tag{64}$$

from which the entropic contribution χ_s to the parameter χ is also accessible. This procedure gained wide acceptance and was used for calculation of δ_2 and (sometimes) χ_s for poly(vinyl acetate) (15); polychloroprene, *cis*-1,4-polybutadiene, poly(butadiene-acrylonitrile), and poly(ethylene-vinylacetate) (16), as well as for poly(ethylene oxide) (17, 18).

Schuster et al. (19) used IGC for a detailed analysis of thermodynamic interactions of polystyrene with alkanes, branched alkanes, and alkybenzenes. They analyzed the contribution of methylene groups, branching, and phenyl groups to sorption and mixing functions, and compared the results obtained using the original Flory–Huggins theory with results of the theory modified by introduction of the core volumes of components (along the lines of the Flory–Prigogine theory).

Fernández–Berridi et al. (20) measured the dependence of parameter χ on temperature for polyisobutylene–benzene and found good agreement with results of vapor pressure osmometry.

Bonner and co-workers (21–23) used IGC for collecting the values of $\Delta\bar{H}_{1,\mathrm{sorp}}$ (see Eq. 24) for a large number of probes. They separated the values into a part corresponding to nonpolar interactions, dipole interactions, and specific interactions. Significant values of the latter quantity were interpreted as an indication that the particular probe may be a good solvent for the polymer. The method was used when searching for suitable solvents for the hard-to-dissolve polymers: polysulfone and *m*-phenylene polybenzimidazole.

Inverse gas chromatography is being used with increasing frequency for the study of polymer blends. DiPaola-Baranyi et al. (24) studied the blends of poly(methyl methacrylate) with poly(vinylidene fluoride) (PVF_2)—a pair of polymers that is well known to be compatible. Blends with a high proportion of PVF_2 exhibited negative values of $\chi'_{23,\mathrm{app}}$ of the order -0.6; however, blends poorer in PVF_2 displayed values close to zero or slightly positive. The dependence of $\chi'_{23\mathrm{app}}$ on the nature of the probe was only moderate. In another study, DiPaola-Baranyi and Degré (25) studied blends of polystyrene with poly(*n*-butyl methacrylate), which are just marginally compatible (they are compatible only if the polystyrene has very low molecular weight). Accordingly, the values of $\chi'_{23,\mathrm{app}}$ were very small; they also displayed some unexplained scatter in the dependence on the blend composition.

Walsh and McKeown (26) used IGC for studying compatibility of polyacrylates and polymethacrylates with poly(vinyl chloride). The $\chi'_{23,\mathrm{app}}$ values varied from highly negative (-3.0) for poly(ethyl methacrylate) to highly positive ($+1.0$) for poly(*n*-butyl acrylate). The dependence on the nature of the probes was also rather significant.

Walsh and co-workers (27–29) also studied blends of chlorinated polyethylene with poly(ethyl methacrylate), poly(butyl acrylate), poly(vinyl chloride), and ethylene–vinyl acetate copolymers. They were able to correlate the $\chi'_{23,\mathrm{app}}$ values with one-phase or two-phase behavior of these blends. In all cases, $\chi_{23,\mathrm{app}}$ varied appreciably with the probe and the range of its values was very large.

Al-Saigh and Munk (8) studied a blend of poly(methyl acrylate) and poly(epichlorohydrin). Their values of $\chi'_{23,\mathrm{app}}$ covered a narrow range around zero. However, the dependence of these values on the nature of the probe was clearly demonstrated.

Inverse gas chromatography is also useful for understanding the behavior of copolymers. DiPaola-Baranyi (30) measured the retention of several probes by copolymers of styrene with *n*-butyl methacrylate, isobutyl methacrylate, and *sec*-butyl methacrylate. For the latter two copolymers, V_g^0 of most probes could be predicted by simple interpolation of homopolymer properties. However, for the copolymer with *n*-butyl methacrylate, the measured retention volumes were higher than the predicted ones.

Block copolymers having a low T_g component and a high T_g component have complex IGC behavior. Galin and Rupprecht (31) studied dimethylsiloxane–styrene copolymers. Ward et al. (32, 33) employed a copolymer of dimethyl siloxane and polycarbonate made from bisphenol A. The behavior of both systems was quite similar. Above the glass-transition temperature of polystyrene or polycarbonate, the V_g^0 and $\chi'_{23,\mathrm{app}}$ values were rather high reflecting the underlying incompatibility of the two polymers. Below the glass transition, the polystyrene or polycarbonate segments formed domains that adsorbed most of the probes on their surface to a degree comparable to the adsorption on the free surface of these glassy polymers: The V_g^0 values were still quite high. However, as the volume fraction of the polystyrene or polycarbonate increased, the phase inversion occurred and the polysiloxane domains were not accessible to the probes any more. The V_g^0 fell sharply as would be expected for glassy polymers.

All the studies quoted above were done on systems displaying more or less standard IGC behavior with Gaussian peaks and negligible to moderate dependence on flow rate and amount of injected probe. Aspler and Gray (34) applied the IGC techniques for measuring water sorption isotherms on several derivatives of cellulose. The chromatographic behavior of these systems was very complex and a long careful analysis was needed for determining the isotherms.

Inverse gas chromatography technique was employed by Senich (35) for measurement of relative diffusivities of several alkanes in high-density polyethylene melts at 150 °C. The measurements were based on a method similar to the one used by Gray and Guillet (12). The constant C (Eq. 46) was obtained from measurement of HETP at high flow rates and interpreted in

Table 2. Selected Applications of IGC

	Reference
Adsorption, gas	
Cellulosics	36
Polyethersulfone	37
Adsorption, water	
Poly(vinylidene chloride)	37
Cross-linking studies	
Epoxy–amine reactions	38
Diffusion coefficients	
Computer simulation of diffusion	39
Polyisobutylene	40
Poly(methyl methacrylate)	41, 42
Polystyrene	42–44
Glass transition	
Poly(allylbenzene)	45
Methacrylate copolymers	46
Poly(styrene-*co*-divinyl benzene)	47
Interfacial studies	
Polyethylene	48
Vinyl acetate–vinyl alcohol copolymer	49
Methodology	
Automated IGC	49
Determination of χ	50–52
Flow rate effects	53
Moment analysis	54, 55
Optimization	56
Retention data correction	57, 58
Surface characterization	
Acrylic-grafted polyethylene film	59
Alkyd coatings	60
Carbon fibers	61–71
Glass fibers	72
Cellulose	73
Poly(phenylene terephthalamide) fibers	74, 75
Textile fibers	76
Thermodynamic parameters	
Compilation of polymer–solvent interaction coefficients	77
Homopolymers	
Polyarylate	78
Polybutadiene	79
Poly(butyl methacrylate)	80
Polycarbonates	81
Poly(dimethyl siloxane)	82

Table 2. (*Continued*)

	Reference
Polyethylene	83–85
Poly(ethylhexyl methacrylate)	85
Polyisobutylene	82
Poly[*N*-(octadecyl)maleimide]	86
Polyoxyethylene	87
Polystyrene	88, 89
Poly(tetramethylene carbonate)	87
Poly(vinyl acetate)	90
Poly(vinyl chloride)	91–93
Poly(vinyl isobutyl ether)	94
Poly(vinyl trimethylsilane)	95
Poly(vinylidene chloride)	96
Poly(vinylidene fluoride)	97
Copolymers	
Block and branched poly(ester ethers)	81
Ethylene-propylene rubber	82, 98
Polyether alcohols	99
Poly(styrene-*co*-divinylbenzene)	100, 101
Blends	
Polycaprolactone–polyepichlorohydrin	102, 103
Poly(dimethyl siloxane)–polyisobutylene	104
Poly(ethyl acrylate)–poly(vinyl propionate)	105
Poly(ethylene oxide)–poly(methyl methacrylate)	106
Polystyrene–poly(butyl methacrylate)	107
Polystyrene–poly(2, 6-dimethyl-1, 4-phenylene oxide)	107–111
Polystyrene–poly(methyl vinyl ether)	110, 112
Poly(vinyl acetate)–poly(butyl methacrylate)	113
Poly(vinyl acetate)–polyepichlorohydrin	114
Poly(vinyl acetate)–poly(methyl acrylate)	115
Poly(vinyl acetate)–poly(vinyl isobutyl ether)	113
Poly(vinyl methyl ether)–phenoxy resin	116
Poly(vinyl propionate)–poly(ethyl acrylate)	117
Poly(vinylidene fluoride)–poly(methyl methacrylate)	107

terms of Eq. 48. Estimated values of A and B were used for more dependable evaluation of the constant C.

For more recent publications on IGC methodology, instrumentation, and applications, the reader should consult Table 2 and the general reference and reviews listing.

6. CONCLUSIONS

In the 1970s, during the first decade of its existence, IGC established itself as a versatile method capable of providing many types of information about polymeric systems. It was employed primarily as a source of qualitative data in the fields of crystallization, glass transition, and surface adsorption. It has demonstrated its utility for obtaining thermodynamic and diffusion data.

In the 1980s, the method matured. The evaluation methods were getting more sophisticated and IGC provided a wealth of thermodynamic data for polymer–solvent interaction, as well as for the study of polymer blends. These achievements were possible because of improvements in experimental techniques and, especially, more careful analysis of experimental data. However, the reliability of the data is still not fully satisfactory. Good thermodynamic data call for an overall accuracy of the IGC measurements of about 1%. In our opinion, only few laboratories may have achieved this accuracy. Once the technical problems are solved, IGC may become the leading method for the collection of thermodynamic data, especially in the field of polymer blends, where good thermodynamic information is difficult to obtain by any means. Another promising field is the study of diffusion coefficients, especially for high-boiling probes that might be difficult to handle in standard diffusion measuring instruments.

NOTATION

a_1	Activity of the probe
A	Eddy diffusion term in the van Deemter equation
B	Longitudinal diffusion term in the van Deemter equation
B_{11}	Second virial coefficient of the probe in the vapor phase
c_1	Concentration of the probe
C_g	Resistance to mass transfer
d_l	Stationary phase thickness
D_g	Diffusion coefficient of the probe in the gas phase
D_l	Diffusion constant of the probe in the stationary phase
f	Pressure drop factor defined by Eq. 52
F	Flow rate
F_0	Flow rate at column outlet
g	Interaction function defined by Eqs. 34a and 43
ΔG_{mix}	Gibbs energy of mixing
$\Delta \bar{G}_{1,sorp}$	Partial molar free energy of sorption
H	Height equivalent to a theoretical plate
$\Delta \bar{H}_{1,sorp}$	Partial molar enthalpy of sorption

$\Delta \bar{H}_{1,\mathrm{vap}}$	Molar enthalpy of vaporization
j	Pressure drop factor defined by Eq. 5
k	Partition ratio
k_{H}	Henry's constant
L	Column length
m	Mass
M	Molecular weight
n	Number of moles
N	Number of theoretical plates
P_{A}	Atmospheric pressure
P	Pressure
P_{i}	Column inlet pressure
P_0	Column outlet pressure
P_1	Partial vapor pressure of probe
P_1^0	Saturated vapor pressure of probe
r_h	Hydrodynamic resistance
R	Gas constant
R_f	Fraction of the probe in the gas phase (retardation factor)
$t_{h/2}$	Width of the peak at half-height in time units
t_m	Retention time of unretained peak (marker)
t_p	Retention time of probe
T	Absolute temperature
T_0	Reference absolute temperature (273.15 k)
T_{g}	Glass-transition temperature
T_{th}	Temperature of the bubble-flow meter
u	Flow velocity
u_0	Flow velocity at column outlet
u_p	Average probe velocity
v	Specific volume
V_c	Column void volume
V_d	Apparatus dead volume
V_g	Specific retention volume
V_g^0	Corrected specific retention volume
$V_{h/2}$	Peak width at half-height in volume units
$V_{h/2}^{\mathrm{d}}$	Peak width at half-height caused by peak spreading within dead volume
V_l	Volume of the liquid (stationary) phase
V_N	Net retention volume
V_o	Void or dead volume
V_R	Retention volume
$\bar{V}$	Partial molar volume
w	Mass fraction

x	Length coordinate along the column
x_{cr}	Fraction of crystalline polymer
Z	Compressibility factor of a gas
γ, γ_g, and γ_l	Structural factors in the van Deemter equation
γ_1^0	Fugacity coefficient of the probe in its standard state
δ	Solubility parameter
μ_1	Chemical potential of the probe
ϕ	Volume fraction
χ	Flory–Huggins parameter
χ_s	Entropic contribution to χ
Ω	Weight fraction activity coefficient

Superscripts and/or Subscripts

c	Column
d	Dead volume
f	Fraction
g	Gas phase
h	Hydrodynamic
$h/2$	Half-height
l	Liquid (stationary, polymer) phase
m	marker
N	Net
p	Probe
R	Retention
∞	Infinite dilution of the probe
0	Refers to pure solute or probe
1	Refers to probe
2 and 3	Refers to polymers used as the stationary phase

GENERAL REFERENCES AND REVIEWS

O. E. Schupp III, "Gas Chromatography," in E. S. Perry and A. Weissberger (Eds.), *Techniques of Organic Chemistry*, Vol. **13**, Interscience, New York, 1968.

J.-M. Braun and J. E. Guillet, "Study of Polymers by Inverse Gas Chromatography," *Adv. Polym. Sci.*, **21**, 107 (1976).

J. E. G. Lipson and J. E. Guillet, "Study of Structure and Interactions in Polymers by Inverse Gas Chromatography," in J. V. Dawkins, (Ed.), *Developments in Polymer Characterization-3*, Applied Science, London, 1982, Chapter 2.

J. S. Aspler, "Theory and Applications of Inverse GC," *Chromatogr. Sci.*, **29**, 399 (1985).

R. Vilcu and M. Leca, "Thermodynamic Characterization of Polymers by Chromatography with Gas Mobile Phase," *Rev. Roum. Chim.*, **34**, 387 (1989).

H. P. Schreiber and D. R. Lloyd, "Overview of Inverse Gas Chromatography," *Am. Chem. Soc. Symp. Ser.*, **391**, 1 (1989).

H. P. Schreiber, "Surface Characterization by Inverse Gas Chromatography," *Adv. Org. Coat. Sci. Technol. Ser.*, **11**, 192 (1989).

P. Munk, P. Hattam, and Q. Du, "Thermodynamic Interactions in Mixtures," *J. Appl. Polym. Sci., Appl. Polym. Symp.*, **43**, 373 (1989).

D. R. Lloyd, T. C. Ward, and H. P. Schreiber (Eds.), *Inverse Gas Chromatography, Am. Chem. Soc. Symp. Ser. 391, Am. Chem. Soc.*, Washington, DC, 1989.

P. Munk, P. Hattam, Q. Du, and A.-A. A. Abdel-Azim, "Determination of Polymer–Solvent Interaction Coefficients by Inverse Gas Chromatography," *J. Appl. Polym. Sci., Appl. Polym. Symp.*, **45**, 289 (1990).

REFERENCES

1. O. Smidsrod and J. E. Guillet, "Study of Polymer–Solute Interactions by Gas Chromatography," *Macromolecules*, **2**, 272 (1969).
2. D. G. Gray and J. E. Guillet, "Gas Chromatography on Polymers at Temperatures Close to the Glass Transition," *Macromolecules*, **7**, 244 (1974).
3. J. Pouchlý, A. Živný, and K. Šolc, "Thermodynamic Equilibrium in the System Macromolecular Coil-Binary Solvent," *J. Polym. Sci., Part C*, **23**, 245 (1968).
4. D. G. Gray and J. E. Guillet, "The Application of the Molecular Probe Technique to a Study of Polymer Crystallization Rates," *Macromolecules*, **4**, 129 (1971).
5. J.-M. Braun and J. E. Guillet, "Studies of Polystyrene in the Region of the Glass Transition Temperature by Inverse Gas Chromatography," *Macromolecules*, **8**, 882 (1975).
6. D. D. Deshpande and O. S. Tyagi, "Gas Chromatographic Behavior of Poly(vinyl acetate) at Temperatures Encompassing T_g: Determination of T_g and χ," *Macromolecules*, **11**, 746 (1978).
7. J. J. van Deemter, F. J. Zuiderweg, and A. Klinkenberg, "Longitudinal Diffusion and Resistance to Mass Transfer as Causes of Nonideality in Chromatography," *Chem. Eng. Sci.*, **5**, 271 (1956).
8. Z. Y. Al-Saigh and P. Munk, "Study of Polymer–Polymer Interaction Coefficients in Polymer Blends Using Inverse Gas Chromatography," *Macromolecules*, **17**, 803 (1984).
9. T. W. Card, Z. Y. Al-Saigh, and P. Munk, "Diffusion in the Bubble Flow Meter in Inverse Gas Chromatography Experiments," *J. Chromatogr.* **301**, 261 (1984).
10. T. W. Card, Z. Y. Al-Saigh and P. Munk, "Inverse Gas Chromatography 2. The Role of 'Inert' Support," *Macromolecules*, **18**, 1030 (1985).
11. P. Munk, Z. Y. Al-Saigh, and T. W. Card, "Inverse Gas Chromatography 3.

Dependence of Retention Volume on the Amount of Probe Injected," *Macromolecules*, **18**, 2196 (1985).

12. D. G. Gray and J. E. Guillet, "Studies of Diffusion in Polymers by Gas Chromatography," *Macromolecules*, **6**, 223 (1973).
13. J. E. G. Lipson and J. E. Guillet, "Study of Structure and Interactions in Polymers by Inverse Gas Chromatography" in J. V. Dawkins (Ed.), *Developments in Polymer Characterization-3*, Applied Science, London, 1982, Chapter 2.
14. G. DiPaola-Baranyi and J. E. Guillet, "Estimation of Polymer Solubility Parameters by Gas Chromatography," *Macromolecules*, **11**, 228 (1978).
15. M. J. Fernández-Berridi, G. M. Guzmán, J. M. Elorza, and L. Garijo, "Study by Gas-Liquid Chromatography of the Thermodynamics of the Interaction of Poly(vinyl acetate) with Various Solvents," *Eur. Polym. J.*, **19**, 445 (1983).
16. J. E. G. Lipson and J. E. Guillet, "Studies of Polar and Nonpolar Probes in the Determination of Infinite-Dilution Solubility Parameters," *J. Polym Sci., Polym. Phys. Ed.*, **19**, 1199 (1981).
17. M. J. Fernández-Berridi, G. M. Guzmán, J. J. Iruin, and J. M. Elorza, "Determination of the Interaction Parameter χ of Poly(ethylene oxide) by Gas–Liquid Chromatography below the Melting Temperature," *Polymer*, **24**, 417 (1983).
18. M. Galin, "Gas–Liquid Chromatography Study of Poly(ethylene oxide)–Solvent Interactions: Estimation of Polymer Solubility Parameter," *Polymer*, **24**, 865 (1983).
19. R. H. Schuster, H. Gräter, and H. J. Cantow, "Thermodynamic Studies on Polystyrene–Solvent Systems by Gas Chromatography," *Macromolecules*, **17**, 619 (1984).
20. M. J. Fernández-Berridi, J. I. Equiazábal, J. M. Elorza, and J. J. Iruin, "Vapor-Pressure Osmometry and Inverse Gas Chromatography in the Analysis of Thermodynamic Properties of Polymer Solutions," *J. Polym. Sci., Polym. Phys. Ed.* **21**, 859 (1983).
21. K. A. Karim and D. C. Bonner, "Thermodynamic Interpretation of Solute–Polymer Interactions at Infinite Dilution," *J. Appl. Polym. Sci.*, **22**, 1277 (1978); "An Improved Concept in Solubility Parameter Theory Applied in Amorphous Polymers. I. Poly(Ethyl Methacrylate)," *Polym. Eng. Sci.*, **19**, 1174 (1979).
22. K. C. B. Dangayach and D. C. Bonner, "Solvent Interactions with Polysulfone," *Polym. Eng. Sci.*, **20**, 59 (1980).
23. K. C. B. Dangayach, K. A. Karim, and D. C. Bonner, "Interactions of Organic Solvents with Aromatic Heterocyclic Polymers. I. *m*-Phenylene Polybenzimidazole," *J. Appl. Polym. Sci*, **26**, 559 (1981).
24. G. DiPaola-Baranyi, S. J. Fletcher, and P. Degré, "Gas Chromatographic Investigation of Poly(vinylidene-fluoride)-Poly(methyl methacrylate) Blends," *Macromolecules*, **15**, 885 (1982).
25. G. DiPaola-Baranyi and P. Degré, "Thermodynamic Characterization of Polystyrene–Poly(*n*-butyl methacrylate) Blends," *Macromolecules*, **14**, 1456 (1981).

26. D. J. Walsh and J. G. McKeown, "Compatibility of Polyacrylates and Polymethacrylates with Poly(vinyl chloride): 2. Measurement of Interaction Parameters," *Polymer*, **21**, 1335 (1980).
27. C. P. Doubé and D. J. Walsh, "Studies of Poly(vinyl chloride)/Solution Chlorinated Polyethylene Blends by Inverse Gas Chromatography," *Eur. Polym. J.*, **17**, 63 (1981).
28. D. J. Walsh, J. S. Higgins, S. Rostami, and K. Weeraperuma, "Compatibility of Ethylene–Vinyl Acetate Copolymers with Chlorinated Polyethylenes. 2. Investigation of the Thermodynamic Parameters," *Macromolecules*, **16**, 391 (1983).
29. C. Zhikuan and D. J. Walsh, "Inverse Gas Chromatography for the Study of One Phase and Two Phase Polymer Mixtures," *Eur. Polym. J.*, **19**, 519 (1983).
30. G. DiPaola-Baranyi, "Thermodynamic Miscibility of Various Solutes with Styrene-Butyl Methacrylate Polymers and Copolymers," *Macromolecules*, **14**, 683 (1981).
31. M. Galin and M. C. Rupprecht, "Gas Chromatographic Study of the Interactions in Styrene–Dimethylsiloxane Block Copolymers and Blends," *Macromolecules*, **12**, 506 (1979).
32. T. C. Ward, D. P. Sheehy, and J. E. McGrath, "Polydimethyl Siloxane–Polycarbonate Block Copolymer Studies Utilizing Inverse Gas Chromatography," *ACS Polym. Div. Prepr.* **21**(2), 70 (1980).
33. T. C. Ward, D. P. Sheehy, J. E. McGrath, and J. S. Riffle, "Inverse Gas Chromatography Studies of Polydimethylsiloxane (PDMS) Polycarbonate (PC) Copolymers and Blends," *ACS Polym. Div. Prepr.* **22**(1), 187 (1981).
34. J. S. Aspler and D. G. Gray, "An Inverse Gas-Chromatographic Study of the Interaction of Water with Some Cellulose Derivatives," *J. Polym. Sci., Polym. Phys. Ed.* **21**, 1675 (1983).
35. G. A. Senich "Migration to and from Plastics," CHEMTECH, **11**, 360 (1981); "Measurements of Diffusion in Polymers by Inverse Gas Chromatography," *ACS Polym. Div. Prepr.* **22**(2), 343 (1981); "Chromatographic Studies of Diffusion in Polymers," *Proc. IUPAC Macromol. Symp. 28th, 1982*, 740.
36. B. Shiyao, S. Sourirajan, F. D. F. Talbot, and T. Matsuura, "Gas and Vapor Adsorption on Polymeric Materials by Inverse Gas Chromatography," *Am. Chem. Soc. Symp. Ser.*, **391**, 59 (1989).
37. P. G. Demertzis and M. G. Kontominas, "Thermodynamic Study of Water Sorption and Water Vapor Diffusion in Poly(vinylidene chloride) Copolymers," *Am. Chem. Soc. Symp. Ser.*, **391**, 77 (1989).
38. M. F. Grenier-Loustalot, G. Mouline, and P. Grenier, "Inverse Gas Chromatography Used to Follow Kinetics of Epoxy-Amine Reactions in the Molten State," *Polymer*, **28**, 2275 (1987).
39. P. Hattam and P. Munk, "Inverse Gas Chromatography. 5. Computer Simulation of Diffusion Processes on the Column," *Macromolecules*, **21**, 2083 (1988).
40. P. Munk, T. W. Card, P. Hattam, M. J. El-Hibri, and Z. Y. Al-Saigh, "Inverse Gas Chromatography. 4. The Diffusion Phenomena on the Column," *Macromolecules*, **20**, 1278 (1987).

41. D. Arnould and R. L. Laurence, "Solute Diffusion in Polymers by Capillary Column Inverse Gas Chromatography," *Am. Chem. Soc. Symp. Ser.*, **391**, 87 (1989).
42. C. A. Pawlisch, J. R. Bric, and R. L. Laurence, "Solute Diffusion in Polymers. 2. Fourier Estimation of Capillary Column Inverse Gas Chromatography Data," *Macromolecules*, **21**, 1685 (1988).
43. C. A. Pawlisch, A. Macris, and R. L. Laurence, "Solute Diffusion in Polymers. 1. The Use of Capillary Column Inverse Gas Chromatography," *Macromolecules*, **20**, 1564 (1987).
44. J. Miltz, "Inverse Gas Chromatographic Studies of Styrene Diffusion in Polystyrene and Monomer/Polymer Interaction," *Polymer*, **27**, 105 (1986).
45. G. Boiteux-Steffan, J. P. Soulie, D. Sage, and J. M. Letoffe, "Multiple Transitions in Poly(allylbenzene): 1. Thermophysical Properties," *Polymer*, **26**, 1443 (1985).
46. A. B. Wojcik, "Porous Bead Aliphatic-Aromatic Methacrylate Copolymers. VII. TGA, DTA, DSC, and Inverse Gas Chromatography Studies," *J. Appl. Polym. Sci.*, **39**, 179 (1990).
47. R. Sanetra, B. N. Kolarz, and A. Wlochowicz, "Determination of the Glass Transition Temperature of Poly(styrene-co-divinylbenzene) by Inverse Gas Chromatography," *Polymer*, **26**, 1181 (1985).
48. F. Chen, "Study of Acceptor-Donor Interactions at the Polymer Interface by Inverse Gas Chromatography Data Analysis," *Macromolecules*, **21**, 1640 (1988).
49. J. E. Guillet, M. Romansky, G. J. Price, and R. van der Mark, "Studies of Polymer Structure and Interactions by Automated Inverse Gas Chromatography," *Am. Chem. Soc. Symp. Ser.*, **391**, 20 (1989).
50. A. C. Su and J. R. Fried, "On the Determination of Polymer–Polymer Interaction Parameter by Inverse Gas Chromatography," *J. Polym. Sci. Part C: Polym. Lett.*, **24**, 343 (1986).
51. M. G. Prolongo, R. M. Masegosa, and A. Horta, "Polymer–Polymer Interaction Parameter in the Presence of a Solvent," *Macromolecules*, **22**, 4346 (1989).
52. I. C. Sanchez, "Relationships Between Polymer Interaction Parameters," *Polymer*, **30**, 471 (1989).
53. O. S. Tyagi and D. D. Deshpande, "Inverse Gas Chromatography of Poly(*n*-butyl methacrylate): Effect of Flow Rate on Specific Retention Volume and Detection of Glass Transition Temperature," *J. Appl. Polym. Sci.*, **34**, 2377 (1987).
54. J. Y. Wang and G. Charlet, "The Use of Moment Analysis in Inverse Gas Chromatography," *Macromolecules*, **22**, 3781 (1989).
55. P. Hattam, Q. Du, and P. Munk, "Computer Simulation of Elution Behavior of Probes in Inverse Gas Chromatography. Comparison With Experiment," *Am. Chem. Soc. Symp. Ser.*, **391**, 33 (1989).
56. A. E. Bolvari, T. C. Ward, P. A. Koning, and D. P. Sheehy, "Experimental Techniques of Inverse Gas Chromatography," *Am. Chem. Soc. Symp. Ser.*, **391**, 12 (1989).

57. M. J. El-Hibri and P. Munk, "Marker Retention in Inverse Gas Chromatography Experiments on Polymers," *Macromolecules*, **21**, 264 (1988).
58. H. Becker and R. Gnauck, "Determination of Dead Times for Inverse Gas Chromatography: Measurements by Use of Methane," *J. Chromatogr.*, **366**, 378 (1986).
59. L. Lavielle and J. Schultz, "Surface Properties of Graft Polyethylene in Contact with Water. I. Orientation Phenomena," *J. Colloid Interface Sci.*, **106**, 438 (1985).
60. M. Laleq, M. Bricault, and H. P. Schreiber, "Component Interactions and Their Influence on the Uniformity of Coating Films," *J. Coat. Technol.*, **61**, 45 (1989).
61. S. Dong, M. Brendle, and J. B. Donnet, "Study of Solid Surface Polarity by Inverse Gas Chromatography at Infinite Dilution," *Chromatographia*, **28**, 469 (1989).
62. M. F. Greiner–Loustalot, Y. Borthomieu, and P. Grenier, "Surfaces of Carbon Fibers Characterized by Inverse Gas Chromatography," *Surf. Interface Anal.*, **14**, 187 (1989).
63. A. E. Bolvari and T. C. Ward, "Adhesion and Acid–Base Interactions via Inverse Gas Chromatography," *Polym. Mater. Sci. Eng.*, **58**, 655 (1988).
64. J. Schultz, L. Lavielle, and C. Martin, "Surface Properties of Carbon Fibers Determined by Inverse Gas Chromatography," *J. Chim. Phys. Phys. -Chim. Biol.*, **84**, 231 (1987).
65. A. E. Bolvari and C. Thomas, "Determination of Fiber-Matrix Adhesion and Acid–Base Interactions," *Am. Chem. Soc. Symp. Ser.*, **391**, 217 (1989).
66. S. P. Wesson and R. E. Allred, "Surface Energetics of Plasma-Treated Carbon Fiber," *Am. Chem. Soc. Symp. Ser.*, **391**, 203 (1989).
67. A. J. Vukov and D. G. Gray, "Properties of Carbon Fiber Surfaces," *Am. Chem. Soc. Symp. Ser.*, **391**, 168 (1989).
68. J. Schultz and L. Lavielle, "Interfacial Properties of Carbon Fiber–Epoxy Matrix Composites," *Am. Chem. Soc. Symp. Ser.*, **391**, 185 (1989).
69. M. Escoubes, B. Chabert, P. Hoffmann, D. Sage, and J. P. Soulier, "Interactions that Can Appear at the Fiber/Matrix Interface During Polyepoxide/Carbon Fiber Composite Making," *J. Colloid Interface Sci.*, **124**, 375 (1988).
70. A. J. Vukov and D. G. Gray, "Adsorption of *n*-Alkanes on Carbon Fibers at Zero Surface Coverage," *Langmuir*, **4**, 743 (1988).
71. J. Schultz, L. Lavielle, and C. Martin, "The Role of the Interface on Carbon Fiber–Epoxy Composites," *J. Adhes.*, **23**, 45 (1987).
72. E. Osmont and H. P. Schreiber, "Surface Characteristics of Glass Fibers," *Am. Chem. Soc. Symp. Ser.*, **391**, 230 (1989).
73. H. L. Lee and P. Luner, "Characterization of AKD Sized Papers by Inverse Gas Chromatography," *Nord. Pulp Pap. Res. J.*, **4**, 164 (1989).
74. P. J. C. Chappell and D. R. Williams, "Surface Characterization of Kevlar by IGC," in F. L. Matthews (Ed.), *Sixth International Conference on Compos. Material*, Vol. 5, Elsevier, London, 1987, p. 5.346.

75. P. J. C. Chappell and D. R. Williams, "Determination of Poly(*p*-phenylene-terephthalamide) Fiber Surface Cleanliness by Inverse Gas Chromatography," *J. Colloid Interface Sci.*, **128**, 450 (1989).
76. A. S. Gozdz and H. D. Weigmann, "Surface Characterization of Intact Fibers by Inverse Gas Chromatography," *J. Appl. Polym. Sci.*, **29**, 3965 (1984).
77. P. Munk, P. Hattam, D. Qiangguo, and A.-A. A Abdel-Azim, "Determination of Polymer–Solvent Interaction Coefficients by Inverse Gas Chromatography," *J. Appl. Polym. Sci., Appl. Polym. Symp. Ed.*, **45**, 289 (1990).
78. J. I. Eguiazabal, M. J. Fernandez-Berridi, J. J. Iruin, and J. M. Elorza, "Chromatographic Determination of Polymer Solubility Parameters," *Polym. Bull.*, **13**, 463 (1985).
79. H. S. Tseng, P. C. Wong, D. R. Lloyd, and J. W. Barlow, "Thermodynamic Interaction in Polybutadiene/Solute Systems by Inverse Gas Chromatography," *Polym. Eng. Sci.*, **27**, 1141 (1987).
80. S. P. Nunes, B. A. Wolf, and H. E. Jeberien, "On the Co-occurrence of Demixing and Thermoreversible Gelation of Polymer Solutions. 2. Thermodynamic Background," *Macromolecules*, **20**, 1948 (1987).
81. A. Edelman and A. Fradet, "Inverse Gas Chromatography Study of Some Triacetin–Polymer Systems," *Polymer*, **30**, 317 (1989).
82. G. J. Price, "Calculation of Solubility Parameters by Inverse Gas Chromatography," *Am. Chem. Soc. Symp. Ser.*, **391**, 48 (1989).
83. H. S. Tseng, R. Douglas, and T. C. Ward, "Solubility of Nonpolar and Slightly Polar Organic Compounds in Low-Density Polyethylene by Inverse Gas Chromatography with Open Tubular Column," *J. Appl. Polym. Sci.*, **30**, 1815 (1985).
84. U. Freytag and H. J. Radusch, "Study of the Thermodynamic Compatibility of High-Density Polyethylene and Low-Density Polyethylene in the Melt State. I. Determination of Solubility Parameters," *Angew. Makromol. Chem.*, **152**, 1 (1987).
85. P. A. Koning, A. E. Bolvari, and T. C. Ward, "Investigations of Acid–Base Interactions in Polymeric Adhesives," *Am. Chem. Soc. Polym. Prepr.*, **27** (2), 131 (1986).
86. J. M. Barrales-Rienda and J. Vidal Gancedo, "Thermodynamic Studies on Poly[*N*-(*n*-Octadecyl) Maleimide] (PMI-18)/Solvent Systems by Inverse Gas Chromatography With Capillary Columns," *Macromolecules*, **21**, 220 (1988).
87. A. Edelman and A. Fradet, "Inverse Gas Chromatography Study of Some Alkyl Trinitrate–Polymer Systems," *Polymer*, **30**, 324 (1989).
88. C. Uriarte, M. J. Fernandez-Berridi, J. M. Elorza, J. J. Iruin, and L. Kleintjens, "Determination of the Interaction Parameter by Inverse Gas Chromatography: An Additional Experimental Test of the Classic Lattice Model," *Polymer*, **30**, 1493 (1989).
89. J. Miltz and V. Rosen-Doddy, "Sorption Isotherms of Styrene on Polystyrene and Monomer–Polymer Interaction," *Eur. Polym. J.*, **22**, 327 (1986).
90. H. S. Tseng, D. R. Lloyd, and T. C. Ward, "Correlation of Organic Solubility in Poly(vinyl acetate)," *Polym. Commun.*, **25**, 262 (1984).

91. P. Hattam, W. Cheng, Q. Du, and P. Munk, "Study of Poly(vinyl chloride) by Inverse Gas Chromatography," *Makromol. Chem., Macromol. Symp.*, **29**, 297 (1989).
92. P. G. Demertzis and M. G. Kontominas, "Interaction of Vinyl Chloride with Poly(vinyl chloride) by Inverse Gas Chromatography: Effect of Monomer Concentration, Plasticizer Content and Temperature," *Dev. Food Sci.*, **12**, 513 (1986).
93. M. G. Kontominas, P. G. Demertzis, and S. G. Gilbert, "Sorption of Vinyl Chloride onto Poly(vinyl chloride) by Classical Partition and Inverse Gas Chromatography: Comparison of Two Methods," *J. Food Process. Preserv.*, **9**, 223 (1985).
94. D. D. Deshpande and O. S. Tyagi, "Inverse Gas Chromatography of Poly(vinyl isobutyl ethers) and Polymer–Solute Thermodynamic Interactions," *J. Appl. Polym. Sci.*, **33**, 715 (1987).
95. Y. P. Yampolskii, N. E. Kaliuzhnyi, and S. G. Durgar', "Thermodynamics of Sorption in Glassy Poly(vinyltrimethylsilane)," *Macromolecules*, **19**, 846 (1986).
96. P. G. Detertzis and M. G. Kontominas, "Interaction Between Vinylidene Chloride and Copolymers of Vinylidene Chloride by Inverse Gas Chromatography," *Lebensm-Wiss. Technol.*, **19**, 249 (1986).
97. C. T. Chen and Z. Y. Al-Saigh, "Characterization of Semicrystalline Polymers by Inverse Gas Chromatography. 1. Poly(vinylidene fluoride)," *Macromolecules*, **22**, 2974 (1989).
98. G. J. Price, K. S. Siow, and J. E. Guillet, "Use of Gas Chromatography to Determine the Degree of Crosslinking of a Polymer Network," *Macromolecules*, **22**, 3116 (1989).
99. H. Becker and R. Gnauck, "Inverse Gas Chromatographic Study of Polyether Alcohols," *J. Chromatogr.*, **410**, 267 (1987).
100. R. Sanetra, B. N. Kolarz, and A. Wlochowicz, "Determination of Thermodynamic Data for the Interaction of Aliphatic Alcohols with Poly(styrene-*co*-divinylbenzene) Using Inverse Gas Chromatography," *Polymer*, **28**, 1753 (1987).
101. R. Sanetra, B. N. Kolarz, and A. Wlochowicz, "The Study of Modified Poly(styrene-*co*-divinylbenzene) by Inverse Gas-Chromatographic Analysis," *Angew. Makromol. Chem.*, **140**, 41 (1986).
102. M. J. El-Hibri, W. Cheng, P. Hattam, and P. Munk, "Inverse Gas Chromatography of Polymer Blends. Theory and Practice," *Am. Chem. Soc. Symp. Ser.*, **391**, 121 (1989).
103. M. J. El-Hibri, W. Cheng, and P. Munk, "Inverse Gas Chromatography. 6. Thermodynamics of Polycaprolactone–Polyepichlorohydrin Blends," *Macromolecules*, **21**, 3458 (1988).
104. H. S. Tseng, D. R. Lloyd, and T. C. Ward, "Correlation of Solubility in Polydimethylsiloxane and Polyisobutylene Systems," *J. Appl. Polym. Sci.*, **30**, 307 (1985).
105. C. Bhattacharya, N. Maiti, B. M. Mandal, and S. N. Bhattacharyya, "Thermodynamic Characterization of Miscible Blends from very Similar Polymers by

Inverse Gas Chromatography. Poly(ethyl acrylate)–Poly(vinyl propionate) System," *Macromolecules*, **22**, 4062 (1989).

106. Y. Murakami, "Studies on Compatibility of Poly(ethylene oxide) and Poly(methyl methacrylate) by Inverse Gas Chromatography," *Polym. J.*, **20**, 549 (1988).
107. G. DiPaola-Baranyi, "Thermodynamics of Polymer Blends by Inverse Gas Chromatography," *Am. Chem. Soc. Symp. Ser.*, **391**, 108 (1989).
108. A. C. Su and J. R. Fried, "Poly(2,6-dimethyl-1,4-phenylene oxide) Blends Studied by Inverse Gas Chromatography," *Am. Chem. Soc. Adv. Chem. Ser.*, **211**, 59 (1986).
109. A. C. Su and J. R. Fried, "Interaction Parameters of Poly(2,6-dimethyl-1,4-phenylene oxide) Blends," *Am. Chem. Soc. Symp. Ser.*, **391**, 155 (1989).
110. S. Klotz, H. Graeter, and H.J. Cantow, "Estimation of Free Energy of Polymer Blends," *Am. Chem. Soc. Symp. Ser.*, **391**, 135 (1989).
111. G. DiPaola-Baranyi, J. Richer, and W. M. Prest, "Thermodynamic Miscibility of Polystyrene–Poly(2,6-dimethyl-1,4-phenylene oxide) Blends," *Can. J. Chem.*, **63**, 223 (1985).
112. J. M. Elorza, M. J. Farnandez-Berridi, J. J. Iruin, and C. Uriarte, "Study of Miscibility of the System Polystyrene/Poly(methyl vinyl ether) by Inverse Gas Chromatography," *Makromol. Chem.*, **189**, 1855 (1988).
113. O. S. Tyagi, S. M. Sajjad, and S. Husain, "Polymer–Polymer Interaction Parameter Determined by Inverse Gas Chromatography," *Polymer*, **28**, 2329 (1987).
114. M. J. El-Hibri and P. Munk, "Recent Development in Inverse Gas Chromatography of Polymer Blends," *Am. Chem. Soc. Polym. Prepr.*, **28**(1), 262 (1987).
115. A. K. Nandi, B. M. Mandal, and S. N. Bhattacharyya, "Miscibility of Poly(methyl acrylate) and Poly(vinyl acetate): Incompatibility in Solution and Thermodynamic Characterization by Inverse Gas Chromatography," *Macromolecules*, **18**, 1454 (1985).
116. C. Uriarte, J. J. Iruin, M. J. Fernandez-Berridi, and J. M. Elorza, "Chromatographic Studies of a Poly(vinyl methyl ether)/Phenoxy Resin Blend Near the Lower Critical Solution Temperature," *Polymer*, **30**, 1155 (1989).
117. B. M. Mandal, C. Bhattacharya, and S. N. Bhattacharyya, "Thermodynamic Characterization of Binary Polymer Blends by Inverse Gas Chromatography," *J. Macromol. Sci., Chem.*, **A26**, 175 (1989).

CHAPTER

6

MEASUREMENT OF MOLECULAR WEIGHTS OF POLYMERS BY OSMOMETRY

JIMMY W. MAYS

Department of Chemistry
University of Alabama at Birmingham
Birmingham, Alabama

and

NIKOS HADJICHRISTIDIS

Division of Chemistry
University of Athens
Athens, Greece

Modern Methods of Polymer Characterization, Edited by Howard G. Barth and Jimmy W. Mays
ISBN 0-471-82814-9

1. INTRODUCTION

1.1. Polymer Molecular Weights

Measurement of polymer molecular weights is a matter of fundamental importance for the characterization of polymeric materials. Unlike low-molecular-weight compounds, polymers have no exact or formula molecular weight. In addition, polymers are usually polydisperse, that is, they exhibit a *distribution* of molecular weights. These factors tend to complicate polymer molecular weight determinations.

A typical molecular weight distribution MWD is depicted in Figure 1. The most important average molecular weights, namely, number-average molecular weight $\bar{M}_n$, weight-average molecular weight $\bar{M}_w$, and z-average molecular weight $\bar{M}_z$ are labelled. The term $\bar{M}_n$ is defined as

$$\bar{M}_n = \frac{\sum N_\mathrm{i} M_\mathrm{i}}{\sum N_\mathrm{i}} \tag{1}$$

where N_i is the number of moles of species i of molecular weight $\bar{M}_\mathrm{i}$. The term $\bar{M}_n$ is merely the weight of the molecules divided by the number of molecules; this definition is consistent with the classic concept of molecular weight for a pure species.

The higher moments of the MWD, $\bar{M}_w$ and $\bar{M}_z$, are given by the equations

$$\bar{M}_w = \frac{\sum N_\mathrm{i} M_\mathrm{i}^2}{\sum N_\mathrm{i} M_\mathrm{i}} \tag{2}$$

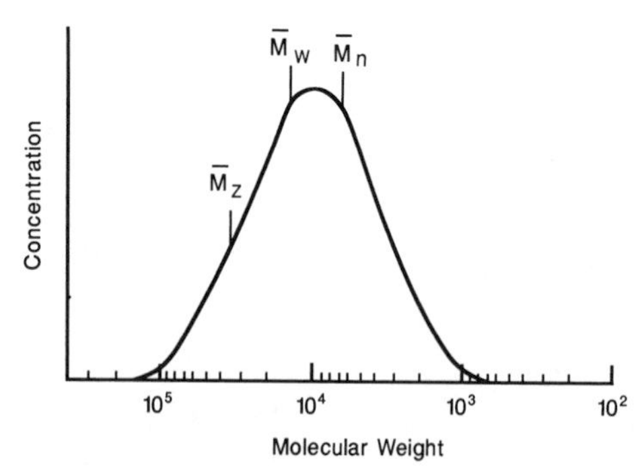

Figure 1. A typical polymer molecular weight distribution and corresponding average molecular weights.

$$\bar{M}_z = \frac{\sum N_i M_i^3}{\sum N_i M_i^2} \tag{3}$$

It is clear from these equations that for $\bar{M}_w$ and $\bar{M}_z$ the molecules of greater mass contribute more to the average than do less massive molecules. Conversely, $\bar{M}_n$ is very sensitive to the presence of low-molecular-weight tails. As an example, Figure 2 illustrates the effect of low- and-high-molecular-weight tails on the average molecular weights. With a small increase in the low-molecular-weight end of the distribution, $\bar{M}_n$ is decreased by 18% as compared to a 2 and 1.1% decrease of $\bar{M}_w$ and $\bar{M}_z$, respectively. At the high-molecular-weight end of the distribution, however, $\bar{M}_w$ and $\bar{M}_z$ are increased by 12 and 7%, respectively, as compared to only a 1% increase in $\bar{M}_n$.

The question now arises as to which average molecular weight should be measured for a particular polymer. The answer usually lies in the circumstances surrounding its manufacture or application. For example, in the field of polymerization kinetics, $\bar{M}_n$ is a critical parameter. The glass-transition temperature T_g of a polymer is also particularly sensitive to low-molecular-weight species. Knowledge of $\bar{M}_n$ is also necessary for evaluating the level of functionalization in polymers with specific end groups. In more applied areas of polymer science, for example, rheology, various average molecular weights *and* the shape and breadth of the distribution are important. Polydispersity ratios such as $\bar{M}_w/\bar{M}_n$ and $\bar{M}_z/\bar{M}_w$ can give some insight into this latter problem. Absolute methods are available, which allow

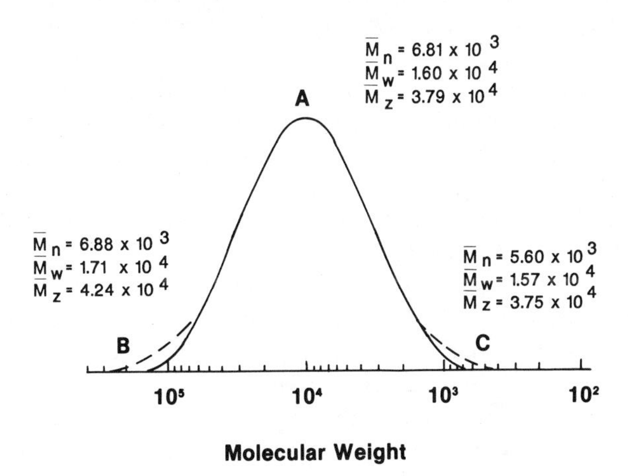

Figure 2. The effect of low- and high-molecular-weight tails on average molecular weights. The addition of a small amount of high-molecular-weight polymer (curve B) to the "base polymer" (curve A) has a large effect on $\bar{M}_z$; $\bar{M}_w$ and $\bar{M}_n$ are relatively unaffected. Conversely, addition of a small amount of low-molecular-weight polymer (curve C) causes a large reduction in $\bar{M}_n$.

the direct measurement of these various averages. Measurement of $\bar{M}_n$ is the primary subject of this chapter.

1.2. Techniques for Measurement of Number-Average Molecular Weights

It is clear from the preceding section that $\bar{M}_n$ is a very important characteristic of a polymer. Fortunately, a number of techniques exists for measuring $\bar{M}_n$. These methods include size exclusion chromatography (SEC), end-group analysis, and colligative properties measurements. Size exclusion chromatography gives not only $\bar{M}_n$ but also MWD and potentially all of the statistical average molecular weights, that is, it gives a complete description of the polymer with respect to molecular weight. This is the reason why SEC is the premier technique used for polymer characterization. The significant disadvantage of SEC is that it is a relative technique and that careful calibration is required; SEC is not an absolute method for measuring polymer molecular weights.

End-group analysis (titration, IR, UV, NMR, labeling, etc.) is also useful for obtaining $\bar{M}_n$, particularly for lower molecular weight species ($< 10{,}000$ g/mol), where the concentration of end groups is higher. A critical requirement is that the nature and number of end groups should be well established. Side reactions frequently lead to small amounts of unanticipated end groups, for example, oxidation of $—NH_2$ groups to $—NO_2$ groups. In the usual case one assumes a certain number of specific end groups per molecule. Cyclization or branching can therefore lead to serious errors in the value of $\bar{M}_n$ that is obtained.

For the reasons given above, absolute number-average molecular weights are mainly derived from the colligative methods, which are based on vapor pressure lowering, freezing point depression (cryoscopy), boiling point elevation (ebulliometry), and osmotic pressure measurements on dilute polymer solutions. These methods are dependent only on the number and not the kind of molecules that are present.

Of the four colligative methods only two are widely used with polymers at present. These techniques are membrane osmometry (MO) and vapor pressure osmometry (VPO). Perhaps the main reason for the continued popularity of MO and VPO is that commercial instruments are readily available. These instruments allow rapid and reliable determinations of $\bar{M}_n$ in a variety of solvents. Also, small sample sizes (< 1 g) are normally adequate for a successful measurement of $\bar{M}_n$. In addition, experimental difficulties (association at low temperatures, supercooling, superheating, local pressure changes, and foaming) have prevented cryoscopy and ebulliometry from becoming more popular.

2. COLLIGATIVE PROPERTIES OF DILUTE POLYMER SOLUTIONS

Colligative properties of dilute solutions, which include vapor pressure lowering, boiling point elevation, freezing point depression, and osmotic pressure, make use of the fact that the addition of solute to a solvent results in a measurable change in the chemical potential of the solvent. The change in chemical potential in terms of activity of the solvent is given by

$$\mu_1 - \mu_1^0 = RT \ln a_1 \tag{4}$$

where μ_1 is the chemical potential of the solvent in the solution, μ_1^0 is the chemical potential of the pure solvent at the same temperature and pressure, a_1 is the activity of the solvent, R is the gas constant, and T is absolute temperature. The chemical potential is a partial molar free energy, and variations with pressure and temperature are described by

$$\left(\frac{\partial \mu_1}{\partial P}\right)_T = \bar{v}_1 \tag{5}$$

$$\left(\frac{\partial \mu_1}{\partial T}\right)_P = \bar{S}_1 \tag{6}$$

where $\bar{v}_1$ is the partial molar volume of the solvent and $\bar{S}_1$ is the partial molar entropy of the solvent.

In a system where solution is in contact with pure solvent, a necessary condition for equilibrium is that the chemical potential of the solvent in both phases must be equal. This necessitates making μ_1 equal to μ_1^0 by changing either the temperature or the pressure of the solution. The amount of change required to restore equilibrium may be calculated from Eqs. 5 or 6. This change is a direct measure of the activity of the solvent in solution and forms the basis of colligative methods for the study of dilute solutions.

Practical equations that enable polymer number-average molecular weights to be calculated from measurements of colligative properties are given below. For membrane osmometry

$$\lim_{c \to 0}\left(\frac{\pi}{c}\right) = \frac{RT}{\bar{M}_n} \tag{7}$$

where π is the osmotic pressure and c is concentration of polymer. For vapor pressure osmometry

$$\lim_{c \to 0} \left(\frac{\Delta P}{c} \right) = \frac{-P^0 V^0}{\bar{M}_n} \tag{8}$$

where ΔP is the vapor pressure differences between solvent and solution, P^0 is the vapor pressure of the solvent, and V^0 is the molar volume of the solvent. For cryoscopy

$$\lim_{c \to 0} \left(\frac{\Delta T_f}{c} \right) = \left(\frac{R T_f^2 M_s}{\rho \Delta H_f} \right) \left(\frac{1}{\bar{M}_n} \right) \tag{9}$$

where ΔT_f is the freezing point depression, ρ is the density of the solvent, M_s is the molecular weight of the solvent, T_f is the solvent freezing point, and ΔH_f is the enthalpy of fusion. For ebulliometry

$$\lim_{c \to 0} \left(\frac{\Delta T_b}{c} \right) = \left(\frac{R T_b^2 M_s}{\rho \Delta H_v} \right) \left(\frac{1}{\bar{M}_n} \right) \tag{10}$$

where ΔT_b is the boiling point elevation, T_b is the solvent boiling point, and ΔH_v is the enthalpy of vaporization.

The effect on the colligative properties brought about by adding polymer of varying molecular weight to pure solvent is given in Table 1. The change in osmotic pressure is largest and, therefore, is the most easily measured. Consequently, MO is the most frequently utilized colligative method for the determination of polymer molecular weights. Membrane osmometry, being the most sensitive, is also applicable to polymers of higher molecular weight.

The very slight changes is vapor pressure given in Table 1 cannot be directly measured with the required accuracy. Nevertheless, an indirect method of measuring the vapor pressure lowering by the technique called vapor pressure osmometry (or vapor phase osmometry, VPO) has enjoyed widespread success. Vapor pressure osmometry is actually a misnomer since

Table 1. Relative Comparison of the Sensitivity of Various Colligative Properties[a]

Property	$\bar{M}_n = 10^4$	$\bar{M}_n = 10^5$	$\bar{M}_n = 10^6$
Osmotic pressure (cm solvent)	30	3	0.3
Vapor pressure lowering (mm Hg)	8×10^{-3}	8×10^{-4}	8×10^{-5}
Freezing point depression (°C)	5×10^{-3}	5×10^{-4}	5×10^{-5}
Boiling point elevation (°C)	2.5×10^{-3}	2.5×10^{-4}	2.5×10^{-5}

[a]A 1% solution of polystyrene in benzene at 25 °C.

the measurement is in no way related to osmometry, but is used because of historical reasons. Vapor phase osmometry is currently the method of choice for measuring $\bar{M}_n$ for polymers where $\bar{M}_n < 20{,}000$ g/mol, since diffusion of low-molecular-weight polymer through the membrane limits the utility of MO in the region of low $\bar{M}_n$.

The combined popularity of MO and VPO has severely limited cryoscopy and ebulliometry as techniques for measuring $\bar{M}_n$. Perhaps the single most important factor is that commercial MO and VPO units are readily available: whereas, commercial cryoscopic and ebulliometric equipment, designed for use with polymers, are less readily available.

It can be shown that colligative properties measure $\bar{M}_n$, rather than some other average molecular weight. Equation 7 may be written for monodisperse polymer

$$\pi = \frac{RTc}{M} \tag{11}$$

For a polydisperse sample

$$\pi = RT \sum \left(\frac{c_i}{M_i} \right) \tag{12}$$

where c_i is the concentration of species i of molecular weight M_i. Furthermore, the concentration of any species c_i is defined as

$$c_i = N_i M_i \tag{13}$$

N_i is the number of molecules of molecular weight M_i. Therefore,

$$\pi = RT \left(\frac{\sum N_i M_i}{\sum M_i} \right) \tag{14}$$

and (from Eq. 1)

$$\frac{\pi}{c} = RT \left(\frac{\sum N_i}{\sum N_i M_i} \right) = \frac{RT}{\bar{M}_n} \tag{15}$$

3. MEMBRANE OSMOMETRY

3.1. Basic Principles

In membrane osmometry, a solution of a polymer is separated from pure solvent by a semipermeable membrane, that is, a membrane that allows solvent to pass through while preventing the passage of solute. Since the chemical potential of the pure solvent is greater than that of the solvent in the solution, there is a tendency for solvent to diffuse from the pure solvent side of the membrane into the solution side. Pressure can be applied to the solution side to prevent solvent flow from taking place. Under conditions where there is no net flow across the membrane, the applied pressure equals the osmotic pressure π. In other words, the osmotic pressure of a solution is the pressure that must be applied to the solution to make the activity of the solvent in the solution equal to the activity of the pure solvent. A device that measures osmotic pressure is called an osmometer. Such a device is depicted in Figure 3.

The van't Hoff relation

$$\pi = mRT \tag{16}$$

where m is the molar concentration of solute, can be used to relate π to the solute molecular weight M by Eq. 11. Equation 11 is an approximate relationship, which becomes exact in the limit of infinite dilution

$$(\pi/c)_{c\to 0} = \frac{RT}{M} \tag{17}$$

The ratio π/c is known as the reduced osmotic pressure.

Since experimental measurements of π/c must be carried out at finite concentrations, the concentration dependence of π must be understood. Flory and co-workers (1–3) found that a virial expansion gives good agreement with

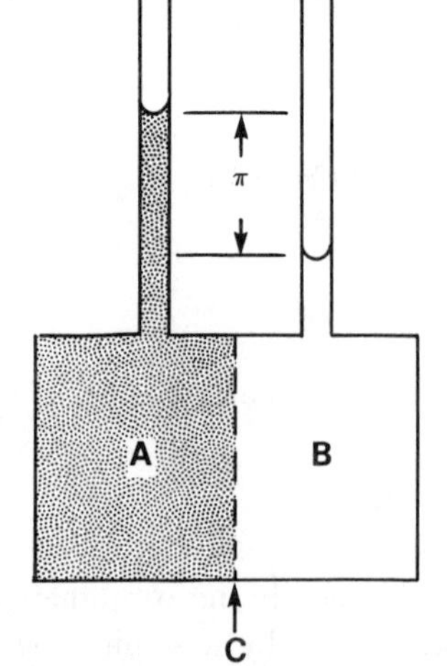

Figure 3. A polymer solution in compartment A is separated from pure solvent in compartment B by a semipermeable membrane C. Solvent diffuses from B to A until the osmotic pressure is just balanced by the hydrostatic pressure. The difference in the heights of the column is the osmotic pressure π.

experimental values

$$\pi/c = (\pi/c)_{c \to 0}(1 + \Gamma_2 c + \Gamma_3 c^2 + \cdots) \tag{18}$$

where Γ_2 and Γ_3 are the second and third virial coefficients, respectively. Equation 18 may also be recast in the form

$$\pi/c = RT(1/\bar{M}_n + A_2 c + A_3 c^2 + \cdots) \tag{19}$$

where A_2 and A_3 are again virial coefficients. The term A_2 is normally expressed in units of mL mol/g^2. The value of A_2 is a measure of polymer–solvent interactions and, hence, solution ideality.

At the theta (θ) temperature, $A_2 = 0$. If the third and higher terms are small, π/c is not a function of concentration at theta conditions. This means that the extrapolation to infinite dilution is avoided by carrying out measurements under theta conditions. Membrane osmometry thus has the potential of determining the θ temperature for a given polymer–solvent pair. However, it should be noted that MO experiments are rarely conducted under θ conditions in actual practice. The potential advantages are offset by possible solubility problems. In addition, θ conditions are frequently not known for a given polymer. Normally, positive solvent values of A_2 are observed (typically $10^{-4} < A_2 < 10^{-3}$) indicating "good" solvent conditions. The value of A_2 in a given solvent decreases with increasing molecular weight except at θ. The empirical relation $A_2 \simeq KM^{-\nu}$, where K is an empirical constant and $\nu \simeq 0.25$ for flexible polymers in good solvents, is often found to hold true.

Frequently, the higher terms in Eqs. 18 and 19 are ignored

$$\frac{\pi}{c} = \frac{RT}{M} + RTA_2 c \tag{20}$$

and a plot of π/c versus c is expected to be linear. When using Eq. 20, upward curvature is often noted at higher concentrations. This curvature is due to contributions from the third and higher virial coefficients (4–6). Stockmayer and Casassa (4) made the useful suggestion that

$$\Gamma_3 = 0.25\Gamma_2^2 \tag{21}$$

This in turn suggests that

$$\left(\frac{\pi}{c}\right)^{0.5} = \left(\frac{\pi}{c}\right)^{0.5}_{c \to 0} [1 + (\Gamma_2/2)c] \tag{22}$$

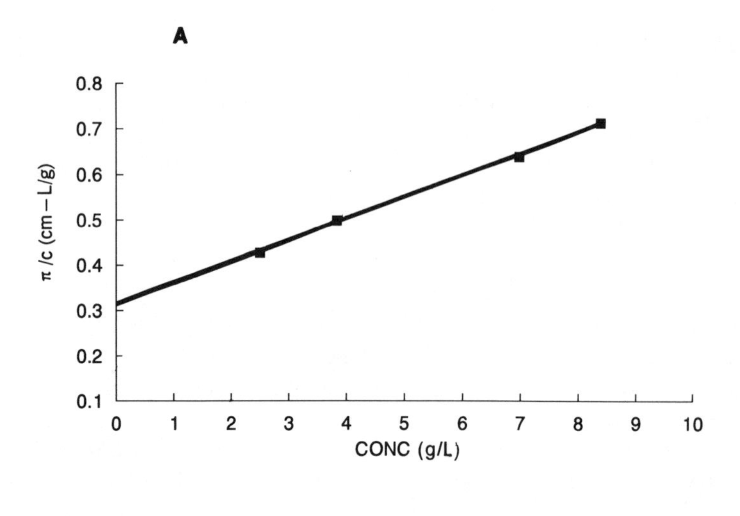

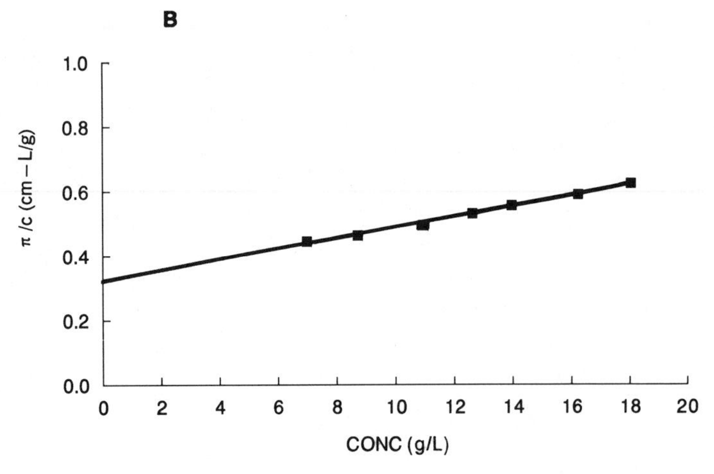

Figure 4. Square-root plots of osmotic pressure data for high-molecular-weight polybutadiene [Fig. 4(*a*) $\bar{M}_n = 338{,}000$] and polystyrene [Fig. 4(*b*) $\bar{M}_n = 279{,}000$]. Note that linear extrapolations to infinite dilution are carried out.

and a plot of $(\pi/c)^{0.5}$ versus c should be linear over a broader range of concentration. Yamakawa (5) considered the virtues and deficiencies of the so-called "square-root plot." He concluded that the square root is generally more reliable than the conventional π/c versus c plot, especially at higher molecular weights. If no curvature were present in the conventional plot, both procedures would yield the same value for $\bar{M}_n$. Where differences exist in $\bar{M}_n$ values obtained by the two procedures, the value arrived at through the use of the square-root plot is to be preferred (4). Flory and co-workers (1–3) opted for use of the square-root plot over 30 years ago.

We strongly suggest, for the above reasons, that the square-root plot *always* be used when $\bar{M}_n$ is evaluted from MO data. Figures 4 and 5 illustrate the way in which the square-root plot better facilitates $\bar{M}_n$ determinations, particularly for polymers with high molecular weights. The correct values of $\bar{M}_n$ for polybutadiene and polystyrene are obtained by linear regression analysis of the $(\pi/c)^{1/2}$ versus c data in Figure 4. The same data, replotted in the conventional manner in Figure 5, show considerable curvature.

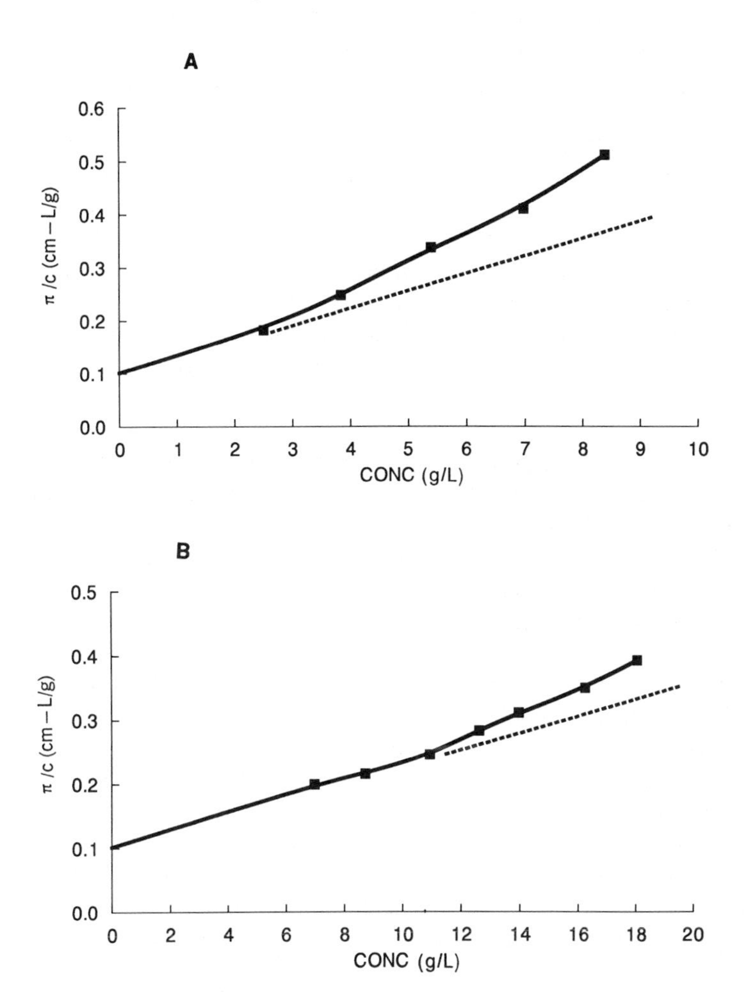

Figure 5. Conventional π/c versus c plots for polybutadiene [Fig. 5(*a*)] and polystyrene [Fig. 5(*b*)] using the data of Figure 4. Note the curvature present at the higher concentrations.

3.2. Practical Aspects

Membrane osmometry is a powerful and widely used method for polymer molecular weight determinations. The technique is both theoretically sound and, in general, experimentally straightforward. This was not always the case.

While Adair (6) in 1925 was probably the first to successfully apply osmometry to the study of macromolecules (proteins), many years passed before accurate results could be routinely obtained. In fact, a round-robin characterization effort (7) sponsored by the International Union of Pure and Applied Chemistry in the 1950s showed membrane osmometry to be quite unsatisfactory for molecular weight measurements on standard polystyrenes. It was even suggested (8) that osmometry be abolished altogether until more suitable membranes could be found.

Membrane osmometry came of age as a routine method with the introduction of automatic osmometers in the early 1960s. These new instruments allowed reliable measurements of π in only a few minutes. The static osmometers (9) required from several hours to even days to reach equilibrium. Static osmometry techniques will not be discussed here, as only the automatic osmometers enjoy widespread use (9). Also, so-called "dynamic osmometry" (10), where at least limited diffusion through the membrane takes place, will not be covered since it is well established (10) that accurate values of $\bar{M}_n$ cannot be readily obtained by this method.

Automatic osmometers are of two basic types. One type measures the osmotic pressure directly using a pressure transducer. The other type uses a servo mechanism to balance the hydrostatic pressure and prevent diffusion of solvent through the membrane. Both types of instruments are capable of measuring pressures very quickly, generally before more than about 10^{-6} mL of solvent has diffused across the membrane.

A schematic illustrating the operation of the Wescan Model 230 Membrane Osmometer is shown in Figure 6. Knauer (15) also sells an osmometer that operates on a similar basis. This method of operation was pioneered in the Melabs osmometer, which is no longer available. In these instruments the membrane separates the solvent and solution compartments. A strain gauge, attached to a stainless steel diaphragm, measures the deflection of the diaphragm caused by the osmotic pressure difference between the solvent and solution. Output is recorded continuously on a recorder.

Both the Knauer and Wescan instruments are capable of giving good performance in a variety of organic solvents and in water. Also, both instruments can be operated at elevated temperatures (ca. 130 °C) for the analysis of polyolefins.

Although no longer available, a great many of the Mechrolab 500 Series osmometers are still in use. Thus, a brief discussion of these units is in order.

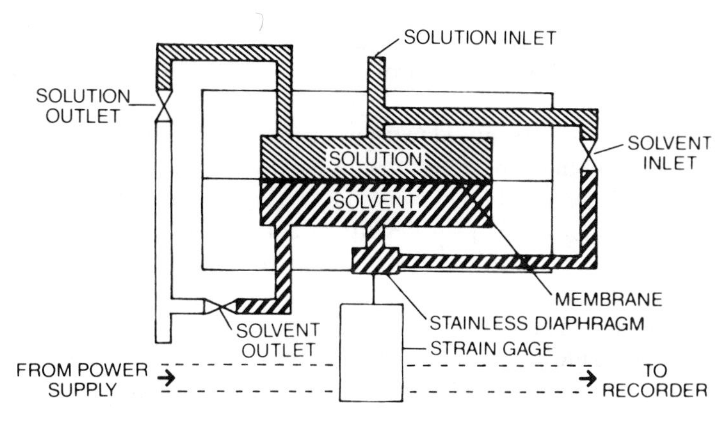

Figure 6. Schematic diagram of the Wescan Model 230 membrane osmometer. Reproduced with permission from Wescan Instruments, Inc.

These instruments use the servo balancing principle and were sold in three models. The three models differ in the range of temperature over which they may be operated.

Two problem sources with the Mechrolab osmometer were bubble loss and lack of access to the solvent side of the membrane without membrane removal. The former problem resulted from the use of a bubble intentionally introduced into a capillary to trigger a photodector. Equilibrium was reached by balancing the upper edge of the bubble just at the level of the light source. The bubble sometimes dissolved in polar solvents or grew larger on standing. It could also be lost if solutions, particularly viscous ones, were introduced too rapidly into the osmometer. Lack of access to the solvent side of the membrane was a problem if polymers of low enough molecular weight to penetrate the membrane were introduced into the instrument. Once polymer had penetrated the membrane, there was no convenient mode of removal other than dismantling the instrument.

In general though, with proper care and experience, reliable osmotic pressure measurements can be carried out using any of the instruments described above. Experience has demonstrated that the prime factor in the performance of any membrane osmometer is the nature of the membrane. As mentioned earlier, diffusion of solute through the membrane must be avoided. Fortunately, the automatic osmometers allow measurement to be made so quickly ($\simeq 10$ min) that diffusion of solute through the membrane is normally not a serious concern. Commercial membranes, suitable for use in a variety of solvents, are available from several suppliers. Most of the membranes used are cellulose based. They are of uniform porosity and free of pinholes, thus avoiding many of the problems encountered with casting one's own membranes. Regenerated cellulose membranes for organic solvents and cellulose acetate membranes for aqueous work can be purchased from Schleicher & Schuell or Wescan.

Recently, Wescan introduced two new membranes (Types 477-007 and 477-008), which combine the desirable combination of fast equilibration and permeability. These membranes consist of two layers. One layer, which is very permeable and spongelike, provides strength and allows the thickness of the low-porosity membrane to be substantially reduced. By reducing membrane thickness, rapid equilibration can be achieved in combination with the small average pore size. The 477-007 membrane has a nominal molecular weight cutoff of 15,000 g/mol and equilibrates in 5 min or less. The 477-008 membrane has a nominal molecular weight cutoff of 5000 g/mol and equilibrates in about 20 min.

The major limitation of the new Wescan membranes is that they may be used only in toluene or other hydrocarbon solvents at or near room temperature, although Burge (11) described similar membranes for aqueous applications with a molecular weight cutoff between 3000 and 5000 g/mol.

A major unsolved problem in membrane osmometry is poor membrane performance at elevated temperatures, which is necessary for maintaining semicrystalline polyolefins in solution. Chiang (12) discussed problems associated with membrane osmometry in solvents such as decalin and trichlorobenzene at elevated temperatures. Chiang suggested using conventional regenerated cellulose membranes, even though they only last several days before deteriorating, because no reliable inorganic membranes are available. The same situation persists even today. Also, with the present interest in high-performance materials, membranes that are chemically resistant to a broad range of organic solvents would be most useful, since many of today's speciality polymers have only limited solubility.

In summary, the advantages and disadvantages of membrane osmometry are listed below. Advantages include:

- Membrane osmometry yields an absolute number-average molecular weight. Calibration with standards is not required. Standards are still useful, however, in examining membrane performance and assuring proper operator technique.
- Low-molecular-weight contaminents (traces of residual solvent, antioxidants, etc.) can be tolerated, since they will readily equilibrate on both sides of the membrane. With other colligative methods, small amounts of low-molecular-weight impurities will lead to the underestimation of $\bar{M}_n$.
- $\bar{M}_n$ is easily determined for block or graft copolymers by MO, since the measurement is independent of chemical heterogeneity. This situation is

in stark contrast to light scattering, where the measurement of $\bar{M}_w$ for block copolymers is a more complicated task (13). Normally, only apparent molecular weights are obtained from light scattering experiments on block copolymers (13) because of the uncertainity associated with determining the specific refractive index increment of the sample and its chemical heterogeneity.

- Membrane osmometry is applicable to polymers having a very broad range of molecular weights. The low-molecular-weight limit is governed by membrane permeability. The high-molecular-weight limit (about 5×10^5) is higher than that of any other colligative method (or end-group analysis).
- Information about polymer–solvent interactions may be derived from A_2 values obtained by MO, for example, θ conditions for a polymer–solvent system may be determined using MO.
- Commercially available osmometers allow the measurements of π for a single solution in about 5–20 min.

Disadvantages of membrane osmometry pertain, largely, to problems with the membranes.

- Erronous results (overestimation of $\bar{M}_n$) are obtained when solute diffuses through the membrane. Consequently, samples of high polydispersity are often not well suited to MO analysis.
- Characterization of polyolefins at elevated temperature (> 100 °C) is difficult because of problems with membrane stability (12, 14).
- Characterization of polyelectrolytes by MO requires special care because of ionic effects. For example, π/c values may be *much* higher for polyelectrolytes solutions in the absence of added salt, and π/c may increase with dilution (15). Most of the difficulties encountered with osmotic pressure measurements on polyelectrolytes are overcome by working in aqueous salt solutions (16) at low concentrations of polymer.

4. VAPOR PRESSURE OSMOMETRY

4.1. Basic Principles

The current method of choice for measuring $\bar{M}_n$ values of less than about 20,000 g/mol is vapor pressure osmometry (VPO). As we pointed out in Section 2, a direct measure of vapor pressure lowering is impractical because of the extreme sensitivity that is required. Why then has VPO enjoyed widespread use with good success while cryoscopy and ebulliometry have not? The answer is that VPO, as currently practiced, does not entail making a direct measure of vapor pressure lowering. Instead, the thermoelectric method originally introduced by Hill (17) in 1930 is employed.

In the thermoelectric method, two matched temperature-sensitive thermistors are placed in a chamber, under controlled temperature, with pure solvent present. The chamber atmosphere is saturated with solvent vapor. If drops of solvent are placed on both thermistors, the thermistors will be at the same temperature. If, however, a drop of solution is placed on one thermistor and solvent is placed on the other, a temperature imbalance is created. Condensation of solvent from the atmosphere onto the solution thermistor causes the solution to increase in temperature until its vapor pressure rises to match that of the pure solvent. The temperature imbalance that is created, although it is very small ($\simeq 10^{-5}$ K), can be related to solute concentration and, ultimately, molecular weight by

$$\Delta T = \frac{-RT^2 N_2}{\Delta H_v} \tag{23}$$

where N_2 is the mole fraction of the solute. The molecular weight of the solute can be calculated knowing its weight concentration. Equation 23 is obtained by combining Raoult's law with the Clapeyron equation. It should be noted that for Eq. 23 to be strictly valid, there should be no heat losses due to condensation or radiation, complete saturation of the vapor, and efficient condensation on the liquid surface. Such conditions are impossible to achieve in actual practice with current methods and instruments. It is therefore customary to calibrate the instrument with substances of known molecular weight rather than attempting a direct calculation of the instrument constant. Also, sensitivity is enhanced by measuring changes in resistance ΔR of the thermistors instead of directly measuring ΔT. For small values of ΔT, $\Delta T \propto \Delta R$.

The working equation is a virial expansion analogous to Eq. 18 for membrane osmometry

$$\frac{\Delta R}{c} = \left(\frac{\Delta R}{c}\right)_{c \to 0} (1 + \Gamma_2 c + \Gamma_3 c^2 + \cdots) \tag{24}$$

where,

$$\left(\frac{\Delta R}{c}\right)_{c \to 0} = \frac{K}{\bar{M}_n} \tag{25}$$

and K is a calibration constant. The same data treatment techniques that were used for MO (Section 3.1) are also applicable to VPO, that is, third and higher terms in Eq. 24 are normally ignored and square-root plots may be successfully employed (18).

4.2. Practical Aspects

Unlike membrane osmometry, VPO normally requires calibration with a standard material of known molecular weight. As pointed out in the prior section, this occurs because true equilibrium values of ΔT are not normally obtained in VPO experiments. It is therefore improper to consider molecular weights obtained from conventional VPO experiments to be absolute molecular weights in a precise sense of the term. Nevertheless, intercomparisons (18, 19) between MO and VPO (with narrow molecular weight distribution standards of appropriate $\bar{M}_n$ values for both techniques) indicate that reliable $\bar{M}_n$ values can be obtained using VPO (Table 2).

Vapor pressure osmometry is usually the method of choice for measuring $\bar{M}_n$ of polymers having $\bar{M}_n < 20{,}000$ g/mol. This section describes techniques and instrumentation used in VPO.

Experience has shown that careful selection of solvent and temperature is critical to the success of the VPO experiment. Nearly all common solvents, including water, can be used with VPO. The temperature should be chosen so that vapor pressure of the solvent will be greater than 50 torr, but not so high

Table 2. $\bar{M}_n$ of Polystyrene Standards by VPO[a]

Sample	Nominal MW	$\bar{M}_n$ (MO)	$\bar{M}_n$ (VPO)
Pressure No. 61222	2,000	2,050[b]	1,970
Pressure No. 80314	9,000		8,540
Pressure No. 41220	17,500	15,100	16,100
			16,400
			17,800[c]
Pressure No. 80317	35,000	30,000	29,100
NBS 1478	37,400	36,000[d]	36,900
Pressure No. 60917	50,000	51,100	49,500
			46,400[c]
Pressure No. 4b	110,000	111,000	108,000
			100,000[c]

[a] Wescan Model 232-A in toluene at 40 or 50 °C.
[b] Value determine using Mechrolab Model 302B VPO.
[c] Data from Ref. 18.
[d] Calculated from $\bar{M}_w = 37{,}400$ and $\bar{M}_w/\bar{M}_n = 1.04$ reported by the National Bureau of Standards (NBS).

as to lead to problems with evaporation from the VPO chamber. Solvent purity is critical, as volatile impurities lead to drifting and slow equilibration. Reagent–grade solvents should be rejected in favor of commercial "distilled-in-glass" solvents or, if distilled-in-glass solvents are unavailable, reagent grade solvents should be dried and distilled before use. A commercial water purification system provides high quality water for work with aqueous systems.

Under certain conditions, stabilizers should be added to the system to prevent oxidation. Examples include tetrahydrofuran (THF), where the addition of a small amount of 2,6-di-*tert*-butyl-4-methylphenol (BHT) is suggested (20), and high-temperature work with polyolefins, where tetrakis [methylene (3,5-di-*tert*-butyl-4-hydroxyhydrocinnamate)] methane (Irganox 1010) has been successfully used.

An additional factor is that greater sensitivity can be achieved by using solvents with low heats of vaporizations (21).

Many substances have been successfully utilized as VPO calibrants. Two requirements that any potential calibrant should meet are (11)

- Total vapor pressure, including any impurities, should be $< 0.1\%$ of the vapor pressure of solvent.
- High purity, preferably $> 99.9\%$, is required.

An additional, although obvious, qualification is that the molecular weight of the calibrant should be known with a high degree of certainty. Some of the materials, which have been succesfully employed as VPO standards are listed in Table 3. Narrow-molecular-weight distribution polystyrene and polyethylene standards are listed as calibrants even though they are not pure substances, that is, there is a finite, although small, polydispersity present. The reason for including polymers in the list of calibrants is that some VPO instruments require calibration with a material that has a molecular weight near that of the sample to be analyzed (22). In other words, for some VPO instruments (the Mechrolab design, in particular), the measured molecular weight will depend somewhat on the molecular weight of the calibrant (23–28).

Vapor pressure osmometers sold commercially can be divided into two basic types—those that employ thermistors and those that use conventional hanging drop thermistors. These two types of thermistor designs are shown in Figure 7. The vertical thermistors automatically control drop size to ensure reproducible response. The hanging drop design requires the operator to manually monitor (via a mirror) and control drop size.

The Mechrolab 300 Series of vapor pressure osmometers were the first commercial instruments to become available (early 1960s). They incorporated the classic hanging drop design. Although these instruments are no longer

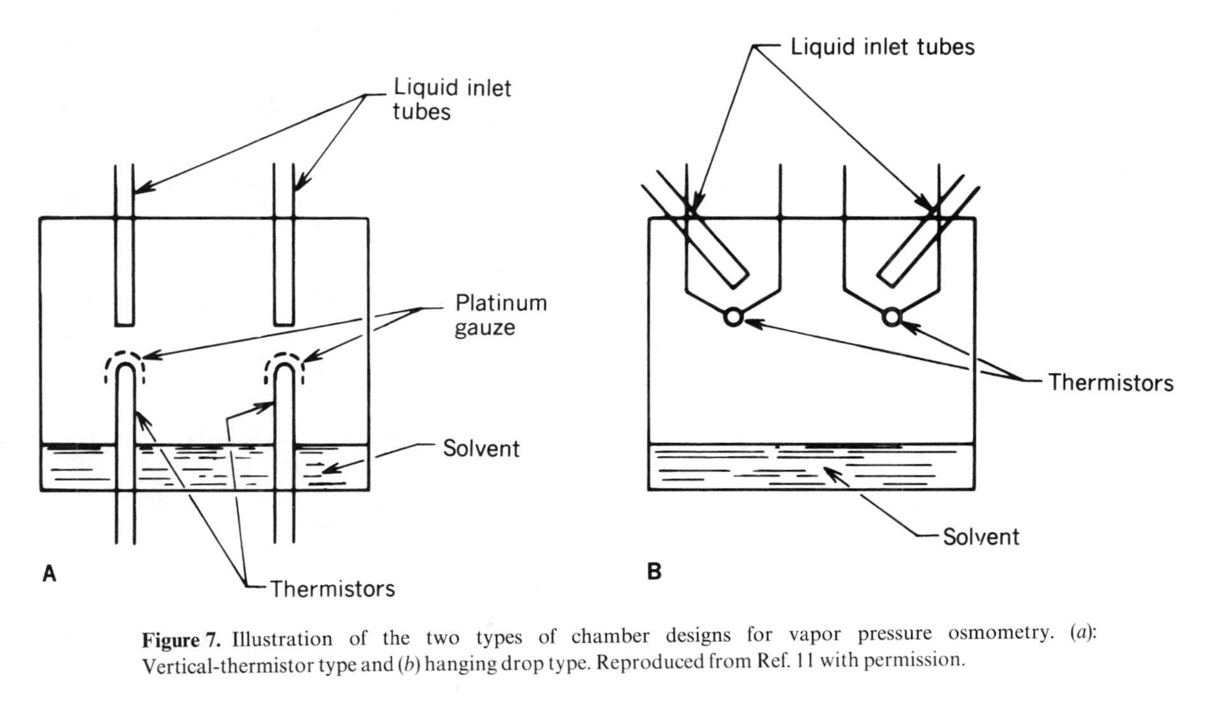

Figure 7. Illustration of the two types of chamber designs for vapor pressure osmometry. (*a*): Vertical-thermistor type and (*b*) hanging drop type. Reproduced from Ref. 11 with permission.

Table 3. Common Calibrants for Use in Vapor Pressure Osmometry

Calibrants	Solubility
Mannitol	Aqueous
Sucrose	Aqueous
Pentaacetylsucrose	Aqueous
Benzil	Organics
Biphenyl	Organics
Pentaerythrityl tetrastearate	Organics[a]
Sucrose octaacetate	Organics[a]
Polystyrene standards	Organics[a]
Polyethylene standards	[b]

[a]These standards are high enough in molecular weight that they are suitable for work at elevated (> 100 °C) temperature.
[b]Soluble in organic solvents only at temperatures in excess of 100 °C.

sold, many units remain in use today. Companies such as Knauer and UIC currently market vapor pressure osmometers (Fig. 8), which are similar in design to the Mechrolab instruments, although these new units, of course, incorporate solid-state electronics. Both the Knauer and Mechrolab units are capable of operating at temperatures up to 130 °C in a variety of solvents.

The measurement time for a single solution is only a few minutes. The upper limit for reasonably accurate (±10%) measurements of $\bar{M}_n$ is usually about 25,000 g/mol. Measured values of $\bar{M}_n$ will depend somewhat on the molecular weight of the calibrant (23–28). Therefore, calibration should be carried out

Figure 8. The UIC VPO unit. This instrument is available in both basic (operator driven) and computer-interfaced designs. The hanging drop thermistor style is employed. Photograph compliments of UIC Inc.

with a material having $\bar{M}_n$ close to that of the sample to be analyzed. With the hanging-drop thermistor design it is very important to make sure that the drops placed on the thermistors are always of uniform size. Otherwise, reproducible readings may not be obtained (29, 30).

In light of difficulties with the hanging drop design and especially, the molecular weight dependence of the calibration constant, Figini and co-workers (27, 31–33) published a series of papers that describe reliable calibration and data interpretation procedures for the Mechrolab instrument. In view of these problems, commercial instruments have been introduced, which utilize vertical thermistors having cups or pieces of platinum gauze that control drop size in a highly reproducible manner (Fig. 7a). Also, this design eliminates the need to visually monitor drop size.

The first commercial instrument that employed vertical thermistors was the Hitachi–Perkin–Elmer Model 115 Molecular Weight Apparatus. However, this instrument is no longer in production.

Currently, two vapor pressure osmometers that utilize platinum gauzes on vertical thermistors are sold by Wescan. These units appear to be improved updated versions of the Hitachi design. The Wescan Model 233 Molecular Weight Apparatus is reportedly able to measure molecular weights as high as 25,000 g/mol in certain solvents; whereas the Wescan Model 232-A Molecular Weight Apparatus claims a much higher upper limit of 100,000 g/mol under favorable conditions.

With the Wescan Model 232-A, saturation of the chamber around the thermistors with solvent vapor is aided by the use of two cylinders of filter paper. Solvent and solutions are injected onto the thermistors through the injection pipes (Fig. 7*a*). Platinum gauzes assure reproducible volume. Equilibration occurs over 5–10 min. Published (18, 37) and unpublished (19) work, using polystyrene and polyethylene standards, has confirmed the manufacturer's claim with regard to the high sensitivity of this device (see Table 2). In fact, molecular weights as high as 400,000 g/mol were reportedly measured for standard polystyrenes (34) using a prototype of similar design. It is our opinion, however, based on personal experience, that extreme care is required to accurately measure molecular weights greater than about 50,000 g/mol.

There is little or no variation of the calibration constant with molecular weight of the calibrant for the Hitachi (35) or the Wescan (11, 18) designs. This constitutes a major advantage of these instruments, and it also suggests that the difference in thermistor design (vertical versus hanging drop) is somehow accountable. Burge (18) considered this matter in detail and reaches the same conclusion. At any rate, it is clear that the Wescan Model 232-A unit offers advantages in both thermistor design and sensitivity relative to other instrument designs.

Mirabella (36) successfully applied the Wescan 232-A to the measurement of $\bar{M}_n$ of polyolefins at about 140 °C. Recent work (37) using NBS standard polyethylenes has shown that molecular weights as high as about 100,000 g/mol can be measured with reasonable accuracy (± 10%). Thus, the Wescan unit operates quite satisfactorily for work at elevated temperatures.

In summary, VPO complements MO especially in the low-molecular-weight region where diffusion of polymer through the membrane is a problem. In this regard, VPO utilizes an "ideal membrane," namely, the vapor phase transport of solvent in the presence of a nonvolatile polymer. The "ideal membrane" idea also manifests itself in the instrument by Pals and Staverman (see Ref. 8). In fact, solute volatility defines the lower limit of molecular weights measurable by VPO. As currently practiced, with instruments of the vertical thermistor design, VPO approaches MO in reliability and ease of operation.

5. MISCELLANEOUS TECHNIQUES

5.1. Cryoscopy

Cryoscopy has not been used much for measuring $\bar{M}_n$ of polymers. Reasons include the lack of commercially available equipment designed for use with polymers and practical difficulties associated with performing cryoscopic measurements on polymer solutions, for example, association at low temperatures. To the authors' knowledge, only one commercial cryoscopic unit is available (from Knauer). It claims to be applicable to materials with molecular weights up to 5000 g/mol. Measurements on a single solution reportedly take only 2 min.

Cryoscopy generally offers greater sensitivity than ebulliometry and should be capable, in theory, of accurately measuring molecular weights up to about 50,000 g/mol. Limitations of the method include the tendency for polymer to precipitate from solution before the freezing point is reached, complications brought about by supercooling, and association of polymers at low temperatures. The application of cryoscopy to polymers has been reviewed by Glover (38).

5.2. Ebulliometry

Similar to the case of cryoscopy, ebulliometry has not encountered widespread use with polymers. Experimental difficulties complicate ebulliometric measurements; difficulties include the tendency of polymer solutions to foam when boiled and errors due to superheating and pressure variations.

Commercial instruments designed for use with polymer solutions are rare. Gallenkamp markets an ebullioscope, and Davison (39) mentions both Japanese and Russian units, which were sold in at least limited quantities in the 1970s. Ebulliometry of polymer solutions has been extensively reviewed (39–42).

5.3. End-Group Analysis

Many polymerizations, especially condensation polymerizations, produce polymers with reactive or otherwise distinguishable end groups. If the polymers obtained are of low molecular weight ($< 10{,}000$ g/mol), as is often the case, the end groups are frequently present in sufficiently high concentrations so that chemical titration or other analytical procedures can be used to measure $\bar{M}_n$.

A critical requirement is that the nature and number of the end groups should be well established. Side reactions frequently lead to small amounts of unanticipated end groups. In the usual case one assumes exactly two end groups per molecule. Cyclization or branching can therefore lead to serious errors in the value of $\bar{M}_n$ that is obtained.

The various techniques of end-group analysis are too numerous and too varied in nature to discuss here. Many of the classical chemical techniques have been reviewed by Price (43).

6. CONCLUSIONS

In conclusion, the combination of MO and VPO allows reliable determination of polymer molecular weights up to about 5×10^5 g/mol. There is presently no reliable technique for measuring absolute number-average molecular weights larger than about 5×10^5 g/mol. Another present limitation is the need for membranes with improved chemical and thermal stability. Such membranes would better facilitate determination of $\bar{M}_n$ for many of today's "high-performance" polymers, which are frequently of limited solubility under conditions suitable for long-term use with current cellulosic membranes.

INSTRUMENT COMPANIES

- Gallenkamp, Bolton Road West, Loughborough LE11 OTR, England, UK
- Dr. Herbert Knauer, Wissenschaftliche Gerate KG, Heuchelheimer Str. 9, D-6380 Bad Homburg v.d.H., Germany

- UIC Incorporated, P.O. Box 863, Joliet, Illinois 60434
- Schleicher & Schuell, Inc., Keene, New Hampshire 03431

REFERENCES

1. T. G. Fox, Jr., P. J. Flory, and A. M. Bueche, "Treatment of Osmotic and Light Scattering Data for Dilute Polymer Solutions," *J. Am. Chem. Soc.*, **73**, 285 (1951).
2. W. R. Krigbaum and P. J. Flory, "Treatment of Osmotic Pressure Data," *J. Polym. Sci.*, **9**, 503 (1952).
3. W. R. Krigbaum and P. J. Flory, "Statistical Mechanics of Dilute Polymer Solutions. IV. Variation of the Osmotic Second Coefficient with Molecular Weight," *J. Am. Chem. Soc.*, **75**, 1775 (1953).
4. W. H. Stockmayer and E. F. Casassa, "The Third Virial Coefficient in Polymer Solutions," *J. Chem. Phys.*, **20**, 1560 (1952).
5. H. Yamakawa, *Modern Theory of Polymer Solutions*, Harper & Row, New York, 1971, Chapter 7.
6. G. S. Adair, "A Critical Study of the Direct Method of Measuring the Osmotic Pressure of Hemoglobin," *Proc. R. Soc. London Ser. A*, **108**, 627 (1925).
7. H. P. Frank and H. F. Mark, "Report on Molecular-Weight Measurements of Standard Polystyrene Samples," *J. Polym. Sci.*, **10**, 129 (1953); "Report on Molecular-Weight Measurements of Standard Polystyrene Samples II. International Union of Pure and Applied Chemistry," *J. Polym. Sci.*, **17**, 1 (1955).
8. G. V. Schultz, Diskussionstagung Bunsengesellschaff, Nov. 1955 [taken from P. T. Pals and A. J. Staverman, "An Osmometer with an Ideal Semipermeable Boundary," *J. Polym. Sci.*, **23**, 69 (1957)].
9. For a critical description of classical osmometers see H. T. Hookway, "Number-Average Molecular Weights by Osmometry," in P. W. Allen (Ed.), *Techniques of Polymer Characterization*, Butterworths, London, 1959.
10. H. G. Elias, "Dynamic Osmometry," in D. McIntyre (Ed.), *Characterization of Macromolecular Structure*, National Academy of Science, Washington, DC, 1968.
11. D. E. Burge, "Molecular Weight Measurements by Osmometry," *Am. Lab.*, (June), 41 (1977).
12. R. Chiang, "Characterization of High Polymers in Solutions—with Emphasis on Techniques at Elevated Temperatures," in B. Ke (Ed.), *Newer Methods of Polymer Characterization*, Interscience, New York, 1964, pp. 503–513.
13. H. Benoit and D. Froelich, "Application of Light Scattering to Copolymers," in M. B. Huglin (Ed.), *Light Scattering from Polymer Solutions*, Academic, New York, 1972.
14. H. Coll and F. H. Stross, "Determination of Molecular Weights by Equilibrium Osmotic—Pressure Measurements," in D. McIntyre (Ed.), *Characterization of Macromolecular Structure*, National Academy of Sciences, Washington, DC, 1968.

15. U. P. Strauss and R. M. Fuoss, "Polyelectrolytes. V. Osmotic Pressures of Poly(4-vinyl-*N*-*n*-butylpyridonium Bromide) in Ethanol at 25 °C," *J. Polym. Sci.*, **4**, 457 (1949).
16. H. Eisenberg, *Biological Macromolecules and Polyelectolyte Solutions*, Clarendon, Oxford, 1976.
17. A. V. Hill, "A Thermal Method of Measuring the Vapor Pressure of an Aqueous Solution," *Proc. R. Soc. London*, Ser. A, **127**, 9 (1930).
18. D. E. Burge, "Calibration of Vapor Pressure Osmometers for Molecular Weight Measurements," *J. Appl. Polym. Sci.*, **24**, 293 (1979).
19. J. W. Mays, unpublished results.
20. D. E. Burge, Debtek Inc., Cambell, Calfornia, personal communication.
21. K. Kamide, T. Terrakawa, and S. Matsuda, "Determination of Number-Average Molecular Weight of Atactic Polystyrene and Cellulose Diacetate by Vapour Pressure Osmometry," *Br. Polym. J.*, **15**, 91 (1983).
22. S. Kume and H. Kobayashi, "Molecular Weight Measurement by Vapour Pressure Osmometry," *Makromol. Chem.*, **79**, 1 (1964).
23. J. Brzezinski, H. Glowala, and A. Kornas-Calka, "Note on the Molecular Weight Dependence of the Calibration Constant in Vapour Osmometry," *Eur. Polym. J.*, **9**, 1251 (1973).
24. B. H. Bersted, "Molecular Weight Determination of High Polymers by Means of Vapor Pressure Osmometry and the Solute Dependence of the Constant of Calibration," *J. Appl. Polym. Sci.*, **17**, 1415 (1973).
25. B. H. Bersted, "Vapor Pressure Osmometry: A Model Accounting for the Solute Dependence of the Calibration Constant," *J. Appl. Polym. Sci.*, **18**, 2399 (1974).
26. C. E. Morris, "Molecular Weight Determinations by Vapor-Pressure Osmometry," *J. Polym. Sci., Symp. Ed.*, **55**, 11 (1976).
27. M. Marx-Figini and R. V. Figini, "On the Molecular Weight Determination by Vapour Pressure Osmometry, 1: Consideration of the Calibration Function," *Makromol. Chem.*, **181**, 2401 (1980).
28. I. Kucharikova, "Some Aspects of the Estimation of $\bar{M}_n$ by Vapor Pressure Osmometry," *Makromol. Chem.*, **181**, 2401 (1980).
29. W. Simon, J. T. Clerc, and R. E. Dohner, "Thermoelectric (Vapormetric) Determination of Molecular Weight on 0.001 Molar Solutions: A New Detector for Liquid Chromatography," *Microchem. J.*, **10**, 495 (1966).
30. A. C. Meeks and I. J. Goldfarb, "Time Dependence and Drop Size Effects in Determination of Number Average Molecular Weight by Vapor Pressure Osmometry," *Anal. Chem.*, **39**, 908 (1967).
31. R. V. Figini, "On the Molecular Weight Determination by Vapor Pressure Osmometry (VPO), 2: Molecular Weight Average Obtained by VPO," *Makromol. Chem.*, **181**, 2049 (1980).
32. R. V. Figini and M. Marx-Figini, "On the Molecular Weight Determination by Vapour Pressure Osmometry, 3: Relationship Between Diffusion Coefficient of the Solute and Non-Colligative Behaviour," *Makromol. Chem.*, **182**, 437 (1980).

33. M. Marx-Figini, M. Tagliabue, and R. V. Figini, "On the Molecular Weight Determination by Vapour Pressure Osmometry," 4: Experimental Proof of the Deviation of $\bar{M}_{vpo}$ from $\bar{M}_n$," *Makromol. Chem.* **184**, 319 (1983).
34. A. H. Wachter and W. Simon, "Molecular Weight Determination of Polystyrene Standards by Vapor Pressure Osmometry," *Anal. Chem.*, **41**, 90 (1969).
35. L. Mrkvicakova and S. Pokorny, "On the Reliability of Molecular Weight Determination by Vapour Pressure Osmometry," *J. Appl. Polym. Sci.*, **30**, 1211 (1985).
36. F. M. Mirabella, "Measurement of Number-Average Molecular Weights and Second Virial Coefficients of Polyolefins Using an Improved Vapor Pressure Osmometer," *J. Appl. Polym. Sci.*, **25**, 1775 (1980).
37. J. W. Mays and E. G. Gregory, "Molecular Weights of Polyethylene and Polypropylene by Vapor Pressure Osmometry at Elevated Temperatures," *J. Appl. Polym. Sci.*, **34**, 2619 (1987).
38. C. A. Glover, "Cryoscopy," in P. E. Slade, Jr. (Ed.), *Polymer Molecular Weights*, Dekker, New York, 1975, Chapter 4.
39. G. Davison, "Ebullioscopic Methods for Molecular Weights," in L. S. Bark and N. S. Allen (Eds.), *Analysis of Polymer Systems*, Applied Science, London, 1982, Chapter 7.
40. D. F. Rushman, "Other Methods for the Determination of Number-Average Molecular Weights," in P. W. Allen (Ed.), *Techniques of Polymer Characterization*, Butterworths, London, 1959, Chapter 4.
41. M. Ezrin, "Determination of Molecular Weight by Ebullimetry," in D. McIntyre (Ed.), *Characterization of Macromolecular-Structure*, National Academy of Sciences Washington, DC, 1968.
42. R. S. Lehrle, "Ebulliometry Applied to Polymer Solutions," in J. C. Robb and F. W. Peaker (Eds.), *Progress in High Polymers*, Vol. 1, Academic, New York, 1961.
43. G. F. Price, "Techniques of End-group Analysis," in P. W. Allen (Ed.), *Techniques of Polymer Characterization*, Butterworths, London, 1959, Chapter 7.

CHAPTER

7

POLYMER CHARACTERIZATION USING DILUTE SOLUTION VISCOMETRY

JIMMY W. MAYS

Department of Chemistry
University of Alabama at Birmingham
Birmingham, Alabama

and

NIKOS HADJICHRISTIDIS

Department of Chemistry
The University of Athens
Athens, Greece

Modern Methods of Polymer Characterization, Edited by Howard G. Barth and Jimmy W. Mays
ISBN 0-471-82814-9

1. INTRODUCTION

Early investigators sought to establish relationships between the viscosity of polymer solutions and the molecular weight of the polymer. In addition to providing a measure of polymer molecular weight, measurements of dilute solution viscosities are now recognized as being useful in the study of polymer branching, polymer dimensions and its temperature dependence, chain flexibility, and association of polymers in solution.

This chapter reviews the current status of viscometric characterization of dilute polymer solutions. Emphasis is placed on reviewing techniques that are useful to experimentalists entering or already active in this field. Instrumentation is also reviewed. Most of the attention will be focused on capillary viscometry, since this technique is by far the most commonly employed means of measuring dilute solution viscosities. A minimum amount of theory necessary for a detailed understanding of the subject is introduced. Readers interested in more detailed theoretical treatments are referred to books on the subject (1, 2). Finally, no attempt is made to discuss the viscosity behavior of concentrated polymer solutions where polymer–polymer interactions become important.

2. ACTUAL AND INTRINSIC VISCOSITIES

The viscosity of a liquid flowing through a capillary is given by Poiseuille's equation

$$\frac{dV}{dt} = \frac{\pi p r^4}{8\eta l_c} \tag{1}$$

where dV/dt is the volume in cubic centimeters (cm^3) of liquid flowing through the capillary per unit time in seconds (s), p is the hydrostatic pressure head in dynes per square centimeter (dyn/cm^2), r is the radius of the capillary in inches (in.), l_c is the length of the capillary in centimeters (cm), and η is the viscosity of the liquid in poise (p). Since a constant volume of liquid is normally used, Eq. 1 can be written as

$$\frac{V}{t} = \frac{\pi \bar{p} r^4}{8\eta l_c} \tag{2}$$

where t is the flow time through the capillary and $\bar{p}$ is the average hydrostatic pressure under which flow occurs. The average pressure $\bar{p}$ is defined as

$$\bar{p} = \rho g \bar{h} \tag{3}$$

where ρ is the density of the liquid in grams per cubic centimeter (g/cm^3), g is the acceleration of gravity, and $\bar{h}$ is the average value of the liquid head. Substituting for $\bar{p}$ in Eq. 2 and rearranging we obtain

$$t = \frac{8V\eta l_c}{\rho g\bar{h}r^4\pi} \tag{4}$$

which simplifies to

$$\eta = At\rho \tag{5}$$

where A is a constant for a particular viscometer and may be evaluated using liquids of known viscosity.

For Eq. 5 to be valid, all of the pressure difference applied across the capillary must be used in overcoming viscous forces, that is, the potential energy of the liquid column should not impart kinetic energy to the efflux. This is, of course, not true and we may rewrite Eq. 5 in the form

$$\frac{\eta}{\rho t} = A - \frac{B}{t^2} \tag{6}$$

where the second term in Eq. 6 corrects for contributions to kinetic energy. The factor B decreases as the length of the capillary increases, as the radius decreases, and as the liquid head $\bar{h}$ decreases. In practice, commercial viscometers are designed so that the kinetic energy constant B is as small as possible (3). Consequently, Eq. 5 can normally be applied. Both Eqs. 5 and 6 assume laminar flow with zero velocity at the walls of the capillary, as well as incompressibility of the liquid (4). For dilute polymer solutions, these are almost always valid assumptions.

Calibration is achieved by plotting η/t versus $1/t^2$ for at least two liquids of known viscosity; B is obtained from the slope and A is the Y intercept.

Thus far the measurement of η, the "actual viscosity" of a polymer solution, has been described. Normally, we are more concerned with measuring the increase in viscosity of the solvent brought about by the presence of polymer molecules. The relative viscosity η_r defined as

$$\eta_r = \frac{\eta}{\eta_0} \tag{7}$$

where η_0 is the viscosity of the pure solvent, is of more practical importance (see

Table 1. Terminology Used in Viscometry

Symbol	Common Name	IUPAC Name	Definition
η_r	Relative viscosity	Viscosity ratio	η/η_0
η_{sp}	Specific viscosity		$\dfrac{\eta - \eta_0}{\eta_0} = \eta_r - 1$
η_{red}	Reduced viscosity	Viscosity number	$\dfrac{\eta_{sp}}{c}$
η_{inh}	Inherent viscosity	Logarithmic viscosity number	$\dfrac{\ln \eta_r}{c}$
$[\eta]$	Intrinsic viscosity	Limiting viscosity number	$\left(\dfrac{\eta_{sp}}{c}\right)_{c \to 0}$ or $\left(\dfrac{\ln \eta_r}{c}\right)_{c \to 0}$

Table 1 for nomenclature). From Eq. 5 it is clear that

$$\eta_r = \frac{t\rho}{t_0 \rho_0} \tag{8}$$

where t_0 is the solvent flow time and ρ_0 is the solvent density. For very dilute solutions of the type discussed here, solvent and solution densities are very nearly equal and

$$\eta_r = \frac{t}{t_0} \tag{9}$$

For infinitely dilute solutions the value of η_r approaches unity, and it is useful to define the specific viscosity η_{sp} as

$$\eta_{sp} = \eta_r - 1 = \frac{t - t_0}{t_0} \tag{10}$$

The specific viscosity is a measure of the increase in viscosity brought about by the addition of polymer. The ratio η_{sp}/c is a measure of the specific capacity of the polymer to increase the relative viscosity (5) and, in the limit of infinite dilution, is known as the intrinsic viscosity (6), $[\eta]$

$$[\eta] = \left(\frac{\eta_{sp}}{c}\right)_{c \to 0} \tag{11}$$

The most general form for the concentration dependence of η_{sp}/c may be written as (7, 8)

$$\frac{\eta_{sp}}{c} = a_1 + a_2 c + a_3 c^2 + \cdots \tag{12}$$

where $a_1 = [\eta]$. In practice, Huggins equation (9)

$$\frac{\eta_{sp}}{c} = [\eta] + k' [\eta]^2 c \tag{13}$$

is more commonly employed. From Eq. 13 a plot of η_{sp}/c versus c will be linear with $[\eta]$ as the intercept. The term k' is derived from the slope of the plot and is commonly referred to as the "Huggins viscosity constant." Actually, k' is not a constant, the value of k' depending on a variety of conditions (2, 10–13), and use of the term "Huggins coefficient" is to be preferred. For polymers in good solvents, k' often has a value of about $\frac{1}{3}$, while larger values of 0.5–1 are typically found in poor solvents (2, 10, 12, 13). Thus the value of k' is a measure of solvent quality. A large number of theoretical treatments have been applied to the calculation of k' (14–22).

Kraemer (6) proposed an alternate expression for evaluating $[\eta]$, that is,

$$\frac{\ln \eta_r}{c} = [\eta] + k'' [\eta]^2 c \tag{14}$$

Here k'' is known as the Kraemer coefficient. For polymers in good solvents, k'' is negative in sign and smaller in magnitude than k' and $k' - k'' = 0.5$. Since Eq. 14 shows less concentration dependence than Eq. 13, many workers give preference to Eq. 14 on this basis. Nevertheless, Eq. 13 is probably more widely used. The preferred procedure is to plot both equations and to take the mutual intercept as $[\eta]$ as shown in Figure 1. Ibrahim and Elias (23) pointed out that the Kraemer equation is not well behaved if $k' > 0.5$. In many cases, Eqs. 13 and 14 do not give mutual intercepts and the use of Eq. 13 is the preferred method.

Numerous other extrapolation procedures have been suggested (24–28), though only one of these procedures deserves mention here. Martin (28)

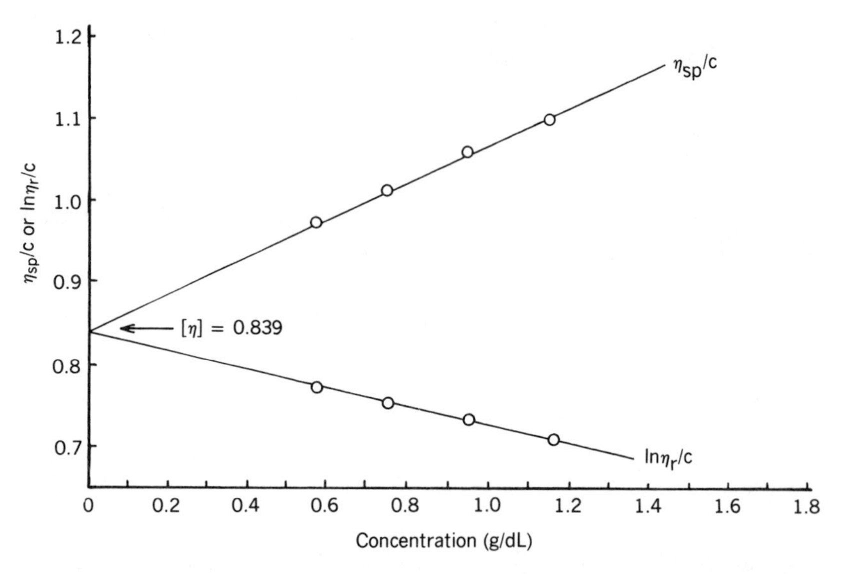

Figure 1. Typical plot of η_{sp}/c and $\ln \eta_r/c$ versus c for a mixed microstructure polyisoprene of $\bar{M}_w = 1.01 \times 10^5$ in THF at 30 °C. The mutual intercept is $[\eta]$.

suggested

$$\log\left(\frac{\eta_{sp}}{c}\right) = \log[\eta] + k'''[\eta]c \tag{15}$$

This equation is particularly useful when regular plots of η_{sp}/c versus c show considerable curvature and are, consequently, difficult to extrapolate to infinite dilution.

3. CALCULATION OF MOLECULAR WEIGHTS FROM INTRINSIC VISCOSITIES

In 1930, Staudinger and Heuer (29) were the first to attempt to relate the viscosity of polymer solutions to the molecular weight of the solute. In essence, Staudinger postulated:

$$[\eta] \propto M \tag{16}$$

where M is the polymer molecular weight. Experiments later demonstrated that Eq. 16 only held true for "stiff" polymers, such as certain cellulosics (6, 30). It was subsequently shown empirically in 1938 by Mark (31) that

$$[\eta] = KM^a \tag{17}$$

where K and a are constants for a given polymer, solvent, and temperature. Equation 17 is now commonly known as the Mark–Houwink (32) equation or even as the Mark–Houwink–Sakurada (M–H–S) (33) equation. The M–H–S relationship holds true for many polymer systems. Flory (34) demonstrated the validity of the M–H–S equation for polyisobutylene fractions in diisobutylene having a molecular weight range of nearly three orders of magnitude. Generally, for flexible polymer molecules, $0.5 \leqslant a \leqslant 0.8$. Higher values of a, equaling or exceeding unity, are observed for certain less flexible (wormlike or rodlike) macromolecules such as cellulosics and polyelectrolytes. Thus, the value of a provides insight into polymer conformation. Values of K and a have been determined for many polymer systems by plotting $\ln[\eta]$ versus $\ln M$ for homologous series of polymers (differing only in molecular weight).

$$\ln[\eta] = \ln K + \ln \bar{M}_w \cdot a \tag{18}$$

where a and K are determined from the slope and intercept, respectively. Here M must be determined by an independent method [preferably light scattering (see below)]. Compilations of K and a values are available (35–38). Once values of these constants are reliably established, M can be calculated based on the value of $[\eta]$ for the polymer.

Caution must be exercised in using M–H–S relationships for calculating M for polymers of low molecular weight ($M \leqslant$ about 20,000 g/mol). Not only will the values of $[\eta]$ be small for polymers of such limited size, but it is well established (39–42) that M–H–S relations are not valid at low molecular weights. This is a consequence of the non-Gaussian behavior of short-chain polymers, that is, coils approach wormlike behavior at low molecular weight (43, 44). Furthermore, recent work by Lodge and co-workers (45, 46) indicates that changes in solvent friction can occur with a change in solution concentration. This can have an impact on measured values of $[\eta]$, particularly at low molecular weights. Altares et al. (39) demonstrated that for anionically synthesized polystyrene in benzene, $[\eta] \propto \bar{M}_n^{0.5}$ when $\bar{M} < 10{,}000$ and $[\eta] \propto \bar{M}_w^{0.75}$ for $25{,}000 < M < 1.5 \times 10^6$. Kow et al. (42) found $[\eta] \propto M^{0.6}$ for a great number of samples of polystyrene and polyisoprene of low molecular weight in THF at 30 °C. In addition, McIntyre et al. (47) suggested that a break in the M–H–S relation may occur at extremely high molecular weights ($M > 10^7$) for polystyrene in benzene, although this finding has been questioned (44). To ensure the validity of the M–H–S equation, one should not extrapolate the relationship beyond the molecular weight range over which it was determined. It is also desirable to use K and a values that are established using narrow-molecular-weight distribution samples, since Eq. 17 is valid only for monodisperse polymers.

Flory (48) showed that the viscosity–average molecular weight $\bar{M}_v$ for polydisperse macromolecules is related to $[\eta]$ by

$$[\eta] = K\frac{\sum N_i M_i^{1+a}}{\sum N_i M_i} = K\bar{M}_v^a \tag{19}$$

where N_i is the number of molecules of molecular weight M_i, and K and a are defined as before. For most polymers the viscosity–average molecular weight lies between $\bar{M}_n$ and $\bar{M}_w$. In addition, it is clear from Eq. 19 that for polymers where $0.5 \leqslant a \leqslant 0.8$, $\bar{M}_v$ is much closer to $\bar{M}_w$ than to $\bar{M}_n$. It is preferable, for this reason, to use values of $\bar{M}_w$ when establishing M–H–S parameters using polydisperse standards. However, either $\bar{M}_n$ or $\bar{M}_w$ values can be employed, but must be specified.

4. DETERMINATION OF MOLECULAR DIMENSIONS FROM INTRINSIC VISCOSITIES

Frictional interactions between the polymer coil and the solvent are primarily responsible for the increase in viscosity of a polymer solution relative to the viscosity of the solvent. Consequently, hydrodynamic models are invoked to describe this behavior. Flory and Fox (49), building upon the earlier theories of Debye and Bueche (50), Kirkwood and Riseman (51), and Brinkman (52), showed that the intrinsic viscosity may be written as

$$[\eta] = K_\theta M^{0.5} \alpha^3 \tag{20}$$

where

$$K_\theta = \Phi_0 \left(\frac{\langle r^2 \rangle_0}{M}\right)^{1.5} \tag{21}$$

α is the expansion factor for the polymer coil (which accounts for excluded volume effects), Φ_0 is a universal constant (for linear flexible chain molecules under θ conditions) designated as the Flory constant, and $\langle r^2 \rangle_0$ is the unperturbed mean-square end-to-end distance of the polymer chain. Under theta (θ) conditions (53), $\alpha = 1$, that is, there are no excluded volume effects, and we may write

$$[\eta]_\theta = K_\theta M^{0.5} \tag{22}$$

Notice that Eq. 20 and 22 are written in the form of the M–H–S equation, but a is assigned a value of 0.5. For conditions of excluded volume $\alpha > 1$ and may be further defined as

$$\alpha^2 = \frac{\langle S^2 \rangle}{\langle S^2 \rangle_0} \tag{23}$$

where $\langle S^2 \rangle$ is the mean-square radius of gyration of the polymer chain under conditions of excluded volume and the subscript "0" denotes that θ state.

Values of the Flory constant Φ_0 have been derived theoretically. Theory (54, 55) gives $\Phi_0 = 2.66 \times 10^{21}$ when $\langle r^2 \rangle_0$ is in centimeters squared and $[\eta]$ is in deciliters per gram (dL/g). Of course, Φ_0 may also be evaluated experimentally since

$$\Phi_0 = \frac{[\eta]_\theta M}{\langle r^2 \rangle_0^{3/2}} \tag{24}$$

$\langle r^2 \rangle_0$ may be calculated from $\langle S^2 \rangle_0$, which is obtained the angular dissymetry of scattered light (56) under θ conditions for a sample of known M and compared with the value of $[\eta]$ under the same conditions of solvent and temperature. The widely accepted value of Φ_0 from careful experimental studies on various polymers (57–64) is $2.5(\pm 0.1) \times 10^{21}$. Interestingly, this is the same value predicted by Flory (65) 35 years ago. With the value and validity of the Flory constant established, $\langle r^2 \rangle_0$ may be calculated from a single measurement of $[\eta]$ at the θ temperature, assuming M is known. Under conditions of excluded volume the viscometric expansion factor α_η may be calculated as

$$\alpha_\eta = ([\eta]/[\eta]_\theta)^{1/3} \tag{25}$$

where $\alpha_\eta < \alpha\,(1)$.

The combined accuracy and simplicity of intrinsic viscosity measurements under θ conditions lead Flory to suggest (66) that $[\eta]_\theta$ measurements are to be much preferred over $\langle S^2 \rangle_0$ measurements (from light scattering dissymmetry) for the evaluation of polymer unperturbed dimensions. The most useful measure of polymer unperturbed dimensions is Flory's characteristic ratio C_∞ defined as (67)

$$C_\infty \equiv \lim_{N \to \infty} \left(\frac{\langle r^2 \rangle_0}{N l^2} \right) \tag{26}$$

where N is the number of main chain bonds and l is the average bond length.

For the hypothetical freely jointed chain, $C_\infty = 1$. For real flexible polymer chains, C_∞ varies between about 4 and 20, with increasing values of C_∞ reflecting greater deviation from freely jointed behavior (see Section 8.2 for a discussion of semiflexible chains). Consequently, this dimensionless parameter is useful for comparing chain flexibility.

Calculation of C_∞ for a polymer is conveniently carried out by evaluating K_θ from intrinsic viscosities under θ conditions (Eq. 22). The ratio $\langle r^2 \rangle_0/M$ is then calculated from Eq. 21. The term C_∞ is then obtained by rearranging Eq. 26 to

$$C_\infty = \frac{\langle r^2 \rangle_0/M}{(N'/M_0)l^2} \tag{27}$$

where N' is the number of main chain bonds per repeating unit and M_0 is the molecular weight of the repeating unit. A number of extrapolation procedures have been advanced (68–73) for estimating K_θ values for flexible polymers from intrinsic viscosity measurements in moderate to good solvents (above the θ temperature). In particular, the Burchard–Stockmayer–Fixman (B–S–F) method (68, 69) yielded reliable estimates of K_θ for a wide variety of chains, especially when the M–H–S exponent is less than 0.7 and molecular weights are between 10^4 and 10^6 g/mol (74).

The B–S–F relationship is given as

$$[\eta] = K_\theta M^{0.5} + 0.51\Phi_0 BM \tag{28}$$

where B is related to the binary cluster integral (71). Equation 28 suggests that a plot of $[\eta]/M^{0.5}$ versus $M^{0.5}$ will yield K_θ as the intercept with B obtained from the slope. Recent work (75) has also shown Eq. 28 to be useful for estimating K_θ from viscosity data obtained well below the θ temperature.

Extensive compilations of C_∞ values are available in the literature (35, 37, 76, 77). Values of C_∞ may also be calculated theoretically using rotational isomeric state models (66, 78, 79).

It is also possible to calculate a viscometric radius R_V by assuming a hard sphere model (80). The intrinsic viscosity is given as

$$[\eta] = \tfrac{5}{2} N_A V/M \tag{29}$$

where N_A is Avogadro's number and V is the volume of the sphere. The radius (in nm) is calculated as

$$R_V = 5.41 \times 10^{-9} ([\eta]M)^{1/3} \tag{30}$$

R_V is slightly larger than the hydrodynamic radius R_H, which is obtained from photon correlation spectroscopy or ultracentrifugation. Recent theory (81) suggests that R_V/R_H equals 1.12 for linear polymers in good solvents. A somewhat larger value of 1.23 is suggested theoretically (82, 83) for θ conditions. Experimental values found in good solvents (84, 85) are in agreement with theory, but the larger values predicted in poor solvents are not found experimentally (86, 87). Instead, $R_V/R_H = 1.0 - 1.1$ is observed over the entire range of solvent conditions (86). Equation 29 may also be applied to branched polymers, where it is found (87, 88) that R_V/R_H is near unity for a variety of star- and comb-branched materials.

5. TEMPERATURE DEPENDENCE OF VISCOSITIES OF DILUTE POLYMER SOLUTIONS

5.1. Effects on the Intrinsic Viscosity

Changes in temperature may alter the intrinsic viscosity by changing $\langle r^2 \rangle_0$ and/or α. Changes of $\langle r^2 \rangle_0$ with temperature are usually small in magnitude and may be positive or negative (89, 90). A negative temperature coefficient $(d \ln \langle r^2 \rangle_0 / dT)$ implies an increase in chain flexibility (decrease in C_∞) with increasing temperature. Conversely, a positive temperature coefficient is indicative of decreasing chain flexibility with increasing temperature.

The temperature dependence of a has been given by Flory (91) as

$$\alpha^5 - \alpha^3 = 2\, C_M \varphi_1 (1 - \theta/T) M^{1/2} \tag{31}$$

where $C_M = (3^3/2^{5/2}\, \pi^{3/2})\, (\bar{v}^2/V_1 N_A)\, (M/\langle r^2 \rangle_0)^{3/2}$

$\bar{v}$ = partial specific volume of polymer

V_1 = molar volume of the solvent

φ = parameter characterizing the entropy of dilution of polymer with solvent

The heat of dilution K_1 is defined as

$$K_1 = \varphi_1 \left(\frac{\theta}{T} \right) \tag{32}$$

In good solvents, $K_1 < 0$ and $\varphi_1 > 0$, so α decreases with an increase in temperature. In poor solvents, $K_1 > 0$ and $\varphi_1 > 0$, so α increases with temperature. In an athermal solvent, $K_1 = 0$ and α is independent of temperature. Normally, an increase or decrease in α will be accompanied by a corresponding increase or decrease in $[\eta]$. Equation 31 also indicates that α is most sensitive to temperature in the vicinity of the θ temperature and that α should change more rapidly with temperature for higher molecular weights. These predictions are borne out by experimental work (92–94).

For a quantitative evaluation of the effects of temperature on $\langle r^2 \rangle_0$, K_θ values should be firmly established by measurements in θ solvents having θ temperatures over the desired range of interest. Care should be taken to use chemically similar solvents in the evaluation of K_θ so that specific solvent effects (95–97) are avoided. So-called mixed θ solvents are best avoided because of preferential adsorption (98–101). The term α_η may then be measured from values of $[\eta]$ obtained at the desired temperatures (102).

The temperature coefficient may also be obtained by measuring $[\eta]$ in a good solvent as a function of temperature and correcting for polymer–solvent interactions (103). Corrections are greatly simplified if an athermal solvent is choosen. In addition, $d\ln C_\infty/dT$ may be derived from thermoelastic experiments on lightly cross-linked bulk polymers (90)

$$\frac{f_e}{f} = T\left(\frac{d\ln \langle r^2 \rangle_0}{dT}\right) \tag{33}$$

where f_e is the energetic component of the total elastic force f. In general, satisfactory agreement exists between values of $d\ln \langle r^2 \rangle_0/dT$ measured by these two techniques (89, 90, 103–105). While McCrum (106, 107) has criticized the thermoelasticity method, Smith and Mark (108) recently showed McCrum's analysis to be flawed.

Accurate information regarding $d\ln \langle r^2 \rangle_0/dT$ is extremely useful in developing a theoretical understanding of polymer chain conformation (109), for example, in the rotational isomeric state model. Knowledge and understanding of $d\ln \langle r^2 \rangle_0/dT$ may also be useful in explaining unusual rheological phenomena (110), such as increased activation energy of flow for certain branched polymers.

5.2. Temperature Dependence of Actual Viscosities

The temperature dependence of the actual viscosity of polymer solutions may be expressed by the well-known equation (111, 112)

$$\eta = A \cdot \exp(Q/RT) \tag{34}$$

where A is a preexponential term with an activation entropy significance, Q is the apparent activation energy of viscous flow, R is the gas constant, and T is the absolute temperature. Both Q and A are dependent on the solvent, molecular weight of the polymer, and concentration. For dilute solutions of polymer, it was empirically shown by Moore and co-worker (111, 112) that

$$Q = Q_0 + K_e \cdot M \cdot c \tag{35}$$

where Q_0 pertains to the pure solvent and K_e is a constant for a given polymer–solvent system. Values of K_e for stiff polymers are always positive. Smaller or even negative values of K_e are found for flexible chain polymers. Also, values of K_e are smaller by 1–2 orders of magnitude for flexible polymers than values of K_e found for stiff polymers. Thus Eq. 35 appears to provide a basis for distinguishing between flexible and stiff polymers and, possibly, a means of at least semiquantitatively evaluating chain flexibility (111, 113).

The preexponential term in Eq. 34 is also known to differ for flexible and for stiff macromolecules. With flexible chains it is empirically observed that (111)

$$A = A_0 + K_\beta \cdot M^\beta \cdot c \tag{36}$$

whereas with stiff chains

$$A = A_0 \exp(K_\gamma \cdot M^\gamma \cdot c) \tag{37}$$

where A_0 refers to pure solvent and K_β, K_γ, β, and γ are constants for a given polymer–solvent system. The significance of these constants is not clear, but they are thought (111, 113) to be related to the entropy of activation of viscous flow. Recently, the equations above have also been shown to hold for dilute solutions of star polystyrene in both good and poor solvents (114).

6. ANALYSIS OF LONG-CHAIN BRANCHING BY VISCOMETRIC STUDIES

The intrinsic viscosity of a polymer in solution is a parameter that is directly dependent on molecular dimensions. For linear polymers $\langle r^2 \rangle$, the mean-square end-to-end distance, is a useful measure of molecular dimensions. With branched polymers $\langle S^2 \rangle$, the mean-square radius of gyration, is used since the concept of end-to-end distance is not applicable to a polymer chain having more than two ends. For linear Gaussian chains,

$$\langle r^2 \rangle_0 = 6 \langle S^2 \rangle_0 \tag{38}$$

where subscripts indicate the unperturbed state; therefore, $\langle S^2 \rangle_0$ may be used to compare dimensions of linear- and branched-polymer chains.

Traditionally, the parameter g,

$$g = \left(\frac{\langle S^2 \rangle_{0,\mathrm{br}}}{\langle S^2 \rangle_{0,\mathrm{lin}}} \right)_M \tag{39}$$

has been used to compare unperturbed dimensions of linear and branched polymers of the same molecular weight. Since polymers with long-chain branching have smaller dimensions than their linear counterparts at constant molecular weight, the value of g will always be less than unity. Often, Eq. 39 has been used to compare dimensions of linear and branched polymers in good solvents. As is known, polymers have larger dimensions in good solvents, and the expansion factor α must be the same for both species for such a comparison to be valid. It is now well established (115, 116) that $\alpha_{\mathrm{br}} < \alpha_{\mathrm{lin}}$, and, consequently, values of g will be smaller in good solvents than under θ conditions. This finding, which contradicts prior theoretical predictions and experimental results (117–120), is rationalized as follows (115). Theoretical (121, 122) and experimental results (123–126) suggest that star molecules have extended configurations compared to corresponding linear materials as a consequence of abnormally high chain segment density near the star's center. When both linear and star molecules are exposed to good solvents they expand, but the star molecule, which was already partially extended under θ conditions, will expand less.

Another consequence of the phenomenon described here is the so-called θ-temperature depression (116, 124), that is, star polymers exhibit lower θ temperatures (defined as the temperature at which $A_2 = 0$) than do linear polymers.

In practice, measurements of $\langle S^2 \rangle$ are difficult to perform with high accuracy, particularly at the θ state where dimensions are smaller. Therefore, g':

$$g' = \left(\frac{[\eta]_{\theta,\mathrm{br}}}{[\eta]_{\theta,\mathrm{lin}}} \right)_M \tag{40}$$

where $[\eta]_\theta$ is the intrinsic viscosity under θ conditions for the linear and branched polymers at constant M and is a more frequently employed parameter than g. It is again common practice to compare values of $[\eta]$ for branched and linear polymers in good solvents, though such a comparison is only justified when $\alpha_{\mathrm{lin}} = \alpha_{\mathrm{br}}$. In practice, values of g and g' obtained for perturbed macromolecules should, to a first approximation, provide reasonable estimates of values obtained for unperturbed molecules since the ratio $\alpha_{\mathrm{lin}}/\alpha_{\mathrm{br}} \simeq 1$.

The value of g' is unity for a linear polymer and decreases with increasing numbers of branches. One way to visualize the effect of increased branching is to consider a star-shaped macromolecule of rather high functionality. The addition of

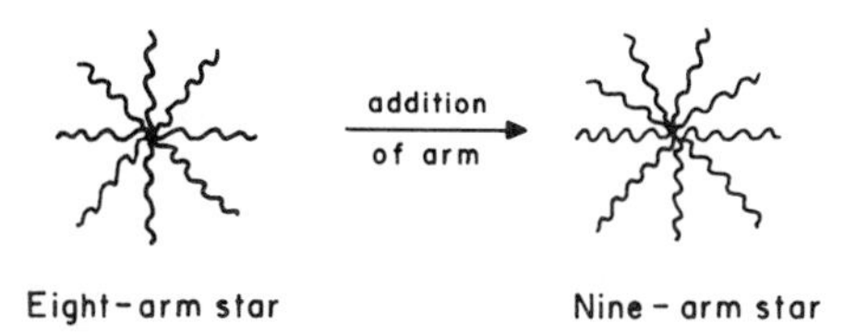

a ninth arm to an eight-arm star polymer increases the molecular weight without appreciably increasing the size. Therefore, the viscosity of the star remains roughly constant, while the viscosity of a linear molecule of corresponding molecular weight increases noticeably because of the 12% increase in M.

Zimm and Kilb (127) and Stockmayer and Fixman (128) theoretically predicted values of g' for star polymers of various functionality (number of arms) f. The model of Stockmayer and Fixman gives

$$g' = f^{3/2}[2 - f + 2^{1/2}(f - 1)]^{-3} \tag{41}$$

while the method of Zimm and Kilb gives

$$g' = (2/f)^{3/2}[0.390(f - 1) + 0.196]/0.586 \tag{42}$$

These equations are for use only with star polymers. Experimentally (129–132), values of g' are found to be smaller than those predicted by Eq. 42, but are larger than values given by Eq. 41.

The best fit to experimental data is obtained by use of the parameter G' (115)

$$G' \equiv \frac{[\eta]_{\text{star}}}{[\eta]_{\text{lin}}} \tag{43}$$

which compares polymers of equal arm molecular weight. The linear polymer is treated as a two-arm star and at θ conditions

$$G' = g'(f/2)^{1/2} \tag{44}$$

Theoretical values of G' may be calculated from Stockmayer–Fixman (128) and Zimm–Kilb (127) theories as a function of f. Better agreement between theory and experiment is obtained using the method of Bauer (133), which is based on the work of Casassa and co worker (134–136). This method gives

$$G' = [\langle X \rangle_{\mathrm{br}} / \langle X \rangle_{\mathrm{lin}}]^3 f \tag{45}$$

where $\langle X \rangle$ is the average span given by

$$\langle X \rangle = 4(nfb^2/6)^{0.5}\psi \tag{46}$$

where a star is composed of f arms having n links of length b. Values of the function ψ have been tabulated (135). Equation 45 gives excellent agreement with experimental values of G' as a function of f (137).

When dealing with more complicated forms of branching, as is to be found in many commercial polymers, the relation (127)

$$g' \simeq g^{0.5} \tag{47}$$

may prove useful, since expressions for g have been derived (138–140) for various types of branching. Scholte (141) summarized these expressions as follows:

1. Star with random branch length.

$$g = \frac{6f}{(f+1)(f+2)} \tag{48}$$

2. Comb with M f-functional branch units and p subschains, $p = (f-1)M + 1$, with random lengths.

$$g = \frac{6p^2 + (f-1)^2 M(M^2 - 1)}{p(p+1)(p+2)} \tag{49}$$

3. Random branching with an average of M-branch units (138).

For trifunctional branch units

$$g = [(1 + M/7)^{1/2} + 4M/9\pi]^{-1/2} \tag{50}$$

For tetrafunctional branch units

$$g = [(1 + M/6)^{1/2} + 4M/3\pi]^{-1/2} \tag{51}$$

Zimm and Stockmayer (138) also studied the effects of polydispersity on the value of g. In addition, Mazur and McCrackin (121) recently developed an expression that takes into account excluded volume effects in star polymers. They arrived at $g = (3f - 2)/f^2 + c$ where $c = 0.05$ for $f \leqslant 9$.

A comment about the Mark–Houwink–Sakurada exponent for branched polymers is perhaps in order at this point. For randomly branched polymers values of $a < 0.5$ are frequently reported. This occurs because the extent of branching generally increases with increasing molecular weight. It is sometimes erroneously assumed that all branched polymers will exhibit diminished values of a. For homologous series of model stars where M is varied and the degree of branching is kept constant, the values of a obtained under θ conditions equal 0.50 within experimental error (11, 88).

7. EXPERIMENTAL TECHNIQUES IN SOLUTION VISCOMETRY

Accurate results from solution viscosity measurements are possible only when solution concentrations are known accurately and when the solutions are free of dust or other extraneous matter. It is also important that the solutions be neither too dilute nor too concentrated. If solutions are too dilute, viscosities of solutions differ little from that of the solvent and accurate measurements are not possible. If solutions are too concentrated, many of the relationships used in treating viscosity data are no longer valid because of intermolecular interactions between polymer molecules. Consequently, the seemingly trivial task of preparing solutions requires considerable thought and care if reliable results are to be obtained.

Solvent purity is one of the first issues to be addressed, though this matter will not be treated in detail here. Instead, the reader is advised to use at least reagent grade solvents for even routine purposes. For more careful, research-oriented purposes, the use of commercial "distilled-in-glass" solvents is probably adequate. For the most demanding applications, published purification techniques (142) may be used to further purify reagent quality solvents just before use. Solvents of very high volatility at the measurement temperature should be avoided.

Polymer solutions may be prepared on either a weight-to-volume or a weight-to-weight basis. The former approach requires the use of calibrated volumetric glassware, while the latter does not. For this reason, we prefer to prepare solutions on a weight-to-weight basis using disposable screw-cap vials or bottles. The container weight and polymer weight must be known very accurately. An appropriate volume of solvent is added to give roughly the desired concentration. After allowing sufficient time for dissolution to occur, the concentration is determined just before use by weighing. In this manner, solvent losses due to evaporation are minimized. The solvent density at the temperature of measurement is used to calculate polymer concentration at the temperature of measurement. Solvent and solution densities are normally equal to a very good approximation.

An ideal initial concentration for solution viscosity measurements is one that will give $\eta_r \simeq 1.4$. Some workers suggest somewhat higher concentrations where $\eta_r \simeq 1.8$. This initial concentration can then be diluted by addition of solvent to prepare additional, more dilute concentrations in a convenient manner. Dilution should ideally be carried out so $\eta_r \simeq 1.1$ for the most dilute solution. By working in a concentration range where $1.1 \leqslant \eta_r \leqslant 1.4$, we have found that Huggins' and Kraemer's equations are applicable and that the data can be treated in a linear fashion, that is, there is no evidence of curvature at the higher concentrations. Also, flow times are significantly different for solutions, relative to solvent, so that timing errors are minimized. As mentioned in Section 2, the Kraemer equation may not be well behaved if k' from the Huggins equation is greater than 0.5 in value (23).

Table 2 gives approximate concentrations for obtaining $\eta_r \simeq 1.4$ for polymers of various degrees of polymerization, DP, in "good" solvents. For "poor" or θ solvents, somewhat higher concentrations should be used. The reader should be aware that higher or lower concentrations may be used with complete success in many cases: The concentrations given in Table 2 are intended only as guidelines. An absolute upper limit for concentration may be given as

$$[\eta] \cdot c \leqslant 1 \tag{52}$$

where $[\eta]$ is in deciliters per gram (dL/g) and c is in grams per deciliter (g/dL). Polymer–polymer interactions become significant at or near this point.

Table 2. Approximate Concentrations for Obtaining $\eta_r \simeq 1.4$ in a Good Solvent (or 1.3 in a θ Solvent) as a Function of Weight-Average Degree of Polymerization (DP_w)

DP_w	Concentration (g/dL)
200	2.5–3.0
500	1.5
1,000	1.0
2,000	0.75
4,000	0.40
6,000	0.25
10,000	0.15
15,000	0.10

Filtration is conveniently accomplished using filter holders, which attach directly to syringes. Membrane filters are available with a variety of pore sizes. Filters of 1-μm pore size are quite adequate for solution viscosity work; there is normally no need to use smaller pore sizes. These syringe–filter combinations may be employed to introduce solvent or solution directly into a clean viscometer. Viscometers should be cleaned by soaking in hot (fresh) chromic acid mixtures for at least overnight.

We prefer to work with Cannon–Ubbelohde dilution viscometers. These and other types of capillary viscometers are shown in Figure 2. The advantage of the Cannon–Ubbelohde design (Fig. 2*a*) is that dilutions of a stock solution are made *in situ*. Therefore, these viscometers are normally faster to use and require less polymer solution than other types of capillary viscometers. By quickly weighing the filtration syringe before and after injection of liquid into the viscometer, dilution and filtration are carried out simultaneously. The normal Cannon–Ubbelohde requires 8 mL of stock solution as a minimum (semimicro designs are available, which require a 1 mL minimum). This original solution may then be diluted by adding aliquots of 2, 3, and 5 mL from the same syringe (with intermittent weighings). Sufficient time should be allowed after each dilution for thermal equilibrium to be attained. It is also desirable to rinse the capillary walls after each dilution to remove any of the polymer solution (of higher concentration) that might have been retained there.

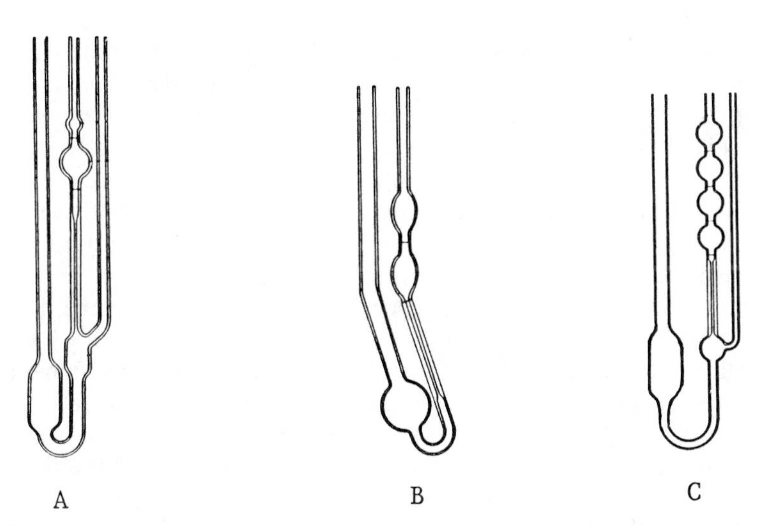

Figure 2. Common types of capillary viscometers: A = Ubbelohde dilution viscometer; B = Cannon–Fenske viscometer; and C = Variable rate-of-shear Ubbelohde viscometer.

The Cannon–Fenske viscometer (Fig. 2*b*) does not allow for dilution inside the viscometer since constant volume must be maintained for each solution, that is, the driving force depends on the height difference between the two tubes. Thus, this viscometer must be emptied and refilled repeatedly during the course of the experiment.

The viscometer shown in Figure 2*c* combines the convenience of the Ubbelohde design with the capacity for extrapolating to zero shear rate. The shear rate is higher in the upper bulb because the shear stress is proportional to the distance between the upper level and the bottom of the capillary (3). The shear rates in the successive bulbs are consequently, lower.

Temperature control is extremely important in solution viscometry. As previously described, the liquid viscosity as a function of temperature is

$$\eta = A \cdot \exp(Q/RT) \tag{34}$$

The values of both A and Q depend on the nature of the solvent, and therefore the dependence of η on T must be known to accurately predict effects of temperature fluctuations on η. As a general guideline, temperature control should be as rigorous as possible. For precise research work, control to $\pm 0.01\,°C$ is desirable. The absolute temperature at which the bath is maintained is generally of less importance than maintenance of constant temperature. When measurements are being carried out under θ or near- θ conditions, absolute temperature must be known. Recall from Section 5.1 that the expansion factor α is most strongly dependent on temperature at or near θ.

The usual experimental arrangement for solution viscometry consists of a battery jar containing water, and equipped with a stirrer and a thermostated heater. For measurements at elevated temperatures, where water can no longer be used, silicone oil is recommended. For work at low temperatures, Cellosolve has been successfully employed.

Care should be taken in mounting the viscometer in a vertical position, since flow times will vary significantly if the capillary is mounted even a few degrees from vertical. Methods for assuring vertical clamping, yet allowing easy removal of the viscometer, have been described (143–146).

The accuracy of the viscosity experiment is dependent on the accurate measurement of flow times. Manual timing will give precision of ± 0.1–0.2 s. Since flow times are generally greater than 100 s, this level of precision is quite adequate and should lead to negligible errors. Automatic timers are available, for example, the Wescan Model 221 Automatic Viscosity Timer, which use two phototransistors to detect passage of the liquid miniscus. Flow times are recorded on a digital readout to ± 0.01 s. This unit is fairly expensive and does not eliminate the need to manually perform dilutions and transfer solutions into the measuring zone. Furthermore, cleaning of the viscometer with chromic acid solution is made more difficult, as care is necessary to avoid getting acid on the timer parts. While the manufacturer claims an improvement in precision of flow time measurements, we have found that careful manual timing is nearly as precise. Nevertheless, the Wescan Model 221 does free the operator from the tiresome task of measuring flow times.

8. SPECIAL TOPICS

8.1. Viscosity of Polyelectrolytes

The viscometric behavior of dilute aqueous solutions of polyelectrolytes is quite different from the behavior of solutions of typical, that is, nonelectrolytic, flexible polymers. This "unusual" behavior is documented in several reviews and books (147–151). The intention here is merely to mention that differences exist between polyelectrolyte and nonelectrolyte viscometric behavior, to briefly describe some of these differences, and to steer the interested reader towards pertinent, more specific information. The presence of ionic groups on a polymer chain normally leads to expansion of the chain due to repulsion among like charges. Also, a high local charge density is created that will repel other similarly charged polymer chains, in addition to strongly affecting the properties of simple ions present in the solution. Expansion of polymer chains naturally leads to an increase in the viscosity of the solution. The addition of simple salts to aqueous solutions of polyelectrolytes shields the charges of the polyions and enables meaningful experimental measurements to be carried out. An additional complication concerning intrinsic viscosities of polyelectrolyte solutions is that the "electroviscous effect" (152, 153), which results from the electrical work required to move a charged particle through a fluid, may need to be taken into account.

With ordinary uncharged polymers, $[\eta]$ may be evaluated by extrapolating η_{sp}/c to zero concentration because the molecular conformation is, to a very good approximation, independent of concentration as long as the solution is sufficiently dilute. With polyelectrolytes, increasing dilution can lead to molecular expansion because of reduced shielding brought on by counterions moving further away from the ionic sites on the polymer chain. The result is that plots of η_{sp}/c versus c tend to be highly curved. Fuoss and Strauss (154, 155) showed that the relation

$$\frac{\eta_{sp}}{c} = \frac{A}{1 + B\sqrt{c}} \tag{53}$$

should be linear. Here $A = [\eta]$ as η_{sp}/c is extrapolated to $\sqrt{c} \to 0$.

As a final caution, polyelectrolyte solutions may show a very strong dependence on the rate of shear (156). Recently, a viscometer has been described (157), which allows accurate viscosity measurements at low shear rates on low ionic strength, dilute polyelectrolyte solutions, down to polymer concentrations below 1 ppm.

8.2. Semiflexible Chains

The bulk of this review has been concerned with the viscometric behavior of flexible polymer chains having random coil conformations. However, macromolecules with more extended conformations are being increasingly studied, especially due to their tendency to form liquid crystalline mesophases. Typical stiff-chain polymers include certain cellulosics, polyisocyanates, aromatic polyamides and polyesters, and polypeptides in helical conformations.

A useful and convenient model for studying the conformations of such chains is the wormlike chain model (158). Wormlike chains are characterized by a contour length L, a diameter d, and persistence length q. The Kuhn statistical segment length l' is equal to $2\,q$. Benoit and Doty (159) showed that q can be derived from the radius of gyration by the equation

$$\langle S^2 \rangle = q^2[\tfrac{1}{3}(L/q) - 1 + (2\,q/L) - (2\,q^2/L^2)[1 - \exp(-L/q)]] \qquad (54)$$

Alternatively, a number of approaches (160–167) have been developed for deriving parameters of wormlike chains from hydrodynamic quantities such as the intrinsic viscosity.

Perhaps the most widely accepted of these latter approaches is the theory of Yamakawa and Fujii (166). Bohdanecky (168) recently developed a simple graphical procedure based on the Yamakawa–Fujii equation

$$[\eta]_\theta = \Phi(l')^{3/2} L^{0.5}/M_L \qquad (55)$$

where Φ is the hydrodynamic parameter and $M_L = M/L$ where M is molecular weight. The value of Φ depends on L_r and d_r where $L_r = L/l'$ and $d_r = d/l'$. The term Φ takes on the limiting value (Φ_0) for flexible unperturbed coils (see Section 4) in the limit of $L_r \to \infty$. Bohdanecky (168) demonstrated that Φ can be given a simple form

$$\Phi = \Phi_0[B_0 + A_0(l'/L)^{0.5}]^{-3} \qquad (56)$$

where B_0 is essentially constant, and A_0 is a function of d_r.

The final form of the Bohdanecky equation is

$$(M^2/[\eta]_\theta)^{1/3} = A_\eta + B_\eta M^{0.5} \qquad (57)$$

where

$$A_\eta = A_0 M_L \Phi_0^{-1/3} \tag{58}$$

and

$$B_\eta = B_0 \Phi_0^{-1/3} (\langle r^2 \rangle_0 / M)_\infty^{-0.5} \tag{59}$$

where the subscript ∞ denotes that the $\langle r^2 \rangle_0/M$ value obtained from B_η is the random coil value. A plot of $(M^2/[\eta]_\theta)^{1/3}$ versus $M^{0.5}$ is expected to be linear, and $(\langle r^2 \rangle_0/M)_\infty$ is derived from the slope. The Kuhn length is calculated as $l' = \langle r^2 \rangle_0 / N l_u$ where $N = M/M_u$, M_u is the molecular weight of the repeating unit, and l_u is the length of the repeating unit projected on the molecular axis.

In a rigorous sense, unperturbed values of the intrinsic viscosity are required by Eq. 57. From a practical perspective, θ conditions are rarely reported for stiff-chain polymers due to their tendency to crystallize and precipitate from solution under these conditions. Bohdanecky (168) considered the effects of excluded volume when applying Eq. 57. Fortunately, chain expansion due to favorable thermodynamic interactions with solvent are normally quite small, that is, $\alpha \simeq 1$, for stiff-chain polymers. Norisuye and Fujita (169) showed that the onset of the excluded volume effect occurs at a critical chain length of about 50 l', a value that far exceeds the length of most such polymers. Thus, $[\eta]$ values obtained under good solvent conditions can usually be used in the Bohdanecky approach without correcting for chain expansion.

Chain diameters can also be evaluated from the Bohdanecky plot if the partial specific volume $\bar{v}$ of the polymer is known. Bohdanecky (168) showed that

$$(d^2/A_0) = (4\Phi_0 / 1.215\pi N_A)(\bar{v}/A_\eta) B_\eta^4 \tag{60}$$

where values of (d_r^2/A_0) have been empirically established (168) as a function of d_r

$$\log(d_r^2/A_0) = 0.173 + 2.158 \log d_r \tag{61}$$

Equation 59 is valid for $d_r \leqslant 0.1$, and d can be calculated directly from Eq. 59 since $d = d_r l'$.

Recently, Reddy and Bohdanecky (170) derived a method for estimating the temperature coefficient of chain dimensions $d \ln \langle r^2 \rangle_0 / dT$ of stiff-chain polymers. Their approach again utilizes the Yamakawa–Fujii wormlike cylinder model (166). Reddy and Bohdanecky (170) showed that $d \ln [\eta]/dT$ is independent of M,

$$d \ln \langle r^2 \rangle_0 / dT = \tfrac{2}{3} d \ln[\eta]/dT \tag{62}$$

Thus, viscosity data can be used to obtain chain dimensions and their variation with temperature, as well as chain diameters for stiff-chain macromolecules.

8.3. Capillary Viscometry at Elevated Temperatures

Certain polymers, for example, stereoregular polyolefins, dissolve only at elevated temperatures near or in excess of their crystalline melting points. In many cases, it is necessary to go as high as 140 °C to ensure complete dissolution of the polymer. The use of elevated temperatures introduces certain complications into the measurement of solution viscosities. Notable complications include

1. Preparing and handling solutions at elevated temperatures without precipitation of the polymer.
2. Filtration.
3. Oxidative degradation.

In view of such difficulties, Chiang (171) described methods for characterizing polymers at elevated temperatures. To obtain reliable viscometric results at temperatures exceeding 100 °C, oxygen should be rigorously excluded during dissolution, handling, and filtration of polymer solutions. Viscometers designed specifically for this purpose have been described in the literature (172, 173). Figure 3 shows the viscometer used by Nakajima et al. (173). Nitrogen

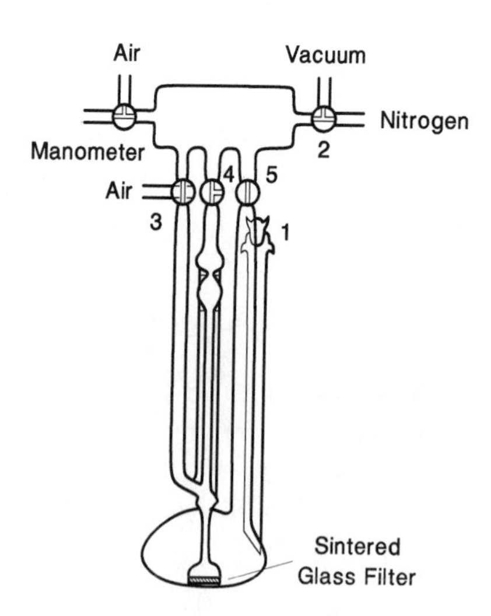

Figure 3. A viscometer designed for use at elevated temperatures. This design allows filtration and measurements to be carried out under an inert atmosphere. Polymer and solvent are introduced through tube (1). The viscometer is then evacuated and filled with nitrogen several times through stopcock (2), and the contents are then heated to the desired temperature for dissolution. Solutions are then pushed into the delivery bulb through the sintered glass filter via nitrogen pressure through stopcock (5) with stopcock (3) closed and with intermittent venting through stopcock (4). Flow times are measured with stopcocks (3) and (4) open to nitrogen atmosphere. Reproduced with permission from the author (173). Copyright © 1966 John Wiley and Sons, Inc.

gas provides an inert atmosphere and is used to force liquid up through the filter and into the capillary. It is also recommended that a small amount of antioxidant be added as a further precautionary measure. Desreux (174) devised a closed viscometer that avoids evaporation of solvent at elevated temperatures. Wagner (175) described the assembly and operation of a low-shear viscometer at elevated temperatures under a blanket of argon for use with ultrahigh-molecular-weight polyethylene.

In industrial laboratories, it is often necessary to obtain viscosity results for many polymer samples, so that the procedures outlined above are no longer feasible due to economics and time constraints. A reasonable procedure in this case is to use additional antioxidant and to allow just enough time for the polymer to fully dissolve before proceeding with the experiment. Useful guidelines for dissolving polyethylene have been given by Wagner (175). Grinshpun and Rudin (176) described the preparation of stabilized, aggregate-free solutions of polypropylene.

8.4. Couette Viscometry

In practice, a major limitation of capillary viscometry is that rather high shear rates, on the order of 100–1000 s^{-1}, are imposed on the polymer solutions as they flow through the capillary. For typical polymers with $M \leqslant 10^6$ g/mol, these shear rates are not usually sufficient to cause degradation of polymer

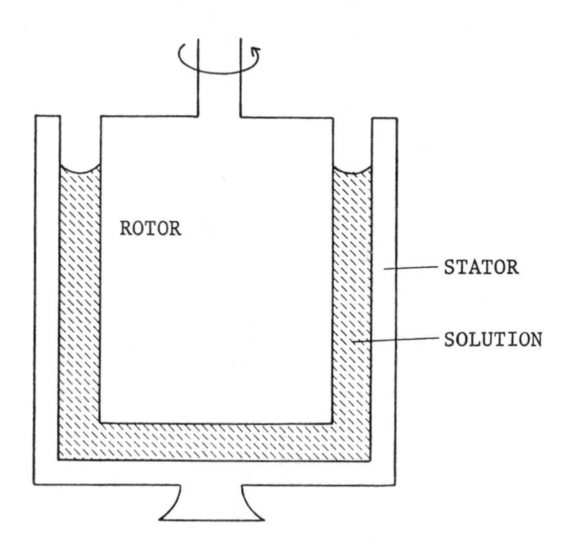

Figure 4. Couette rotational viscometer with a stationary outer cylinder (stator) and rotating inner cyclinder (rotor). Alignment along the mechanical axis is critical.

chains. For polymers with $M > 10^7$ g/mol, for example, certain types of DNA, dextran, and some high-molecular-weight synthetic polymers, shear degradation is essentially unavoidable in capillary viscometers. Rotation viscometers having low rotational speeds and narrow gaps between rotor and stator can produce exceptionally low linear shear rates ($< 1\ s^{-1}$), which will not degrade even ultrahigh-molecular-weight macromolecules. Also, the need to correct

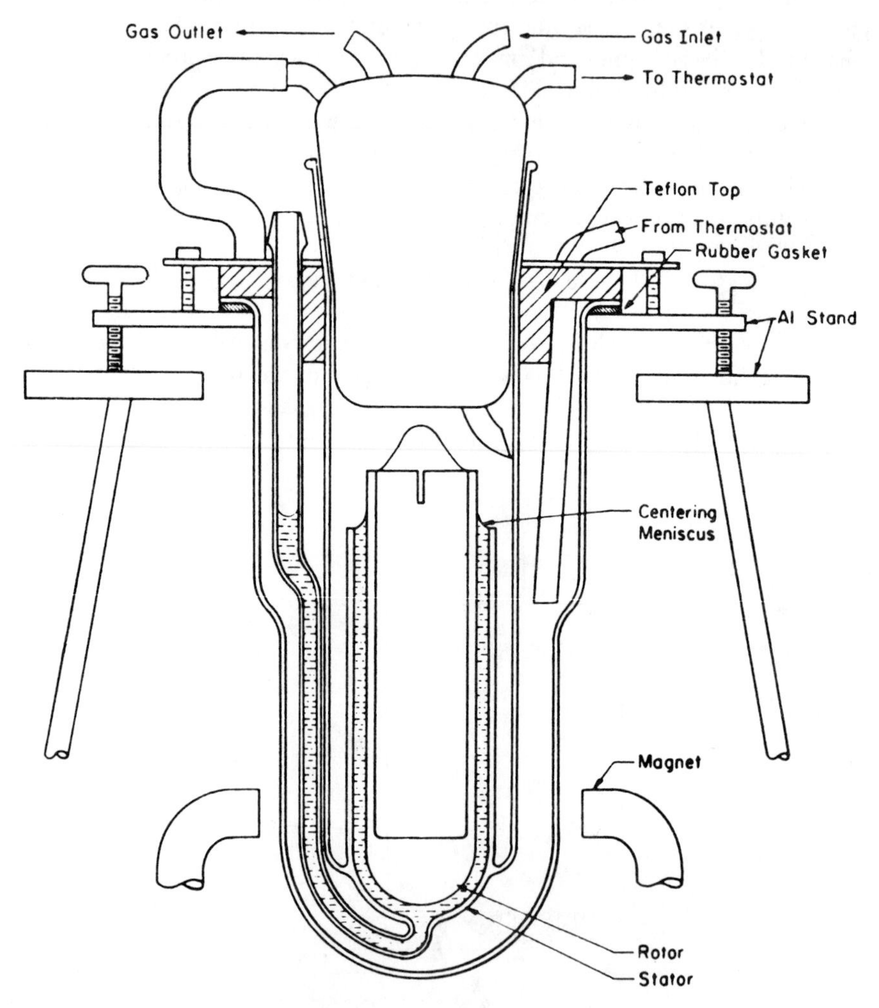

Figure 5. Zimm–Crothers viscometer as employed by Berry (58). Alignment is achieved automatically since the free floating inner cylinder is supported by its own buoyancy. Reproduced with permission from the publisher (58).

for shear rate effects, which may be required when using capillary viscometry, is effectively eliminated.

Two types of rotation viscometers are of considerable importance and will be briefly described here. The Couette viscometer, shown in Figure 4, achieves centering through the use of a mechanical axis. Alignment is critical due to the narrow gap between the stator and the rotor. The Couette viscometer depicted in Figure 4 is shown as having the outer cylinder stationary, with the inner cylinder driven at a constant speed. The resulting torque is measured by some appropriate device, and this torque is related to the viscosity of the liquid.

Zimm and Crothers (177) introduced a simplified rotating cylinder viscometer that has, appropriately, become known as the Zimm–Crothers viscometer. Other workers (58, 178–179) have since constructed and employed similar devices with considerable success. The basic design is depicted in Figure 5. The Zimm–Crothers viscometer uses a freely floating inner cylinder that is supported by its own buoyancy and held in place by surface forces, eliminating the need for careful alignment along a mechanical axis as is necessary with the Couette design. A constant torque is applied by the interaction of a steel pellet in the bottom of the rotor with a rotating applied magnetic field. A constant shear stress is applied to the liquid, and the rate of rotation is inversely proportional to the viscosity of the liquid. With no mechanical devices attached to the moving cylinder, energy dissipation is entirely due to viscous forces. The relative viscosity is simply

$$\eta_r = \frac{P - P_n}{P_0 - P_n} \tag{63}$$

where P_n is the period of revolution of the magnet, and P and P_0 refer to the polymer solution and solvent, respectively.

Slagowski (179) described the construction and use of a Zimm–Crothers type of viscometer in considerable detail. Wagner (175) also recently described the construction and operation of such a unit, specifically designed for work at elevated temperatures. Commercial low-shear viscometers are available from Bohlin Reologi AB, Contraves AG, and Rheometrics, Inc.

8.5. Automatic Viscometers

Despite the low cost and relative simplicity of ordinary, manually conducted, capillary viscometry, there have been repeated efforts to develop automatic viscometers. Potential advantages of automatic viscometers include

- Reduced analysis time.
- Enhanced sensitivity and improved accuracy.

- Computer interfacing capabilities.
- Utility as a detector for size exclusion chromatography (SEC).

One of the first advances in the area of automated viscometry was the development of photoelectric devices, which sense the passage of a liquid miniscus and eliminate the need for manual timing. As early as 1933, Jones and Talley (180) reported the use of such a device. Wescan currently markets the Model 221 Automatic Viscosity Timer, which makes use of this photoelectric principle (see Section 7).

Unfortunately, automated timing does not eliminate the manual tasks of loading the viscometer, performing dilutions, and forcing solutions into the measuring zones. As of this writing EcoPlastics Limited plans to market a viscometer which, in addition to automatic timing, offers automated liquid transfer and microprocessor capabilities. Publications describing the prototype upon which this instrument is based have appeared (181, 182).

The Schott AVS system is a fully automated viscometer, which is currently available. The measuring sequence is such that siphoning of the liquid, observation of the liquid miniscus with simultaneous time measurement, repetition of the measuring process, rinsing, filling, sample changing, and so on, are fully automated. A computer can be connected to the unit for system control and data evaluation. The Schott AVS can also be operated at elevated temperatures. This unit is capable of measuring flow times reproducibly to ± 0.01 s, thus allowing measurements on extremely dilute solutions.

A somewhat different approach to automating viscosity measurements has been taken by Ouano (183). Utilizing an SEC arrangement with a concentration detector, Ouano measured the pressure drop ΔP across a capillary at constant flow rate for solvent and polymer solution. The intrinsic viscosity may then be calculated for any eluent fraction (at low c) as

$$[\eta]_{\mathrm{i}} = \frac{1}{c_{\mathrm{i}}} \ln\left(\frac{\Delta P_{\mathrm{i}}}{\Delta P_{\mathrm{o}}}\right) \tag{64}$$

where ΔP_{i} and ΔP_{o} refer to solution and solvent, respectively. Lesec and coworkers (184–186) utilized a similar instrument with apparent success.

A major problem that had to be overcome with viscometers of the latter type involve reducing pulses from the pumping system to an acceptable level. Obviously, a viscometer of this type is extremely sensitive to variations in the flow rate. This difficulty was reportedly overcome by introducing high-sensitivity filters just after the pumping system (184). Water Chromatography Division of Millipore Corp. has also recently incorporated a single-capillary viscometer detector into their Model 150C liquid chromatograph (187).

The most recent development in automated solution viscosity measurements is the Model 100 Differential Viscometer (DV) from Viscotek. Here the specific viscosity of an extremely dilute polymer solution is derived directly from a measure of the difference in pressure across a wheatstone bridge arrangement of four matched capillaries. During the elution of the sample, three arms (capillaries) of the bridge contain the mobile phase, while the fourth arm (capillary) contains the eluting solute. The resulting imbalance is then measured as a pressure drop from which the intrinsic viscosity is determined (188).

Viscotek has recently introduced a different type of differential viscometer, Model Y-500, based on a design first reported by Yau et al. (189). This instrument consists of an analytical capillary and a reference capillary in series, separated by a delay volume. The pressure drop across each of the capillaries is measured with a pressure transducer, the signals of which are processed using a differential log amplifier. The resulting detector output is $\ln \eta_r$. The intrinsic viscosity is obtained by dividing this quantity by the polymer concentration. The reference capillary serves to cancel out any flowrate and temperature fluctuations. Both models can be used offline for batch measurements as well as for an online SEC detector. A sensitivity of approximately 1×10^{-4} specific viscosity units, at a signal-to-noise ratio of 4, has been reported. The manufacturer claims the following advantages.

- Extrapolation to infinite dilution is not necessary, since η_{sp}/c may be measured at extremely low concentrations.
- High sensitivity and precision are obtained without the usual precise control of temperature, since differential pressure measurements tend to cancel these differences.
- Capillaries have high length-to-diameter ratios, eliminating the need for kinetic energy and end effect corrections.
- Measurements of η_{sp}/c in about 3 min.
- Utility as an SEC detector, since the differential design eleminates the need for rigorous pulse damping.

The manufacturer also points out some apparent disadvantages.

- The DV is more complex and more expensive than ordinary glass viscometers.
- The pressure transducer of model 100 DV must maintain calibration in order to obtain accurate data.
- Corrosive solvents should be avoided to prevent corrosion of stainless steel parts.

The potential of the DV as a viscometric detector for SEC is of considerable value. Knowledge of $[\eta]$ as a function of elution volume enables average molecular weights to be calculated for "unknown" samples using the universal calibration procedure (190).

9. CONCLUSIONS

A wealth of information concerning polymer structure and properties can be gleaned from viscosity measurements on dilute polymer solutions. While the methods used to measure viscosities normally require only inexpensive equipment, valuable insight into polymer size, molecular weight, and branching can be obtained. Thus, it comes as no surprise that viscometry remains one of the most common and useful means of polymer characterization. Recent advances in automation assure that viscometry will retain this prominent place among polymer characterization techniques.

NOTATION

a	Mark–Houwink–Sakurada exponent
a_1, a_2	Virial coefficients
A	Preexponential term
A, B	Constants used in viscometer calibration
A_0, B_0	
A_η, B_η	Parameters in Bohdanecky treatment of wormlike chains
B	Excluded volume parameter
c	Concentration
C_M	Parameter occurring in thermodynamic relations for dilute polymer solutions
C_∞	Characteristic ratio
d	Chain diameter
d_r	Reduced chain diameter, d/l'
f	Total elastic force
f	Functionality (number of arms) of a star polymer
f_e	Energetic component of the total elastic force
g	Acceleration of gravity
g	Parameter relating reduction in radius of gyration of a branched polymer compared to a linear polymer with the same molecular weight, $\langle S^2 \rangle_{br}/\langle S^2 \rangle_l$
g'	Parameter relating reduction in intrinsic viscosity of a branched polymer compared to a linear polymer with the same molecular weight, $[\eta]_{br}/[\eta]_l$

G'	Ratio of intrinsic viscosity of a star polymer to the intrinsic viscosity of a linear chain with twice the arm molecular weight
$\bar{h}$	Average value of the liquid head
k'	Huggins coefficient
k''	Kraemer coefficient
k'''	Constant in Martin's equation
K	Mark–Houwink–Sakurada constant, dL/g
K_θ	Unperturbed chain dimensions, dL $mol^{0.5}/g^{3/2}$
K_1	Heat of dilution
K_β, K_e, K_γ	Empirical viscosity constants in the treatment of Moore
l	Average bond length
l_c	Length of capillary
l_u	Length of the repeating unit projected on the molecular axis
l'	Kuhn statistical segment length
L	Chain contour length
L_r	Reduced contour length, L/l'
M	Molecular weight, g/mol
$\bar{M}_n$	Number-average molecular weight
M_u	Molecular weight of the repeating unit
$\bar{M}_w$	Weight-average molecular weight
N	Number of main chain bonds
N_A	Avogadro's number
N_i	Number of molecules of molecular weight M_i
N'	Number of main chain bonds per repeating unit
p	Hydrostatic pressure head, dyn/cm^2
$\bar{p}$	Average hydrostatic pressure under which flow occurs
P	Period of revolution with polymer solution in the Zimm–Crothers viscometer
P_n	Period of revolution of the magnet (no solvent or solution) in a Zimm–Crothers viscometer
P_o	Period of revolution with solvent in the Zimm–Crothers viscometer
ΔP_i	Pressure drop across a capillary for a polymer solution of concentration c_i
ΔP_o	Pressure drop across a capillary for solvent
q	Persistence length
Q	Apparent activation energy of viscous flow for a polymer solution
Q_o	Activation energy of viscous flow for solvent
r	Radius of capillary
$\langle r^2 \rangle_o$	Unperturbed mean-square end-to-end distance
R	Gas constant

R_H Hydrodynamic (Stokes) radius
R_V Viscometric (Einstein) radius
$\langle S^2 \rangle$ Mean-square radius of gyration
$\langle S^2 \rangle_o$ Mean-square radius of gyration at θ condition
t Time
t Flow time for polymer solution
t_0 Solvent flow time
T Temperature
V Volume
V_1 Molar volume of solvent
$\langle x \rangle$ Average span defined by Eq. 46
α Chain expansion factor, $(\langle S^2 \rangle / \langle S^2 \rangle_0)^{0.5}$
α_η Chain expansion factor, $([\eta]/[\eta]_\theta)^{1/3}$
β Empirical constant
α Empirical constant
η, η_0 Viscosity and viscosity of solvent, poise (P)
η_r Relative viscosity, η/η_0
η_{sp} Specific viscosity, $\eta_r - 1$
$[\eta]$ Intrinsic viscosity, usually in dL/g
$[\eta]_\theta$ Intrinsic viscosity at θ condition
θ Theta condition or temperature
$\bar{v}$ Partial specific volume of the polymer
ρ, ρ_0 Density of solution or solvent, g/cm^3
φ Parameter characterizing the entropy of dilution of polymer with solvent
Φ, Φ_0 Hydrodynamic parameter relating the intrinsic viscosity to the molecular dimension. The subscript denotes the value for flexible linear chains at θ condition.

INSTRUMENT COMPANIES

- Bohlin Reologi AB, Science Park Ideon, Lund S-22370, Sweden
- Cannon Instrument Company, P. O. Box 16, State College, Pennsylvania, 16804
- Contraves AG, Schaffhauserstr. 580, CH-8052, Zurich, Switzerland
- EcoPlastic Limited, 518 Gordon Baker Rd., Willowdale, Toronto, Canada M2H 3B4
- Rheometrics, Inc., One Possumtown Rd., Piscataway, New Jersey 08554
- Schott America, 3 Odell Plaza, Yonkers, New York 10701
- Viscotek Corporation, 1032 Russell Dr., Porter, Texas 77365

- Water Chromatography, Division of Millipore Corp., Maple St., Millford, Massachusetts
- Wescan Instruments, Inc., 3018 Scott Blvd., Santa Clara, California 95050

REFERENCES

1. H. Yamakawa, *Modern Theory of Polymer Solutions*, Harper & Row, New York, 1971.
2. M. Bohdanecky and J. Kovar, *Viscosity of Polymer Solutions*, Elsevier, Amsterdam, 1982.
3. M. R. Cannon, R. E. Manning, and J. D. Bell, "Viscosity Measurement: The Kinetic Energy Correction and a New Viscometer," *Anal. Chem.*, **32**, 355 (1960)
4. L. H. Cragg and H. van Oene, "Shear Dependence in the Viscometry of High Polymer Solutions: A New Variable-Shear Capillary Viscometer," *Can. J. Chem.*, **39**, 203 (1961).
5. P. J. Flory, *Principles of Polymer Chemistry*, Cornell University Press, Ithaca, New York, 1953, p. 309.
6. E. O. Kraemer, "Molecular Weights of Celluloses," *Ind. Eng. Chem.*, **30**, 1200 (1938).
7. F. Eirich and J. Riseman, "Some Remarks on the First Interaction Coefficient of the Viscosity-Concentration Equation," *J. Polym. Sci.*, **4**, 417 (1949).
8. H. L. Frisch and R. Simha, "The Viscosity of Colloidal Suspensions and Macromolecular Solutions," in F. R. Eirich (Ed.), *Rheology*, Vol. 1, Academic, New York 1956, Chapter 14.
9. M. L. Huggins, "The Viscosity of Dilute Solutions of Long-Chain Molecules. IV. Dependence on Concentration," *J. Am. Chem. Soc.*, **64**, 2716 (1942).
10. Z. Xu, R. Qian, N. Hadjichristidis, and L. J. Fetters, "Molecular Weight Dependence of Huggins Coefficient Under θ Conditions," *J. China Univ. Sci. Technol.*, **14**, 228 (1984).
11. B. J. Bauer, L. J. Fetters, W. W. Graessley, N. Hadjichristidis, and G. Quack, "Chain Dimensions in Dilute Polymer Solutions: A Light Scattering and Viscometric Study of Multi-armed Polyisoprene Stars in Good and Theta Solvents," *Macromolecules*, **22**, 2337 (1989).
12. M. Bohdanecky, "The Effect of Polydispersity on the Huggins Constant," *Coll. Czech. Chem. Commun.*, **31**, 4095 (1966).
13. M. Bohdanecky, "The Huggins Viscometric Constant of Some Linear Polymers in a Series of Solvents," *Coll. Czech. Chem. Commun.*, **35**, 1972 (1970).
14. J. Riseman and R. Ullman, "The Concentration Dependence of the Viscosity of Solutions of Macromolecules," *J. Chem. Phys.*, **19**, 578 (1951).
15. H. Yamakawa, "Concentration Dependence of Polymer Chain Configurations in Solution," *J. Chem. Phys.*, **34**, 1360 (1961).

16. J. M. Peterson and M. Fixman, "Viscosity of Polymer Solutions," *J. Chem. Phys.*, **39**, 2516 (1963).
17. T. Sakai, "Huggins Constant k' for Flexible Chain Polymers," *J. Polym. Sci., Part A* 2, **6**, 1535 (1968).
18. T. Sakai, "Huggins Constant k', for Chain Polymers at the Theta Point," *Macromolecules*, **3**, 96 (1970).
19. T. Ogasa and S. Imai, "Theory on the Mechanical Properties of Dilute Polymer Solutions," *J. Chem. Phys.* **54**, 2989 (1971).
20. K. F. Freed and S. F. Edwards, "Huggins Coefficient for the Viscosity of Polymer Solutions," *J. Chem. Phys.*, **62**, 4032 (1975).
21. M. Muthukumar and K. F. Freed, "Huggins Coefficient for Polymer Solutions with Excluded Volume," *Macromolecules*, **10**, 899 (1977).
22. M. Muthukumar, "The Series Expansion for the Concentration Dependence of Relaxation Times for Dilute Polymer Solutions," *J. Chem. Phys.*, **79**, 4048 (1983).
23. F. Ibrahim and H. G. Elias, "Uber eine Fehlerguelle bei der Berechnung von Staudinger-Indices (Grenzviskositatszahl, intrinsic viscosity)," *Makromol. Chem.*, **76**, 1 (1964).
24. F. Baker, "The Viscosity of Cellulose Nitrate Solutions," *J. Chem. Soc.*, **103**, 1653 (1913).
25. G. V. Schulz and F. Blaschke, "Eine Gleichung zur Berechnung der Viscositatszahl fur sehr kleine Konzentrationen," *J. Prakt. Chem.*, **158**, 130 (1941).
26. G. V. Schulz and G. Sing, "Uber den Anstieg der spezifischen Viscositat makromolekularer Losungen im Bereich kleiner Konzentrationen," *J. Prakt. Chem.*, **161**, 11 (1943).
27. W. Heller, "Treatment of Viscosity Data on Polymer Solutions (An Analysis of Equations and Procedures)," *J. Colloid Sci.*, **9**, 547 (1954).
28. A. F. Martin, ACS Meeting, Memphis, TN, 1942; see Ref. 27 for a summary of Martin's results.
29. H. Staudinger and W. Heuer, "Uber hochpolymere Verbindungen, 33. Mittielung: Beziehungen Zwischen Viscositat und Molekulargewicht bei Poly-styrolen," *Chem. Ber.*, **63**, 222 (1930); H. Staudinger and R. Nodzu, "Uber hochpolymere Verbindungen, 36. Mitteil: Viscositats-Untersuchungen an Paraffin-Losungen," *Chem. Ber.*, **63**, 721 (1930).
30. E. O. Kraemer and W. D. Lansing, "The Molecular Weights of Cellulose and Cellulose Derivatives," *J. Phys. Chem.*, **39**, 153 (1935).
31. H. Mark, *Der Feste Korper*, Leipzig, 1938 (from P. F. Onyon, "Viscometry," in P. W. Allen (Ed.), *Techniques of Polymer Characterization*, Butterworths, London, 1959, Chapter 6).
32. R. Houwink, "Zusammenhang zwischen viscosimetrisch und osmotisch bestimmten Polymerisationsgraden bei Hochpolymeren," *J. Prakt. Chem.*, **157**, 15 (1940).
33. I. Sakurada, "Genugt eine einzige Konstante fur die Darstellung der

Konzentration-abhangigkeit der Viscositat von hochpolymeren Verbindungen?," *Kolloid Z.*, **82**, 345 (1938).

34. P. J. Flory, "Molecular Weights and Intrinsic Viscosities of Polyisobutylenes," *J. Am. Chem. Soc.*, **65**, 372 (1943).
35. M. Kurata and Y. Tsunashima, "Viscosity-Molecular Weight Relationships and Unperturbed Dimensions of Linear Chain Molecules," in J. Brandrup and E. H. Immergut (Eds.), *Polymer Handbook*, 3rd ed., Wiley, New York, 1989, Chapter VII.
36. M. Kurata and W. H. Stockmayer, "Intrinsic Viscosities and Unperturbed Dimensions of Long Chain Molecules," *Adv. Polym. Sci.*, **3**, 196 (1963).
37. R. Jenkins and R. S. Porter, "Unperturbed Dimensions of Stereoregular Polymers," *Adv. Polym. Sci.*, **36**, 1 (1980).
38. G. Meyerhoff, "Die Viscosimetrische Molekulargewichtsbestimmung von Polymeren," *Adv. Polym. Sci.*, **3**, 59 (1961).
39. T. Altares, Jr., D. P. Wyman, and V. R. Allen, "Synthesis of Low Molecular Weight Polystyrene by Anionic Techniques and Intrinsic Viscosity—Molecular Weight Relations over a Broad Range in Molecular Weight," *J. Polym. Sci.*, A2, 4533 (1964).
40. U. Bianchi, M. Dalpiaz, and E. Patrone, "Viscosity-Molecular Weight Relationship for Low Molecular Weight Polymers: 1. Polydimethylsiloxane and Polyisobutylene," *Makromol. Chem.*, **80**, 112 (1964).
41. T. G. Fox, J. B. Kinsinger, H. F. Mason, and E. M. Schuele, "Properties of Dilute Polymer Solutions: I Osmotic and Viscometric Properties of Solutions of Conventional Poly(methyl methacrylate)," *Polymer*, **3**, 71 (1962).
42. C. Kow, N. Hadjichristidis, and L. J. Fetters, unpublished results; see also, C. Kow, "Synthesis of Uniform and Non-Uniform Polyisoprene Networks," Ph.D. Thesis, University of Akron, 1982, p. 121.
43. K. Huber, S. Bantle, P. Lutz, and W. Burchard, "Hydrodynamic and Thermodynamic Behavior of Short-Chain Polystyrene in Toluene and Cyclohexane at 34.5°C," *Macromolecules*, **18**, 1461 (1985).
44. Y. Einaga, Y. Miyake, and H. Fujita, "Intrinsic Viscosity of Polystyrene," *J. Polym. Sci., Polym. Phys. Ed.*, **17**, 2103 (1979); see also: H. Fujita, "Some Unsolved Problems on Dilute Polymer Solutions," *Macromolecules*, **21**, 179 (1988).
45. R. L. Morris, S. Amelar, and T. P. Lodge, "Solvent Friction in Polymer Solutions and its Relation to the High Frequency Limiting Viscosity." *J. Chem. Phys.*, **89**, 6523 (1988).
46. E. D. von Meerwall, S. Amelar, M. A. Smeltzly, and T. P. Lodge, "Solvent and Probe Diffusion in Aroclor Solutions of Polystyrene, Polybutadiene, and Polyisoprene," *Macromolecules*, **22**, 295 (1989).
47. D. McIntyre, L. J. Fetters, and E. Slagowski, "Polymers: Synthesis and Characterization of Extremely High Molecular Weight Polystyrene," *Science*, **176**, 1041 (1972).
48. P. J. Flory, *Principles of Polymer Chemistry*, Cornell University Press, Ithaca, New York, 1953, p. 313.

49. P. J. Flory and T. G. Fox, Jr., "Treatment of Intrinsic Viscosities," *J. Am. Chem. Soc.*, **73**, 1904 (1951).
50. P. Debye and A. M. Bueche, "Intrinsic Viscosity, Diffusion and Sedimentation Rate of Polymers in Solutions," *J. Chem. Phys.*, **16**, 573 (1948).
51. J. G. Kirkwood and J. Riseman, "The Intrinsic Viscosities and Diffusion Constants of Flexible Macromolecules in Solution," *J. Chem. Phys.*, **16**, 565 (1948).
52. H. C. Brinkman, "A Calculation of the Viscous Force Exerted by a Flowing Fluid on a Dense Swarm of Particles," *Appl. Sci. Res.*, **A-1**, 27 (1947).
53. P. J. Flory, *Principles of Polymer Chemistry*, Cornell University Press, Ithaca, New York, 1953, Chapter 10.
54. C. W. Pyun and M. Fixman, "Perturbation Theory of the Intrinsic Viscosity of Polymer Chains," *J. Chem. Phys.*, **44**, 2107 (1966).
55. Recent renormalization group calculations [Y. Oono and M. Kohmoto, *J. Chem. Phys.*, **78**, 520 (1983)] have yielded $\Phi_o = 2.36 \times 10^{21}$.
56. P. Debye, "Molecular-Weight Determination by Light Scattering," *J. Phys. Coll. Chem.*, **51**, 18 (1947).
57. D. McIntyre, A. Wims, L. C. Williams, and L. Mandelkern, "Conformation and Frictional Properties of Polystyrene in Dilute Solutions," *J. Phys. Chem.*, **66**, 1932 (1962).
58. G. C. Berry, "Thermodynamic and Conformational Properties of Polystyrene: II. Intrinsic Viscosity Studies on Dilute Solutions of Linear Polystyrenes," *J. Chem. Phys.*, **46**, 1338 (1967).
59. K. Kawahara, T. Norisuye, and H. Fujita, "Excluded Volume Effects in Dilute Polymer Solutions: II. Limiting Viscosity Number," *J. Chem. Phys.*, **49**, 4339 (1968).
60. K. Takashima, G. Tanaka, and H. Yamakawa, "Further Test of the Two-Parameter Theory of Dilute Polymer Solutions: Poly(*p*-bromostyrene)," *Polymer J.*, **2**, 245 (1971).
61. A. Yamamoto, M. Fujii, G. Tanaka, and H. Yamamoto, "More on the Analysis of Dilute Solution Data: Polystyrenes Prepared Anionically in Tetrahydrofuran." *Polymer J.*, **2**, 799 (1971).
62. M. Fukuda, M. Fukutomi, Y. Kato, and T. Hashimoto, "Solution Properties of High Molecular Weight Polystyrene," *J. Polym. Sci., Polym. Phys. Ed.*, **12**, 871 (1974).
63. Y. Miyaki, Y. Einaga, H. Fujita, and M. Fukuda, "Flory's Viscosity Factor for the System Polystyrene + Cyclohexane at 34.5 °C," *Macromolecules*, **13**, 588 (1980).
64. N. Hadjichristidis, Z. Xu, L. J. Fetters, and J. Roovers, "The Characteristic Ratios of Stereoirregular Polybutadiene and Polyisoprene," *J. Polym. Sci., Polym. Phys. Ed.*, **20**, 743 (1982).
65. P. J. Flory, *Principles of Polymer Chemistry*, Cornell University Press, Ithaca, New York, 1953, p. 617.

66. P. J. Flory, *Statistical Mechanics of Chain Molecules*, Interscience, New York, 1969, pp. 35–39.
67. P. J. Flory, *Statistical Mechanics of Chain Molecules*, Interscience, New York, 1969, p. 11.
68. W. Burchard, "Uber den Einfluß der Losungsmittel aud die struktur linearer Makromekule," *Makromol. Chem.*, **50**, 20 (1960).
69. W. H. Stockmayer and M. Fixman, "On the Estimation of Unperturbed Dimensions from Intrinsic Viscosities," *J. Polym. Sci., Part. C1*, 137 (1963).
70. M. Kurata and W. H. Stockmayer, "Intrinsic Viscosities and Unperturbed Dimensions of Long Chain Molecules," *Fortschr. Hochpolym. Forsch.*, **3**, 196 (1963).
71. H. Yamakawa, *Modern Theory of Polymer Solutions*, Harper & Row, New York, 1971, Chapter 7.
72. G. Tanaka, "Intrinsic Viscosity and Friction Coefficient of Flexible Polymers," *Macromolecules*, **15**, 1028 (1982).
73. K. Kamide and W. R. Moore, "Analysis of Intrinsic Viscosity–Molecular Weight Data," *J. Polym. Sci. Part B*, **2**, 809 (1964).
74. W. H. Stockmayer, *Br. Polym. J.*, **9**, 89 (1977).
75. J. S. Lindner, N. Hadjichristidis, and J. W. Mays, "Application of Extrapolation Procedures to Viscosity Data Obtained Below the Theta Temperature," *Polym. Commun.*, **30**, 174 (1989).
76. P. J. Flory, *Statistical Mechanics of Chain Molecules*, Interscience, New York, 1969, Chapter 2, Table 1.
77. J. W. Mays and N. Hadjichristidis, "Characteristic Ratios of Polymethacrylates," *J. Macro. Sci., Revs.*, **C28** (3 & 4), 371 (1988).
78. M. V. Volkenstein, *Configurational Statistics of Polymeric Chains*, Interscience, New York, 1963.
79. P. J. Flory, "Foundations of Rotational Isomeric State Theory and General Methods for Generating Configurational Averages," *Macromolecules*, **7**, 381 (1974).
80. P. J. Flory, *Principles of Polymer Chemistry*, Cornell University Press, Ithaca, New York, 1953, p. 353.
81. Y. Oono, "Crossover Behavior of Transport Properties of Dilute Polymer Solutions: Renormalization Group Approach. III," *J. Chem. Phys.*, **79**, 4629 (1983).
82. C. W. Pyun and M. Fixman, "Intrinsic Viscosity of Polymer Chains," *J. Chem. Phys.*, **42**, 3838 (1965).
83. C. W. Pyun and M. Fixman, "Frictional Coefficient of Polymer Molecules in Solution," *J. Chem. Phys.*, **41**, 937 (1964).
84. N. S. Davidson, L. J. Fetters, W. G. Funk, N. Hadjichristidis, and W. W. Graessley, "Measurements of Chain Dimensions in Dilute Polymer Solutions; A Light Scattering Study of Linear Polyisoprene in Cyclohexane," *Macromolecules*, **20**, 2614 (1987).

85. J. S. Lindner, W. W. Wilson, and J. W. Mays, "Properties of Poly(α-methylstyrene) in Toluene: A Comparison of Experimental Results with Predictions of Renormalization Group Theory," *Macromolecules*, **21**, 3304 (1988).
86. J. S. Lindner, W. W. Wilson, and J. W. Mays, unpublished results.
87. J. Roovers and P. M. Toporowski, "Hydrodynamic Studies on Model Branched Polystyrenes," *J. Polym. Sci., Polym. Phys. Ed.*, **18**, 1907 (1980).
88. N. Khasat, R. W. Pennisi, N. Hadjichristidis, and L. J. Fetters, "Dilute Solution Behavior of Asymmetric Three-Arm and Regular Three- and Twelve-Arm Polystyrene Stars," *Macromolecules*, **21**, 1100 (1988).
89. P. J. Flory, *Statistical Mechanics of Chain Molecules*, Interscience, New York, 1969, Chapter 2, Table 2.
90. J. E. Mark, "Thermoelastic Properties of Rubberlike Networks and Their Thermodynamic and Molecular Interpretation," *Rubber Chem. Technol.*, **46**, 593 (1973).
91. P. J. Flory, *Principles of Polymer Chemistry*, Cornell University Press, Ithaca, New York, 1953, p. 600.
92. T. G. Fox, Jr., and P. J. Flory, "Intrinsic Viscosity—Molecular Weight Relationships for Polyisobutylene," *J. Phys. Colloid. Chem.*, **53**, 197 (1949).
93. T. G. Fox, Jr., and P. J. Flory, "Intrinsic Viscosity—Temperature Relationships for Polyisobutylene in Various Solvents," *J. Am. Chem. Soc.*, **73**, 1909 (1951).
94. T. G. Fox, Jr., and P. J. Flory, "Intrinsic Viscosity Relationships for Polystyrene," *J. Am. Chem. Soc.*, **73**, 1915 (1951).
95. T. A. Orofino, "Dilute Solution Properties of Polystyrene in θ-Solvent Media: II. An Analysis of Environmental Effects," *J. Chem. Phys.*, **45**, 4310 (1966).
96. T. A. Orofino and J. W. Mickey, Jr., "Dilute Solution Properties of Linear Polystyrene in θ-Solvent Media," *J. Chem. Phys.*, **38**, 2512 (1963).
97. J. W. Mays, N. Hadjichristidis, and L. J. Fetters, "Solvent and Temperature Influences on Polystyrene Unperturbed Dimensions," *Macromolecules*, **18**, 2231 (1985).
98. A. Dondos and H. Benoit, "Unperturbed Dimensions of Polymers in Binary Solvent Mixtures," *Eur. Polym. J.*, **4**, 561 (1968).
99. A. Dondos and H. Benoit, "Influence de la Temperature Sur les Dimensions Non Perturbees des Polymers Dissous Dans des Melanges de Solvants," *Eur. Polym. J.*, **6**, 1439 (1970).
100. A Dondos and H. Benoit, "The Influence of Solvents on Unperturbed Dimensions of Polymer in Solution," *Macromolecules*, **4**, 279 (1971).
101. A. Dondos and H. Benoit, "The Relationship Between the Unperturbed Dimensions of Polymers in Mixed Solvents and theThermodynamic Properties of the Solvents Mixture," *Macromolecules*, **6**, 242 (1973).
102. See, for example, P. J. Flory, *Principles of Polymer Chemistry*, Cornell University Press, Ithaca, New York, 1953, pp. 623–625.
103. S. Bluestone, J. E. Mark, and P. J. Flory, "The Interpretation of Viscosity-

Temperature Coefficients for Poly(oxyethylene) Chains in a Thermodynamically Good Solvent," *Macromolecules,* **7**, 325 (1974).

104. J. E. Mark, "Thermoelastic Results on Rubberlike Networks and Their Bearing on the Foundations of Elasticity Theory," *J. Polym. Sci. Part D*, **11**, 135 (1976).
105. J. E. Mark, "The Rubber State," in J. E. Mark, A. Eisenberg, W. W. Grassley, L. Mandelkern, and J. L. Koenig (Eds.), *Physical Properties of Polymers*, American Chemical Society, Washington, 1984, Chapter 1.
106. N. G. McCrum, "The Determination of $\ln \bar{r}_o^2/dT$ by the Method of Temperature Induced Creep," *Polymer Commun.*, **25**, 213 (1984).
107. N. G. McCrum, "The Determination of $d\ln \bar{r}_o^2/dT$ by the Method of Temperature Induced Creep in Tension and Torsion," *Polymer,* **27**, 47 (1986).
108. K. J. Smith, Jr., and J. E. Mark, "Criticisms of the Recently Proposed Analysis and Methods of Thermoviscoelasticity," *Polymer*, **29**, 292 (1988).
109. P. J. Flory, *Statistical Mechanics of Chain Molecules*, Interscience, New York, 1969, p. 39.
110. W. W. Graessley, "Effect of Long Branches on the Temperature Dependence of Viscoelastic Properties in Polymer Melts," *Macromolecules*, **15**, 1164 (1982).
111. R. Moore, "Viscosities of Dilute Polymer Solutions," in A. D. Jenkins (Ed.), *Progress in Polymer Science*, Vol. 1, Pergamon, Oxford, 1967, Chapter 1.
112. W. R. Moore and A. M. Brown, "Viscosity–Temperature Relationships for Dilute Solutions of Cellulose Derivatives. I. Temperature Dependence of Solution Viscosities of Ethyl Cellulose," *J. Colloid* Sci., **14**, 1 (1959).
113. N. Hadjichristidis, "Viscosity and Intrinsic Viscosity/Temperature Relationships for Dilute Solutions of Poly (2-biphenylyl methacrylate) and Poly(4-biphenylyl methacrylate)," *Makromol. Chem.*, **184**, 1043 (1983).
114. M. Liouni, C. Touloupis, N. Hadjichristidis and J. W. Mays, "Viscosity–Temperature Relationships for Linear and 12-Arm Star Polystyrenes in Dilute Solution," *J. Appl. Polym. Sci.*, **37**, 2699 (1989).
115. B. J. Bauer and L. J. Fetters, "Synthesis and Dilute-Solution Behavior of Model Star-Branched Polymers," *Rubber Chem. Technol.*, **51**, 406 (1978).
116. S. Bywater, "Preparation and Properties of Star-Branched Polymers," *Adv. Polym. Sci.*, **30**, 89 (1979).
117. E. F. Casassa, "Statistical Thermodynamics of Polymer Solutions: III. The Second Virial Coefficient for Branched Star Molecules," *J. Chem. Phys.*, **37**, 2176 (1962).
118. G. C. Berry and T. A. Orofino, "Branched Polymers: III. Dimensions of Chains with Small Excluded Volume," *J. Chem. Phys.*, **40**, 1614 (1964).
119. G. C. Berry and E. F. Casassa, "Thermodynamic and Hydrodynamic Behavior of Dilute Polymer Solutions," *J. Polym. Sci., Part D*, **4**, 1 (1970).
120. G. C. Berry, "Thermodynamic and Conformational Properties of Polystyrene: III. Dilute Solution Studies on Branched Polymers," *J. Polym Sci., Part A2*, **9**, 687 (1971).

121. J. Mazur and F. McCrackin, "Configurational Properties of Star-Branched Polymers," *Macromolecules*, **10**, 326 (1977).
122. M. Daoud and J. P. Cotton, "Star Shaped Polymers: A Model for the Conformation and Its Concentration Dependence," *J. Phys.*, **43**, 531 (1982).
123. See, for example, B. J. Bauer and L. J. Fetters, "Synthesis and Dilute-Solution Behavior of Model Star-Branched Polymers," *Rubber Chem. Technol.*, **51**, 424 (1978) and references therein.
124. B. J. Bauer, N. Hadjichristidis, L. J. Fetters, and J. E. L. Roovers, "Star-Branched Polymers: 5. The Temperature Depression for 8- and 12-Arm Polyisoprenes in Dioxane," *J. Am. Chem. Soc.*, **102**, 2410 (1980).
125. K. Huber, W. Burchard, S. Bantle, and L J. Fetters, "Monte Carlo Calculations in Comparison to Neutron Scattering Studies: 2. Global Dimensions of 12-Arm Stars," *Polymer*, **28**, 1990 (1987).
126. K. Huber, W. Burchard, S. Bantle, and L. J. Fetters, "Monte Carlo Calculations in Comparison to Neutron Scattering Studies: 3. On the Structure of 12-Arm Star Molecules," *Polymer*, **28**, 1997 (1987).
127. B. H. Zimm and R. W. Kilb, "Dynamics of Branched Polymer Molecules in Dilute Solutions," *J. Polym. Sci.*, **37**, 19 (1959).
128. W. H. Stockmayer and M. Fixman, "Dilute Solutions of Branched Polymers," *Ann. N. Y. Acad. Sci.*, **57**, 334 (1953).
129. J. E. L. Roovers and S. Bywater, "Preparation and Characterization of Four-Branched Star Polystyrene," *Macromolecules*, **5**, 384 (1972).
130. J. E. L. Roovers and S. Bywater, "Preparation of Six-Branched Polystyrene. Thermodynamic and Hydrodynamic Properties of Four- and Six-Branched Star Polystyrenes," *Macromolecules*, **7**, 443 (1974).
131. J. Roovers, N. Hadjichristidis, and Lewis J. Fetters, "Analysis and Dilute Solution Properties of 12- and 18-Arm-Star Polystyrenes," *Macromolecules*, **16**, 214 (1983).
132. J. W. Mays, N. Hadjichristidis, and L. J. Fetters, "Star-Branched Polystyrenes: An Evaluation of Solvent and Temperature Influences on Unperturbed Chain Dimensions," *Polymer*, **29**, 680 (1988).
133. B. J. Bauer, "Bulk and Solution Properties of Star-Branched Polymers," Ph.D. Thesis, University of Akron, 1981.
134. E. F. Casassa, "Equilibrium Distribution of Flexible Polymer Chains Between a Macroscopic Solution Phase and Small Voids," *J. Polym. Sci., Polym. Lett. Ed.*, **5**, 773 (1967).
135. E. F. Casassa and Y. Tagami, "An Equilibrium Theory for Exclusion Chromatography of Branched and Linear Polymer Chains," *Macromolecules*, **2**, 14, (1969).
136. E. F. Casassa, "Comments on Exclusion of Polymer Chains from Small Pores and Its Relation to Gel Permeation Chromatography," *Macromolecules*, **9**, 182 (1976).
137. B. J. Bauer "Bulk and Solution Properties of Star-Branched Polymers," Ph.D. Thesis, University of Akron, Ohio, 1981, pp. 87–88.

138. B. H. Zimm and W. H. Stockmayer, "The Dimensions of Chain Molecules Containing Branches and Rings," *J. Chem. Phys.*, **17**, 1301 (1949).
139. T. A. Orofino, "Branched Polymers: II. Dimensions in Non-interacting Media," *Polymer*, **2**, 305 (1961)
140. M. Kurata and M. Fukatsu, "Unperturbed Dimension and Translational Friction Constant of Branched Polymers," *J. Chem. Phys.*, **41**, 2934 (1964).
141. T. G. Scholte, "Characterization of Long-Chain Branching in Polymers," in J. F. Dawkins (Ed.), *Development in Polymer Characterization*, Vol. 4, Applied Science, London, 1983, pp. 3–4.
142. J. A. Riddick, W. B. Bunger, and T. K. Sakano, *Organic Solvents: Physical Properties and Methods of Purification*, 4th ed., Wiley, New York, 1986.
143. J. Duke and D. Prem, "Dilute Solution Viscometry: Its Use and Interpretation with GR-S Polymers," *Rubber Age*, **80**, 83 (1956).
144. W. Cooper and R. K. Smith, "Apparatus for Viscosity Determination," *Chem. Ind.*, **936**, 1185 (1957).
145. R. B. Brock, "Apparatus for Viscosity Determinations," *Chem. Ind.*, 1185 (1957).
146. N. C. Billingham, *Molar Mass Measurements in Polymer Science*, Wiley, New York, 1977, Chapter 7.
147. C. Tanford, *Physical Chemistry of Macromolecules*, Wiley, New York, 1961.
148. H. Morawetz, *Macromolecules in Solution*, 2nd ed., Wiley, New York, 1975.
149. P. J. Flory, *Principles of Polymer Chemistry*, Cornell University Press, Ithaca, New York, 1953, p. 635.
150. V. Crescenzi, "Some Recent Studies of Polyelectrolyte Solutions," *Adv. Polym. Sci.*, **5**, 358 (1968).
151. H. Eisenberg, *Biological Macromolecules and Polyelectrolytes in Solution*, Oxford University Press, Oxford, 1976.
152. F. Booth, "The Electroviscous Effect for Suspensions of Solid Spherical Particles," *Proc. R. Soc. London Series A*, **203**, 533 (1950).
153. J. Stone-Masui and A. Watillon, "Electroviscous Effects in Dispersions of Monodisperse Polystyrene Latices," *J. Colloid Interface Sci.*, **28**, 187 (1968).
154. R. M. Fuoss and U. P. Strauss, "Polyelectrolytes: II. Poly-4-vinylpyridonium Chloride and Poly-4-vinyl-*N*-*n*-butyl-pyridonium Bromide," *J. Polym. Sci.*, **3**, 246, (1948); "Electrostatic Interaction of Polyelectrolytes and Simple Electrolytes," *J. Polym. Sci.*, **3**, 602 (1948).
155. R. M. Fuoss, "Polyelectrolytes", *Disc. Faraday Soc.*, No. 11, 125 (1951).
156. H. Eisenberg, "Viscosity Behaviour of Polyelectrolyte Solutions at Low and Medium Rates of Shear," *J. Polym. Sci.*, **23**, 579 (1957).
157. J. Cohen, Z. Priel and Y. Rabin, "Viscosity of Dilute Polyelectrolyte Solution," *J. Chem. Phys.*, **88**, 7111 (1988).
158. O. Kratky and G. Porod, "Rotgen-Untersuchung Geloster Fadenmolecule," *Rec. Trav. Chim.*, **68**, 1106 (1949).

159. H. Benoit and P. Doty, "Light Scattering from Non-Gaussian Chains," *J. Phys. Chem.*, **57**, 958 (1953).
160. A Peterlin, "Viscosity and Sedimentation of Linear Macromolecules Exhibiting Partial Solvent Immobilization," *J. Polym. Sci.*, **5**, 473 (1950).
161. A. Peterlin, "Gradient Dependence of Intrinsic Viscosity of Flexible Linear Macromolecules," *J. Chem. Phys.*, **33**, 1799 (1960).
162. Y. E. Eizner and O. B. Ptitsyn, *Vysokomol. Soedin*, **4**, 1725 (1962).
163. J. E. Hearst, "Rotary Diffusion Constants of Stiff-Chain Macromolecules," *J. Chem. Phys.*, **38**, 1062 (1963).
164. J. E. Hearst, "Shear Dependence of the Intrinsic Viscosity of Rigid Distributions of Segments with Cylindrical Symmetry," *J. Chem. Phys.*, **42**, 4149 (1965).
165. S. F. Kurath, C. A. Schmitt, and J. J. Bachhuber, "Hydrodynamic Behavior of Fully Acetylated Guaran. Test of the Eizner–Ptitsyn Theory for the Semirigid Macromolecule," *J. Polym. Sci., Part A*, **3**, 1825 (1965).
166. H. Yamakawa and J. Fujii, "Intrinsic Viscosity of Wormlike Chains. Determination of the Shift Factor," *Macromolecules*, **7**, 128 (1974).
167. H. Yamakawa and T. Yoshizaki, "Transport Coefficient of Helical Wormlike Chains. 3. Intrinsic Viscosity," *Macromolecules*, **13**, 633 (1980).
168. M. Bohdanecky, "New Method for Estimating the Parameters of the Wormlike Chain Model from the Intrinsic Viscosity of Stiff-Chain Polymers," *Macromolecules*, **16**, 1483 (1983).
169. T. Norisuye and H. Fujita, "Excluded-Volume Effects in Dilute Polymer Solutions. XIII. Effects of Chain Stiffness," *Polym. J.*, **14**, 143 (1982).
170. G. V. Reddy and M. Bohdanecky, "Analysis of the Temperature Quotient of the Intrinsic Viscosity of Stiff-Chain Polymers," *Macromolecules*, **20**, 1393 (1987).
171. R. Chiang, "Characterization of High Polymers in Solutions—with Emphasis on Techniques at Elevated Temperatures," in B. Ke (Ed.), *Newer Methods of Polymer Characterization*, Interscience, New York, 1964, Chapter XII.
172. A. Ciferri, "Conformational Energy of Chain Molecules. Part 2. Intrinsic Viscosity–Temperature Coefficient for Athermal Polydimethylsiloxane Solutions," *Trans. Faraday Soc.*, **57**, 853 (1961).
173. A. Nakajima, F. Hamada, and S. Hayashi, "Unperturbed Chain Dimensions of Polyethylene in Theta Solvents," *J. Polym. Sci., Part C*, **15**, 285 (1966).
174. V. Desreux, Univ. Liege, Liege, Belgium, private communication.
175. H. L. Wagner, "Viscosity and Molecular Weight Distribution of Ultrahigh Molecular Weight Polystyrene Using a High Temperature Low Shear Rate Rotational Viscometer," *J. Appl. Polym. Sci.*, **36**, 567 (1988).
176. V. Grinshpun and A. Rudin, "High Temperature GPC of Polypropylene." *J. Appl. Polym. Sci.*, **30**, 2413 (1985).
177. B. H. Zimm and D. M. Crothers, "Simplified Rotating Cylinder Viscometer for DNA," *Proc. Natl. Acad. Sci.*, **48**, 905 (1962).
178. A. R. Sloniewsky, G. T. Evans, and P. Ander, "Modified Zimm–Crothers Rotating-Cylinder Viscometer," *J. Polym. Sci.Part A2*, **6**, 1555 (1968).

179. E. L. Slagowski, "Characterization of Chain Conformation and Thermodynamics of Extremely High Molecular Weight Polystyrene," Ph.D. Thesis, University of Akron, Ohio, 1972.

180. G. Jones and S. K. Talley, "The Viscosity of Aqueous Solutions as a Function of the Concentration," *J. Am. Chem. Soc.*, **55**, 624 (1933).

181. T. Kilp, B. Houvenaghel-Defoort, W. Panning, and J. E. Guillet, "Automatic Recording Capillary Viscometer for the Study of Polymeric Reactions," *Rev. Sci. Instrum.*, **47**, 1496 (1976).

182. M. Breton and D. Gustafson, "A Simple, Cost-Effective Automatic Capillary Viscometer System," *J. Polym. Sci., Polym. Phys. Ed.*, **21**, 1559 (1983).

183. A. C. Ouano, "Gel-Permeation Chromatography: VII. Molecular Weight Detection of GPC Effluents," *J. Polym. Sci., Part A1*, **10**, 2169 (1972).

184. D. Lecacheux, J. Lesec, and C. Quivoron, "High-Temperature Coupling of High-Speed GPC with Continuous Viscometry: I. Long-Chain Branching in Polyethylene," *J. Appl. Polym. Sci.*, **27**, 4867 (1982).

185. D. Lecacheux, J. Lesec, C. Quivoron, R. Prechner, R. Panaras, and H. Benoit, "High-Temperature Coupling of High-Speed GPC with Continuous Viscometry: II. Ethylene-Vinyl Acetate Copolymers," *J. Appl. Polym. Sci.*, **29**, 1569 (1984).

186. J. Lesec, P. Lecacheux, and G. Marot, "Continuous Viscometric Detector in Size Exclusion Chromatography," *J. Liq. Chromatogr.*, **11**, 2571 (1988).

187. J. L. Ekmanis, R. A. Skinner, and N. F. Waldhauser, "GPC Analysis of Polymers with an On-line Viscometric Detector," 2nd International Symposium on Polymer Analysis and Characterization, April 1989, Austin, Texas, Abstract A9. Also see *J. Appl. Polym. Sci., Appl. Polym. Symp.,* 1991 (in press).

188. M. A. Haney, "The Differential Viscometer, II. On-line Viscosity Detector for Size-Exclusion Chromatography," *J. Appl. Polym. Sci.*, **30**, 3037 (1985).

189. W. W. Yau, S. D. Abbott, G. A. Smith, and M. Y. Keating, "A New Stand-Alone Capillary Viscometer Used as a Continuous SEC Detector," *Am. Chem. Soc. Symp. Ser.*, **352**, 80 (1987).

190. Z. Grubisic, P. Rempp, and H. Benoit, "A Universal Calibration for Gel Permeation Chromatography," *J. Polym. Sci., Polym. Lett. Ed.*, **5**, 753.

CHAPTER

8

POLYMER CHARACTERIZATION USING THE ULTRACENTRIFUGE

PETR MUNK

Department of Chemistry and Biochemistry and Center for Polymer Research
University of Texas at Austin
Austin, Texas

1. TRADITIONAL METHODS IN SEDIMENTATION

Sedimentation analysis in the hands of its founder, Theodor Svedberg, played an extremely important role in the early days of polymer science, in the 1920s.

Modern Methods of Polymer Characterization, Edited by Howard G. Barth and Jimmy W. Mays
ISBN 0-471-82814-9

Svedberg introduced both (a) the sedimentation velocity method, which together with measurements of diffusion coefficients yielded the first dependable molecular weights of macromolecular materials, and (b) the sedimentation equilibrium method, yielding both the weight-average $\bar{M}_w$ and the z-average $\bar{M}_z$ molecular weights. The sedimentation molecular weights finally put to rest all objections to the concept of macromolecules being really high-molecular-weight substances, and brought to life the art of macromolecular characterization.

Since the time of Svedberg, sedimentation analysis provides absolute values of molecular weight with very good accuracy, is applicable to a broad range of molecular weights (from 200 to 10,000,000 and more) and provides precise information about hydrodynamic properties of macromolecules, as well as valuable thermodynamic data. It requires only a minimum amount of material (often 20 μg of protein is all that is needed for measurement of molecular weight and polydispersity).

Sedimentation measurements, however, are tedious, ultracentrifuges are prone to malfunction (especially during experiments lasting many days), and the evaluation techniques are sometimes quite time consuming. With the advent of newer methods of polymer characterization, such as light scattering, size exclusion chromatography (SEC), rheological, and spectroscopic methods, sedimentation analysis is being used less often. This is perfectly understandable since the newer methods often provide the required information with much less effort. However, the strength of the sedimentation method *survives*. With the exception of light scattering, it is the *only* method that gives absolute molecular weight, uses minute amounts of materials, and is extremely flexible.

We will review neither the mathematical and physical background of sedimentation analysis nor will we describe the basic features of the ultracentrifuge, which are adequately covered in the monographs and review articles quoted in the General References. After a short review of sedimentation velocity and sedimentation equilibrium, we will describe a few experimental developments, which recently made the method less tedious and more precise. We will then present several instances in which the ultracentrifuge was used in solving some difficult characterization problems.

1.1. Sedimentation Velocity

Traditionally, the method of sedimentation velocity has been used for two major purposes: as a criterion of purity of biochemical materials and for the measurement of sedimentation coefficients. This coefficient serves for the characterization of the material and for the calculation of molecular weight and hydrodynamic properties. In the simplest case, the sedimentation coefficient is related to other quantities as

$$s_0 = M(1 - \bar{v}\rho)/N_A f_0 \quad (1)$$

where the subscript zero refers to data extrapolated to zero concentration, M is the molecular weight, $\bar{v}$ the partial specific volume, ρ the density of the solution, f is the friction coefficient, and N_A is Avogadro's number. Equation 1 is used either for calculating the friction coefficient (when M is known, e.g., from light scattering experiments) or for calculating molecular weight if f_0 is known from the measurement of the diffusion coefficient D.

$$f_0 = kT/D_0 \quad (2)$$

Here, k is Boltzmann's constant and T is absolute temperature. In fact, carefully measured data may lead to values of molecular weight surpassing in accuracy the data obtained by other methods (1). The friction coefficients may serve as a basis for a detailed analysis of basic hydrodynamic constants (2).

Sedimentation velocity experiments are used as a check of purity of the macromolecular solutes. In a paucidisperse system (e.g., mixture of two or more proteins), each component is resolved. The individual peaks may be either fully or partially separated. Even a shoulder on the peak may reveal the presence of an impurity. In a polydisperse system, a broad peak is considered to be an indication of polydispersity. However, both diagnostic tests should be treated with caution, since they may be disturbed by two unwanted phenomena: self-sharpening of the peaks and convection within the cell.

Self-sharpening is a result of higher solute concentration. The sedimentation coefficient decreases with increasing solute concentration. It follows that those molecules that are trailing the boundary as a result of either diffusion polydispersity move within a region of low concentration. They move faster than the molecules in the plateau region and are "catching up." Consequently, the peak is narrower than expected. Sometimes an impurity with slightly lower sedimentation constant may escape detection. This may be remedied by using a lower concentration, for which the self-sharpening effect is smaller.

At very low concentrations, however, convection in the cell may occur easily. The viscosity of most solutions studied by an ultracentrifuge is rather low. From the view of macroscopic hydrodynamics, they may behave as ideal fluids (zero viscosity). It is a well-known fact in the hydrodynamics of ideal fluids that a vortex, once formed, lasts forever (or for a long time when the viscosity is small but finite). In the cell of the ultracentrifuge, a vortex may be formed by the smallest mechanical irregularity of the rotor movement. Generally, the heavier rotors are less prone (but not immune) to the formation of vortices. The vortices lead to mixing of the solution in the whole cell or in a part of it negating the centrifugal processes. Even in the smoothest run of the ultracentrifuge, there are sufficient opportunities to start a convected run. In our experience, a convected run cannot be salvaged and must be stopped. Fortunately, there is an effect that works against vortex formation: a density gradient. In the presence of a density gradient, the convection would cause the denser liquid to rise above the less dense one, which is against the rules of mechanical equilibrium. Thus, the density gradients stabilize the run against convections. Of course, the stabilizing gradient must be large enough. Under otherwise comparable conditions, low-velocity runs are more prone to convection than the high-velocity ones.

The stabilizing gradient is frequently provided by the macromolecular solute itself, and the convection almost never occurs within the sedimenting peak. However, in the concentration-depleted region behind the sedimenting peak, as well as in the plateau region, the density gradient may be insufficient and a vortex may develop in these portions of the cell. Even a slow vortex equalizes the concentration within its region, and also "chews" on the neighboring peak. In our experience, a depleted meniscus region almost always indicates convection. One test for convection is the radial dilution law: When the peak area decreases during the sedimentation faster than predicted, convection should be considered. Convection is usually not very troublesome in biochemistry. Biochemical materials are usually dissolved in buffers. The salts forming the buffer redistribute themselves in the cell creating a density gradient that prevents convection. However, synthetic polymers dissolved in single solvents are very prone to convection. We have seen many photographs from velocity experiments allegedly proving a narrow distribution of molecular weights, which were the result of such insidious convection. When a mixed solvent is used for such an experiment, it may form a stabilizing density gradient similar to those formed by the salts in buffered solutions.

Provided that the problems with peak sharpening and convection can be avoided, the shape of the sedimenting boundary may be interpreted in terms of molecular weight distribution. For this purpose, the effect of diffusion must be eliminated. The distance traveled by a molecule because of diffusion is proportional to the square root of time $t^{1/2}$, while the distance sedimented is proportional to time. It follows that the shape of the boundary must be extrapolated toward $1/t$ (or better, toward $1/t^{1/2}$) to eliminate the effect of diffusion (3). If the sedimentation coefficients depend on concentration, another extrapolation toward vanishing concentration is necessary. These procedures are described in detail by Williams (4) and by Fujita (5).

1.2. Sedimentation Equilibrium

Traditionally, sedimentation equilibrium experiments were used for the measurement of molecular weights. The analysis is reasonably simple for solutions that are thermodynamically ideal. The weight-average molecular weight $\bar{M}_w$, as well as the centrifuge average $\bar{M}_z$, may be obtained from a single equilibrium experiment. The ratio of $\bar{M}_z/\bar{M}_w$ provides a measure of the polydispersity of the sample. Technically, for the evaluation of these experiments, the equilibrium concentration at the column meniscus must be known. It is either established using the law of mass conservation from the original concentration of the sample, or the speed of the rotor can be selected high enough to make the concentration at meniscus negligibly small. The latter method is slightly less precise, but has an advantage in not requiring a separate experiment for measurement of the original concentration.

During the process of sedimentation, a sample of a polydisperse macromolecular material redistributes itself in the cell. The heavier molecules sediment faster and, at equilibrium, they settle closer to the cell bottom. This feature complicates the experiments designed for obtaining the average molecular weight. However, this redistribution presents an opportunity for studying the polydispersity of polymer samples, as well as the association equilibria of protein solutions. The situation is complicated further by nonideality of the solutions. Extensive literature in this field was reported by Adams and co-workers (6,7) and by Williams and co-workers (8). A detailed analysis requires measurements on several solutions with different initial concentrations of the polymer. For each solution, sedimentation equilibrium has to be established at several rotor speeds. An appropriate analysis will yield both the distribution of molecular weights and the value of the second virial coefficient. However, the work involved is considerable. A similar analysis applied to protein solutions may yield information about the type of equilibrium involved (e.g., monomer–dimer–tetramer) as well as the values of the equilibrium constants involved.

2. TECHNICAL IMPROVEMENTS

Major drawbacks of the ultracentrifuge methods are the length of time needed for the experiments (especially the equilibrium experiments) and the lengthy collection and analysis of the data. However, remedies for both problems are readily available, as described in this section.

2.1. Multicell Experiments

The time needed for the establishment of the sedimentation equilibrium is generally rather long. While a few hours are sufficient for a solution column 1 mm high, 1–2 days are needed for 3-mm columns, and several days may be required for longer columns. For many types of experiments, the experimental data must be collected for a sufficiently long column. In this case, the experiment has to be run for a long time. The only time-saving device is in performing a large number of experiments simultaneously in the same rotor. For such an experiment multicell rotors are needed. While Beckmann manufactures four-cell rotors, which may be operated at speed up to 52,000 rpm, Heraeus-Christ produces custom-made titanium rotors with up to eight cells and maximum speed of 60,000 rpm (9).

In the older designs of the optical system, the images of individual cells were separated using wedged windows. However, modern modulated light beams may be synchronized with the rotor movement and the cells may be illuminated separately. The modulation of the UV light is used together with the scanning device (10). For visible light and Schlieren and/or interference optics, a stroboscopic light source (11) or a modulated laser (9, 12, 13) are used.

2.2. Computer Assisted Analysis of Interference Photographs

The distribution of concentrations within the sedimentation cell is obtained most conveniently from scans using UV light of an appropriate wavelength (14). While the method is very useful for biochemical materials dissolved in aqueous buffers, it can seldom be used for polymers in organic solvents. On the one hand, many polymers may not have any absorption at available wavelengths, on the other hand, solvents with carbonyl function, halogenated solvents, aromatic solvents, and so on, may not have enough light transmission in the UV. For example, the absorption maximum of carbonyl compounds is around 250 nm, yet in neat form in a 12-mm layer, they are virtually opaque for most wavelengths in the UV region. Under such circumstances, the measurement of concentration profile must be based on the refractive index. The Schlieren optics is very sensitive but quantitative measurements can rarely be performed. Consequently, most researchers use the interference optics for evaluation of sedimentation equilibrium experiments.

Manual measurement of interference photographs is very tedious. About a hundred fringe positions should be read for each photograph with a precision of at least 5 μm on the plate. Several photographs must be evaluated for every run. De Rosier et al. (15) coupled a Nikon profile projector with computer driven motors moving the mechanical stage. The projected image was scanned by a photomultiplier across the fringe pattern. The Fourier transform analysis of the scan yielded the position of the central fringe, fringe interval, fringe contrast, and the maximum of the envelope of the intensity of the fringe pattern. The data were collected on magnetic tape in a format suitable for subsequent computer analysis. Using this instrument, we were able to read the position of the fringe with a standard deviation of about 2 μm. Most of the deviation was caused by the actual distortion of the fringes on the photograph (caused by imperfections of the optical system) and by photographic noise. Collection of the data for one equilibrium experiment together with extensive computer analysis required a few hours for a semiskilled operator as compared to several days of a skilled operator to manually collect the same amount of data. Recent progress in image analysis and computers allow even faster data collecting routines.

3. NEWER APPLICATIONS OF THE ULTRACENTRIFUGE

3.1. Refinements in the Analysis of Sedimentation Velocity Experiments

In sedimentation velocity experiments, the concentration of the sedimenting material in the plateau region decreases as the boundary travels along the column (radial dilution law). Since the sedimentation coefficient is concentration dependent, it changes during the experiment. Moreover, at high rotor velocities, an appreciable pressure gradient is generated within the cells. The pressure at the meniscus is atmospheric while at the bottom it is of the order of 200 atm. At high pressures, the compression of liquids leads to changes in density, viscosity, and thermodynamic quality of the solvent. All these changes influence the sedimentation coefficient, which, as a result, depends on the position in the cell, rotor speed, and so on. Consequently, the dependence of the logarithm of the boundary position on the time of sedimentation (linear in a textbook case) is curved even in ideal experiments. In older studies of biochemical materials dissolved in aqueous buffers, the deviations from linearity were relatively insignificant because of the low compressibility of water and the small values of the coefficient (k_s) characterizing the concentration dependence of the sedimentation coefficient. However, for polymers dissolved in organic solvents, the error caused by neglecting the curvature of the plots may easily reach 20–30%.

Obviously, careful analysis is necessary when high-precision experimental data are needed. Vidakovic et al. (16) pointed out that the nonlinear term in the above quoted dependence (logarithm of the radial distance of the boundary on time) has two contributions: A smaller contribution for the concentration dependence, which is proportional to the fourth power of the angular velocity ω, and a larger contribution for the pressure dependence, which is proportional to the sixth power. Hence, from experiments at different concentrations and different rotor speeds, both terms can be evaluated.

Mulderije (17–19) devoted extensive effort to improvements of the analysis of sedimentation velocity experiments. He has developed a precise method for finding peak maxima from the steep sections of the Schlieren pattern and their fine interference structure (19). For velocity experiments at higher rotor speeds, a linear approximation of the pressure effect is not sufficient and a quadratic approximation is necessary (19). For measurement of small differences in the sedimentation coefficients of two samples (e.g., solutions of the same polymer at two different concentrations), Mulderije (18) developed a method comparing peak positions recorded at the same time—preferably in a two-cell experiment in a single run. This procedure eliminates the need for correcting for the acceleration of the rotor, the pressure effect, and so on. The method is especially valuable for calculating the coefficients of the dependence of sedimentation coefficient on concentration and temperature (18).

In a typical velocity experiment designed for calculating the distribution of sedimentation coefficients in a polydisperse sample, the continuous radial distribution of concentration (i.e., the peak shape) is measured at several times (i.e., several photographs are taken). Mulderije (17) showed that equivalent information may be obtained by continuous monitoring of the polymer concentration at one or more locations in the cell. In these experiments, the concentration is measured by following the change of the intensity of the light passing through the cell, caused either by the light absorption (UV scanner is conveniently used for these experiments), or by the light scattered by large sedimenting particles. This method is advantageous especially in cases when the effects of diffusion and interaction vanish, wherein the averages of the sedimentation coefficients are easily obtained. The method is also more powerful than the customary one when analyzing the distribution of slow components within the sedimenting peak.

Mächtle (20) used the same principle for studying dispersions of larger particles (e.g., latexes in the range of 20–2000-nm diameters). In a multicell rotor he measured the same sample dispersed in H_2O, D_2O, and a 1:1 mixture of H_2O/D_2O. These three liquids have densities 1.0, 1.1, and 1.05 g/mL, respectively. From the comparison of the three runs he was able to calculate the density of the particles. Knowing the density he calculated the friction coefficient and the particle radius from the time in which the boundary passed the detector (i.e., from the sedimentation coefficient). Finally, from the particle radius and turbidity (provided by the detector) he obtained the refractive index of the particle with the use of Mie theory of light scattering. Multimodal distributions were successfully studied by this method.

3.2. Sedimentation Velocity of Semidilute Solutions

Originally, the method of sedimentation velocity was employed mainly for studying the properties of individual macromolecules or particles. It was soon discovered that the sedimentation behavior depends on the concentration of the solution because of the hydrodynamic and thermodynamic interactions of the dissolved molecules. However, the analysis (using the model of slightly interacting particles) was geared toward elimination of these unwanted effects so that the properties of individual particles could be deduced. This model is reasonable for compact particles (proteins and latexes) up to rather high concentration, but it is applicable for coiled molecules of linear polymers only at rather low concentrations.

At some relatively low overlap concentration c^*, the molecular coils fill all the available volume. At concentrations $c > c^*$ the individual coils overlap, form many contacts, and become entangled. This change is accompanied by a fundamental change in the sedimentation behavior of the polymer solution. At low concentrations, the effect of hydrodynamic interaction is mainly intramolecular: The polymer coil is essentially impermeable to the solvent and the latter flows mainly in regions in-between the individual coils. However, at concentrations above c^*, there is no space between the coils, and the solvent has to flow through the entangled network of all polymer coils. At a given concentration of polymer, the hydrodynamic properties of the network are no longer dependent on the length (molecular weight) of the individual molecules. Therefore, the dependences of the sedimentation coefficient on concentration for samples of different molecular weights should merge into a single dependence above the overlap concentration. This behavior was observed experimentally (Fig. 1).

Mijnlief and Jaspers (21) pointed out that the sedimentation of an entangled polymer network is physically equivalent to flow of the solvent through a porous medium. Simple porous media are customarily characterized by a parameter called permeability, which is independent of the nature of the liquid flowing through it and of temperature and pressure. Mijnlief and Jaspers (21) showed that the permeability of the polymer network k is related to its sedimentation coefficient s. Their relation (after a slight change in notation) reads

$$k = \eta s / c(1 - \bar{v}\rho) \tag{3}$$

where c and $\bar{v}$ are the concentration and specific volume of the polymer, respectively. The terms η and ρ are, respectively, the viscosity and density of the solvent. Permeabilities of a sample of poly(α-methylstyrene) (PAMS) at a single concentration, which is well above the overlap concentration, are

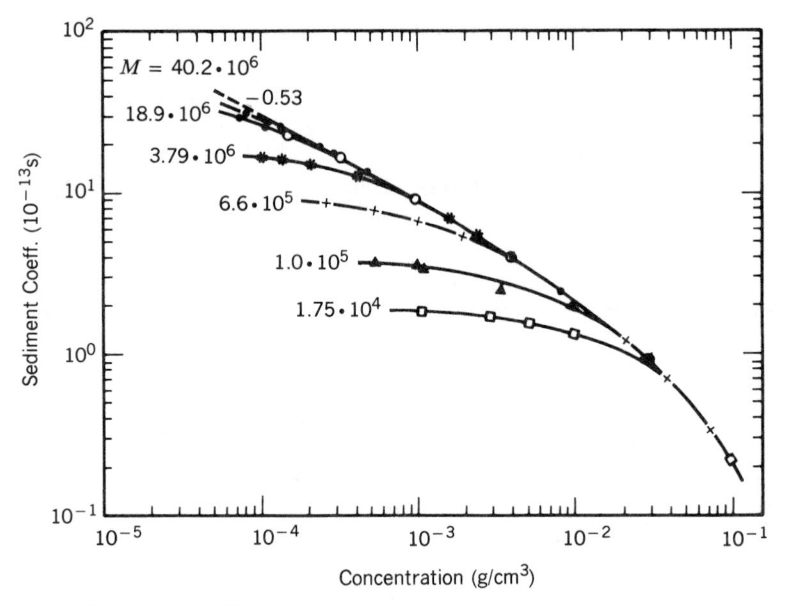

Figure 1. Sedimentation coefficient s versus concentration c plot for polystyrene samples with narrow molecular weight distribution in toluene at 20 °C. The numerical values at the curves denote the molecular weight M; -0.53 is the limiting slope of the envelope [Huang and Meyerhoff (27)]. Reproduced with permission from Hüthig & Wèpf Verlag.

plotted in Figure 2 as a function of temperature for a good solvent (toluene) and a poor solvent (cyclohexane). The difference of permeability for the two solvents, and its change with temperature for the poor solvent indicate that the structure of the entangled network (of a given average density of polymer segments) depends on the thermodynamic quality of the solvent. Seemingly, the segments of the polymer are more evenly distributed in the space in good solvents than in the poor solvents. A more even distribution means narrower channels for the solvent flow and lower permeabilities.

More recently, the structure of the entangled networks in the semidilute region (at concentrations above the overlap concentration) has been studied in considerable detail. The "scaling laws," first introduced by deGennes and Brochard (22, 23) gained wide acceptance. According to their model, the polymer network is characterized by a set of contact points between polymer chains. When the solvent and the polymer are in relative motion (as in sedimentation), the hydrodynamic interactions among polymer segments do not extend beyond the contact points. The average length of the polymer chains between two contact points is called a screening length: It decreases with increasing concentration of the solution. The section of the chain between

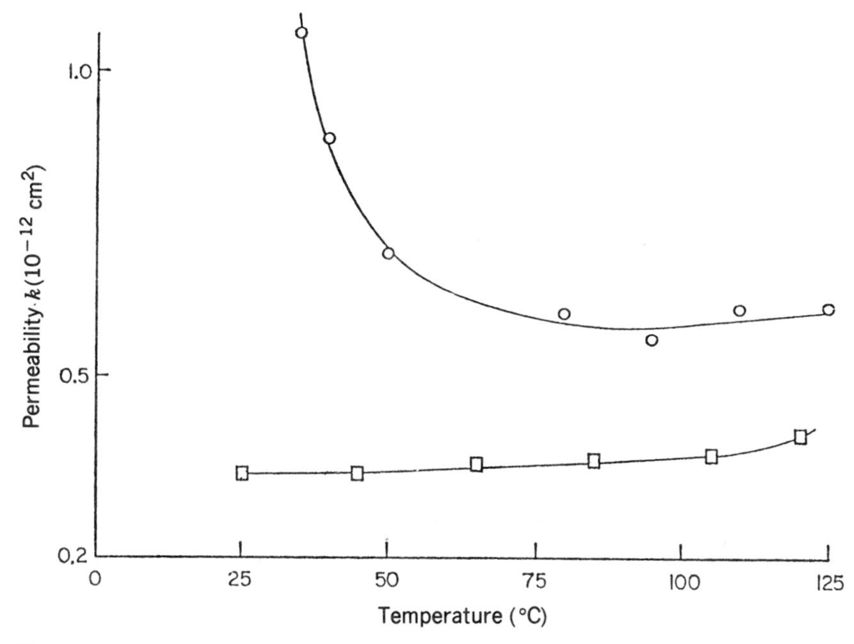

Figure 2. Permeability k in solutions of PAMS ($M = 6.5 \times 10^6$) as a function of temperature. Concentration: 0.0164 g/mL; ○, in cyclohexane; □ in toluene [Mijnlief and Jaspers (21)]. Reproduced with permission from the Royal Society of Chemistry.

the contact points is called a "blob," and it obeys the conformational rules of polymer chains (which depend, of course, on the thermodynamic interactions between the polymer and the solvent). The hydrodynamic interaction plays its full role within the blob, which moves as a more-or-less impermeable particle through the solvent. The hydrodynamic radius of the blob (for a given concentration of a given polymer sample) depends on the thermodynamic quality of the solvent. The scaling theory casts all the experimental dependences in the form of power laws and predicts the applicable exponents, but does not predict the proportionality constants. For sedimentation, the theory predicts that in the semidilute region the sedimentation coefficient is independent of the molecular weight of the polymer, and is proportional to $c^{-0.5}$ for good solvents and to $c^{-1.0}$ for θ (pseudoideal) solvents. These predictions were confirmed experimentally with reasonably good precision (16, 24–27).

It should be noted that the semidilute region (when defined by the above relations) is rather narrow, and the logarithmic dependence of s versus c is strongly curved in the region between 0.01 and 0.1 g/mL. The exponent approaches -2.0 at the latter concentration for good as well as for poor solvents (24, 27).

3.3. Molecular Weight of Polymers in Nonideal Solutions

The measurement of molecular weight by means of sedimentation equilibria, which is straightforward for monodisperse polymers in pseudoideal solutions, becomes rather involved for polydisperse polymers in nonideal solutions. Scholte (28–30) developed a method for measurement of molecular weight distribution (MWD) for polymers in pseudoideal (θ) solutions. The method requires measurement of sedimentation equilibrium at several (5–7) rotor speeds. Because of the nature of the polymer distribution in the sedimentation cell at equilibrium, polymers with different MWD may yield very similar concentration profiles. For example, a mixture of two monodisperse polymers differing by a factor of two in molecular weight yields an essentially identical concentration profle as a polymer with unimodal MWD of appropriate width and average molecular weight. It follows that any method for evaluation of MWD from sedimentation equilibria may yield only broad outlines of MWD but not a detailed function. Adams and co-workers (6, 7, 31, 32) generalized Scholte's method for polymers in nonideal solutions. The generalized method requires knowledge of the second virial coefficient B. This coefficient may be obtained either from light scattering experiments or from the sedimentation equilibrium itself. In the latter case, sedimentation equilibria have to be measured for several different starting concentrations of polymer at several rotor speeds for each concentration, a rather tedious procedure.

Under these circumstances, an alternative procedure, which would evaluate the molecular weight, polydispersity, and second virial coefficient from a single sedimentation equilibrium experiment would be attractive. The method cannot deliver all this information in a single run without limiting the accuracy of all the experimental values. Nevertheless, the results may be fully satisfactory under a number of circumstances.

1. For characterizing a new polymer.
2. For determining the molecular weight of a sample known to be monodisperse (e.g., a protein).
3. For measuring thermodynamic parameters.

The evaluation techniques for sedimentation equilibria developed by Munk and co-workers (33–35) take advantage of the fact that the equilibrium concentrations in the cell may cover a rather broad range. Hence, the dependence of experimental data on concentration of the polymer, which is needed for the analysis of the nonideality, may be obtained from a single equilibrium run (33–35).

The condition for equilibrium in an ultracentrifuge for each component of the solute is

$$M_i(1 - \bar{v}_i\rho)\omega^2 r = \sum_k (\partial\mu_i/\partial c_k)_{T,P,c_{j\neq k}}(dc_k/dr) \tag{4}$$

which for a homogeneous solute reduces to

$$M_2(1 - \bar{v}_2\rho)\omega^2 r = (\partial\mu_2/\partial c_2)_{T,P}(dc_2/dr) \tag{5}$$

Here, index 1 refers to the solvent, index 2 to the homogeneous solute, and indices i, j, k to the components of heterogeneous solute. The terms M_i, $\bar{v}_i$, μ_i, and c_i are, respectively, the molecular weight, partial specific volume, chemical potential, and concentration (in g/mL) of the i component. The distance from the rotational axis is r, ω is the angular velocity of the rotor, ρ is the density of the solution, P is the pressure, and T is the absolute temperature. In the relations applying to the heterogeneous solutes, $c = \sum_i c_i$ will be the total concentration of solutes.

For a homogeneous solute, the partial derivative in Eq. 5 reads

$$(\partial\mu_2/\partial c_2)_{T,P} = RT(1 - c_2\bar{v}_2)(1/c_2 + 2BM_2 + 3CM_2^2 c_2 + \cdots) \tag{6}$$

where B and C are the osmotic virial coefficients, RT has its usual significance. For most solutions, the volume changes in mixing are negligible. In such a case the following relation holds

$$(1 - \bar{v}_2\rho) = (1 - \bar{v}_2\rho_1)(1 - c_2\bar{v}_2) \tag{7}$$

where ρ_1 is the density of pure solvent.

For further calculations it is convenient to define an experimentally accessible quantity M_{app} as

$$1/M_{app} \equiv (1 - \bar{v}_2\rho_1)\omega^2 c_2 r/RT(dc_2/dr) \tag{8}$$

Obviously, the apparent molecular weight M_{app} is a function of the position within the cell r. Combination of Eqs. 5–8 yields

$$1/M_{app} = 1/M_2 + 2Bc_2 + 3CM_2c_2^2 + \cdots \tag{9}$$

Thus, the plot of $1/M_{app}$ versus c_2 yields for a homogeneous solute $1/M_2$ as an intercept and the second virial coefficient as the initial slope. For slightly nonideal solutions, the whole dependence is linear. For stronger nonidealities it curves up because of the third virial term. In the latter case, the plot of $1/M_{app}^{1/2}$ versus c_2 exhibits usually much better linearity allowing for more confident extrapolation. The procedure is similar to the one used for treatment of osmotic and light scattering data (36).

Thus, for monodisperse samples, a single equilibrium run provides unambiguously both the molecular weight and the virial coefficient. The above analysis is especially useful for samples of biochemical origin known to be homogeneous (37).

For heterogeneous solutes, it is customary to assume that all the interactions among molecules of different size may be expressed by the same set of virial coefficients (B, C,...). Moreover, all the specific volumes are assumed to have the same value $\bar{v}$. In this case, an equation analogous to Eq. 9 may be derived as

$$1/M_{app} = 1/\bar{M}_w + 2\,Bc + 3\,C\bar{M}_z c^2 + \cdots \tag{10}$$

where $\bar{M}_n$, $\bar{M}_w$, and $\bar{M}_z$ are the molecular weight averages defined in the usual way. However, in this case $\bar{M}_w$ is a function of r and, because of this, it depends on the local concentration c. As a result, $\bar{M}_w$ increases with increasing concentration. This effect leads to an upturn in the plot of $1/M_{app}$ versus c in the region of low concentrations. This upturn ruins the validity of the extrapolation procedure. Thus, a different procedure was developed for heterogeneous samples. A tangent is drawn to the plot of $1/M_{app}$ versus c or (for strongly nonideal systems) to the plot of $1/M_{app}^{1/2}$ versus c. The intercept of the tangent is the local value of $\bar{M}_z/\bar{M}_w^2$ in the former case, and $(\bar{M}_z + \bar{M}_w)/2\,\bar{M}_w^{3/2}$ in the latter case. The slope of the tangent has two terms: one related to the virial coefficient and the other to heterogeneity. The procedure is advantageous for polymer samples of moderate heterogeneity (fractionated polymers, narrow distribution polymers from living polymerization, etc.) brought to equilibrium at conditions when the concentration at meniscus is still a sizeable fraction (10–30%) of the concentration at the bottom. For such experiments, the whole experimental dependences are usually linear within experimental error, and there is no problem with choosing the point for drawing the tangent (33).

For equilibrium experiments at higher rotor speeds leading to depletion of the polymer from the meniscus region, the redistribution of the solute in the cell leads to a rather significant dependence of $\bar{M}_w$ on the position in the cell. It could be shown (34, 35) that

$$d\ln \bar{M}_w/d\ln c = (\bar{M}_z/\bar{M}_w) - 1 \equiv u \tag{11a}$$

$$d\ln \bar{M}_n/d\ln c = 1 - (\bar{M}_n/\bar{M}_w) \tag{11b}$$

where the polydispersity parameter u is defined by the identity in Eq. 11a. This relation is valid for all polymer solutions, from ideal to strongly nonideal, provided only that a single set of virial coefficients can describe all the interactions. It is obviously the dependence of $\bar{M}_w$ on *concentration*, which is the same for systems of different ideality. The dependence of $\bar{M}_w$ on the *radial position* in the cell is different for different nonidealities because the dependence of the concentration on the position is different.

In the following, we will show that for one type of unimodal molecular weight distribution—the Schulz–Zimm distribution—a single sedimentation equilibrium experiment may provide the average molecular weight, the width of the distribution, and the values of thermodynamic parameters (35).

The Schulz–Zimm distribution is characterized by two parameters $\bar{M}_w$ and z:

$$dc = c[b^{z+1}/\Gamma(z+1)]M^z \exp(-bM)\,dM \tag{12}$$

$$b \equiv (z+1)/\bar{M}_w \tag{12a}$$

$$u = 1/(z+1) \tag{12b}$$

where $\Gamma(x)$ is the gamma function of argument x.

If the MWD at any point in the sedimentation cell can be described by Eq. 12, it is possible to show by modification and integration of Eq. 2 that the MWD at any other point in the cell is described also by Eq. 12 with a different value of $\bar{M}_w$, but with the same value of z. Thus, the parameter u in Eq. 11 is constant throughout the cell and Eq. 11 may be integrated to yield

$$\bar{M}_w = Kc^u \tag{13}$$

where K is a constant.

Once the parameters K and u are known, the molecular weight averages of the original sample $\bar{M}_w^{\text{or}}$ and $\bar{M}_z^{\text{or}}$ may be calculated from the relations

$$\bar{M}_w^{\text{or}} = K \int_{r_m}^{r_b} c^{1+u} r\,dr \Big/ \int_{r_m}^{r_b} c^u r\,dr \tag{14}$$

$$\bar{M}_z^{\text{or}} = K(1+u) \int_{r_m}^{r_b} c^{1+2u} r\,dr \Big/ \int_{r_m}^{r_b} c^{1+u} r\,dr \tag{15}$$

The integration limits r_m and r_b are the radial distances of meniscus and bottom, respectively. Equations 14 and 15 are written for sector-shaped cells and the integrals are computed numerically from the experimental concentration profile. The polydispersity factor of the original sample u^{or} is given as

$$u^{or} = \bar{M}_z^{or}/\bar{M}_w^{or} - 1 \tag{16}$$

It is interesting to note that for sedimentation equilibria of pseudoideal solutions at rotor speeds sufficient for depletion of the meniscus region, the distribution of the original sample is again of the Schulz–Zimm type. The characteristic parameter z^{or} is related to z as

$$z^{or} = z - 1 \tag{17}$$

Thus, MWD of the original sample is broader than that of the redistributed one. For nonideal solutions and nonnegligible meniscus concentrations, the difference between the characteristic parameters for the local and overall MWD is even less, while the unimodal character remains qualitatively the same.

Parameters K and u are calculated as follows. Equation 10 may be rewritten in a form

$$1/M_{app} = (1/\bar{M}_w) f(c, \bar{M}_w) \tag{18}$$

where the function $f(c, \bar{M}_w)$, which characterizes thermodynamically the given polymer–solvent system, contains a limited number of parameters. For example, $f(c, \bar{M}_w)$ for a system obeying Eq. 10 reads

$$f(c, \bar{M}_w) = [1 + 2B\bar{M}_w c + 3c\bar{M}_w^2(1 + u)c^2 + \cdots] \tag{19}$$

In the above cited study (35), we found that the function $f(c, \bar{M}_w)$ may be approximated in a broad region of concentrations and nonidealities by an empirical relation

$$f(c, \bar{M}_w) = (1 + \alpha\bar{M}_w c)^2/(1 + \beta\bar{M}_w c) \tag{20a}$$

$$B \equiv \alpha - (\beta/2) \tag{20b}$$

Combining Eq. 13 with Eq. 18 we obtain

$$1/M_{app} = K^{-1}c^{-u} f(c, Kc^u) \tag{21}$$

or

$$1/M_{app} = K^{-1}c^{-u}(1 + \alpha K c^{u+1})^2/(1 + \beta K c^{u+1}) \tag{22}$$

when Eq. 20a is used for $f(c, \bar{M}_w)$.

When analyzing the experimental dependence of M_{app} on c, a nonlinear least-square fit is used to find the adjustable parameters K, u, and the thermodynamic parameters of $f(c,\bar{M}_w)$, for example, α and β. The values pertinent to the original sample are then calculated from Eqs. 14–16.

The method yielded reasonably good results for narrow fractions and for moderately broad distributions of the most probable distribution type. It accounted for the behavior of very nonideal solutions up to $\bar{M}_w/M_{app} \sim 30$. For still larger nonidealities, it failed. However, such huge nonidealities are never encountered in experiments designed for a measurement of molecular weight and its distribution. The method also fails for very broad distributions and for bimodal distributions.

There is a similarity between our method [in its simpler version applicable to pseudoideal systems where $f(c, \bar{M}_w) = 1$] and the method of Donnelly (38, 39). Donnelly's method is based on the inversion of Laplace transforms and it is quoted as a method yielding a complete MWD. However, the method is unambiguous only for experiments characterized by two parameters (P and Q in Donnelly's notation). The resulting MWDs are two-parameter distributions of a specific type. The same is true for our method, which is based on the Schulz–Zimm distribution. The similarity of both methods is further stressed by the inability of both methods to handle bimodal distributions. For such distributions, the Laplace transform method is easily disturbed even by the smallest experimental errors.

3.4. Thermodynamic Data from Sedimentation Equilibria

Most sedimentation equilibrium studies consider the thermodynamic nonideality as a nuisance, which has to be corrected when molecular weight and MWD are measured. In a reversal of this approach, the equilibrium measurements may be used for obtaining extensive and rather precise information about the thermodynamic properties themselves. Scholte et al. (40, 41) developed a method for the measurement of chemical potentials and their dependence on polymer concentration from the sedimentation equilibria. From chemical potential, other thermodynamic data are obtained easily. The method employed Schlieren optics and several runs at different starting concentrations using short solution columns. However, a run with a long column yielded equivalent data. Thermodynamic parameters were measured up to 80% wt concentration. The authors have shown a clear difference in the shape of thermodynamic dependences for polystyrene in toluene (a good solvent) and in cyclohexane (a θ solvent). Moreover, they have shown that the dependence of thermodynamic parameters on concentration is molecular weight dependent up to at least 40% wt concentration of polystyrene in cyclohexane. Murakami et al. (42) used the same method for demonstrating

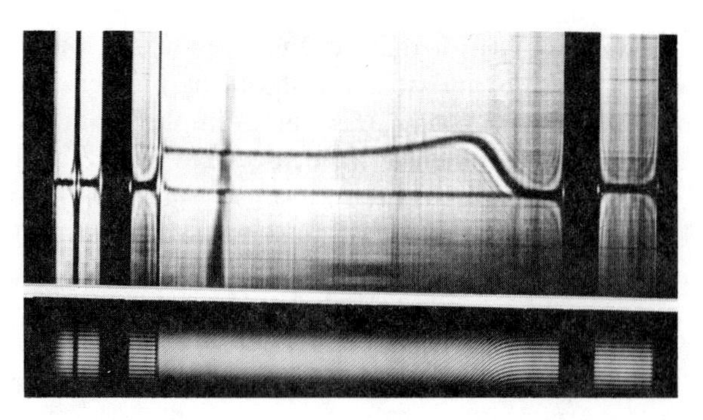

Figure 3. Interference and Schlieren photographs of a sedimentation equilibrium run with a long column. Polystyrene in bromobenzene, $M = 181{,}000$; 13,000 rpm, after 14 days [Chu and Munk (34)]. Reprinted with permission from S. G. Chu and P. Munk, *Macromolecules*, **11**, 101 (1978). Copyright © 1978 American Chemical Society.

the peculiar behavior of poly(chloroprene) in butanone. This system exhibits θ behavior in a broad temperature region between 15 and 35 °C.

Chu and Munk (34) used this method for studying a flotating system polystyrene–bromobenzene. They used interference optics instead of Schlieren (Fig. 3) gaining much better precision. To obtain this precision, they had to use a much narrower range of concentrations—up to about 7 volume percent. The theoretical approach and the evaluation analysis by the latter authors is equivalent to the principles developed by Scholte (40, 41). However, instead of first evaluating the somewhat difficult-to-handle chemical potentials, the interaction parameters are evaluated directly improving the precision of the results. We will therefore present the method of Chu and Munk.

The starting relation is again Eq. 4. The derivatives $(\partial \mu_i / \partial c_k)_{T,P,c_{j \neq k}}$ are obtained from a generalized relation of Flory–Huggins for the Gibbs function of mixing ΔG_{mix}

$$\Delta G_{\text{mix}}/RT = n_1 \ln \phi_1 + \sum_i n_i \ln \phi_i + n_i \phi_{\text{p}} g \tag{23}$$

where ϕ values are the volume fractions, ϕ_{p} is the total volume fraction of the polydisperse polymer, and g is a complicated function of the polymer concentration and of the MWD of the polymer. The phenomenological function g characterizes the thermodynamic interactions within the system. For polymer samples with narrow MWD it is permissible to assume that for a given sample, g is a function only of ϕ_{p} and is not influenced by the redistribution of molecular weights along the column in sedimentation equilibrium experiments.

The required derivatives then read

$$(\partial\mu_i/\partial c_k)_{P,T,c_{j\neq k}}/RT = \delta_{ik}/c_i + \sigma_i\bar{v}(1 - 1/\sigma_K) - 2F\phi_1\sigma_i\bar{v} \tag{24}$$

where $\sigma_i \equiv V_i/V_1$; V values are molar volumes; $\delta_{ik} = 1$ for $i = k$ and $\delta_{ik} = 0$ for $i \neq k$; $\bar{v}$ is the specific volume, which is the same for all molecular weights. The interaction function F is related to the well-known Flory–Huggins parameter χ as

$$F = \chi + (\phi_p/2)(\partial\chi/\partial\phi_p)_{P,T} \tag{25}$$

χ is related to the interaction function g as

$$\chi = g - \phi_1(\partial g/\partial\phi_p)_{P,T} \tag{26}$$

Obviously, in a special case when g is independent of concentration ϕ_p, all three parameters g, χ, and F are equal to each other. Substituting Eq. 24 into Eq. 4, summing over k, using the definition of number-average molecular weight $\bar{M}_n$, and realizing that the molecular weight averages are functions of the position in the cell and that, consequently, for a given experiment, they are functions of the local value of the total concentration of the polymer c, we obtain after some manipulation

$$c_i M_i B = dc_i/dr \tag{27}$$

$$B \equiv (1 - \bar{v}\rho_1)\omega^2 r\phi_1/RT + (dc/dr)[(\bar{v}/\bar{M}_n)(1 - d\ln\bar{M}_n/d\ln c) - (\bar{v}^2/V_1)(1 - 2F\phi_1)] \tag{28}$$

Summation of both sides of Eq. 27 over all polymer species yields for B

$$B = (dc/dr)/c\bar{M}_w \tag{29}$$

Combination of Eqs. 11b, 28, and 29 yields after some manipulation the expression for F as

$$2F = \frac{1}{\phi_1} - \left[\frac{(1 - \bar{v}\rho_1)\omega^2 r}{RT(dc/dr)} - \frac{1}{c\bar{M}_w}\right]\frac{V_1}{\bar{v}^2} \tag{30}$$

In Eq. 30, F is expressed as a function of experimentally accessible quantities and of the position dependent value $\bar{M}_w$. However, in the region of high concentrations and high molecular weight, the contribution of the term

$1/c\bar{M}_w$ to the value of F is very small. For example, for typical polymer–solvent systems and for $c = 0.05$ g/mL, $\bar{M}_w = 10^6$, this contribution is about 0.001. Thus, in such a case almost any estimate of $\bar{M}_w$ would be good enough. Even in the region of lower concentrations and molecular weights, a reasonable estimate is easy to make; for example, calculations related to Eq. 13 may yield satisfactory data. Once F is known as a function of the concentration, the concentration dependent parameter χ is evaluated from the relation

$$\chi = \frac{2}{c^2} \int_0^c F c dc \tag{31}$$

Chu and Munk (34) reported that polystyrene samples of different molecular weight in bromobenzene exhibited $\chi(c)$ functions, which did not display a tendency to merge into a single "master" function even at volume fractions of polymer $\phi_p \equiv c\bar{v}$ equal to 0.07 (Fig. 4). Rietveld et al. (41) report similar behavior for polystyrene in cyclohexane up to volume fractions $\phi_p = 0.32$.

The sedimentation equilibrium may also be used for obtaining another type of thermodynamic information, the preferential adsorption onto polymer in mixed solvents (43, 44).

The preferential adsorption is most conveniently described in terms of osmotic equilibrium. Let us consider the following hypothetical experiment.

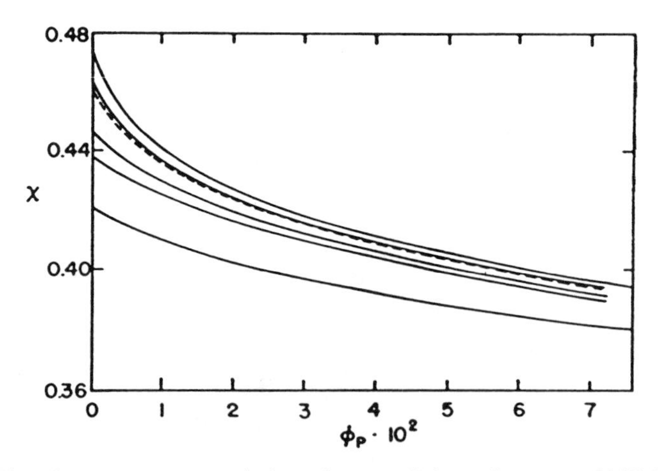

Figure 4. Plot of parameter χ versus ϕ_p for polystyrene in bromobenzene at 20 °C. Molecular weights of polystyrene (from bottom to top): 35,500; 114,000; 181,000; 366,000 (broken line); 596,000; 1,750,000. Reprinted with permission from S. G. Chu and P. Munk, *Macromolecules*, **11**, 101 (1978). Copyright © American Chemical Society.

Some amount of polymer is added to the solution side of the osmotic chamber while the equilibrium with the other unchanged part is preserved. This is possible only when the pressure on the solution side is changed and some amount of one of the solvents is added together with the polymer. The relative amount of the additions $(\partial m_2/\partial m_3)_\mu$ is the measure of the preferential adsorption. Here, subscript 1 refers to the principal solvent, 2 to the other solvent (to be added or subtracted), 3 to the polymer, m_i is the molality of the ith component. The subscript μ refers to the fact that at the preserved equilibrium the chemical potentials of the two solvents are unchanged. As a result of the preferential adsorption, the density increment of the polymer at constant composition of the mixed solvent $(\partial\rho/\partial c_3)_{m_2}$ differs from the increment at preserved osmotic equilibrium $(\partial\rho/\partial c_3)_\mu$. In fact, the difference of these increments is related to the parameter of preferential adsorption λ and to the derivative $(\partial m_2/\partial m_3)_\mu$ as

$$\lambda \equiv -\bar{v}_2\phi_1\frac{M_2}{M_3}\left(\frac{\partial m_2}{\partial m_3}\right)_\mu = \frac{[(\partial\rho/\partial c_3)_\mu - (\partial\rho/\partial c_3)_{m_2}]}{(d\rho/d\phi_1)} \tag{32}$$

Here, ϕ_1 is the volume fraction of solvent 1 in the absence of polymer, M values are molecular weights and $\bar{v}$ values are specific volumes. The derivative $d\rho/d\phi_1$ describes the dependence of the density of the mixed solvent on its composition.

Customarily, the density increments are related to partial specific volumes of the polymer as

$$(\partial\rho/\partial c_3)_{m_2} = 1 - \bar{v}_3\rho_0 \tag{33}$$

$$(\partial\rho/\partial c_3)_\mu = 1 - \bar{v}_3^*\rho_0 \tag{34}$$

where $\bar{v}_3$ is the usual specific volume, while $\bar{v}_3^*$ is a quantity defined by Eq. 34; ρ_0 is the density of the mixed solvent.

Combination of Eqs. 32–34 gives

$$\lambda = (\bar{v}_3 - \bar{v}_3^*)\rho_0/(d\rho/d\phi_1) \tag{35}$$

While $\bar{v}_3$, ρ_0, and $d\rho/d\phi_1$ are obtained by routine measurements of density, $\bar{v}_3^*$ is obtained from sedimentation equilibria. A routine but lengthy analysis of the sedimentation equilibrium of polymers in mixed solvents yields an equivalent of Eqs. 8 and 9 in the form

$$\frac{\omega^2 c_3 r}{RT(dc_3/dr)} = \frac{1}{(1 - v_3^*\rho_0)M_3} + \frac{2Bc_3}{1 - \bar{v}_3^*\rho} + \cdots \tag{36}$$

Provided that M_3 is known (e.g., from an experiment using a single solvent), the intercept and initial slope of the dependence of the experimental left-hand side of Eq. 36 on the concentration c_3 yield $\bar{v}_3^*$ and the second virial coefficient B.

Both the parameter λ and coefficient B are related to the mixing function ΔG_{mix}. Conversely, when λ and B are measured as a function of the solvent composition, the parameters of ΔG_{mix} may be evaluated. Munk and co-workers (43, 44) measured several polystyrene samples in mixtures of benzene–cyclohexane and ethyl acetate–cyclohexane. They employed the function ΔG_{mix} in the form

$$\Delta G_{\text{mix}}/RT = n_1 \ln \phi_1 + n_2 \ln \phi_2 + n_3 \ln \phi_3 + g_{12} n_1 \phi_2 + g_{13} n_1 \phi_3 + g_{23} n_2 \phi_3 + g_T n_1 \phi_2 \phi_3 \tag{37}$$

where the significance of the symbols is similar to Eq. 23. The solvent–solvent interaction function g_{12} could be deduced from the thermodynamic data for the solvent mixture alone. Chu and Munk (43) developed a procedure for calculation of the difference $(g_{13} - g_{23} V_1/V_2)^0$ and the ternary functions g_{T}^0 and $\chi_T^0 \equiv g_T^0 - (dg_T/d\phi_3)^0$. Here, the superscript zero denotes data at vanishing concentration of polymer. For both experimental systems, the ternary functions g_T^0 and χ_T^0 were very significant. This indicates a need for modification of the simple models of polymer solutions, which rely mainly on binary interactions.

Alternatively, (using an approach known from light scattering measurements), λ may be obtained also from the refractive increments $(\partial n/\partial c_3)_\mu$ and $(\partial n/\partial c_3)_{m_2}$ as

$$\lambda = [(\partial n/\partial c_3)_\mu - (\partial n/\partial c_3)_{m_2}]/(dn/d\phi_1) \tag{38}$$

where n is the refractive index of the solution. The quantity $(\partial n/\partial c_3)_\mu$ can be advantageously obtained by another ultracentrifuge method as we will describe in Section 3.6. It is very satisfying to find that the two completely unrelated methods yielded quite similar values of λ (44).

Rietveld (45) developed still another approach for obtaining thermodynamic data using ultracentrifugation. It is well known that the fine details of the thermodynamics of polymer solutions may be obtained from phase equilibria. Below the critical temperature, polymer solutions separate into two phases. From the composition of the two phases and its dependence on temperature, detailed information may be obtained about interaction coefficients. However, the separation of the phases may be rather difficult in the vicinity of the critical temperature. The centrifugal force is used with advantage for this purpose. A polymer solution of known composition is introduced into the cell and the temperature is lowered to the required value while the centrifuge is running at low speed ($\sim$ 120 rpm). After the two finely dispersed phases come to equilibrium, the speed is increased to about 1000 rpm and the two phases are rapidly separated. The cell is observed by the standard Schlieren optics. From the position of the separating meniscus, it is possible to calculate the volume ratio of the two phases. In an interesting experimental twist, the temperature of the rotor is then increased above the critical temperature. The two phases become miscible and the phase boundary is transformed into a typical concentration boundary that starts spreading due to the diffusion. Once the boundary spreads sufficiently, its area in the Schlieren photograph is a measure of the concentration difference between the two original phases. With the concentration difference and volume ratio known, it is easy to calculate the compositions of the two phases.

The main drawback of the method is the poor stability of the standard ultracentrifuges at very low velocities. Low velocities are mandatory, since high velocities would increase the pressure appreciably. This, in turn, would change the values of the interaction parameters. Another drawback is relatively insensitive temperature control of the centrifuges. Rietveld (45) constructed a low-speed centrifuge with carefully controlled stability and fine temperature control. The overall design was based on a Spinco Model E ultracentrifuge. The modified centrifuge performed as expected.

3.5. Sedimentation Equilibrium in a Density Gradient

The single experimental method that contributed most to the early understanding of the structure of nucleic acids and of the mechanism of their genetic function was sedimentation equilibrium in a density gradient (46, 47). For biochemical research, the method was developed to a high degree of sophistication. It allowed the study of miniscule differences in the density of nucleic acids (48) (the density differences were a manifestation of differences in chemical composition), it contributed to the understanding of the hydration of biochemical materials (49), and it provided the molecular weight of nucleic acids even in the region 10^6–10^7 g/mol, and higher (50). Furthermore, the amount of material needed for analysis was extremely low: Typically a few micrograms per experiment. We refer the reader to two excellent reviews of the method as applied to nucleic acids and proteins (49, 51).

It soon became obvious that the above mentioned features would also be extremely helpful in the study of synthetic polymers. Specifically, the density differences seem to offer a sensitive way for detection of polymeric impurities differing in chemical composition, as well as for an analysis of polymer blends and mixtures of polymers (e.g., grafted copolymer and the contaminating homopolymers). It seems that even the study of chemical heterogeneity of copolymers could be measured by this method. This property is difficult to determine by any other method. Furthermore, the measurement of molecular weights in the density gradient offers some special features, which may compensate for the long time requirements of the method. Namely, the method may yield several averages of molecular weight. The number-average molecular weight $\bar{M}_n$ is useful in the analysis of polymerization kinetics. For high-molecular-weight polymers, $\bar{M}_n$ is difficult to determine, although for lower molecular weight polymers, $\bar{M}_n$ can be measured by osmometry (see Chapter 6). Under favorable circumstances, $\bar{M}_n$ may be measured by density gradient sedimentation up to the region of 10^7 g/mol. Molecular weight measurements by any method are extremely difficult in this molecular weight region.

Recognizing these advantages, Hermans and Ende in the early 1960s presented the theory of sedimentation equilibrium in a density gradient for polymers (52) and copolymers (53–56). They treated combined effects of nonideality, polydispersity, and chemical heterogeneity. They also published several experimental studies dealing with most of the above-mentioned possibilities of the method (57–59). Their studies are reviewed in Ref. 60. Since that time, the method was almost never mentioned again in the literature in connection with synthetic polymers (61–63).

What went wrong? Why did a method so successful for nucleic acids not get the same acceptance in the synthetic polymer field? In our opinion, the main reason was that several fortunate features of nucleic acid solutions were absent in the polymer solutions.

1. Nucleic acids have a very strong absorption band at 260 nm. Solutions of most cesium salts, which were used as the density-gradient forming medium, are optically clear at this wavelength. Consequently, UV optics with a scanner could be used to provide an unambiguous record of the concentration within the cell. Moreover, because of the high extinction coefficient of DNA, concentrations even in the middle of the DNA band were so low that nonideality effects could be handled confidently.

2. Water, the principal solvent for DNA measurements, has rather low compressibility. Thus, the bothersome pressure-compressibility effects were less prominent than those for most organic solvents.

3. Many samples of DNA (mainly viral DNA) were strictly monodisperse and chemically homogeneous, simplifying the analysis.

4. Because of the very high molecular weight of DNA, the effects of heterogeneity and of molecular weight were clearly separated.

5. The thermodynamic behavior of solutions of cesium chloride and other

cesium salts has been known with good precision from electrochemical studies, greatly facilitating the evaluation of the density gradient.

Synthetic polymers lack most of these convenient features. Many of the important water-soluble polymers (polyacrylamide, polyacrylic acid, and polysaccharides) have no absorption band in the usable part of the UV spectrum. Similarly, only rarely does a polymer soluble in organic solvents exhibit absorption in a spectral region where density-gradient forming solvent components are optically clear. (Most dense solvents have chemical groups, e.g., bromine, which absorb most of the UV.) Consequently, the polymer bands in the ultracentrifuge cell have to be detected by means of their refractive properties using Schlieren or interference optics. Schlieren optics is very sensitive for the detection of the polymer bands, but is not sufficiently precise for a meaningful analysis of the band shape. The interference optics, which was not employed by Hermans, Ende and their co-workers (52–61), may have sufficient sensitivity and precision. Nevertheless, it requires the use of higher concentrations than were used in DNA studies. Consequently, efficient procedures must be used for the treatment of nonideality. Such procedures are feasible, and consist of using solvent mixtures, which are thermodynamically rather poor, and of measuring the sedimentation equilibrium of the same cell load at several rotor velocities. The latter procedure produces a rather convenient technique for establishing a concentration scale suitable for extrapolation to vanishing concentration.

Before treating the newer and tentative developments in density gradient ultracentrifugation, the physical and matematical background of the method will be reviewed.

When a mixture of two solvents differing in density is subjected to a centrifugal field, the solvent components are redistributed in the cell and (at equilibrium) form a density gradient. When a macromolecular component with a density intermediate between the solvent densities at meniscus and bottom of the cell is added to the mixture, the buoyancy forces would cause it to travel toward a particular position in the cell where the buoyant density of the macromolecules is equal to the density of the liquid.

3.5.1. Measurement of Molecular Weight

It was shown by Meselson et al. (46) that the shape of the band is Gaussian provided that, 1. the polymer is homogeneous in composition and molecular weight, 2. the density of the solvent is a linear function of radius within the band, and 3. the behavior of the polymer is thermodynamically pseudoideal. The width of the Gaussian band is inversely proportional to the molecular weight of the polymer, to the square of the rotor velocity, and to the density gradient. The density gradient itself is also proportional to the square of rotor velocity. It also depends on the density and thermodynamic behavior of the solvent components (64). Thus, when the density gradient is known, the molecular weight may be calculated from the width of the band.

When the polymer is polydisperse, the band is actually a superposition of many Gaussian curves. From the shape of such bands, Meselson et al. (46) calculated the molecular weight averages $\bar{M}_n$ and $\bar{M}_w$ using the expressions

$$\bar{M}_n = -\beta \int c(r)dr \Big/ \int (r - r_0)^2 c(r)\, dr \tag{39}$$

$$\bar{M}_w = -\beta \int \left(\frac{dc/dr}{r - r_0} \right) dr \Big/ \int c(r)dr \tag{40}$$

$$\beta = RT/\bar{v}(d\rho/dr)_{\mu_0}\omega^2 r_0 \tag{41}$$

Here, $c(r)$ is the position dependent concentration of polymer, $\bar{v}$ is the specific volume of the polymer, r_0 is the radial position of the band center, and integrals extend over the whole bands. These expressions are valid for polydisperse systems where all polymer components have the same specific volume, where the density gradient $d\rho/dr$ is independent of the position in the cell, and for thermodynamically pseudoideal systems. When considering nonlinear density gradients, Eqs. 39–41 should be modified to read (65)

$$\bar{M}_n = \frac{2RT}{\bar{v}\omega^2} \int c(r)d(r^2) \Big/ \int (r^2 - r_0^2)(\rho - \rho_0)c(r)d(r^2) \tag{42}$$

$$\bar{M}_w = -\frac{2RT}{\bar{v}\omega^2} \int \left[\frac{dc/d(r^2)}{\rho - \rho_0} \right] d(r^2) \Big/ \int c(r)d(r^2) \tag{43}$$

Because real polymer solutions are always more or less nonideal, Hermans and Ende (52) also studied the effect of nonideality. They found that the apparent molecular weight averages are reduced by nonideality. The reduction of the weight-average molecular weight $\bar{M}_w$ was larger than the reduction of the number average $\bar{M}_n$. Consequently, the apparent ratio $\bar{M}_w/\bar{M}_n$ may be less than unity. Hermans and Ende (52) recommend performing the analysis at several concentrations of polymer and extrapolating the apparent values to vanishing concentration of polymer. Recently, we were able to show (66) that the concentration dependence of $\bar{M}_{n,\mathrm{app}}$ and $\bar{M}_{w,\mathrm{app}}$ (these quantities are equal to right-hand sides of Eqs. 42 and 43, respectively) may be expressed by means of the second virial coefficient B (as presented in Eqs. 6 and 36)

$$1/\bar{M}_{n,\mathrm{app}} = 1/\bar{M}_n + B\bar{c} + \cdots \tag{44}$$

$$1/\bar{M}_{w,\mathrm{app}} = 1/\bar{M}_w + 2Bf\bar{c} + \cdots \tag{45}$$

Here, $\bar{c}$ is the average polymer concentration in the band defined as

$$\bar{c} \equiv \int c^2 d(r^2) \Big/ \int c d(r^2) \tag{46}$$

and f is a function of the polydispersity of the polymer, which is equal to unity for monodisperse samples and is larger than unity for polydisperse samples.

The recommended experimental procedure consists of equilibrating the same cell load sequentially at several rotor speeds; evaluating $\bar{M}_{n,\mathrm{app}}$, $\bar{M}_{w,\mathrm{app}}$, and $\bar{c}$ for each speed (the $\bar{c}$ value is decreasing with decreasing rotor speed); plotting these data according to Eqs. 44 and 45; and evaluating $\bar{M}_n$, $\bar{M}_w$, B, and f from the intercepts and slopes of these plots.

For the application of Eqs. 42 and 43 it is obviously necessary to be able to evaluate precisely the density profile in the cell. The density profile is a function not only of the nature of the two solvent components and of the rotor speed, but also of the pressure (which depends on the column length). For polymers dissolved in organic solvents, the thermodynamics of which is usually not known experimentally, the detailed profile has to be obtained directly from the sedimentation equilibria in density gradients themselves. We recently developed a method for this purpose in which the equilibrium is sequentially established for the same solution of a marker polymer at several rotor speeds and the position of the peak is measured accurately (67). The procedure is then repeated for the same marker polymer dissolved in several solvent mixtures differing in their initial composition. From the data collected in the above way, we were able to evaluate the density ρ in form of a relationship

$$\rho = a_0 + b_0 P_0 + (a_1 + b_1 P_0)(u - u_0) + (a_2 + b_2 P_0)(u - u_0)^2 \tag{47}$$

where a_0, a_1, a_2, b_0, b_1, b_2 are constants characteristic for the solvent mixture, P_0 is the pressure at the location of the center of the marker polymer peak, and $u \equiv r^2\omega^2/2$. Subscript zero refers to the position of the marker peak center. Equations for calculating P_0 for particular experimental conditions were also given. The method is quite lengthy, but has to be applied only once to every solvent system of interest. The constants of Eq. 47 are applicable at all experimental conditions.

In the above section we outlined the so-called two-component theory of the equilibrium in a density gradient. In this theory, the density of the solvent mixture is supposed to vary with the position in the cell, but all other properties (thermodynamic interactions with polymer) are the same within the whole band. In an actual three-component case, the preferential adsorption of one solvent component onto polymer changes its buoyant density and, consequently, determines the position of the band in the cell. Moreover, the preferential adsorption may vary with the composition of the solvent mixture, that is with the position within the band. In other words, in the expression for buoyance $(1 - \bar{v}^*\rho)$ not only the density ρ but also the partial specific volume at constant chemical potential $\bar{v}^*$ are functions of radius (68).

For the three-component case, the expression $(\rho - \rho_0)$ in Eqs. 42 and 43 should be replaced by $(\rho - 1/\bar{v}^*)$. Here, $1/\bar{v}^*$ is equal to ρ_0 at the center of the band, but not elsewhere. The term $\bar{v}^*$ may be measured as a function of solvent composition using pycnometric procedures on a polymer solution brought to osmotic equilibrium against a mixed solvent. However, these procedures are not easy and the establishment of the dependence of $\bar{v}^*$ on the solvent density is one of the more difficult problems. Usually it is assumed that

$$(\rho - 1/\bar{v}^*) \equiv a(\rho - \rho_0) \tag{48}$$

where a is a constant found by a calibration experiment. In the notation of Meselson et al. (46), β in Eq. 41 is replaced by β_{eff}. The latter quantity is found by equilibrating two samples of identical DNA in the same cell: one from bacteria in ordinary growing medium and the other from bacteria grown in the same medium in which the source of nitrogen was made from the isotope ^{15}N. This isotopic substitution makes DNA denser by a precisely known amount, which serves for the calibration (69).

3.5.2. *Mixtures of Polymers and Copolymers–Heterogeneity of Copolymers*

When a mixture of several polymer species is subjected to density-gradient sedimentation, each species forms a band. If the densities of the polymers are sufficiently different and the molecular weights sufficiently high (to obtain narrow bands), the bands are separated and the analysis of the mixture is straightforward.

Most copolymers are not only polydisperse in molecular weight, they are also heterogeneous in chemical composition. Measurement of this heterogeneity is quite difficult. Hermans and Ende (54, 56) tried to obtain the heterogeneity from a detailed analysis of the shape of the band in the density-gradient centrifugation. As we have seen above, the shape is influenced by many factors other than heterogeneity and the analysis could be ambiguous.

In our opinion, the most straightforward exploitation of the density-gradient equilibrium would consist of double detection. The concentrations of the two monomers of the copolymer should be detected by two methods differing in their response toward the two monomers. Ideally, this detection is performed by UV measurements at two different wavelengths or by a combination of UV measurement with refractometric measurement. However the optical absorption of the solvents usually makes this approach difficult.

An alternative method involves performing two equilibrium experiments in two solvent mixtures differing in refractive index in such a way that in one mixture, one monomer is invisible (i.e., isorefractive), and the other monomer is invisible in the other mixture. We have performed a similar experiment on some mixtures of copolymers (Fig. 5 and 6), which demonstrate this approach quite clearly.

Density-gradient centrifugation can also be applied to suspensions of larger particles in nonsolvents, which will form a thin band appearing as a strip on the Schlieren photograph. This method was used to show that a polystyrene, which was prepared by freeze-drying of a very dilute solution in cyclohexane (and presumably composed of spheres formed by a single collapsed macromolecule), has a different density than the same sample of polystyrene

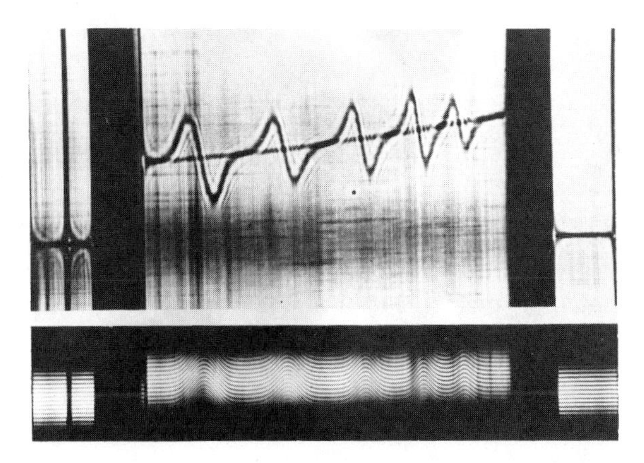

Figure 5. Schlieren and interference photograph at sedimentation equilibrium in a density gradient at 60,000 rpm. Solvent mixture: $CBrClFCBrF_2$/dibutyl succinate. Mixture of equal amounts of polystyrene, poly(methyl methacrylate) and three copolymers of styrene and methyl methacrylate (with the weight fraction of styrene 0.25, 0.50, and 0.75, respectively). The denser poly(methyl methacrylate) bands near the bottom of the cell (right side), the less dense polystyrene bands close to the top of the cell (left side). In this solvent mixture the refractive increment of poly(methyl methacrylate) is about 35% of that of polystyrene. The total amount of each polymer is proportional to the area under the corresponding peak in the interference photograph. The proportionality constant is proportional to the refractive increment.

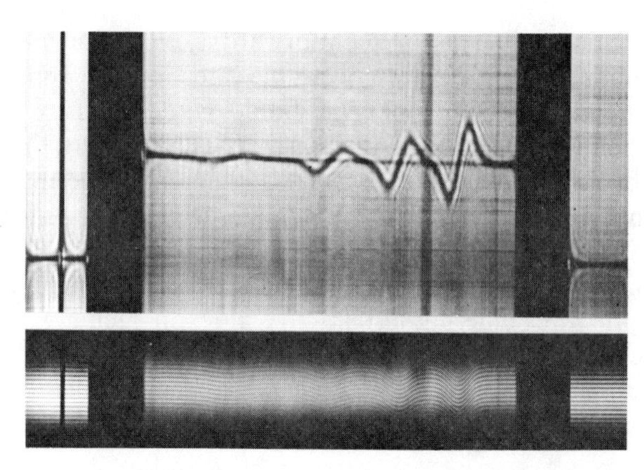

Figure 6. Schlieren and interference photograph at sedimentation equilibrium at 60,000 rpm. Solvent mixture *m*-dibromobenzene/2-ethylnaphtalene. The same mixture of polymers as in Figure 5. In this mixture of solvents, polystyrene is almost invisible because of its low refractive index increment. The refractive index increment of poly(methyl methacrylate) is negative.

dispersed in the same medium (ethanol–lithium iodide–water) by sonication (70).

3.5.3. *Selection of Solvents*

Whenever the refractometric optics (usually interference) is used, the refractive indexes of solvent components become important. At any particular point in the cell, the difference between the refractive indexes of the polymer compartment and reference compartment of the sedimentation cell (measured as a shift in the fringe position) is proportional to the local concentration of polymer. The proportionality constant is the refractive increment of the polymer measured at constant chemical potential of the solvent components (to take care of the preferential adsorption) at the local composition of the solvent mixture. It is well known that the refractive increment is roughly proportional to the difference of refractive indexes between polymer and the solvent (or solvent mixture). Thus, for a given polymer, it depends on the refractive index of the solvent or, in the case of a solvent mixture, on the composition of the solvent. In a density-gradient experiment, the composition of the solvent mixture is changing considerably with the position in the cell. Thus, the proportionality constant between polymer concentration and fringe shift is changing with the position in the cell. Moreover, the change of solvent composition leads to a change of preferential adsorption onto polymer leading to an additional variation of the effective refractive increment.

To avoid the computational and experimental difficulties connected with the variation of the refractive increment, isorefractive solvent pairs are recommended. Such a choice minimizes the variation of the increment. Moreover, when isorefractive solvent pairs are used, the deflection of the light beam out of the optical system at high rotor speeds is avoided. This deflection was frequently the speed-limiting factor when aqueous solutions of CsCl or similar salts were used for density-gradient ultracentrifugation.

The search for convenient solvents is one of the more difficult problems. Each solvent pair must satisfy a number of conditions, and it may be difficult to satisfy all of them simultaneously. For a homopolymer (measurement of molecular weight and polydispersity), the following conditions apply.

1. The solvents are isorefractive.

2. They should be solvents with about equal solvent power for the polymer to minimize the effects of preferential adsorption. The solvent power should be relatively poor to reduce nonideality and to facilitate necessary extrapolations.

3. The refractive index should be reasonably different from the polymer to allow measurements at low concentrations (good visibility of the polymer).

4. The viscosity of the solvents should be low to reduce equilibration times. Solvent mixtures with viscosity of the order of 1 cP require an equilibration time of about 1 day. At higher viscosities the times are proportionally longer.

5. The solvents must be capable of forming a steep density gradient to allow even polymers with relatively low molecular weight to band at several rotor velocities. This condition may be relaxed when only polymers with very high molecular weight are studied (e.g., DNA and polyacrylamide). It is easy to show that the steepest gradients are produced by solvents, which differ appreciably in density. It is also important that the less dense solvent has a density sufficiently lower than the polymer. Thermodynamic properties of the mixture also play an important role. Generally, the more endothermic the mixing (positive deviation from Raoult's law), the steeper the density gradient (67).

For copolymers and/or mixtures of polymeric materials, the above conditions must be satisfied for two pairs of solvents. Condition 2 must be satisfied for both homopolymers, and may be rather difficult to accomplish. Condition 5 is rather severe because two different homopolymers usually differ significantly in density. Condition 3 is replaced by the condition that one pair must have a refractive index less than and the other greater than both homopolymers. There is an additional condition; condition 6.

6. The preferential adsorption onto the two homopolymers must be such as not to expand excessively the difference in the effective densities of the two homopolymers. Such an unfavorable situation may occur if the less dense

solvent is preferentially adsorbed onto the less dense homopolymer, and the denser solvent onto the denser homopolymer.

Recently, we made an extensive search for suitable solvent pairs. At the beginning the main problem was the relative scarcity of solvents with high density and low refractive index, as well as of solvents with low density and high refractive index. The first group of solvents was eventually found among bromofluorocarbons. Both halogens increase the density considerably. Generally, the fluorocarbons have very low refractive index ($n_D = 1.3$ and lower) and are immiscible with most other materials. However, the substitution by bromine alleviates both problems. The fine tuning of the ratio of Br/F serves for changing the refractive index. To date, two mixtures of this type were developed:

1. Cyclohexane ($n_D^{20} = 1.426$) with $CFBrClCF_2Br$ ($n_D^{20} = 1.431$), which is a suitable solvent mixture for polystyrene.
2. Methyl acetate ($n_D^{20} = 1.361$) with $CF_2BrCF_2CF_2Br$ ($n_D^{20} = 1.360$), which is suitable for poly(methyl methacrylate).

Another perspectively convenient system is toluene ($n_D^{20} = 1.497$) and $CHBr_2Cl$ ($n_D^{20} = 1.497$).

The solvents with a low density and a high refractive index were found among the derivatives of naphthalene and chinolin. Unfortunately, these compounds usually interact very strongly with polybrominated compounds (used as the denser component). Also their density is not sufficiently low ($\rho = 1.0$) and their viscosity is high (3–5 cP).

For the study of styrene–methyl methacrylate copolymers, we have developed two reasonable, but not completely satisfactory solvent mixtures: $CBrClFCBrF_2$ ($n_D^{20} = 1.431$) with dibutyl succinate ($n_D^{20} = 1.430$) and *m*-dibromobenzene ($n_D^{20} = 1.608$) with 2-ethyl naphthalene ($n_D^{20} = 1.600$).

3.6. Diffusion Experiments

Synthetic boundary cells are routinely used for the formation of a sharp boundary between two liquids (one of them is usually a macromolecular solution). This procedure is beneficial in sedimentation velocity experiments with solutes having high diffusion coefficients. It circumvents the difficulties related with the slow separation of the boundary from the meniscus. In the sedimentation equilibrium experiments, the procedure is used (in connection with interference optics) for measurement of the solution concentration.

After the boundary is formed, it starts spreading because of diffusion processes. (It may also sediment, but sedimentation could usually be suppressed by choosing a low speed of the rotor: typically 4000–6000 rpm.) In our studies (44, 71, 72) we followed this diffusion process and were able to derive a wealth of information. Besides the diffusion coefficient of the polymer, we were able to measure the mutual diffusion coefficient in a mixture of two liquids and the refractive increments of polymers at constant chemical potential in a mixed solvent $(\partial n/\partial c_3)_\mu$ (the quantity needed in Eq. 38).

For a meaningful analysis of these experiments it is necessary to recognize that diffusion takes place in a sector-shaped short (~1 cm) cell. Thus, the standard approaches for analysis of diffusing boundaries (rectangular infinite cell) are not valid. We resorted to comparison of actual experimental data with computer-simulated diffusion data for the same cell with the same initial position of the boundary. We will describe these procedures in some detail.

The experiments in the ultracentrifuge at low rotor speed are prone to convection inside the cell. The convections are caused by mechanical disturbances, intermittent radiation from the heating element in the rotor chamber, and by an occurrence of a negative density gradient in some part of the cell. For studies of polymers in mixed solvents we prevented convections by routinely adding a small amount of the lighter solvent to the solvent sector of the cell. This excess solvent created a so-called solvent boundary, which helped to stabilize the run. As the diffusion in the cell progressed, the stabilizing effect of the solvent boundary slowly disappeared. However, when this happened, the polymer boundary was already spread sufficiently to prevent convections. Of course, in experiments with single solvents, this strategy could not be employed. Such runs proved to be the most difficult. To prevent the disturbances of temperature, the rotor temperature was brought to about 0.5 °C below the desired temperature. Both refrigeration and temperature controls were then disconnected before the start of the run. The typical rise in temperature during the run was less than 1 °C.

The interference photographs from the ultracentrifuge are usually evaluated by means of so-called fringe count Δf, that is, the difference in the fringe position (corrected for base line and expressed as the number of fringes) for two selected points in the cell. The fringe count is related to the difference of refractive indexes Δn at these two points as

$$\Delta f = h\,\Delta n/\lambda_0 \tag{49}$$

where h is the thickness of the cell and λ_0 is the wavelength of the light. One of the goals of the measurement is to find (for points on either side of the solvent–solution boundary) that part of Δf, which corresponds to solvated polymer at the time when the boundary was formed. However, the fringe count obtained from the photographs corresponds to the superposition of the polymer boundary and solvent boundary at some later time.

For any diffusing entity, for which the diffusion constant D is independent of concentration within the range of concentration covered in the cell, the concentration profile in the cell is a unique function of the boundary conditions (length of the sector-shaped cell), initial conditions (position and size of the sharp boundary), and the product $D \cdot t$, where t is the time. In our procedure, the initial position of the boundary is estimated from the Schlieren photographs. The experimental concentration profile is sampled at three pairs of points, each of them placed more or less symmetrically around the initial boundary. The fringe count is measured for each pair and each interference photograph. Obviously, the fringe count decreases with time: faster for the inner pair, slower for the outer pair. Then, the diffusion of a unit-sized boundary with equivalent original position in an equivalent cell is simulated in a computer. The concentration profile is computed as a function of $D \cdot t$ and is sampled at the three pairs of points, which are equivalent to the experimental positions. Thus, for each pair (subscript j) a function $F_j(D \cdot t)$ is obtained in a form of a table. The experimental fringe count Δf_j is now expressed as

$$\Delta f_j = \Delta f_{\mathrm{p}}^0 F_j[D_{\mathrm{p}} \cdot (t - \Delta t)] + \Delta f_{\mathrm{s}}^0 F_j[D_{\mathrm{s}} \cdot (t - \Delta t)] \tag{50}$$

Here, the subscript p refers to polymer boundary and s to the solvent boundary. The superscript zero refers to the original sharp boundary, t is the time from the start of the run, Δt is the time, which takes into account the period needed for the original formation of the boundary, as well as the original disturbances of the boundary. A nonlinear least-square procedure could now be applied to fit all three sets of experimental points simultaneously to Eq. 50. The five adjustable parameters, Δf_{p}^0, Δf_{s}^0, D_{p}, D_{s}, Δt, could be obtained this way. However, it proved to be more precise to first measure the diffusion constant for the solvent boundary in a separate experiment. This experiment was essentially the same as the one described above, but the polymer was absent. The boundary was formed between two solvent mixtures differing by about 2% in their composition. In this case, the experimental fringe count was fitted to a simpler expression

$$\Delta f_j = \Delta f_{\mathrm{s}}^0 F_j[D_{\mathrm{s}} \cdot (t - \Delta t)] \tag{51}$$

and the diffusion constant D_{s} was obtained with good precision. Once D_{s} was known, the data for the polymer sample were fitted to Eq. 50. Four quantities were then obtained from the least-square analysis. Δf_{p}^0, Δf_{s}^0, D_{p}, Δt. It was gratifying to find that the value of Δt was usually quite close to the observed time for the initial forming of the boundary.

The experiment described above could be used for three different purposes.

1. The original size of the polymer boundary Δf_p^0 may be converted to the difference of refractive indexes Δn according to Eq. 49 and to $(\partial n/\partial c_3)_\mu$, if the concentration c_3 is known. This analysis may eventually lead to values of λ (Eq. 38) and other thermodynamic quantities (44).

2. The diffusion constant of the polymer D_p is an important characteristic of the polymer–solvent system; its value at vanishing concentration of polymer is usually required for physicochemical calculation. The method described above produces integral diffusion coefficients, that is, values averaged over the concentration range within the boundary. Our analysis has shown that for typical initial polymer concentrations (0.004 g/mL), the integral value is a good approximation of the limiting value for polymers in moderately good solvents. For poor solvents, the measured values are low by a few percent; for good solvents, the diffusion constant is overestimated. However, under many circumstances (especially for biochemical samples, which are usually not strongly nonideal) the diffusion coefficient may be combined either with sedimentation coefficient or intrinsic viscosity to yield a rather good approximation of the molecular weight (71).

3. The diffusion experiment in the absence of the polymer proved to be a surprisingly precise method for evaluating the mutual diffusion coefficient of the two solvents D_s. The coefficient D_s is a sensitive function of the thermodynamic properties of the solvent mixture. We applied the method to two solvent mixtures, for which good thermodynamic data were available (cyclohexane–benzene and cyclohexane–ethyl acetate). We found that the data are best described by the theory of Hartley and Crank (73), which can be presented as

$$D_s \eta/Q = (kTV_2/\sigma_2 V_1)\phi_1 + (kT/\sigma_1)\phi_2 \tag{52}$$

$$Q \equiv (\partial \ln a_1/\partial \ln c_1)_{P,T} \tag{53}$$

Here, the thermodynamic factor Q is defined by Eq. 53, a_1 and c_1 are the activity and concentration of solvent 1, respectively, V_i values are the molar volumes of the two solvents, σ_i values are their friction factors presumed to be independent of concentration, and η is viscosity of the mixture. According to Eq. 52, the ratio $D_s\eta/Q$ is a linear function of the volume fraction ϕ_1. This linearity did correspond reasonably well with our data.

These results suggest a new approach for measuring thermodynamic properties of liquid mixtures. The thermodynamic interaction of the two liquids (ΔG_{mix}) is postulated as an expression with several unknown parameters. The thermodynamic factor Q is evaluated from this expression. The unknown parameters are then obtained by a least-squares fit of experimental data to Eqs. 52 and 53 (72).

If the validity of the Hartley–Crank theory is confirmed for a larger number of mixtures, the measurements of diffusion coefficient and viscosity may be used for evaluation of the thermodynamic interaction functions.

4. CONCLUSIONS

As we have seen, ultracentrifugation has been and most probably will remain a source of precise information in the field of molecular weight and MWD, as well as in the area of hydrodynamic studies. The measurement of thermodynamic properties is very promising and deserves more experimental attention.

However, we see promise in the use of the less familiar techniques: ultracentrifugation in a density gradient and diffusion measurements. Especially intriguing is the possibility of measurement of the number-average molecular weight for high-molecular-weight samples, and the double-detection techniques for measurement of chemical heterogeneity of copolymers. The relatively simple diffusion experiment may find broad use in biochemistry. Its solvent version presents a challenging approach to the thermodynamics of solvent mixtures.

The ultracentrifuge *is* alive and well.

NOTATION

a	Activity
B	Second osmotic virial coefficient
B	Variable defined by Eq. 28
C	Third osmotic virial coefficient
c	Concentration
c^*	Chain overlap concentration
D	Diffusion coefficient
D_0	Diffusion coefficient at infinite dilution
F	Interaction function (Eq. 25)
f	Frictional coefficient
f	Function of polydispersity (Eq. 45)
f_0	Frictional coefficient at infinite dilution
Δf	Fringe count
G	Gibbs free energy
g	Interaction function (Eq. 23, Eq. 37)
h	Centrifuge cell thickness

K	Constant from Eq. 13
k	Boltzmann constant
k	Permeability of entangled polymer network
M	Molecular weight
M_{app}	Apparent (experimental) molecular weight
$\bar{M}_n$	Number-average molecular weight
$\bar{M}_w$	Weight-average molecular weight
$\bar{M}_z$	z-Average molecular weight or centrifuge average molecular weight
m	Molality
n	Refractive index of the solution
$\partial n/\partial c$	Refractive index increment
N_A	Avogadro's number
P	Pressure
Q	Thermodynamic factor (Eq. 53)
R	Gas constant
r	Distance from the rotational axis
r_b	Radial distance of the bottom of the cell
r_m	Radial distance of the meniscus
r_o	Radial distance of the band center
s	Sedimentation coefficient
s_o	Sedimentation coefficient at infinite dilution
t	Time
T	Absolute temperature
u	Polydispersity parameter; also $u \equiv r^2\omega^2/2$ (Eq. 47)
V	Molar volume
$\bar{v}$	Partial specific volume
z	Parameter in the Schulz–Zimm distribution
α, β	Parameters of Eq. 20
β	Variable defined by Eq. 41
$\Gamma(x)$	Gamma function of x
δ_{ik}	Kronecker delta
η	Viscosity
λ	Preferential adsorption parameter
λ_0	Wavelength of light
μ	Chemical potential
ρ	Solution density
σ	Friction factor (Eq. 52)
ϕ	Volume fraction
ϕ_p	Total volume fraction of the polydisperse polymer
χ	Flory–Huggins parameter
ω	Angular velocity of the rotor

GENERAL REFERENCES

T. Svedberg and K. O. Pedersen, *The Ultracentrifuge*, Clarendon, Oxford, 1940.

H. K. Schachman, *Ultracentrifugation in Biochemistry*, Academic, New York, 1959.

H. Fujita, *Mathematical Theory of Sedimentation Analysis*, Academic, New York, 1962.

J. J. Hermans and H. A. Ende, "Density-Gradient Centrifugation," in B. Ke (Ed.), *Newer Methods of Polymer Characterization*, Interscience, New York, 1964, p. 525.

D. A. Yphantis (Ed.), "Advances in Ultracentrifugal Analysis," *Ann. N. Y. Acad. Sci.*, **164**, 1 (1969).

J. H. Coates, "Ultracentrifugal Analysis," in S. J. Leach (Ed.), *Physical Principles and Techniques of Protein Chemistry, Part B*, Academic, New York, 1970.

J. W. Williams, *Ultracentrifugation of Macromolecules: Modern Topics*. Academic, New York, 1972.

C. H. Chervenka, *A Manual of Methods for the Analytical Ultracentrifuge*, Beckman Instruments, Palo Alto, California, 1973.

J. E. Hearst and C. W. Schmid, "Density Gradient Sedimentation Equilibrium," *Methods Enzymol. Part D*. **27**, 111 (1973).

H. Fujita, *Foundations of Ultracentrifugal Analysis*, Wiley, New York, 1975.

J. B. Ifft, "Sedimentation Equilibrium of Proteins in Density Gradients," in N. Catsimpoolas (Ed.) *Methods of Protein Separation*, Vol. 1, Plenum, New York, 1975, p. 193.

J. J. H. Mulderije, "Studies in Ultracentrifugation and Diffusion of Random Coil Polymers," Thesis, Rijksuniversiteit, Leiden, 1978.

C. R. Cantor and P. R. Schimmel, *Biophysical Chemistry, Part II, Techniques for the Study of Biological Structure and Function*," Freeman, San Francisco, 1980, Chapter 11.

REFERENCES

1. V. Petrus, B. Porsch, B. Nyström, and L.-O. Sundelöf, "Absolute Determination of Molar Masses of Standard Polystyrenes by Means of Sedimentation—Diffusion Measurements in Cyclohexane," *Makromol. Chem.*, **184**, 295 (1983).
2. M. Bohdanecký, V. Petrus, B. Porsch, and L.-O. Sundelöf, "Hydrodynamic Constants and Frictional Properties of Polystyrene in Dilute Solution," *Makromol. Chem.*, **184**, 309 (1983).
3. K. E. Van Holde and W. O. Weischet, "Boundary Analysis of Sedimentation—Velocity Experiments with Monodisperse and Paucidisperse Solutes," *Biopolymers*, **17**, 1387 (1978).
4. J. W. Williams, *Ultracentrifugation of Macromolecules: Modern Topics*, Academic, New York, 1972.
5. H. Fujita, *Foundations of Ultracentrifugal Analysis*, Wiley, New York, 1975, p. 166ff.

6. E. T. Adams, Jr., W. E. Ferguson, P. J. Wan, J. L. Sarquis, and B. M. Escott, "Some Modern Aspects of Ultracentrifugation," *Sep. Sci.*, **10**, 175 (1975).
7. P. J. Wan and E. T. Adams, Jr., "Molecular Weights and Molecular-Weight Distributions from Ultracentrifugation of Nonideal Solutions," *Biophys. Chem.*, **5**, 207 (1976).
8. H. Kim, R. C. Deonier, and J. W. Williams, "The Investigation of Self-Association Reactions by Equilibrium Ultracentrifugation," *Chem. Rev.*, **77**, 659 (1977).
9. W. Mächtle and U. Klodwig, "A New 8-Cell Interference and Schlieren Optics Multiplexer for the Analytical Ultracentrifuge Based Upon a Modulable Laser and a High Speed 8-Hole-Rotor with 60,000 rev/min." *Makromol. Chem.*, **180**, 2507 (1979).
10. J. Floßdorf, "Erweiterte Meßmoglichkeiten in der analytischen Ultrazentrifugation durch die Verwendung eines neuartigen Kollimators," *Makromol. Chem.*, **181**, 715 (1980).
11. W. Mächtle and U. Klodwig, "Stroboskop-Multiplexer zur fotografischen Registrierung in analytischen Ultrazentrifugen beim Einsatz von Vielzellenotoren," *Makromol. Chem.*, **177**, 1607 (1976).
12. J. L. Sarquis and E. T. Adams, Jr., "Self-Association of β-lactoglobulin C in Acetate Buffers," *Biophys. Chem.*, **4**, 181 (1976).
13. J. Floßdorf, H. Schillig, and K.-P. Schindler, "Intermittierende Laserbeleuchtung für die Interferenz– und Schlieren-optik einer analytischen Ultrazentrifuge," *Makromol. Chem.*, **179**, 1617 (1978).
14. R. H. Crepeau, S. J. Edelstein, and M. J. Rehmar, "Analytical Ultracentrifugation with Absorption Optics and a Scanner-Computer System," *Anal. Biochem.*, **50**, 213 (1972).
15. D. J. DeRosier, P. Munk, and D. J. Cox, "Automatic Measurement of Interference Photographs from the Ultracentrifuge," *Anal. Biochem.*, **50**, 139 (1972).
16. P. Vidakovic, C. Allain, and F. Rondelez, "Sedimentation of Dilute and Semidilute Polymer Solutions at the θ Temperature," *Macromolecules*, **15**, 1571 (1982).
17. J. J. H. Mulderije, "Studies in Ultracentrifugation and Diffusion of Random Coil Polymers," Ph.D. Thesis, Rijksuniversiteit, Leiden, 1978.
18. J. J. Mulderije, "Friction Coefficient of Polymer Molecules in Dilute Solution near the θ Point. 1. A Rapid Method for Determining Small Differences in Sedimentation Coefficients: Polystyrene in Cyclohexane," *Macromolecules*, **13**, 1207 (1980).
19. J. J. H. Mulderije, "Linear and Second-Order Pressure Effects in Velocity Centrifugation of Random Coil Polymers. Polystyrene in Cyclohexane," *Macromolecules*, **15**, 506 (1982).
20. W. Mächtle, "Charakterisierung von Dispersionen durch gekoppelte H_2O/D_2O-Ultrazentrifugenmessungen," *Makromol. Chem.*, **185**, 1025 (1984).
21. P. F. Mijnlief and W. J. M. Jaspers, "Solvent Permeability of Dissolved Polymer Material. Its Direct Determination from Sedimentation Measurements," *Trans. Faraday Soc.*, **67**, 1837 (1971).
22. F. Brochard and P. G. deGennes, "Dynamical Scaling for Polymers in Theta Solvents," *Macromolecules*, **10**, 1157 (1977).

23. P.-G. deGennes, "Scaling Concepts in Polymer Physics," Cornell University Press, Ithaca, NY, 1979.
24. J. Roots and B. Nyström, "Test of 'Scaling Laws' Describing the Concentration Dependence of Osmotic Pressure, Diffusion and Sedimentation in Semidilute Macromolecular Solutions," *Polymer*, **20**, 149 (1979).
25. B. Nyström, J. Roots, and R. Bergman, "Sedimentation Velocity Measurements Close to the Upper Critical Solution Temperature and at θ-Conditions: Polystyrene in Cyclopentane Over a Large Concentration Interval," *Polymer*, **20**, 157 (1979).
26. G. Pouyet and J. Dayantis, "Velocity Sedimentation in the Semidilute Concentration Range of Polymers Dissolved in Good Solvents," *Macromolecules*, **12**, 293 (1979).
27. T. Huang and G. Meyerhoff, "The Sedimentation Behaviour of Polystyrene in Various Solvents in Semidilute Solutions," *Makromol. Chem.*, **185**, 2459 (1984).
28. Th. G. Scholte, "Molecular Weights and Molecular Weight Distributions of Polymers by Equilibrium Ultracentrifugation. Part II. Molecular Weight Distribution," *J. Polym. Sci. Part A2*, **6**, 111 (1968).
29. Th. G. Scholte, "Determination of the Molecular Weight Distribution of Polymers from Sedimentation–Diffusion Equilibria," *Ann. N.Y. Acad. Sci.*, **164**, 156 (1969).
30. Th. G. Scholte, "Determination of the Molecular Weight Distribution of Polymers from Equilibria in the Ultracentrifuge," *Eur. Polym. J.*, **6**, 51 (1970).
31. E. T. Adams, Jr., P. J. Wan, D. A. Soucek, and G. H. Barlow, "Molecular Weight Distributions from Sedimentation Equilibrium Experiments," *Adv. Chem. Ser.*, **125**, 235 (1973).
32. D. A. Soucek and E. T. Adams, Jr., "Molecular Weight Distributions from Sedimentation Equilibrium of Nonideal Solutions," *J. Colloid Interface Sci.*, **55**, 571 (1976).
33. P. Munk and M. E. Halbrook, "Sedimentation Equilibrium of Polymers in Good Solvents," *Macromolecules.*, **9**, 568 (1976).
34. S. G. Chu and P. Munk, "A Sedimentation Equilibrium Study of Thermodynamic Properties of Polystyrene in Bromobenzene," *Macromolecules*, **11**, 101 (1978).
35. P. Munk, "Sedimentation Equilibrium of Highly Nonideal Solutions of Polydisperse Polymers," *Macromolecules*, **13**, 1215 (1980).
36. P. J. Flory, *Principles of Polymer Chemistry*, Cornell University Press, Ithaca, N.Y. 1953.
37. P. Munk and D. J. Cox, "Sedimentation Equilibrium of Protein Solutions in Concentrated Guanidinium Chloride. Thermodynamic Nonideality and Protein Heterogeneity," *Biochemistry*, **11**, 687 (1972).
38. T. H. Donnelly, "The Direct Estimation of Continuous Molecular Weight Distributions by Equilibrium Ultracentrifugation," *J. Phys. Chem.*, **70**, 1862 (1966).
39. T. H. Donnelly, "Some Capabilities and Limitations of the Laplace Transform Method for the Direct Estimation of Continuous Molecular Weight Distributions from Equilibrium Ultracentrifugation," *Ann. N.Y. Acad. Sci.*, **164**, 147 (1969).
40. Th. G. Scholte, "Determination of Thermodynamic Parameters of Polymer–

Solvent Systems from Sedimentation–Diffusion Equilibrium in the Ultracentrifuge," *J. Polym. Sci. Part A2.*, **8**, 841 (1970).

41. B. J. Rietveld, Th. G. Scholte, and J. P. L. Pijpers, "Thermodynamic Parameters of Polystyrene Solutions in Cyclohexane, Determined from Sedimentation-Diffusion Equilibria in the Ultracentrifuge. Dependence on Concentration, Temperature and Molecular Weight," *Br. Polym. J.*, **4**, 109 (1972).
42. H. Murakami, T. Norisuye, and H. Fujita, "Ultracentrifugal Evaluation of Chemical Potentials for the System Poly(chloroprene)–Methyl Ethyl Ketone," *Polymer J.*, **7**, 248 (1975).
43. S. G. Chu and P. Munk, "Thermodynamic Properties of Polystyrene in Mixed Solvents Studied by Sedimentation Equilibrium," *Macromolecules*, **11**, 879 (1978).
44. T. M. Aminabhavi and P. Munk, "Preferential Adsorption onto Polystyrene in Mixed Solvent Systems," *Macromolecules*, **12**, 607 (1979).
45. B. J. Rietveld, "Construction and Application of a Low-speed Centrifuge for Thermodynamic Investigation of Macromolecular Systems," *Br. Polym. J.*, **6**, 181 (1974).
46. M. Meselson, F. W. Stahl, and J. Vinograd, "Equilibrium Sedimentation of Macromolecules in Density Gradients," *Proc. Natl. Acad. Sci. USA.*, **43**, 581 (1957).
47. M. Meselson and F. W. Stahl, "The Replication of DNA in Escherichia Coli," *Proc. Natl. Acad. Sci. USA.*, **44**, 671 (1958).
48. N. Sueoka, J. Marmur, and P. Doty, "II. Dependence of the Density of Deoxyribonucleic Acids on Guanine-Cytosine Content," *Nature* (*London*), **183**, 1429 (1959).
49. J. B. Ifft, "Sedimentation Equilibrium of Proteins in Density Gradients," in N. Catsimpoolas (Ed.). *Methods of Protein Separation*, Vol. 1., Plenum, New York, 1975, p. 193.
50. C. W. Schmid and J. E. Hearst, "Density-Gradient Sedimentation Equilibrium of DNA and the Effective Density Gradient of Several Salts," *Biopolymers*, **10**, 1901 (1971).
51. J. E. Hearst and C. W. Schmid, "Density Gradient Sedimentation Equilibrium," *Methods Enzymol.* (*Part D*), **27**, 111 (1973).
52. J. J. Hermans and H. A. Ende, "Density Gradient Centrifugation of a Polymer-Homologous Mixture," *J. Polym. Sci. Part C*, **1**, 161 (1963).
53. J. J. Hermans and H. A. Ende, "Analysis of Copolymers by Means of Density Gradient Centrifugation," *J. Polym. Sci. Part C*, **4**, 519 (1963).
54. J. J. Hermans, "Information Regarding both Molecular-Weight Distribution and Density Distribution in a Polymer Subjected to Density-Gradient Centrifugation," *J. Chem. Phys.*, **38**, 597 (1963).
55. J. J. Hermans, "Density Gradient Centrifugation of a Mixture of Polymers Differing in Molecular Weight and Specific Volume," *J. Polym. Sci. Part C*, **1**, 179 (1963).
56. H. A. Ende and J. J. Hermans, "Analysis of Copolymers by Means of Density Gradient Centrifugation. II. Comparison with Kinetic Requirements," *J. Polym. Sci. Part A*, **2**, 4053 (1964).

57. R. Buchdahl, H. A. Ende, and L. H. Peebles, "Detection of Structural Differences in Polymers by Density Gradient Ultracentrifugation. II. Detection of Microgel," *J. Polym. Sci. Part C*, **1**, 143 (1963).
58. R. Buchdahl, H. A. Ende, and L. H. Peebles, "Detection of Structural Differences in Polymers by Density Gradient Ultracentrifugation III. Tacticity," *J. Polym. Sci. Part C.*, **1**, 153 (1963).
59. H. A. Ende and V. Stannett, "Density Gradient Centrifugation of a Graft Copolymer," *J. Polym. Sci. Part A*, **2**, 4047 (1964).
60. J. J. Hermans and H. A. Ende, "Density-Gradient Centrifugation," in B. Ke (Ed.), *Newer Methods of Polymer Characterization*, Interscience, New York, 1964, p. 525.
61. A. Nakazawa and J. J. Hermans, "Study of Compositional Distribution in a Styrene-Methyl Acrylate Copolymer by Means of Density-Gradient Centrifugation," *J. Polym. Sci. Part A2*, **9**, 1871 (1971).
62. A. Nakazawa, N. Donkai, t. Kotaka, and H. Inagaki, "A Study of Preferential Solvation on Styrene–Methyl Methacrylate Copolymers of Varying Architecture and Composition by Conventional and Density Gradient Sedimentation Equilibrium Methods," *Br. Polym. J.*, 200 (1977).
63. J. Lamprecht, Cl. Strazielle, J. Dayantis, and H. Benoit, "Etude par diffusion de la lumière et par équilibre de sédimentation en gradient de densité de l' hetérogenéité en composition des copolymères statistiques," *Makromol. Chem.*, **148**, 285 (1971).
64. J. B. Ifft, D. H. Voet, and J. Vinograd, "The Determination of Density Distributions and Density Gradients in Binary Solutions at Equilibrium in the Ultracentrifuge," *J. Phys. Chem.*, **65**, 1138 (1961).
65. P. Munk and G. Meyerhoff, "The Use of the Sedimentation Equilibrium in a Density Gradient for the Study of Polymers and Copolymers," 27th International Symposium on Macromolecules, Strasbourg, p. 742, 1982.
66. P. Munk, unpublished results.
67. P. Munk, "Measurement of an Equilibrium Density Gradient of a Solvent Mixture in an Ultracentrifuge," *Macromolecules*, **15**, 500 (1982).
68. J. E. Hearst and J. Vinograd, "A Three-Component Theory of Sedimentation Equilibrium in a Density Gradient," *Proc. Natl. Acad. Sci. USA.*, **47**, 999 (1961).
69. G. Cohen and H. Eisenberg, "Deoxyribonucleate Solutions: Sedimentation in a Density Gradient, Partial Specific Volumes, Density and Refractive Index Increments, and Preferential Interactions," *Biopolymers*, **6**, 1077 (1968).
70. G. Pouyet, A. Kohler, and J. Dayantis, "Density Difference Determinations of Ordinary and Freeze-Dried Polystyrenes in Bulk in the Analytical Ultracentrifuge," *Macromolecules*, **14**, 1126 (1981).
71. T. M. Aminabhavi and P. Munk, "Measurement of Diffusion Coefficients of Polymer Solutions Using the Ultracentrifuge," *Macromolecules*, **12**, 1194 (1979).
72. T. M Aminabhavi and P. Munk, "Diffusion Coefficients of Some Nonideal Liquid Mixtures," *J. Phys. Chem.*, **84**, 442 (1980).
73. G. S. Hartley and J. Crank, "Some Fundamental Definitions and Concepts in Diffusion Processes," *Trans. Faraday Soc.*, **45**, 801 (1949).

CHAPTER

9

LOW-ANGLE LASER LIGHT SCATTERING (LALLS) OF MACROMOLECULES

JEFFREY S. LINDNER

Diagnostic Instrumentation and Analysis Laboratory
Mississippi State University
Mississippi State, Mississippi

and

SHYHCHANG S. HUANG*

Hercules Incorporated
Research Center
Wilmington, Delaware

* Present address: B. F. Goodrich Company, Research and Development Center, 9921 Brecksville Rd., Brecksville, OH 44141

Modern Methods of Polymer Characterization, Edited by Howard G. Barth and Jimmy W. Mays
ISBN 0-471-82814-9

1. INTRODUCTION

Static light scattering is one of the most widely used polymer characterization methods available. Assuming that the refractive index increment of the polymer–solvent system is known, a single experiment yields the weight-average molecular weight $\bar{M}_w$ and the second virial coefficient A_2. Furthermore, if the average coil radius is larger than about $\lambda/20$, where λ is the wavelength of light, then a measure of the mass-average size of the chain, the z-average radius of gyration $\langle S^2 \rangle_z^{1/2}$, is possible. The ability to obtain these three fundamental chain parameters is not possible with any other characterization method.

The three quantities, $\bar{M}_w$, A_2 and $\langle S^2 \rangle_z^{1/2}$ describe the chain in some detail; however, both the expansion of the coil and the thermodynamic properties of the polymer solution will depend on the polymer–solvent interactions. As a result A_2 and $\langle S^2 \rangle_z^{1/2}$ will be solvent dependent. In most cases, and if aggregates are not present in solution, the determination of $\bar{M}_w$ relates directly to the primary chain structure and polydispersity.

The terms $\bar{M}_w$, A_2 and $\langle S^2 \rangle_z^{1/2}$ are static properties of the solvated macromolecule. In other words, these properties do not depend on the Brownian dynamics of the chain arising from the frictional forces between segments of the chain and the solvent. The transport properties do, however, depend on the thermodynamic interactions and the primary chain structure. Thus, methods such as photon correlation spectroscopy (PCS) (determination of the translational diffusion coefficient) (Chapter 10) and viscometry (Chapter 7) provide complementary information. When results from these three methods are combined (for a series of homologous chains in a specific solvent) a complete understanding of the solution properties is possible. The reader is referred to some recent theoretical and experimental studies (1–5).

A number of specific considerations are required in static light scattering measurements. For example, the scattering from polyelectrolytes and copolymers can lead to interpretation difficulties; such situations have been evaluated (see Sections 2.3 and 2.4) and, provided that caution is exercised, few problems can be expected. From a strictly experimental standpoint the elimination (minimization) of dust must be fully achieved. The incorporation of lasers in light scattering photometers has allowed instrument designers to minimize the scattering volume. Consequently, solution clarification can be more readily accomplished. Another result of the reduced volume is that precious samples of biological (and other) origin can be more readily studied.

In this chapter we focus on a special class of static light scattering (low-angle laser light scattering, LALLS). Since this technique was introduced in the 1970s, a resurgence in light scattering studies has occurred. The LALLS method relies on the fact that the intramolecular interference of the polymer chain essentially vanishes at small forward scattering angles. Therefore, determination of the z-average radius of gyration is not possible from this technique. The rapid determination of $\bar{M}_w$ and A_2 is quite valuable, however, and no other measurement can directly provide $\bar{M}_w$ in less time.

Additional consideration of the LALLS method arises from its use as an online detector for size exclusion chromatography (SEC). The sensitivity of the light scattering detector and the extremely small sample volume permit the determination of the scattering intensity of the macromolecule following separation by SEC. In principle, a molecular weight distribution can be obtained without recourse to direct calibration of the SEC system. In practice, numerous experimental difficulties of LALLS measurements, *along with* those of SEC, must be taken into account.

2. THEORETICAL

2.1. Background

The characterization of macromolecules by static light scattering is a well-established technique. A number of monographs and review articles have appeared (6–15). Pertinent equations are described below.

When an electromagnetic wave interacts with a particle, a dipole is created with the electrons of the particle aligning with the plane of the field while the nuclei attain the opposite alignment. The induced moment can be expressed in terms of the polarizability α,

$$\alpha = \frac{c(\partial n/\partial c)_\mu \bar{n}_0}{2\pi N} \tag{1}$$

where c and N are the concentration and number of particles, respectively, and n_0 is the refractive index of the medium. The term $(\partial n/\partial c)_\mu$ is known as the specific refractive index increment and accounts for the finite change of the medium refractive index due to the presence of the scatterers. The subscript μ indicates constant chemical potential.

The ratio of the light scattered i_s to the incident light intensity I_0 for a collection of small, noninteracting identical particles illuminated with a vertically polarized field is given by

$$\frac{i_s}{I_0} = \frac{16\pi^4 V\alpha^2 \sin^2\theta_v}{\lambda_0^4 r^2} \tag{2}$$

where V corresponds to scattering from unit volume, θ_v is the angle of observation with respect to the plane of polarization, r is the distance from the scattering center to the observer (detector), and λ_0 is the wavelength of light *in vacuo*. For vertically polarized light, $\theta_v = 90°$; thus substitution of Eq. 1 into Eq. 2 yields

$$\frac{i_s}{I_0} = \frac{4\pi^2 Vc^2(\partial n/\partial c)_\mu^2 \bar{n}_0^2}{\lambda_0^4 r^2 N} \tag{3}$$

The total number of scatterers can be expressed as

$$N = \frac{cN_A}{M} \tag{4}$$

where N_A is Avogadro's number. Therefore,

$$\left(\frac{i_s}{I_0}\right)\left(\frac{r^2}{V}\right) = \frac{4\pi^2 \bar{n}_0^2(\partial n/\partial c)_\mu^2 cM}{\lambda_0^2 N_A} \tag{5}$$

The term of the left-hand side is commonly denoted as the Rayleigh ratio, R_θ. It is also helpful to group constants on the right side of Eq. 5 since these are either known for the actual light scattering experiment or can be conveniently measured,

$$K = \frac{4\pi^2 \bar{n}_0^2(\partial n/\partial c)_\mu^2}{\lambda_0^4 N_A} \tag{6}$$

This allows a simplified expression

$$\frac{Kc}{\Delta R_\theta} = \frac{1}{M} \tag{7}$$

where ΔR_θ is the difference in the Rayleigh ratios of the solution and the solvent.

Equation 7 is a limiting equation in that intermolecular interactions have been completely neglected. As a result, the expression is only valid at the limit of $c = 0$. Since intermolecular interactions exist at finite concentrations, an expression must be derived that accounts for these interactions.

Debye (16) and Einstein (17) expanded the local polarizability in the form

$$\alpha = \alpha' + \delta\alpha \tag{8}$$

where α' is the magnitude of the polarizability and $\delta\alpha$ is the fluctuation in α. From thermodynamic arguments it can be shown that at constant temperature T, pressure P and number of scatterers N, the fluctuations in α are related to the chemical potential and therefore to the osmotic pressure of the solution.

The corresponding change in a property can be expressed by a virial expansion. Equation 7 can then be written as follows:

$$\frac{Kc}{\Delta R_\theta} = \frac{1}{M} + 2A_2c + 3A_3c^2 + \cdots \tag{9}$$

where A_2 and A_3 are known as the second and third virial coefficients. The virial coefficients describe the departure from the limiting expression caused by intermolecular interactions at finite concentration. In most work the third virial coefficient is small and can be neglected. This term cannot be overlooked in all circumstances, however, since curvature in the concentration dependence of the ratio $c/\Delta R_\theta$ can occur at moderate solute concentrations. In most cases, the scattering data are strictly linear up to reasonably large concentrations. These concentrations are difficult to estimate since the third term, and thus the scattering intensity, depends on the molecular weight and concentration. Studies on the third virial coefficient have been reported (18, 19).

A considerable amount of theoretical work has addressed the meaning of the second virial coefficient. Approximate closed expressions have been presented by Flory (20) and Tanford (8). The term A_2 has also been calculated based on the average segment density around the molecular center of mass (21, 22). Various perturbative approaches are presented by Yamakawa (18).

The approaches of Flory (20) and Tanford (8) based on the probability of locating chains in unoccupied volumes of the solution lead to

$$A_2 = \frac{N_A u}{2M^2} \tag{10}$$

where M is the molecular weight of the solute and u is a parameter describing the excluded volume. The volume excluded by a particle depends on the shape or conformation of the solvated chain. The parameter u can be expressed in terms of a regular geometry. The volume of a sphere V_s is defined as

$$V_s = \tfrac{4}{3}\pi R_T^3 \tag{11}$$

where R_T is known as the thermodynamic radius of the sphere. Addition of a second particle to the solution contributes the same volume as the first but can only approach within a distance of $2R$ of the first. The excluded volume can be written as follows:

$$u = \tfrac{32}{3}\pi R_T^3 \tag{12}$$

The second virial coefficient for a spherical particle is thus,

$$A_2 = \frac{16\pi R_T^3 N_A}{3M^2} \tag{13}$$

which indicates that A_2 can be used to estimate an average size.

In the 1940s Flory (20) and (independently) Huggins (23) presented theories of dilute polymer solutions. A major result of these work was that polymer molecules could behave ideally ($A_2 = 0$). This condition can be realized either by a specific polymer–solvent pair at a particular temperature or by employing a solvent–nonsolvent combination and is referred to as the Flory theta (θ) temperature or simply as the θ state. At the θ condition the excluded volume vanishes and a reliable estimate of the particle radius from A_2 is impossible. Any number of polymer chains can be added to the volume and all will mix or interact identically. From a somewhat naive point of view, the attractive and repulsive forces between the macromolecule and the solvent and those between any other solution components cancel (24).

The second virial coefficient can therefore be described in terms of the forces between the solvent and the polymer chain; in particular, the polymer–polymer and solvent–polymer interactions. If the polymer–solvent interactions are large the coil will be expanded and the solvent is considered "good"; in this case A_2 is positive. Conversely, if the polymer–polymer forces dominate the interactions in the solution, the chains will collapse either intermolecularly and/or intramolecularly. The term A_2 and the excluded volume will be negative. If the attractive forces are dominant, incipient phase separation can occur. A solvent that yields a negative A_2 value for a given polymer chain is denoted as "poor."

Like many other quantities that characterize the macromolecule, such as $\langle S^2 \rangle_z$, the intrinsic viscosity, and the translational diffusion coefficient, the term A_2, for a chemically homogenous series of fractions, can be expressed in terms of the molecular weight,

$$A_2 = k\bar{M}_{\omega}^{b} \tag{14}$$

Here the prefactor k and exponent b depend on the solvent. In good solvents the value of b normally ranges from -0.20 to -0.30 (25) in agreement with theoretical calculations (26, 27).

Equation 9 is valid for any type of scatterer as long as the principal radial axis of the particle is less than about 5% of the incident wavelength. If the scatterer is larger than $\lambda_0/20$, scattering from two centers, P_1 and P_2, on the particle will exhibit a difference in phase, Figure 1 (8). This phase difference, arising from the difference between the scattering path length of the two centers, is 0 at a scattering angle of 0° (assuming that i_s could be separated from I_0 at this angle) and increases as the observation angle is increased. This

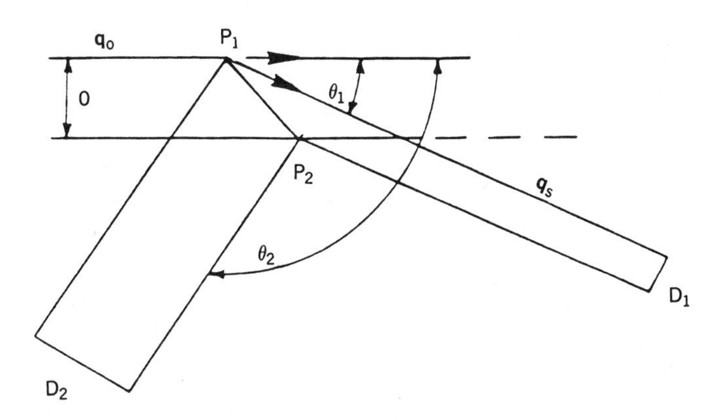

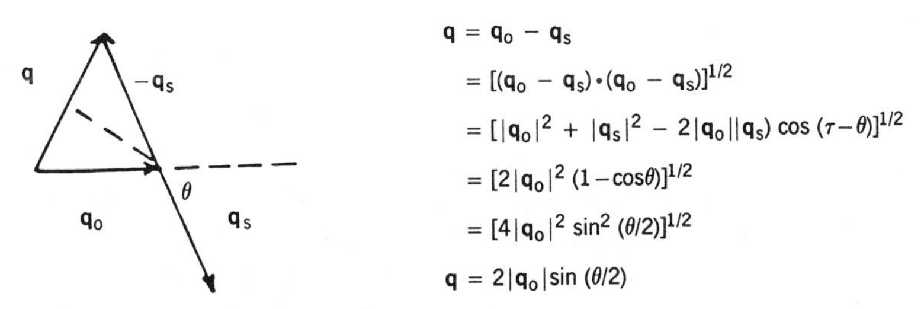

Figure 1. Geometry for determination of the scatter wave vector $\mathbf{q}$ and associated vector notation. An incident vertically polarized beam denoted by $\mathbf{q}_0 = 2\pi n_0/\lambda_0$ is scattered either by a particle at P_1 or P_2 to detectors D_1 and D_2 at the respective angles θ_1 and θ_2. From Ref. 31.

reduction in the scattering intensity is denoted by $P(\theta)$, the particle scattering or form factor.

$$P(\theta) = \frac{\text{scattering intensity at finite angle}}{\text{scattering intensity at } 0^\circ} = \frac{R_\theta}{R_0} \tag{15}$$

Thus for large particles the scattering experiment involves not only extrapolation to zero concentration, Eq. 7, but extrapolation to a scattering angle of 0°.

The formal expression for $P(\theta)$ was derived by Gunier (28), Tanford (8), and Flory (20). All of these studies agreed with the result obtained by Debye (29). Neglecting terms larger than second order, $P(\theta)$ is given by

$$\lim_{\mathbf{q} \to 0} P(\theta) = 1 - \mathbf{q}^2 \langle S^2 \rangle / 3 \tag{16}$$

Here $\mathbf{q}$ is the scattered wave vector derived in Figure 1*b* with $\mathbf{q}_0 = 2\pi/n_0$,

$$\mathbf{q} = \frac{4\pi n_o}{\lambda_0} \sin\frac{\theta}{2} \tag{17}$$

and $\langle S^2 \rangle$ is the squared radius of gyration of the particle

$$\langle S^2 \rangle = \sum_{i=1}^{N} M_i r_i \Big/ \sum_{i=1}^{N} M_i \tag{18}$$

with M_i the mass of the scattering center and r_i the distance of the scattering center from the particle center of mass.

Incorporation of the corrections for intermolecular and intramolecular, $P(\theta)$ interaction was shown by Zimm (30) to lead to

$$\frac{Kc}{\Delta R_\theta} = \frac{1}{M}\left[1 + \frac{16\pi^2 \bar{\eta}_0^2}{3\lambda_0^2} \langle S^2 \rangle \sin^2\left(\frac{\theta}{2}\right)\right] + 2A_2 c + \cdots \tag{19}$$

The ratio $c/\Delta R_\theta$ can therefore be plotted against $\sin^2(\theta/2) + K_s\, c$ where K_s is a constant chosen to separate, graphically, the individual scattering curves at the different concentrations. An example of a scattering diagram is given in Figure 2 (31). Two extrapolations, one to zero concentration at constant angle, yields the values of $c/\Delta R_{\theta = x}$ at $c = 0$ from which the squared radius of gyration is obtained; the second extrapolation is at fixed concentration but as a function of $\sin^2(\theta/2)$. The resulting values of $c/\Delta R_{\theta = 0}$ are then at a scattering angle of 0° and Eq. 7 can be evaluated directly.

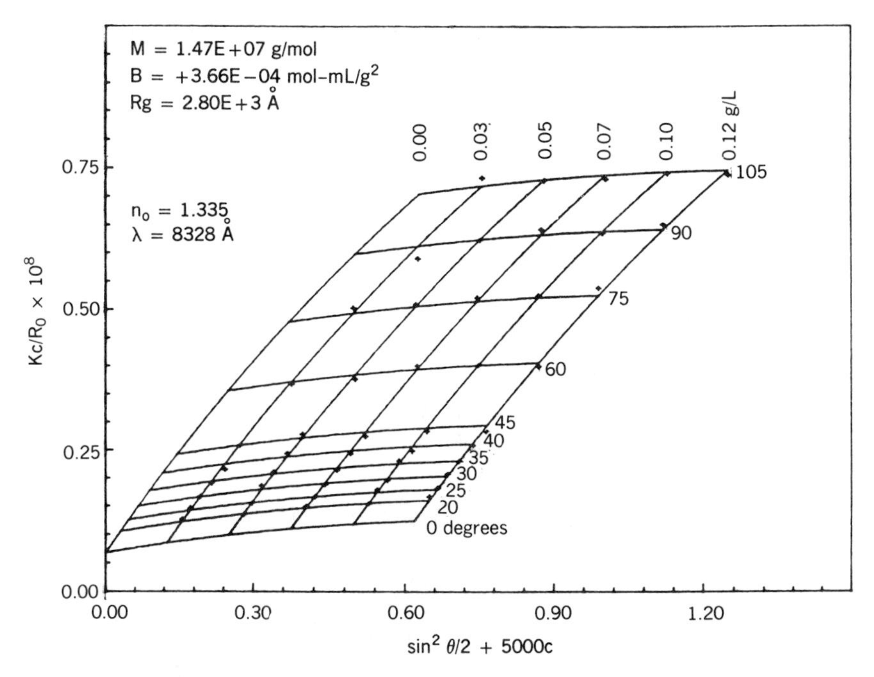

Figure 2. Total scattering diagram for a dextran-g-acrylamide copolymer in water at 25 °C. The terms B and R_g denote the second virial coefficient, A_2, and radius of gyration $\langle S^2\rangle^{1/2}$, as defined in the text. From Ref. 31.

2.2. Macromolecular Polydispersity

The preparation of a polymer normally generates, owing to kinetic considerations, a distribution of molecular weights (polydispersity). Consequently, the values of M and $\langle S^2\rangle$ must relate to some moment of the molecular weight distribution (MWD). Based on Eq. 19 it is possible to average over the specific quantity. For the case of M

$$\lim_{c\to 0} \Delta R_\theta = K \sum c_i M_i \tag{20}$$

The concentration of the ith–polymer chain can be expressed as

$$c_i = n_i M_i \tag{21}$$

where n_i is the number of moles of particles of M_i molecular weight. The definition of the number- and weight-average molecular weights, $\bar{M}_n$ and $\bar{M}_w$, can be rearranged as

$$\sum c_i \sum M_i = \sum c_i \bar{M}_n = c\bar{M}_n \tag{22a}$$

and

$$\bar{M}_w \sum c_i = \sum c_i M_i = c\bar{M}_w \tag{22b}$$

where all the summations run from $i = 1$ to N and c is the total solute concentration. Comparison of Eq. 22b with 20 reveals that the scattering from a polydisperse system yields $\bar{M}_w$. In a similar manner the squared radius of gyration can be shown to be a z-average quantity, $\langle S^2 \rangle_z$.

The formal working Eq. 19, is then

$$\frac{Kc}{\Delta R_\theta} = \frac{1}{\bar{M}_w}\left[1 + \frac{16\pi^2 \bar{n}_0^2}{3\lambda_0^2}\langle S^2 \rangle_z \sin^2\left(\frac{\theta}{2}\right)\right] + 2A_2 c + \cdots \tag{23}$$

and is applicable to large, chemically regular, macromolecules of any dispersity with intermolecular interactions.

2.3. Polyelectrolytes

All of the expressions presented thus far assume that the scattering elements or segments are all chemically similar. Certain experimental problems, both practical and of an interpretational nature, can arise when the polymer chain carries ionic sites on constituent repeating units. These situations have been discussed by Huglin (12), Eisenberg (14), Kratochvíl (11) and others (9).

The molecular conformation of a macromolecule that contains charged groups along its backbone will depend on the number of such groups, as well as the nature of the solvent. In salt-free media the electrostatic repulsions between like charges will lead to molecular expansion and the polyelectrolyte will assume a rodlike or wormlike chain conformation.

The extremely asymmetric molecule will normally exhibit the particle scattering factor of a rod, which makes the extraction of size information difficult. Additional terms also describe the concentration dependence of the osmotic pressure (32). The molecular weight is no longer measured but instead a quantity known as the practical osmotic coefficient Ψ_p is obtained. This quantity is the ratio of the osmotic pressure of the unshielded chain to that of an electrically neutral chain, and its value decreases approximately exponentially with an increase in charge density (concentration). In salt-free solutions the light scattering plot will reflect the above trend.

The above situation can be avoided by the addition of electrolyte. The presence of counterion to the ionized sites on the polyelectrolyte will result in shielding of electrostatic forces. In this case, however, the solution now contains three components. Instead of a normal correspondence between the polymer and solvent, as observed in a binary solution, the solvent sheath can contain more of either the simple salt or the solvent. In a ternary solution one of the solvent components can be preferentially absorbed onto the polymer chain. Casassa and Eisenberg (33) described this problem formally. In essence an apparent molecular weight is obtained. This quantity is sometimes denoted as M_2 or M_{app} and will be in excess of the true moleculer weight; the second virial coefficient and z-average radius of gyration will also be apparent quantities.

The concept of preferential adsorption can be summarized in thermodynamic terms by the expression

$$\gamma_1 = \left(\frac{\partial m_1}{\partial m_2}\right)_\mu \tag{24}$$

where m_1 and m_2 are the molalities of solvent and polymer, respectively, and γ_1 is known as the preferential solvation coefficient. An increase in the solute concentration must correspond to an increase in the concentration of component 1 if the activity (chemical potential) is to remain constant. This situation can be eliminated by exhaustive dialysis of the polyion solution against the mixed solvent (34). It should also be noted that changes in the specific refractive index increment are observed on comparing undialyzed and dialyzed solutions (35). Kratochvíl estimated the ratio M_{app}/M_w for different macromolecules at various degrees of preferential solvation (36). As long as the added electrolyte concentration is high and the counterion of the salt is the same as the ionic moiety of the polyelectrolyte, the resulting error in $\bar{M}_w$ will be about 5%.

2.4. Copolymers

A copolymer dissolved in a single solvent will exhibit an average specific refractive index increment that depends on the copolymer composition and the $\partial n/\partial c$ values of the two components. This average value of $\partial n/\partial c$, denoted as v, can be written as

$$v = W_{\text{A}}(\partial n/\partial c)_{\text{A}} + W_{\text{B}}(\partial n/\partial c)_{\text{B}} \tag{25}$$

where W_{A} and W_{B} are the weight fractions of A and B units in the copolymer. The application of Eq. 25 depends on the chemical heterogeneity of the copolymer which, in turn, depends on the mode of polymerization. If a block copolymer is effectively monodisperse (synthesized by anionic means) then Eq. 25 yields apparent molecular weights that are in excellent agreement with those obtained from membrane osmometry (37).

Random copolymers normally exhibit a compositional distribution, as well as a MWD (38, 39). The weight fractions are therefore average values for all the species in the distribution; thus, v is also an average value and M_{app} is obtained.

Benoit and Froelich (40) showed that M_{app} is related to M by

$$M_{app} = (v_A v_B / v^2) M + [v_A (v_A - v_B) / v^2 W_A M_A] - [v_B (v_A - v_B) / v^2] W_B M_B \quad (26)$$

where v_x is the $\partial n/\partial c$ value of component x and M_A and M_B are the molecular weights of all the repeating units of A and B. Equation 26 can be solved simultaneously from three different sets of experimental data to yield M, M_A, and M_B. If v_A is zero, then M_B can be determined provided v_B is known. In a like manner, M_A can be obtained directly if v_B is zero. A polymer–solvent pair in which $\partial n/\partial c$ is zero is said to be isorefractive, that is, no light will be scattered from the particular component. Isorefractive solvents have therefore been used in the characterization of block copolymers (41). It should also be noted that according to Eq. 26 if both v_A and v_B are large, then M_{app} approaches the true value.

2.5. Low-Angle Laser Light Scattering

Low-angle laser light scattering (LALLS) is a restricted application of static light scattering. At small forward angles the term $\sin^2 \theta/2$ in the working expression (Eq. 19) becomes quite small and the particle scattering factor, $P(\theta)$, is essentially unity. Since the third- and higher-order terms in Eq. 9 are normally small the working expression becomes

$$\frac{Kc}{\Delta R_\theta} = \frac{1}{\bar{M}_w} + 2A_2C \quad (27)$$

An alternate form of Eq. 27 was derived by Berry (42)

$$\left(\frac{Kc}{\Delta R_\theta}\right)^{1/2} = \left(\frac{1}{\bar{M}_w}\right)^{1/2} (1 + 2A_2C)^{1/2} \quad (28)$$

The square-root plot tends to linearize any slight undulations in the scattering data. The analysis of light scattering data has been reviewed by Yamakawa (18) as well as in Refs. 9, 11, 12 and 15.

Special considerations that apply to copolymers and polyelectolytes also apply in LALLS measurements. Finally, in some cases the assumption that $P(\theta) = 1$ may not be correct. Such situations would arise from ultrahigh-molecular-weight macromolecules, generally greater than 1×10^7 g/mol in molecular weight, or for those systems that aggregate or associate intermolecularly in solution. Recently, a conventional laser-based scattering design was introduced for both stand-alone and online SEC applications (see Section 3.1). With this unit there is a potential for determining $\langle S^2 \rangle$ as a function of the MWD.

3. INSTRUMENTATION

3.1. Commercially Available LALLS Photometers

Presently, there are five commercially available laser based photometers that allow the measurement of scattering intensities at small ($< 15°$) angles. Three of these units are true LALLS photometers operating at angles typically less than 7°. These instruments include the KMX-6 and CMX-100 photometers, originally manufactured by Chromatrix and currently available from LDC/Analytical and the LS-8000 distributed in Japan by Tosoh Corporation.

Recently Wyatt Technology introduced the Dawn-F photometer which, depending on solvent refractive index, can allow measurements as low as 9°. Otsuka Electronics manufactures the DLS-700 dynamic light scattering spectrometer. This unit allows both PCS and static intensity measurements to angles as low as 5° depending on cell path length. These latter units, however, should be classified as classical light scattering instruments since intensity measurements at large angles are, based on the design, possible. Specifications of these photometers are compared in Table 1.

Innovations in low-angle scattering spectrometers are not limited to strictly commercial ventures. The prism light scattering spectrometer, developed by Chu et al. (43), allows intensity measurements at angles as small as 2°. Cannel and co-workers (44) also designed an accurate light scattering photometer with 18 fixed scattering angles ranging from 2.6°–163°. Other units (45, 46), primarily developed for Mie scattering applications (small angle near-forward scattering) could readily be converted to allow determination of $\bar{M}_w$ and A_2.

The Chromatix KMX-6 instrument, based on Kaye's design (47, 48) was initially marketed in the mid-1970s. The incident beam is vertically polarized and the source is a low power helium–neon (HeNe) laser operating in the TEM_{∞} mode at $\lambda_0 = 632.8$ nm. For the majority of synthetic macromolecules the effects of absorption and fluorescence are negligible at this wavelength.

Table 1. Commercially Available Laser Light Scattering Photometers

	KMX-6[a]	CMX-100[a]	LS-8000[b]	Dawn F[c]	DLS-700[d]
Cell volume (μL)	10	10.5	30	33	700–2000 Max (12-mm cell) 5000 Max (25-mm cell)
Scattering Volume (nL)	35	40	100	250	0.4, 2, 10 ($\theta = 90°$)
Angle (degrees)	2–3 3–4 3–7 4.5–5.5 6–7 90 175	5.1–6.1	5	Simultaneously monitors 18 angles of which 15 are used in data analysis. Scattering angle depends on solvent refractive index	12-mm cell 10–150 25-mm cell 5–150

[a] LDC Analytical, Riviera Beach, FL.
[b] Tosoh Corporation, Tokyo, Japan.
[c] Wyatt Technology, Santa Barbara, CA.
[d] Otsuka, Corp., Osaka, Japan. (US distributor: Otsuka Electronics U.S.A., Ltd., Fort Collins, CO.)

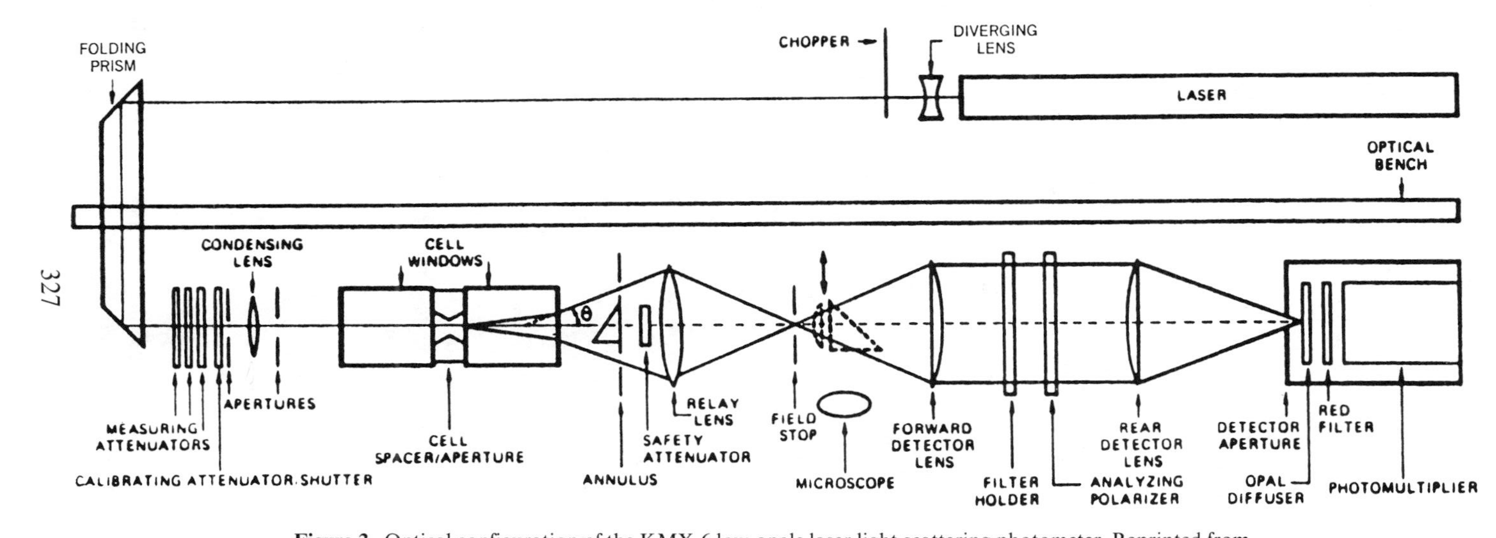

Figure 3. Optical configuration of the KMX-6 low-angle laser light scattering photometer. Reprinted from Ref. 80, p. 448 by courtesy of Marcel Dekker, Inc.

Corrections for these effects may be required for certain polymers, most notably those containing heme groups and highly conjugated systems.

The optical configuration of the KMX-6 is diagrammed in Figure 3. The laser beam is expanded and then modulated by the chopper. A prism is used to route the beam to the scattering cell. Attenuators or neutral density filters of known transmission are used to control the incident light intensity. The laser beam is focused down to a spot diameter of 0.08 mm at the sample volume using the condensing lens.

The sample cell assembly consists of two thick rectangular fused-silica windows between which a spacer is inserted. The insert can be either Teflon® or polycarbonate or stainless steel for chromatographic or high-temperature studies. A small hole located through the center of the spacer defines the cell volume. Two strategically located vertical channels allow the sample solution to flow into and out of the scattering volume.

The receiving optics of the instrument consist of an annular assembly, a beam dump, relay and focusing lenses, a field stop and detector aperture, and locations for mounting a polarizer or interference filter. The annuli are all located on an indexed wheel and serve to define the various solid scattering angles. A small aperture located at 0° is used for I_0 measurements and a vertical slit is available for measurements at 90°.

A small beam trap or dump with an inclined surface is located in the center of each annulus. This component absorbs–reflects the incident beam at 0° thereby protecting the photomultiplier tube from excessive flux. To ensure a safe intensity for the measurements at 0°, a fixed attenuator of high optical density is placed directly after the annulus wheel.

The scattered radiation, defined by the particular annulus selected, is collected by the relay lens and focused on a small field stop. The field stop defines the effective path length l (see below) of the chosen experimental configuration. A microscope objective with a prism assembly is located after the field stop. This feature allows visual observation of the scattering cell. Dust particles passing through the laser beam, air bubbles, and stray scattering from dirty cell window can be observed. The periscope is especially useful at the beginning of the scattering experiment. Lenses are used following the field stop to focus the scattered radiation on the photomultiplier aperture.

The Rayleigh ratio is calculated from

$$R_\theta = \left(\frac{G_\theta}{G_0}\right)\frac{D}{\sigma l} \tag{29}$$

where G is the response at angles θ and 0, D is the transmittance of the attenuator used in measuring the incident power, σ is the solid angle that is defined by the annular geometry, and l is the effective optical path length. The

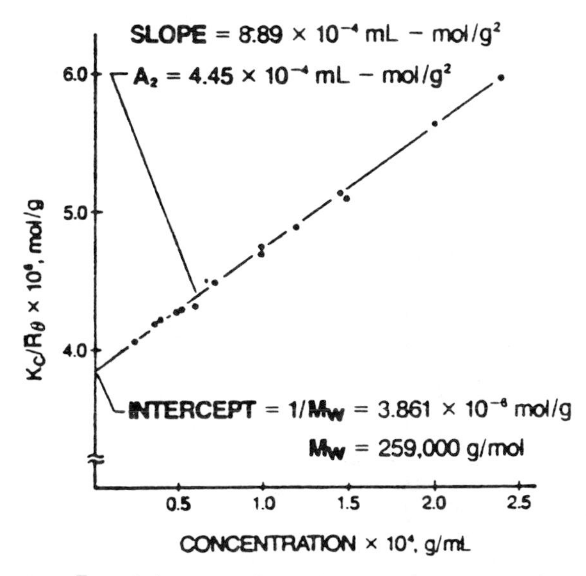

Figure 4. Plot of $Kc/\bar{R}_\theta(\Delta R_\theta)^{-1}$ versus c for SRM 706 polystyrene in toluene. Experimental conditions: KMX-6 LALLS photometer, $\partial n/\partial c = 0.110\,\text{mL/g}$. $\lambda_0 = 632.8\,\text{nm}$, $T = 23\,°\text{C}$ and annulus $= 6°–7°$. Reprinted from *American Laboratory*, **8**(5), 95 (1976). Copyright © 1976 by International Scientific Communication, Inc., Ref. 49.

factor $1/\sigma l$ is obtained from geometrical considerations. No secondary calibration is required with the instrument.

Measurements of R_θ for the polymer solution and the solvent allow calculation of the excess Rayleigh ratio. Assuming that the value of the spcific refractive index increment at $\lambda_0 = 632.8\,\text{nm}$ is known, the data can be evaluated from Eq. 27. A representative plot is given in Figure 4 for NBS 706 polystyrene in toluene (49). The resulting value of 2.59×10^5 g/mol for $\bar{M}_w$ is in excellent agreement with that obtained by other measurements, 2.58×10^5 g/mol.

The KMX-6 can be interfaced with an SEC system (50). Knowledge of the solute concentration of the chromatographed macromolecule along with knowledge of the behavior of the second virial coefficient as a function of $\bar{M}_w$ permits, in principle, the online determination of the MWD. This technique is further discussed in Sections 4 and 5.

The model CMX-100 photometer is a version of the KMX-6 specifically designed for SEC applications although it can be used for offline measurements as well. The optical configuration of the instrument is shown in Figure 5. A beam splitter is used in the CMX-100 to normalize the scattering intensities to the fluctuations in laser beam power and also for determination

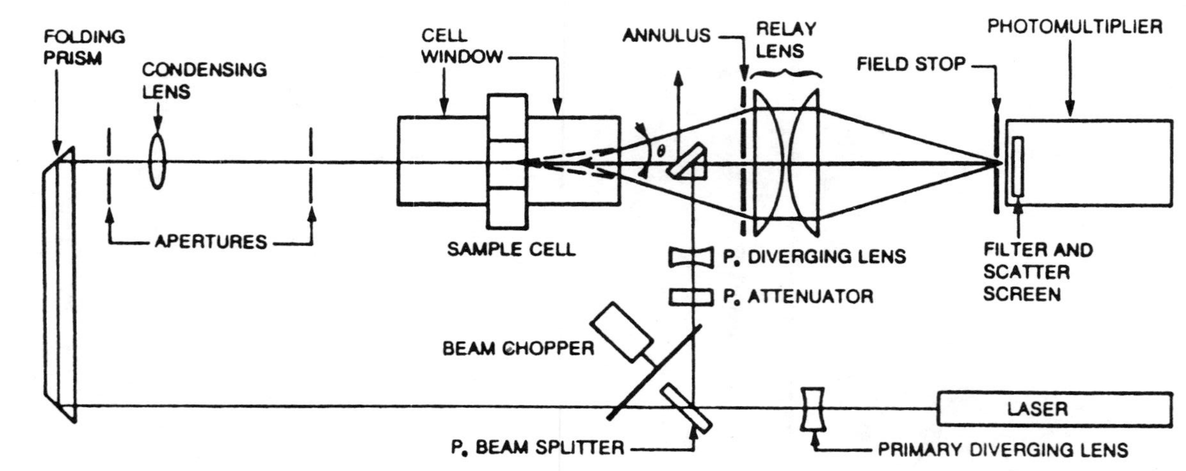

Figure 5. Optical configuration of the CMX-100 LALLS photometer. By permission of LDC Analytical, Riviera Beach, FL.

of G_θ. Modulation of the beam by the chopper allows, as in the KMX-6, quantification of the dark counts or background readings from the phototube. Timing circuitry is used to switch between measurements of scattered light G_θ, incident light G_θ, and dark current. A feedback circuit is used to vary the voltage applied to the photomultiplier, thereby compensating for drifts in incident laser beam power. Circuitry is also used to suppress the dark current.

The sample cell is somewhat different from that of the KMX-6. Two thick windows are used; the cell spacer, however, is 5-mm thick and the cell volume is 10.5 μL.

Although the configuration of the CMX-100 and KMX-6 are similar there are some notably differences: A fixed aperture of 0.2-mm limits the angular range (solvent refractive index dependent) of the CMX-100 to 5.1°–6.1° in air. The photomultiplier tube is located directly after the field stop, thus effectively preventing the insertion of a filter polarizer. Size limitations also dictate that a periscope not be incorporated. A thumbwheel assembly allows the cell to be moved perpendicular to the incident beam.

Determination of the intensities for the incident beam, G_0, and the scattered radiation, G_θ, is accomplished using a switch. Calculation of the Rayleigh ratio R_θ is the same as for the KMX-6.

The optical configuration of the LS-8000 (Fig. 6), is similar to Kaye's design (47, 48). As in the CMX-100, the intensity of the incident beam is continually monitored to compensate for power fluctuations. This measurement, however, is performed using a separate photomultiplier tube located after the scattering cell. (Thus fluctuations in the solution scattering can be observed, inadvertently, as fluctuations in incident beam power.) To determine accurately the molecular weight, a secondary calibration using a polymer of known concentration and $\bar{M}_w$ is required.

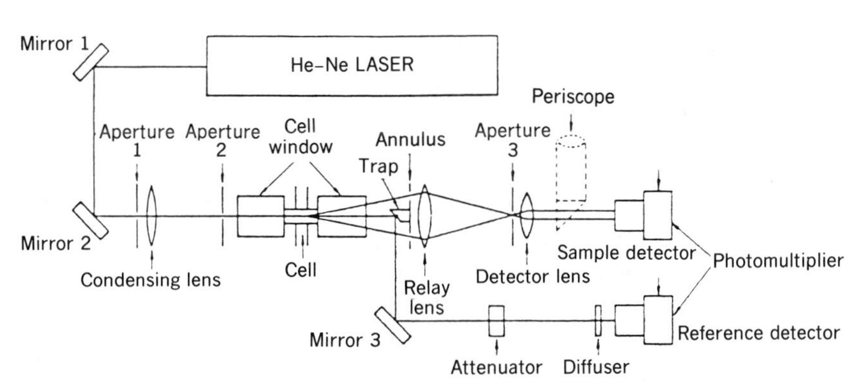

Figure 6. Schematic diagram of the LS-8000 LALLS photometer. By permission of Otsuka Electronics USA Ltd., Fort Collins, CO.

When the LS-8000 is used as an online SEC detector a concentration detector, such as a differential refractometer (DRI), is used to monitor the concentration at each elution volume increment. If the responses at the i elution volume increment from the DRI and LALLS detectors are denoted as h_i^{RI} and h_i^{LS}, respectively, then

$$h_i^{\mathrm{RI}} = k_1 c_i \tag{30}$$

and

$$h_i^{\mathrm{LS}} = k_2 \Delta R_{\theta,i} \tag{31}$$

where c_i and $\Delta R_{\theta,i}$ are the respective concentration of the solute and the excess Rayleigh ratio of the solute at elution volume, i, and k_1 and k_2 are system response constants.

In a typical SEC experiment the concentration of polymer solution injected is, after passing through the columns, diluted considerably. Thus, in the working expression for light scattering it is sometimes permissible to neglect the contribution of A_2 at each elution volume increment

$$\Delta R_{\theta,i} = K c_i \bar{M}_i \tag{32}$$

and

$$h_i^{\mathrm{LS}} = k_2 K c_i \bar{M}_i \tag{33}$$

Substitution of Equation 30 for c_i and rearranging yields

$$\frac{h_i^{\mathrm{LS}}}{h_i^{\mathrm{R1}}} = K\left(\frac{k_2}{k_1}\right)\bar{M}_i \tag{34}$$

Measurement of the LS and RI signal intensities along with knowledge of the specific refractive index increment allows determination of $\bar{M}_i$. It must be cautioned, however, that $(\partial n/\partial c)_\mu$ is only pertinent for a given macromolecule in a particular solvent. Thus the determination of k_2 must correspond to the polymer–solvent system used in the chromatographic analysis.

The main problem in the above approach is the fact that standards of known $\bar{M}_w$ are not available for most polymers. One approach to solving this problem would be fractionation of the sample followed by static measurements of $(\partial n/\partial c)_\mu$ and $\bar{M}_w$ using stand alone instruments. Alternatively, the factor $k_2 K(k_1)^{-1}$ can be estimated through measurement of $(\partial n/\partial c)_\mu$ on the original polymer sample and the use of a UV detector as the SEC concentration detector. Denoting the ratio $k_2 K(k_1)^{-1}$ as B

$$B = B_0\left[\left(\frac{\partial n}{\partial c}\right)\bigg/\left(\frac{\partial n}{\partial c}\right)_0\right] \tag{35}$$

where the subscript 0 corresponds to the standard polymer and $\partial n/\partial c$, and B corresponds to the unknown sample. The ratio of the extinction coefficients, ε_0 and ε, from a UV detector completes the correction term

$$B = B_0\left(\frac{\varepsilon_0}{\varepsilon}\right)\left[\left(\frac{\partial n}{\partial c}\right)\Big/\left(\frac{\partial n}{\partial c}\right)_0\right]^2 \tag{36}$$

Although not a true LALLS unit, the Dawn model F light scattering photometer (51, 52), is capable of measuring intensities at up to 15 angles simultaneously, thus allowing, in principle, the measurement of $\langle S^2 \rangle$.

The flow cell of the DAWN-F consists of a 2-mm channel drilled through a cylindrical cell body. A vertically polarized low-power HeNe laser beam is directed along the length of the channel and the cell is sealed using windows. Eighteen photodiodes are located around a cylindrical scattering cell, Figure 7. Other cell configurations, such as one for batch processing, (Dawn Model B) are available.

Light scattered at an angle from the illuminated sample volume ΔV is refracted at the liquid–glass interface and is measured with the photodiode detector. The measurement angle, as compared to that of the incident beam in air, must be corrected for the refractive index of the solvent using Snell's law.

The lack of prisms, mirrors, and other optical components simplifies the optical configuration. Unlike the KMX-6 and CMX-100, which are calibrated strictly on geometrical considerations, the DAWN-F unit requires calibration using solvents of known Rayleigh ratio or by using a standard of known $\bar{M}_w$ and small radius of gyration, so that $P(\theta) = 1$ at all scattering angles. Prior to

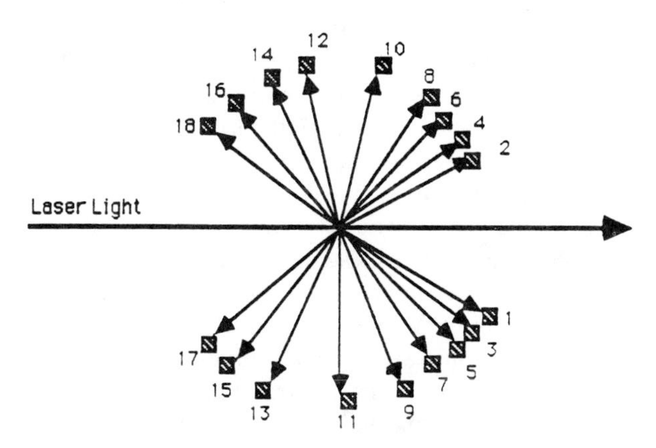

Figure 7. Detector configuration for the Dawn Model F photometer. By permission of Wyatt Technology Corporation, copyright © 1989, 1990.

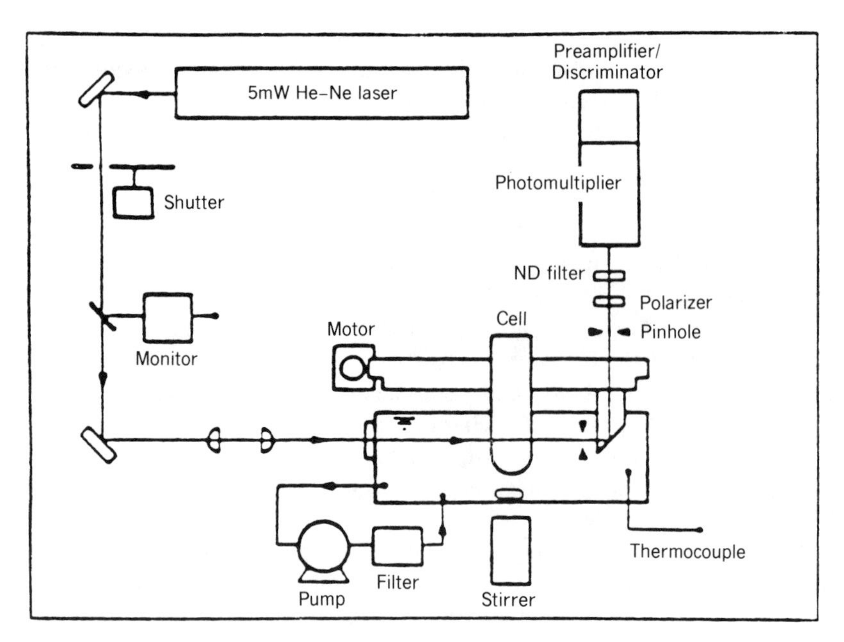

Figure 8. Block diagram of the Otsuka DLS-700 light scattering spectrometer. By permission of Otsuka Electronics USA. Ltd., Fort Collins, CO.

actual calibration of the instrument it is necessary to normalize the response of the photodiode detectors. This is accomplished by relating the signal at each scattering angle to that signal observed at 90°. As in the calibration of the instrument response it is important to perform the detector normalization with a point scatterer (no angular dependence). In practice, the normalization can be performed on the polymer–solvent system used for calibration at the same time the calibration measurement is made.

A block diagram of the Otsuka DLS-700 photometer is shown in Figure 8. A 5-mW HeNe Laser is used as the light source. The monitor is used to normalize all of the scattered intensities to variations in incident laser power. The vertically polarized beam is routed to the scattering cell by fixed mirrors and focused to a small diameter in the center of the cylindrical cell. The cell, which can either be 12- or 25-mm round, is located in the center of a glass vat, which is filled with a refractive index matching fluid. Because the refractive indexes of the cell and the fluid are the same, minor imperfections on the outer surface of the cell do not contribute to stray light.

The scattered radiation is collected at the angle of interest by the prism and focused on a photon counting photomultiplier tube. A pinhole is used to define the collection geometry. Provisions in the design of the receiving train allow for the insertion of a polarizer or other optics such as a narrow band pass (laser line) or neutral density filter. Positioning of the photomultiplier tube assembly is controlled by a steppermotor. An angular range from $5°$–$150°$, depending on cell path length, with an accuracy of $0.1°$ is reported by the manufacturer.

The signals from the monitor detector and the pulses from the photomultiplier tube are routed to a system microcomputer. The computer can either process the incoming pulse train in terms of an average intensity or compute the time correlation function, representing the Brownian motion of the macromolecules, by the time interval method (software correlator). Temperature control of the fluid in the vat, which is critical for translational diffusion measurements, is performed by circulating fluid from an external bath around the outside and below the vat. Temperature control to $\pm 0.2\,°C$ is reported and the temperature range depends on the refracting index matching fluid.

Although the DLS-700 is not a true LALLS unit, the angular range and ability to obtain information on the many parameters available from both static and dynamic light scattering should be noted.

3.2. Refractometers

Knowledge of the specific refractive index increment is critical for the proper determination of $\bar{M}_w$ and A_2. It should be recalled that $(\partial n/\partial c)_\mu$ appears to the second power in the definition of the optical constant K. Some aspects of $(\partial n/\partial c)_\mu$ and its dependence on incident wavelength and molecular weight are briefly discussed in Section 4.2. The wavelength dependence of the refractive index increment is not negligible. Consequently, it is advisable to determine this parameter at the wavelength selected for the intensity measurements.

Measurements of $(\partial n/\partial c)_\mu$ can be accomplished using either a differential refractometer or by interferometric means. The two units described below are examples of commercial instruments that permit determination of $(\partial n/\partial c)_\mu$ by the two principles at the standard operating wavelengths ($\lambda_0 = 633$ nm) of commercially available LALLS photometers.

The optical configuration of the Chromatix KMX-16 laser-based differential refractometer is given in Figure 9. The radiation from a 0.5 mW–HeNe laser is first expanded and collimated prior to traversing the sample cell. The beam then passes through an angle multiplier and a telescope to an adjustable mirror that reflects the beam back through the sample cell to the detector. The detector consists of two separated sensors, which allows determination of the exact position of the beam.

The sample cell is divided into two 3–mL compartments by a window positioned at Brewster's angle. One of the cell compartments contains only solvent and is denoted as the reference side; the other side of the cell either

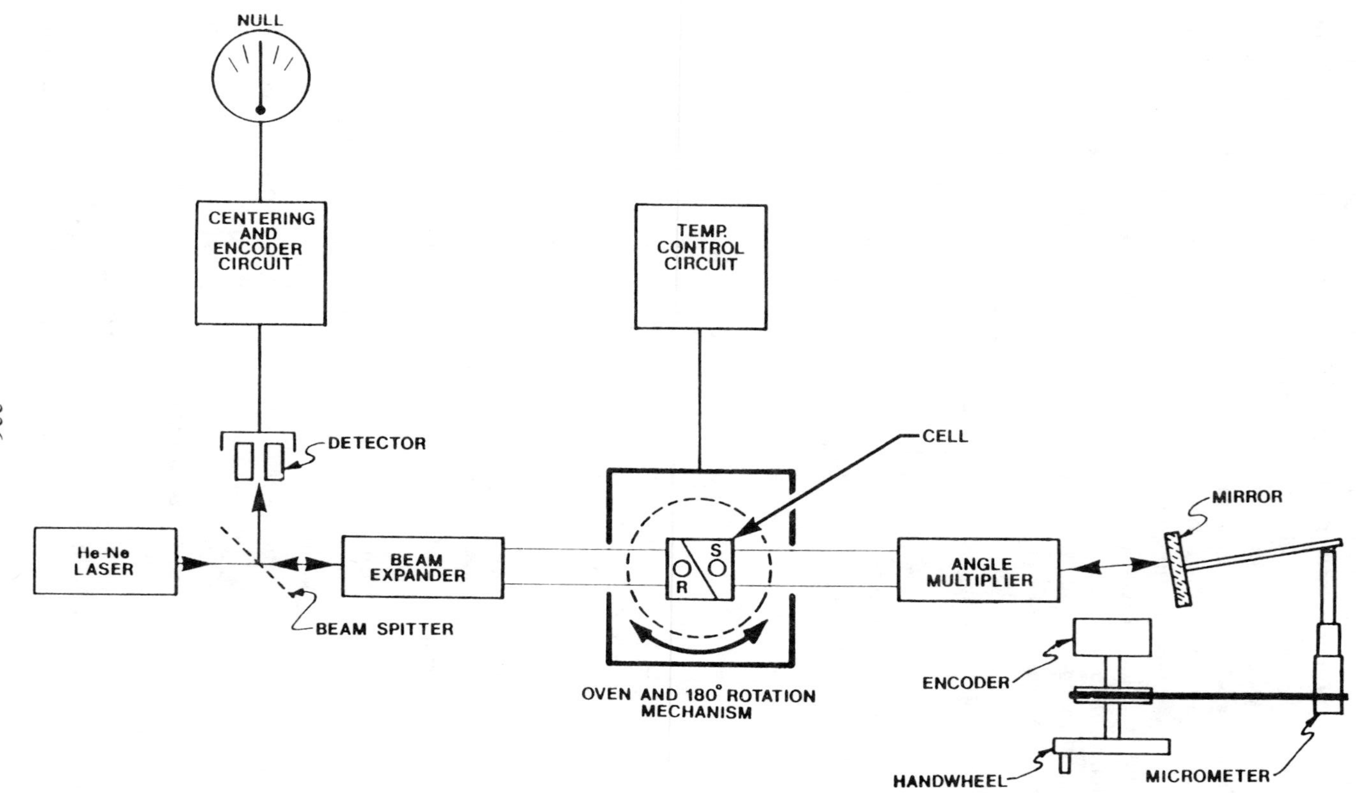

Figure 9. Optical and cell configuration of the KMX-16 differential refractometer. By permission of LDC Analytical, Riviera Beach, Florida.

contains solvent for the determination of the deflection characteristics of the optics or the polymer solution under study for measurement of the excess refractive index, Δn. The sample cell is enclosed in a metal heat transfer block, which is contained in an oven. Temperature can be controlled to 165 °C or to below ambient using an external bath. The sample cell assembly can be rotated 180°.

The deviation of the beam due to the difference in refractive indexes in the two cells is measured by adjusting the calibrated mirror to obtain a null reading on the position sensor. The extent of movement of the mirror is related to the change in refractive index by calibration to a material such as NaCl or KCl with a known $(\partial n/\partial c)_\mu$ (53).

In normal operation, measurements are made through both sides of the cell, L and R. The deviation, ΔX, due to the pressure of the solute is calculated as follows:

$$\Delta X = (L - R)_c - (L - R)_0 \tag{37}$$

where $(L - R)_c$ is at some known concentration and $(L - R)_0$ is that for the pure solvent. The beam deviation is related to the refractive index by

$$\Delta n = k\Delta X \tag{38}$$

where k is a calibration constant.

The specific refractive index increment is that value of $\Delta n/c$ at $c = 0$. This relationship is approximated by (54)

$$\frac{\Delta n}{c} = \left(\frac{\partial n}{\partial c}\right)_\mu + \alpha_1 c + \cdots \tag{39}$$

where α_1 is a constant for a given polymer–solvent system. Extrapolation of values of $\Delta n/c$ at different solute concentrations yields the specific refractive index increment.

Values of $(\partial n/\partial c)_\mu$ for many different polymer–solvent systems are compiled in Refs. 12 and 55. Those given in the *Polymer Handbook* (56) are generally for wavelengths other than 632.8 nm.

In the determination of $(\partial n/\partial c)_\mu$ using the KMX-16 the solute concentrations generally range from 1.0 to 5.0 mg/mL. Differences of 10^{-4}–10^{-3} refractive index units in Δn were reported (55). The KMX-16 was noted by the manufacturer to be accurate to 10^{-6} refractive index units.

An example of an interferometric refractometer is the Optilab 903 distributed by Wyatt Technology. The optical configuration of the instrument is given in Figures 10*a* and 10*b*. As compared to the KMX-16, the Optilab 903

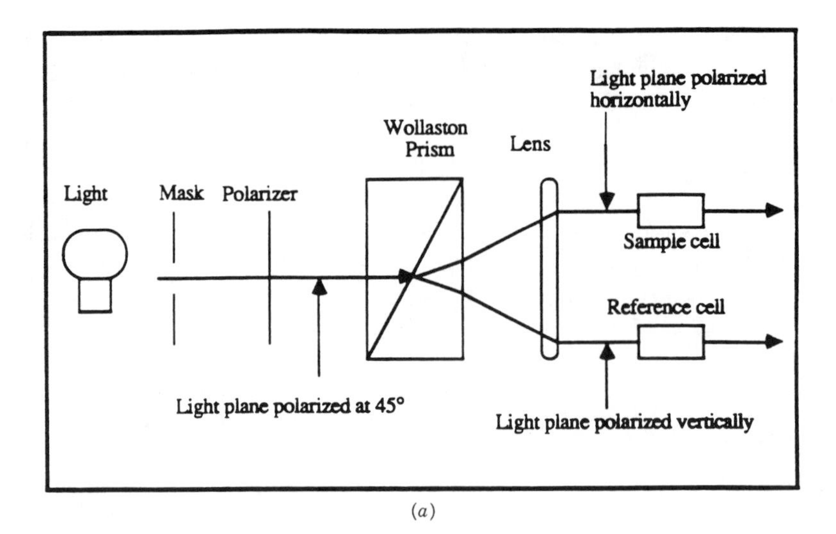

(*a*)

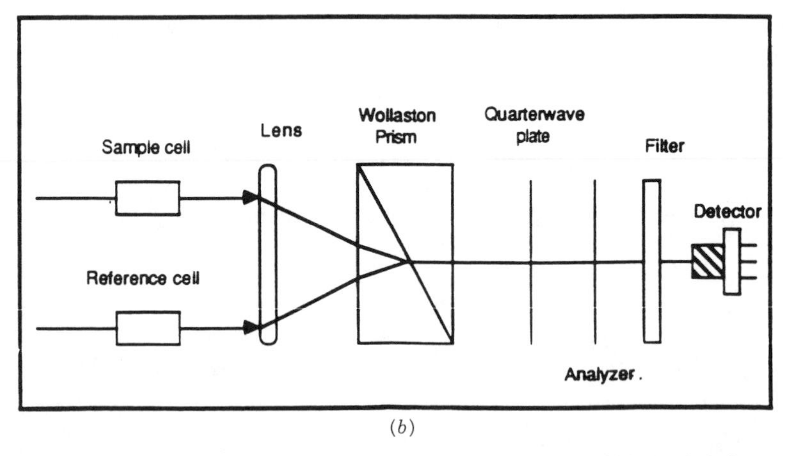

(*b*)

Figure 10. Source side (*a*) and receiving (*b*) optics for the Wyatt/Optilab 903 introferometric refractometer. By permission of Wyatt Technology Corporation, copyright © 1989, 1990.

measures the difference in phase, not the beam displacement, between light passing through a reference cell and a separate solution cell. The light source is masked and passed through a polarizer placed 45° with respect to the incident beam. A Wollaston prism splits the primary beam into horizontal and vertically polarized components of equal intensity. A lens is used to focus the beams to the respective cells.

Prior to traversing the cells, the two beams are in phase with one another. A phase difference, however, will occur after the beams pass through the cell since the refractive indexes in each cell are different. For the solution cell the effective wavelength can be written as follows:

$$\lambda_s = \lambda_0 / n_s \tag{40}$$

and the number of wavelengths present in the cell will equal l/λ_s where l is the cell path length. After passing through the cell, the beams will be in a common refractive index (air) and thus have the same wavelength. The phase difference can then be written as

$$l/\lambda_s - l/\lambda_r = l(1/\lambda_s - 1/\lambda_r) \tag{41}$$

where the subscript r denotes the reference cell.

In the Optilab 903, the phase difference is related in angular units from the fact that a change in one wavelength yields a 360° or 2π radian difference in the waves. Thus, the phase difference expressed in angular terms can be written as follows:

$$\phi = 2\pi l(1/\lambda_s - 1/\lambda_0) \tag{42a}$$

$$= 2\pi l(\Delta n/\lambda_0) \tag{42b}$$

where Eq. 40 has been used. A second lens and Wollaston prism (Fig. 10) are used to recombine the beams. Following passage through the quarterwave plate, a plane polarized beam is obtained. This beam, however, has been rotated by a factor of $\phi/2$. The analyzer located at an angle of $90°-\beta$ with respect to the axis of the initial polarizer, Figure 10*b*, is used to isolate the deviation in phase.

In design the angle β is selected such that at $\phi = 0$ the incident beam intensity is attenuated to 35°. The relationship to the transmission of the incident beam is given by

$$I/I_0 = \cos(90 - \beta) = \sin\beta = 0.35 \tag{43}$$

The angle β will generally reside within the limits

$$\phi/2 < \beta < 90° \tag{44}$$

and the transmission is written as

$$I/I_0 = \sin(\beta + \phi/2) \tag{45}$$

The phase difference is related to the difference in the refractive indexes of the solute and solvent using Eq. 42b. Isolation of the 633 nm wavelength is accomplished using a bandpass filter located directly before the photodiode detector. Calibration of the refractometer is performed in the same manner as for the KMX-16.

Although both the KMX-16 and the Optilab 903 measure $(\partial n/\partial c)_\mu$, there are some notable differences. The KMX-16 is a stand-alone unit, requiring at least 2 mL, whereas the Optilab 903 requires less volume (12, 7 or 1.4 μL). The cell configuration of the Optilab unit allows unidirectional transport of the fluid through the measuring volume. Temperature in this instrument can be controlled up to 85 °C using an external bath as compared to that of the KMX-16, which allows measurements to 165 °C.

According to the manufacturer, the Optilab 903 refractometer is accurate to 1×10^{-5} refractive index units (riu) for the 1.4-μL cell and to 2×10^{-7} riu for the 12-μL cell. The noise level was noted at less than 2×10^{-9} units. The ability to use the refractometer either as a stand alone or online unit for SEC at the proper wavelength for LALLS measurements is an advantage.

4. STATIC LOW-ANGLE LASER LIGHT SCATTERING

4.1. Solvent–Sample Preparation

All of the instruments for LALLS measurements have extremely small scattering volumes. Even with the associated connecting tubing for the transport of the solution into and out of the cell, the volume necessary for the experimental determination of R_θ is less than 1 mL. These volume considerations should be contrasted to earlier light scattering cells, where volumes greater than 10 mL were normally required. This drastic reduction in volume is a direct result of the incorporation of lasers as compared to the isolation of a particular wavelength from a mercury or tungsten lamp with lower flux density.

The parameter required for the analysis of the scattering characteristics of a solute is the excess Rayleigh ratio $\Delta R_\theta = R_{\theta,\text{soln}} - R_{\theta,\text{solv}}$. Any spurious contamination by, for example, dust particles, in either the solution or the solvent will lead to incorrect ΔR_θ values. Consequently, the parameters $\bar{M}_w$, A_2 and $\langle S^2 \rangle_z$ will not represent the contribution of the solute alone. It is, therefore, of importance to minimize or eliminate as best as possible, the amount of dust in the solvent and the solution. Also, it is important to eliminate any air bubbles trapped in the cell.

The scattering from dust is somewhat predictable and the outcome of its presence can be discussed in the context of ΔR_θ and Eq. 19. Figure 11 is a plot

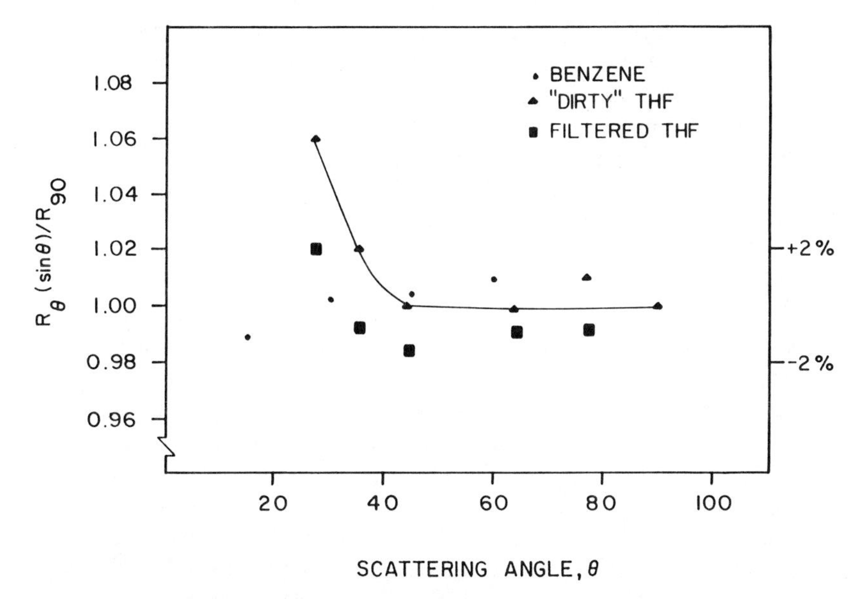

Figure 11. Plot of the ratio $R_\theta(\sin\theta)/R_{90}$ versus scattering angle (degrees) for benzene and THF. The factor $\sin\theta$ in the numerator accounts for that fraction of light viewed at the given scattering angle. The strong upturn at low angles for the unfiltered THF corresponds to scattering from dust. A slope of near zero degrees, corresponds to a clean sample and proper angular alignment of the spectrometer. Results from J. S. Lindner, W. W. Wilson, and J. W. Mays, unpublished results.

of the normalized Rayleigh ratio, defined as $R_\theta \sin\theta/R_{90}$, against the scattering angle (57). The factor $\sin\theta$ accounts for that fraction of scattered light viewed by the detector at angle θ. The three curves shown are for the solvents benzene and tetrahydrofuran (THF). The benzene had been scrupulously cleaned using the closed-loop filtration method discussed below. Also included are data for unfiltered and filtered THF. The flat angular response of the benzene indicates that the solvent is essentially free of dust, that the spectrometer is properly aligned and that stray light from the cell walls is negligible. For the first THF measurement, which was conducted directly after those for benzene, the upturn at the lower scattering angles is indicative of dust contamination. In general, and from experience, dust particles are of large size and possess highly irregular shapes. Note that if measurements could have been extended to small forward angles the dust present in the unfiltered THF solution would lead to drastically increased values of R_θ.

In a similar manner the presence of dust in the polymer solution will lead to increased R_θ values. If the solvent is clean, then the ΔR_θ value will be larger than that of the pure solvent alone and $\bar{M}_w$ will be overestimated. Both the concentration and angular dependencies will then have larger slopes and

$\langle S^2 \rangle_z$ and A_2 will be larger than that of the pure solute. The presence of dust leads to strong downward curvature at low concentrations in a plot of $Kc/\Delta R_\theta$ against c. As the polymer concentration increases, the scattered intensity increases and at moderate or high concentrations the contribution of the dust can diminish relative to that of the solute.

The purification of solvents and solutions has received considerable attention. For batch units the solvent and solutions can be centrifuged at low speed to separate dust particles. The centrifugate is then transferred (pipeted) into the light scattering cell (58). Alternately, the entire cell has been centrifuged and then moved to the holder of the spectrometer (59).

The designs of most of the commercial photometers (Section 3) feature or can feature flow-through cells. This is a direct requirement for their use as SEC detectors, but can also be used to great advantage in static LALLS measurements. Dust is present in solvents and solutions, in glassware and on the inside of the light scattering cell. Through proper cleaning it is possible to minimize the dust in the cell. For example, the two windows for the KMX-6 cell assembly can be readily cleaned by washing the window faces in the solvent followed by the use of spectroscopic grade methanol and optical paper. Some dust will most likely remain. With the method described below, the need to prepare individual concentrations of the polymer solution, and thus to use separate vials for each concentration, is eliminated.

The closed-loop filtration concept is presented in Figure 12 (60). A reservoir is connected to a pump via appropriate tubing (Teflon, etc.) specifically chosen so as not to interact with the solvent. Filters of appropriate pore diameter, whose selection is based on experience and the estimated size of the macromolecule, are located downstream of the pump and prior to the light

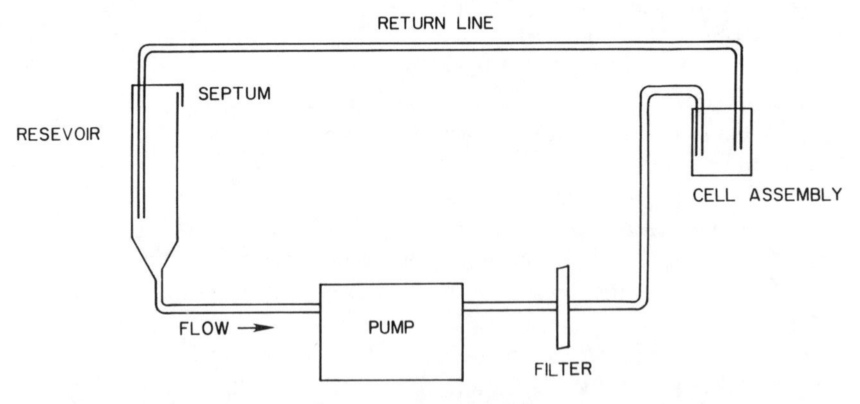

Figure 12. Block diagram of the closed-loop filtration apparatus. Initially developed by Wilson (60).

scattering cell. Filters are also selected based on their chemical compatibility with the solvent. We have found that Nuclepore™ filters (Nuclepore Corp. Pleasanton, Calif.) work quite adequately for aqueous solutions, while the recently introduced Anopore™ filters (Anotec Separation, Ltd., Oxon, England) can be used for both organic and aqueous solutions. Acrodisc™ filters (Gelman Sciences, Ann Arbor, Michigan) are also available in aqueous and chemically resistant varieties. Following passage through the filters, the solvent (solution) is routed to the cell, passes through the scattering volume, and out of the cell. The solution is then directed to the reservoir, which is sealed using a septum in order to prevent evaporation and contamination.

Initially, a known volume of either spectroanalyzed grade or freshly distilled solvent is pipeted into the reservoir and the pump started. The solvent is allowed to filter continuously until the output from the photomultiplier tube (PMT) is stable. Recall also that for most instruments scattering within the cell can be observed. The pump is then stopped and the measurement of the Rayleigh ratio is performed.

At this stage the calculated Rayleigh ratio of the solvent can be compared to literature values (12, 61, 62). Alternately, if the Rayleigh ratio is unknown at the wavelength of measurement, the stability of the photomultiplier output will reveal the cleanliness of the sample. If the solution is dust free then the output of the PMT will be extremely stable, $< 2\%$ output fluctuation. If, however, the Rayleigh ratio does not agree with the literature value to within about 2 or 3%, additional filtration is required.

The closed-loop filtration method can then be used for the actual scattered intensity determination for the polymer solution. If the stock concentration is made sufficiently large (dependent on molecular weight) then only a small amount of solution, on the order of tens of microliters, can be added to a low-volume loop to realize a suitable concentration. Equilibration of the concentration can be followed by monitoring the output of the PMT. At this stage the excess Rayleigh ratio can be determined, and along with the concentration and the value of the optical constant, K, a preliminary apparent molecular weight can be calculated. This, along with the Rayleigh ratio of the solvent, allows some checks on the experimental protocol.

Measurements of R_θ for a given polymer are then performed from lower to higher concentrations. Note also that the cell does not have to be cleaned or moved after each concentration. One potential drawback to the closed-loop technique occurs if the radius of the polymer chain is greater than about one third that of the pore diameter. (Information concerning the radius of the macromolecule can be estimated from viscometry.) Alternatively, the experiments can be performed with a filter of a larger pore size. As a general rule for synthetic macromolecules, if the molecular weight is less than approximately 1×10^6 g/mol a 0.2-μm filter can be used. If the molecular weight is larger than

1×10^6 g/mol then a larger pore size (usually 0.45 μm) must be used to prevent accumulation of the solute on the filter surface. Trapping of the solute by the filter leads to a Rayleigh ratio that will be less than that anticipated and the term $Kc/\Delta R_\theta$ will exhibit a radical increase.

The closed-loop filtration method has been applied to solutions of hemoglobin (63) and other proteins (64) as well as number of water-soluble macromolecules including dextrans, polyacrylamides, and polyelectrolytes (31). More recently it has been applied to the study of poly(α-methylstyrene) (4), polyisobutylene (65), and other nonpolar macromolecules.

4.2. Determination of the Specific Refractive Index Increment

A number of other sources of error must be avoided in conducting light scattering measurements. These have been described in the monographs and review articles referenced in Section 2. For example, an error in $\partial n/\partial c$ of 2% leads to a corresponding error in $\bar{M}_w$ of 4%. Fortunately the determination of $\partial n/\partial c$ is accurate to about 0.001 mL/g when properly performed; therefore, the error arising from $\partial n/\partial c$, if this term has been determined in the same solvent and at the same wavelength as that for the light scattering experiment, is generally less than 2%.

In some instances the value of $\partial n/\partial c$ has been determined at two or more different wavelengths than those employed for the actual measurements. In this case a reasonable estimate can be obtained from (66),

$$(\partial n/\partial c)_{\lambda 2} = (\partial n/\partial c)_{\lambda 1}(k' + k''/\lambda^2) \tag{46}$$

where k' and k'' are polymer–solvent dependent constants. An example of the comparison of the theoretical and experimental determination of $\partial n/\partial c$ for poly(α-methylstyrene) in toluene is given in Figure 13 (4). The agreement between the measured and calculated values is adequate.

Some additional comments relating to $(\partial n/\partial c)_\mu$ concern the known dependence of this parameter on molecular weight as given by (67)

$$\left(\frac{\partial n}{\partial c}\right) = \left(\frac{\partial n}{\partial c}\right)_\infty + \frac{K}{\bar{M}_n} \tag{47}$$

Here $(\partial n/\partial c)_\infty$ is the specific refractive index increment at infinite molecular weight and K is a chemical composition–solvent dependent factor. Typically $(\partial n/\partial c)_\mu$ approaches an asymptotic value at molecular weights greater than about 2×10^4 g/mol.

The only other consideration of $(\partial n/\partial c)_\mu$ is the determination of this parameter for biopolymers and synthetic polyelectrolytes. As discussed in

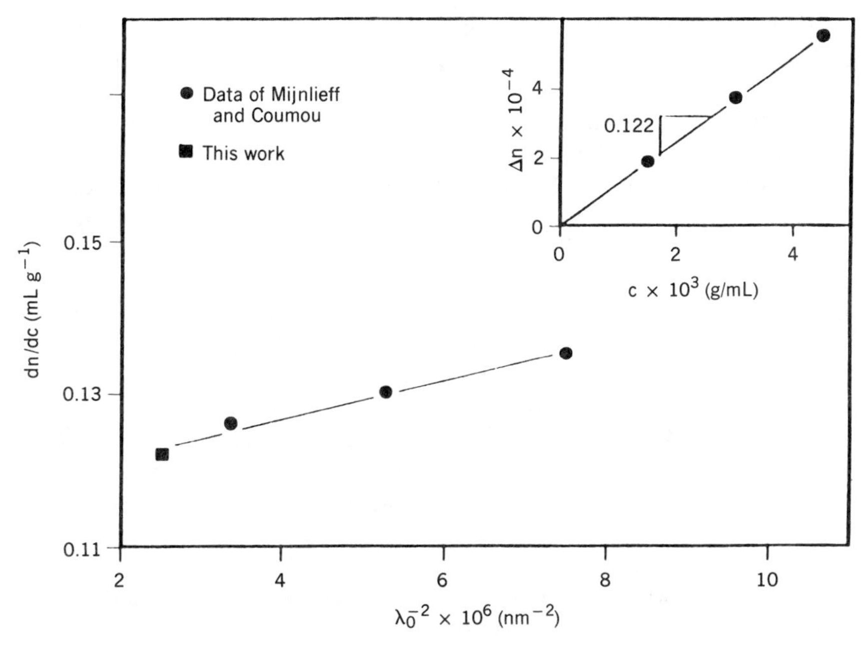

Figure 13. Specific refractive index increment, $\partial n/\partial c$ versus λ_0^{-2} for poly(α-methylstyrene) in toluene at 25 °C. (Circles in the main plot correspond to the data from Ref. 34 in Ref. 4.) Reprinted with permission from J. S. Lindner, W. W. Wilson and J. W. Mays, "Properties of Poly(α-methylstyrene) in Toluene: A Comparison of Experimental Results with Prediction of Renormalized Group Theory," *Macromolecules*, **21**, 3304 (1988), Copyright © 1988, American Chemical Society.

Section 2, the presence of charged moities on these types of macromolecules results in preferential solvation or adsorption of ions on the chain. In some cases it may not be possible, owing to limitations in sample quantity, to perform the measurement of $(\partial n/\partial c)_\mu$. In any such case the molar mass obtained is *apparent*. Sample scarcity should not be such a critical problem for synthetic polyelectrolytes. With any polyelectrolyte it is necessary to dialyze the polymer solution against the solvent until a constant chemical potential is obtained, if a reliable $(\partial n/\partial c)_\mu$ value is to be obtained.

4.3. Other Sources of Error

Aside from errors in $(\partial n/\partial c)_\mu$ there are few other major considerations. One important factor involves the calibration of the response of the photometer. Methods for this have ranged from the use of pure solvents of known Rayleigh ratio, to the use of extremely well-characterized nearly monodisperse polymer standards (61, 68). In practice no calibration is required for LALLS instruments; that is, these instruments have been calibrated based on geometrical considerations by the manufacturers. For users of other systems, the work of Kratochvíl et al. (68) and those references presented at the beginning of Section 2 should be consulted.

As remarked in Section 3.1, some macromolecules absorb or fluoresce at wavelengths employed for light scattering measurements. These additional processes can substantially hinder determination of the true scattering intensity. Casassa and Berry (15) described the effects of absorption and fluorescence on the light scattering method and an excellent discussion on fluorescence effects is contained in the monograph by Kratochvíl (11). In most, but not all, polymer solutions the extinction coefficient at $\lambda_0 = 632.8$ nm is weak or of no consequence. If however, the solution is colored then the process responsible should be investigated. The treatment of absorption or fluorescence and the effects upon the light scattering method are well documented, and reference is made to those works cited previously (6–15).

5. SIZE EXCLUSION CHROMATOGRAPHY–LOW-ANGLE LASER LIGHT SCATTERING

5.1. Background

An important application of LALLS is as a direct molecular weight detector following SEC separation. In conventional SEC, the determination of the moments of the molecular weight distribution is only possible by means of a calibration of the system. For this purpose standards of homogeneous chemical composition, which span a wide range in molar mass are required. The conventional calibration is therefore specific to a given macromolecule in a given solvent. The hydrodynamic volume of a chain may be considerably different from that of the calibrating standards; thus, any attempt to analyze a macromolecule dissimilar to the calibration standards results in the determination of apparent molecular weight averages.

In many cases, standards having the same chemical composition as the polymer under analysis are unavailable. Placing a LALLS photometer between the chromatographic columns and an appropriate concentration detector (differential refractometer, UV detector, etc.) (Fig. 14) (69) permits the determination of the excess Rayleigh ratio ΔR_θ, and solute concentration at each elution volume increment. In principle then, absolute molecular weights can be determined for the sample at different elution volumes without recourse to secondary calibration. In addition to the potential determination of the distribution, SEC–LALLS has also been shown (see below) to aid in the

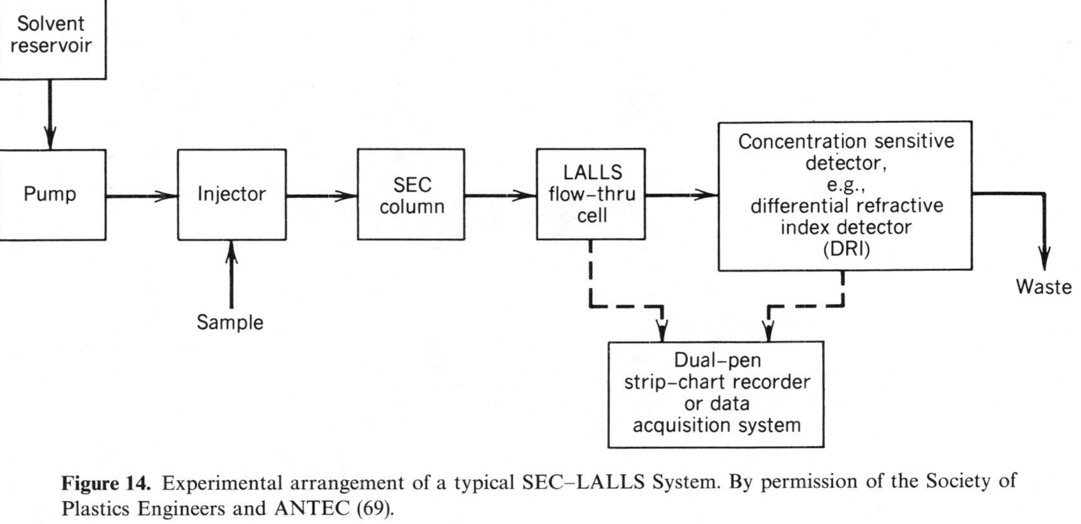

Figure 14. Experimental arrangement of a typical SEC–LALLS System. By permission of the Society of Plastics Engineers and ANTEC (69).

analysis of branching and the determination of trace amounts of ultrahigh-molecular-weight components.

In the typical SEC–LALLS experiment (Fig. 14) the concentration detector is placed after the LALLS photometer in order to prevent damage to the concentration detector cell by back pressure. Outputs from the LALLS detector and the concentration detector can be displayed on a strip-chart recorder or alternately routed to a data acquisition–processing system. LDC/Analytical and Viscotek provide commercial SEC–LALLS data analysis systems. Wyatt Technology offers a software package for use with their DAWN-F multiangle unit.

5.2. Sources of Error

The same factors that impinge on the static light scattering experiment also affect the analysis of the SEC effluent using LALLS detection. Of added importance are the diluted regimes at either ends of the MWD.

As in static light scattering measurements the presence of dust can have deleterious effects in SEC–LALLS. It is therefore important to employ high-purity solvents and to prepare the samples in such solvents. Although the SEC columns can act as a depth filter and remove particulates in the solvent, often times the column packing material itself can be a source of debris, mainly packing fines. In aqueous SEC using silica-based columns, silica dissolution appears to be a major source of particle spikes observed in LALLS tracings. Column conditioning may help to reduce these spikes.

Figure 15 is a plot of LALLS response against elution volume for two standard polystyrenes in THF (70). Each of the spikes on the chromatogram corresponds to the passage of a particle through the scattering volume. Clearly, the presence of these particles introduces an added uncertainty in the data analysis process. To minimize the extent of particles, it may be possible to incorporate an inline filter of low-dead volume before the LALLS detector.

Berkowitz (70) developed an electronic process for the removal of the spike noise. The additional panels in Figure 15 (B–D, F–H) illustrate the filtering characteristics of the hardware. Analysis of the corrected chromatograms, D

→

Figure 15. Panel A is the strip-chart recording of LALLS output for NBS 706 in the presence of particles (spike noise); Panel B is the corrected LALLS trace following first level spike detection; Panel C is the standard deviations after the second level of spike detection, and Panel D is the corrected final strip-chart recording. Panels E–H are the same stage of electronic spike filtering but for a near monodisperse polystyrene, NBS 705. Reprinted with permission from S. A. Berkowitz, "Rejection of Spike Noise from Size Exclusion Chromatography/Low-Angle Laser Light Scattering Experiments," *Anal. Chem.*, **58**, 2571 (1986). Copyright © 1986 American Chemical Society.

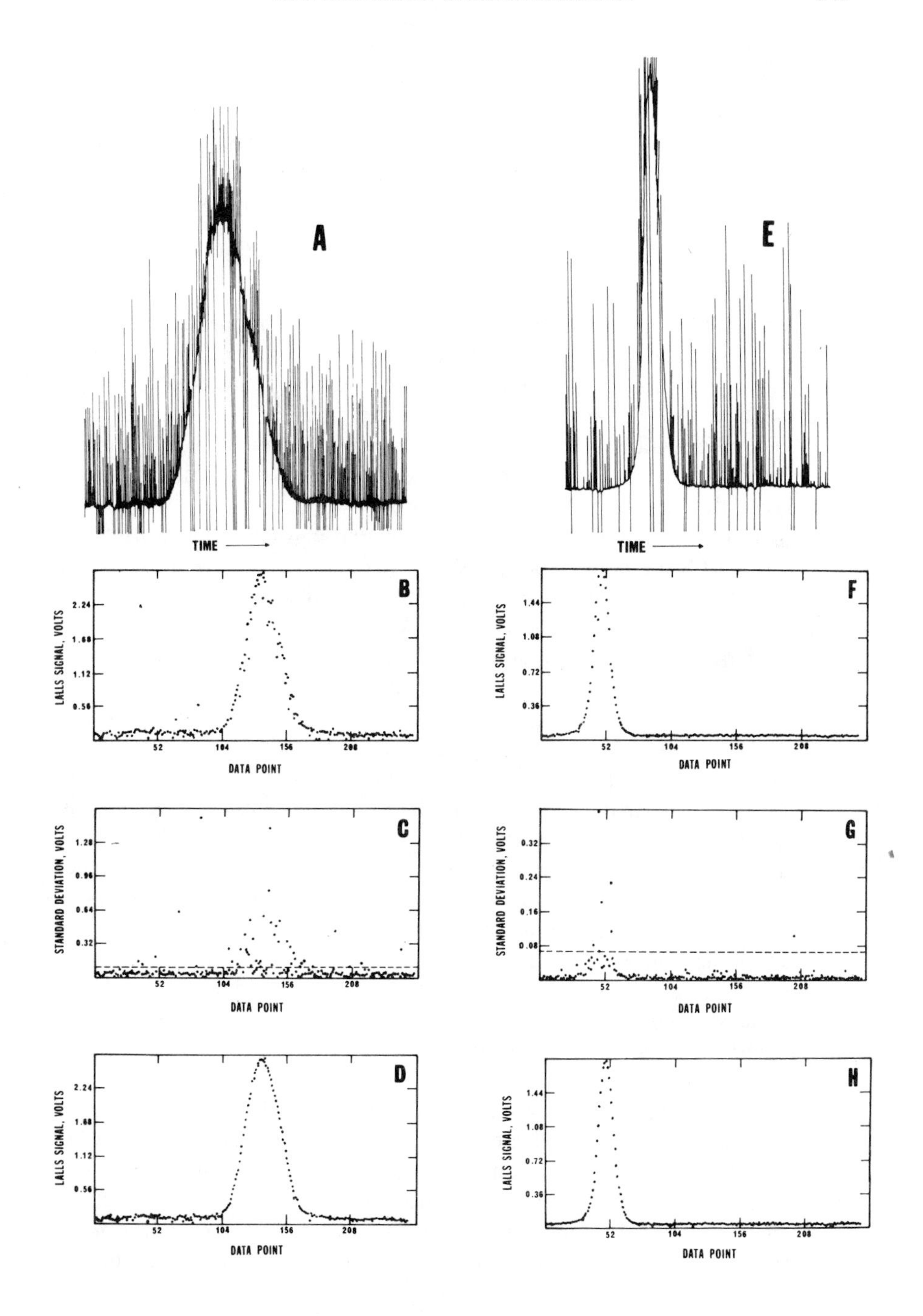
A
TIME
E
TIME
B
LALLS SIGNAL, VOLTS
2.24
1.68
1.12
0.56
52
104
156
208
DATA POINT
F
1.44
1.08
0.72
0.36
C
STANDARD DEVIATION, VOLTS
1.28
0.96
0.64
0.32
G
0.32
0.24
0.16
0.08
D
H

and H, yielded errors of less than 3.2% for $\bar{M}_w$ and less than 14% for $\bar{M}_n$. Because of the limiting characteristics of the LALLS detector (see below), the errors in $\bar{M}_n$ are not excessive and the electronic filter appears to be a reasonable alternative if particles cannot be minimized by physical filtration. In the work cited above it was also noted that the use of commercial software packages without the electronic filter can yield $\bar{M}_w$ errors as large as 39% depending on the presence of particles.

The errors in the concentration and LALLS detectors and the associated effects of these errors on the number- and weight-average molecular weights have been studied by Procházka and Kratochvíl (71, 72). In essence, the low concentrations and small Rayleigh ratios of the late eluting components lead to large errors in this low-molecular-weight regime. Thus, the technique provides a reasonable value of $\bar{M}_w$, but the determination of an accurate value of $\bar{M}_n$ is only possible with narrow distribution macromolecules. The error associated with the determination of $\bar{M}_n$ was shown by model calculations to increase with increasing sample polydispersity.

The above errors dictate the information that can be obtained from the use of LALLS in conjunction with a SEC. In those instances where conventional calibration of the SEC is possible, the determination of $\bar{M}_w$ provides a check on the calibration procedure.

The analysis of Procházka and Kratochvíl (71, 72) assumed that $(\partial n/\partial c)_\mu$ is constant for each elution of volume of the chromatogram. From the previous discussion regarding the molecular weight dependence of this parameter, it is possible that $(\partial n/\partial c)_\mu$ will not, in certain instances, be constant with elution volume. The imposed error on the moments of the MWD will depend on the departure of $(\partial n/\partial c)_\mu$ in the low-molar-mass regime and will therefore affect the lowest moment, $\bar{M}_n$, of the MWD. Note that if the $(\partial n/\partial c)_\mu$ for the lower molecular weight components is less than that of the higher molecular weight species, the optical constant K will be reduced, M at these elution volumes will increase, and the value of $\bar{M}_n$ will be overestimated. The corresponding sample polydispersity index $\bar{M}_w/\bar{M}_n$ will then be underestimated. Normally, a constant value of $(\partial n/\partial c)_\mu$ is employed for SEC–LALLS data analysis. The molecular weight dependence of $(\partial n/\partial c)_\mu$ should be considered if the polymer under analysis has components less than about 2×10^4 g/mol in molecular weight.

Because of band broadening, macromolecules eluting from an SEC column at each elution volume increment are not monodisperse. As a result, the $\bar{M}_n$ that is measured will be overestimated, unless band broadening corrections are applied. If high-resolution columns are used, this error can be minimized.

Figure 16 shows the concentration and LALLS detector responses for a hypothetical macromolecular solution containing a very small amount of ultrahigh-molecular-weight component (69). The high-molecular-weight

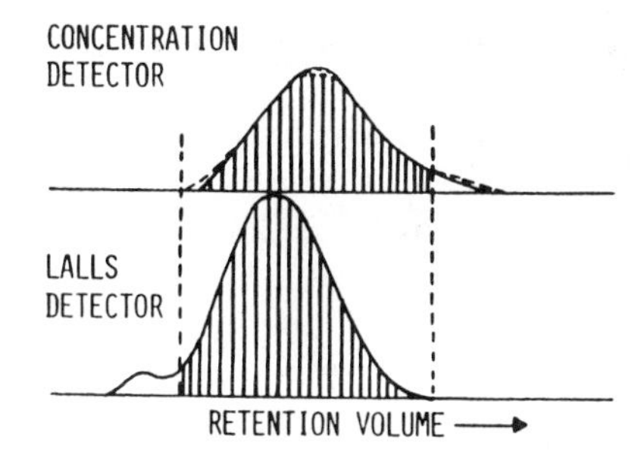

Figure 16. Concentration and LALLS chromatograms for a hypothetical macromolecule with low concentrations of high- and low-molecular-weight components. The areas within the dashed lines indicate the reliable regime of the chromatograms for data analysis. After Barth and Huang (69). Used by permission of the Society of Plastics Engineers.

portion of the distribution is too dilute to be detected with the concentration detector and the low-molecular-weight tail is observed in the concentration detector but not by the LALLS detector. The determination of molecular weight M at the ends of the chromatogram corresponds to regimes of high error, and deviations in the linearity of the calibration of M versus elution or retention volume must be considered. It is possible to assume that the responses at the center of the distributions can be linearly extrapolated as shown in Figure 17. Thus, the unobserved low M portion of the LALLS response curve may be accounted for by considering the concentration data only. This amounts to an improved estimate for $\bar{M}_n$; however, the ultrahigh-molecular-weight components cannot be accounted for and thus $\bar{M}_w$ will be

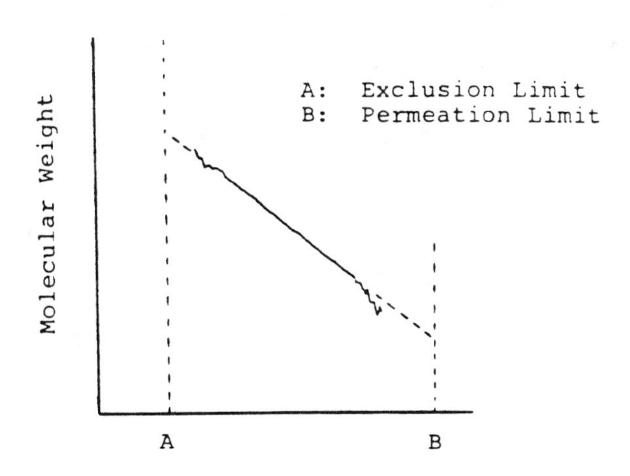

Figure 17. Plausible extrapolation of the column calibration curve to estimate regimes that are outside of the dashed lines in Figure 16.

underestimated. In any case, the presence of any ultrahigh-molecular-weight components can be qualitatively observed.

An estimate of $\bar{M}_w$ can be obtained by ratioing the area of the LALLS curve against the concentration curve (73). The LALLS response at each elution volume v_i (no band broadening) can be derived from Eqs. 20 and 21 as

$$h_i = \Delta R_{\theta,i} = K n_i M_i^2 \tag{48}$$

where h_i is the LALLS response. The total area, A, of the LALLS response curve can be written as

$$A_{LS} = \sum h_i = K \sum n_i M_i^2 \tag{49}$$

while that for the concentration detector is

$$A_c = K' \sum n_i M_i \tag{50}$$

The ratio of the two areas is then

$$\frac{A_{LS}}{A_c} = \frac{K \sum n_i M_i^2}{K' \sum n_i M_i} = K'' \bar{M}_w \tag{51}$$

Assuming that the $\bar{M}_w$ of a lower molecular weight sample for the same type of polymer can be measured, then K'' can be calculated.

5.3. Branching Determination

One of the most important applications of SEC–LALLS is the determination of the degree of long-chain branching with respect to molecular size (see Chapters 2 and 3). Figure 18 shows calibration curves of molecular weight against elution volume V_e for a chemically homogeneous linear and branched polymer (subscripts l and b, respectively) (69). According to the universal calibration method, at identical elution volume increments i,

$$([\eta]_b M_b)_i = ([\eta]_l M_l)_i \tag{52}$$

Here

$$[\eta]_l = K_i M_l^a \tag{53}$$

thus,

$$[\eta]_{b_i} = \frac{M_{l_i}}{M_{b_i}} [\eta]_{l_i} = \left(\frac{K_i M_l^{a+1}}{M_b} \right)_i \tag{54}$$

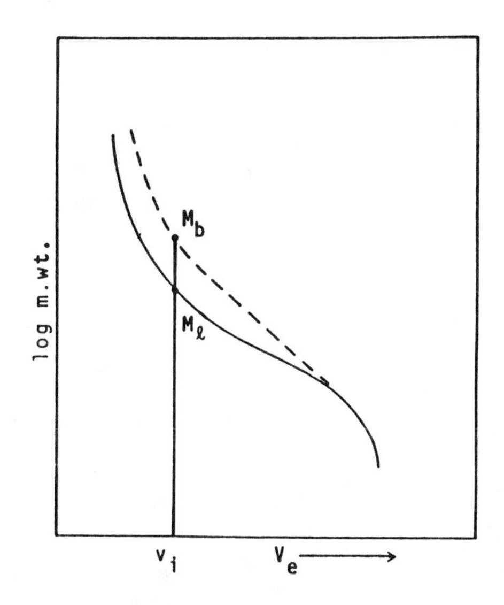

Figure 18. SEC calibration curves for a linear (subscript l) and branched (subscript b) chemically homogeneous chains. After Barth and Huang (69). By permission of the Society of Plastics Engineers.

The reduction in the viscosity arising from branching for identical molar mass samples is given by a branching index g'

$$g' = \left(\frac{[\eta]_{b_i}}{[\eta]_{l_i}}\right)_{M_i} = K_l\left(\frac{M_{l_i}^{a+1}}{M_{b_i}}\right)(K_l M_{l_i}^{a})^{-1} \tag{55}$$

so

$$g' = \left(\frac{M_{l_i}}{M_{b_i}}\right)^{a+1} \tag{56}$$

Application of Eq. 56 is briefly discussed in Section 6 and is also covered in Chapters 2 and 3.

6. SELECTED APPLICATIONS

Table 2 lists selected applications of organosoluble macromolecules which have been characterized using static LALLS or SEC–LALLS. Table 3 continues this compilation for water-soluble macromolecules and bipolymers.

Polystyrene and polyolefins have been studied extensively by LALLS and SEC–LALLS. As a test on the accuracy of static light scattering measurements from different laboratories, Dumelow et al. (75) conducted a round-robin study. Results of this work indicated that deviations in $\bar{M}_w$ values determined by LALLS were less than those observed using other methods.

Table 2. Selected LALLS Applications of Organosoluble Macromolecules

Class	Macromolecule	Static LALLS (Refs.)	SEC–LALLS (Refs.)
Synthetic	Polystyrene	48, 74–76	50, 77–84
	Polyolefins	78, 85, 86	78, 83, 85–93
	Poly(vinyl acetate)	94	94, 95
	Polycarbonate		78, 96
	Polycholoroprene		92
	Epoxy resin		81
	Polyamide	97	97–99
	Poly(ethylene terephthalate)		100
	Polyimide	101	
	Poly(vinylidene fluoride)	102	
	Phenolic resin	103	103
	Polysulfone		77
	Polybutadiene, polyisoprene	104, 105	
	Poly(vinyl chloride)		106
	Poly(methyl methacrylate)		50, 92
	Poly(vinyl alcohol)		107
	Polyquinolines	108	108
Natural products and derivatives	Cellulose		109, 110
	Nitrocellulose		111
	Lignite	112, 113	113
Copolymers	Styrene based	114	115, 116
	Others	117	118–120
	Polyester		121, 122
	Acrylics	123	78

Polystyrene has also been used to study the potentially deleterious effect of shear degradation through SEC columns (83, 84).

Polyolefins, such as low- and high-density polyethylene and polypropylene, have received considerable attention owing to their commercial importance. Such polymers are only soluble in solvents at high temperatures (135–145 °C). This places some additional constraints on the SEC analysis. Factors such as column temperature, as well as detector temperature and general solution preparation, become quite important.

The analysis of long-chain branching (82) is most important for polyolefins (Chapter 3). The determination of long-chain branching with respect to molecular size has been studied using SEC–LALLS and viscosity (87, 89, 91, 93, 94).

Table 3. Selected LALLS Applications of Water-Soluble Macromolecules

Class	Macromolecules	LALLS (Refs.)	SEC–LALLS (Refs.)
Synthetics	Poly(ethylene oxide)	124	125, 126
	Poly(ethylene glycol)		
	Polyacrylamide	31, 78	125, 127, 128
	Sodium polyacrylate		129
	Poly(vinyl pyrrolidone)		126
	Poly(vinyl alcohol)		126
Natural polymers and derivatives	Dextrans		78, 80, 125, 126, 129–131
	Hydroxyethyl starch		132
	Amylose		133, 134
	Polysaccharides		135
	Guar gum		136, 137
	Chitosan		138
	Cellulose derivatives		139, 140
Biopolymers	Peptides	130	141
	Na, K ATPase		142, 143
	Aspergillus oryzae β amalase		144
	NAD^+ Kinase	145	
	Bovine eye lens protein		146
	Ovalbumin		147, 148
	Serum proteins		149
	Proteins (general)	130, 150	150–154
	SDS interactions/ membrane proteins	159	155–158
	Heparin		159
	Scleroglucan		160
	DNA fragments		161
	Hemoglobin	162	

A comparison of static LALLS and SEC–LALLS measurements for polyolefins (86) is given in Table 4. The results indicate that the $\bar{M}_w$ determined by static LALLS is considerably larger than that found using SEC–LALLS. This result was interpreted to arise from the poor signal-to-noise ratio of the high-molecular-weight end of the distribution. Similar results have been observed by MacRury and McConnell (85).

The bulk polymerization of vinyl acetate also yields a branched structure. The MWD, molecular weight averages (94), and the branching index with respect to molecular size (82) have been studied via SEC–LALLS. The effect of

Table 4. Comparison of SEC and LALLS Measurements of Polyethylenes

	SEC/LALLS			LALLS	
Sample	$\bar{M}_n$	$\bar{M}_w$	$\bar{M}_z$	$\bar{M}_w$	Type
NBS 1476	28,400	93,100	3,722,000	214,000	LDPE
D	43,200	176,300	1,519,000	189,000	LDPE
E	21,800	96,000	686,500	115,000	LDPE
F	18,350	151,600	719,000	167,000	LDPE
G	28,300	233,000	1,312,000	251,000	LDPE
H	38,000	145,300	700,000	161,200	LDPE
I	17,200	45,900	225,900	58,100	LDPE
C	46,400	185,300	606,700	217,400	LLDPE
A	46,000	144,000	406,000	233,600	HDPE
B	60,400	144,500	372,000	208,300	LDPE

shear on the branching distribution has also been investigated (95). Other branched polymers that have been analyzed by LALLS include polycarbonate and epoxy resin (96, 122).

The SEC–LALLS of nylon-6 was first reported using hexafluoroisopropanol (HFIP) as the mobile phase on μ-Bondagel columns (98). Hexafluoroisopropanol with electrolyte to suppress the polyelectrolyte effect has also been employed on an unmodified silica gel for successful SEC–LALLS measurements (99).

Pastuska et al. (97) determined the $\bar{M}_w$ and A_2 of several nylon samples in HFIP and also in a mixture of phenol and n-propanol (70:30) at 25 °C. The $\bar{M}_w$ values obtained in either solvent agreed within experimental error. The agreement of $\bar{M}_w$ indicates that preferential solvation mechanisms are absent in the solvent mixture. The SEC–LALLS studies in benzyl alcohol at 100 °C and the solvent mixture at 65 °C yielded similar molecular weights.

Cotts (108) studied the dilute solution properties of three polyquinolines using viscometry, static LALLS and SEC–LALLS. Chain aggregation was found in chloroform and was observed to arise from protonation of the quinolines by HCl present in this solvent. Solutions in N-methylpyrrolidone (NMP) were non-aggregating thereby permitting analysis. The Mark–Houwink–Sakurada equation was obtained from values of $\bar{M}_w$ and $[\eta]$ determined by static LALLS and intrinsic viscosity measurements, respectively.

It was further shown that the SEC–LALLS measurements and the universal calibration method could be employed for determination of $\bar{M}_w$ and

[η] if the chromatograms were properly corrected for axial dispersion. Calibration of the SEC columns was accomplished using nearly monodisperse polystyrene standards in NMP. Thus the determination of [η] is possible when M has been determined by online SEC–LALLS. It therefore appears that reasonable values of the exponent and prefactor of the MHS expression can be obtained using this method. Some caution, as was noted (108), is required in the consideration of dispersion effects. It is also important to recognize that the solvent employed for universal calibration was the same as for the SEC–LALLS determination of the polyquinolines.

As stated in Section 2, the determination of the true molecular weight of a random copolymer requires light scattering measurements in at least three

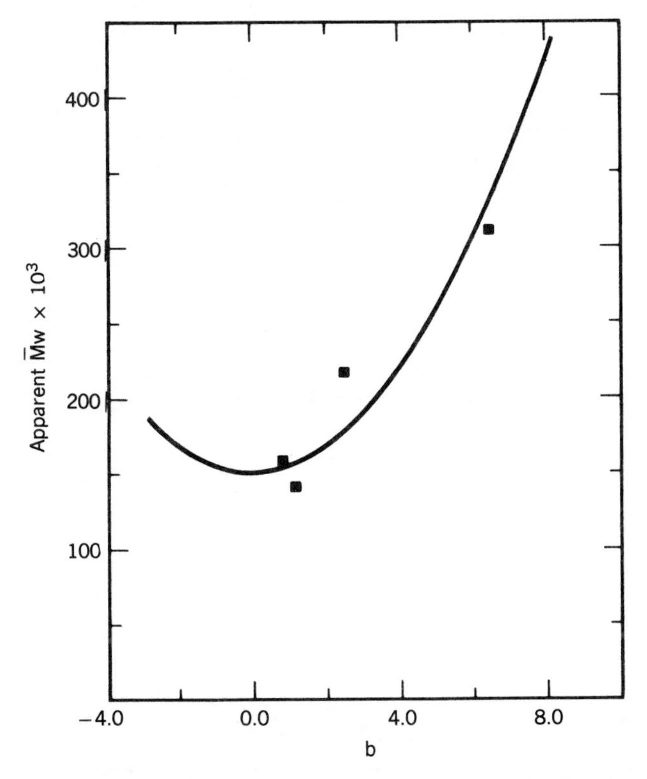

Figure 19. Determination of the true molecular weight for a styrene–butylacrylate copolymer. The plot of apparent $\bar{M}_w$ versus b, where $b = N_a r_B/r^2$ is a parabola with the minimum the true molecular weight. Reprinted by permission of John Wiley & Sons. Inc., After F. B. Malihi, C. Y. Kuo, and T. Provder, "Determination of the Absolute Moleculars Weight of a Styrene-Butyl Acrylate Emulsion Copolymer by LALLS and GPC/LALLS," *J. Appl. Polym. Sci.*, **29**, 925 (1984). Copyright © 1984 John Wiley & Sons, Inc.

different solvents. Malihi et al. (114) determined $\bar{M}_w$ for a styrene–butylacrylate (30:70) emulsion copolymer from measurements in toluene, dimethylformamide (DMF), methyl ethyl ketone, and THF. For the LALLS studies, the total copolymer and the polystyrene and polybutylacrylate specific refractive index increments were determined in all solvents. According to theory, the apparent molecular weight M_{app} is related to the true molecular weight by Eq. 26. Equation 26 is a parabola and the value of M_{app} at the minimum of the curve is the true $\bar{M}_w$ value. Figure 19 illustrates the experimentally determined values. The minimum was determined by curve fitting to be 1.51×10^6 g/mol. The two determinations close to the minimum correspond to measurements in THF and in methyl ethyl ketone. These solvents exhibited $(\partial n/\partial c)_\mu$ values of 0.112 and 0.127 mL/g, respectively, as compared to the lower values, 0.021 and 0.061 mL/g observed in toluene and DMF. The SEC–LALLS values of the copolymer in THF using μ-Styragel columns yielded an $\bar{M}_w$ value of 1.4×10^6 g/mol, in excellent agreement with the value 1.39×10^6 g/mol measured in the same solvent using static LALLS. These values are within 8% of the $\bar{M}_w$ determined from curve fitting.

In some cases SEC–LALLS studies are quite involved. Olson and Diehl (113) studied humic acid obtained from low-rank lignite coals using LALLS

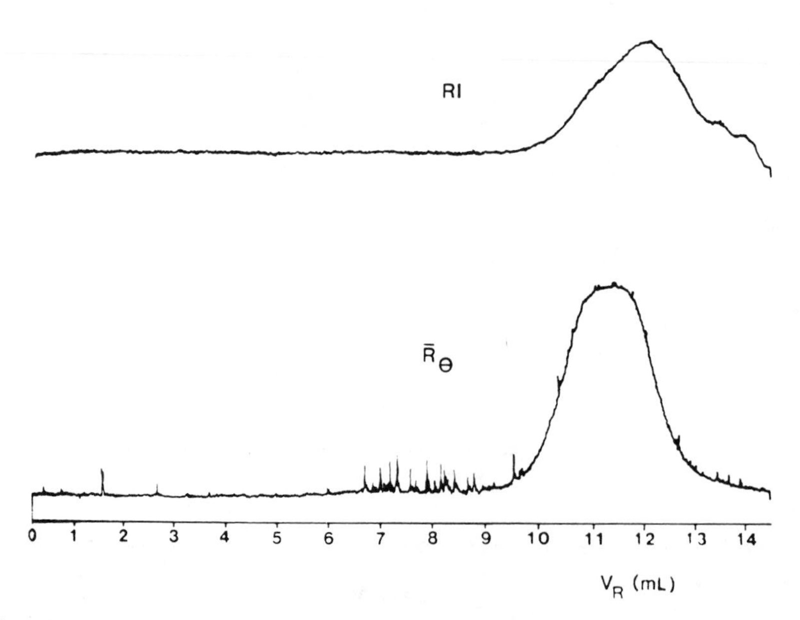

Figure 20. DRI and LALLS tracings for a reduced benzyl humate from lignite coal. Experimental conditions, 10^3 Å PL Gel, mobile phase THF at 22 °C. By permission of Elsevier Science Publishers, Physical Sciences and Engineering Div. (Olson and Diehl, Ref. 113).

and SEC–LALLS. The humic acid contains arenes as well as nitrogen heterocycles and semiquinone-type functional groups. These structures lead to enhanced absorption at 663 nm, thereby interfering with the proper determination of ΔR_θ. Determination of $\bar{M}_w$ by SEC alone is hindered by the lack of suitable calibration standards. Other characterization methods like membrane osmometry and vapor pressure osmometry are limited in terms of the molecular weight range. The presence of groups such as phenols, amines, quinones, and carboxylic acids in the humic fraction leads to adverse column interactions in polar solvents. To minimize optical absorption at 633 nm and minimize absorption of the polymer on the stationary phase, Olson and Diehl (113) prepared methyl, acetyl, and benzyl derivatives of the humic acids followed by reduction. The RI and LALLS tracings for the benzyl derivative in THF are shown in Figure 20. The corresponding $\bar{M}_w$

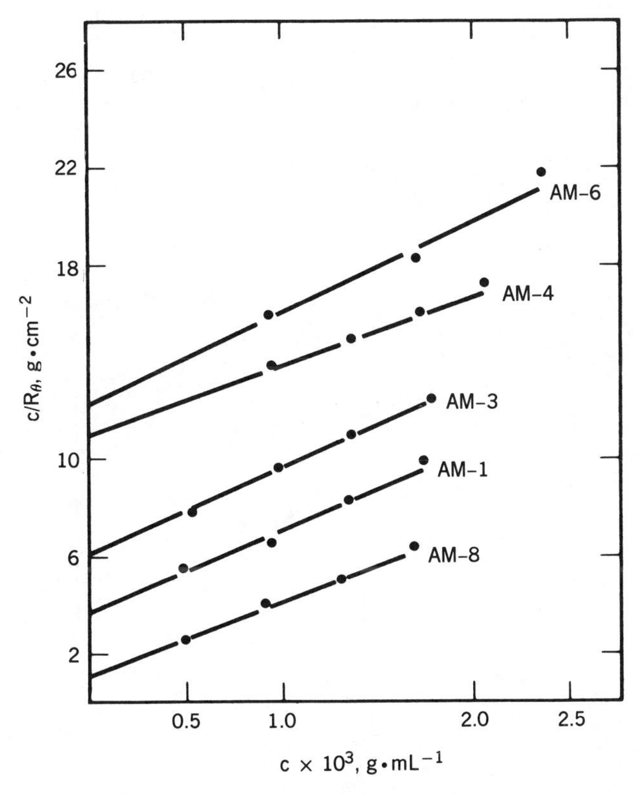

Figure 21. LALLS results for a series of polyacrylamides in water at 25 °C. The values of R_θ have been corrected for the solvent contribution. The linearity of the concentration dependence indicates the lack polyelectrolyte character of the chains. After Ref. 31.

Table 5. Molecular Characteristics for Sodium Polyacrylate Fractions

Sample	$\bar{M}_w \times 10^{-6}$ 0.3 N NaCl	$\bar{M}_w \times 10^{-6}$ 1.0 N NaCl	SEC–LALLS $\bar{M}_w \times 10^{-6}$
NaPA F1	3.27	3.37	
NaPA F2	1.92	1.95	
NaPA F3	1.30	1.10	1.19
NaPA F4	0.700		
NaPA F5	0.619	0.580	0.56
NaPA F6	0.386	0.340	0.368
NaPA F7	0.210	0.203	0.195
NaPA F8	0.095	0.092	0.105

Used by permission of the publishers, Butterworth & Co., (Publishers) Ltd. ©, Ref. 129.

value from SEC–LALLS was 1.7×10^6 g/mol while the static measurements provided a value of 1.8×10^6 g/mol. A major accomplishment of this work was the ability to determine the $\bar{M}_w$ of a component isolated from coal. Although some degradation of the polymer cannot be ruled out, a qualitative means of studying humic acids was developed.

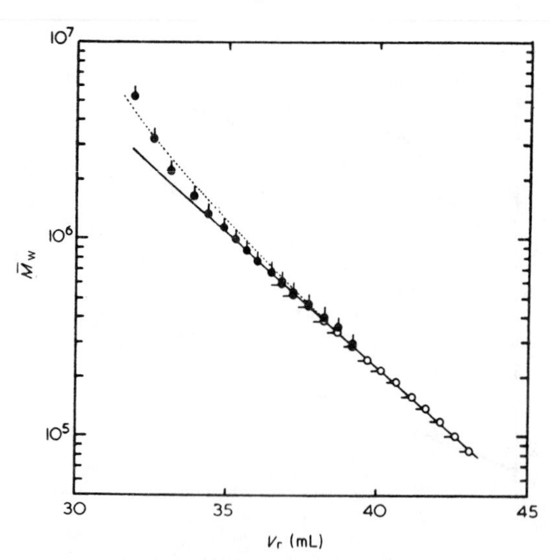

Figure 22. Calibration plot of $\bar{M}_w$ from GPC–LALLS versus elution volume for a series of sodium polyacrylates in 0.3 N NaCl. The upward curvature at small elution volumes corresponds to increased branching at higher molecular weights. By permission of Butterworth and Co. (Publishers) Ltd. ©. Figure from Ref. 129.

Light scattering investigations in pure water are difficult since water has a small Rayleigh ratio, and it is often quite difficult to eliminate dust from aqueous solutions. The closed-loop filtration process described earlier can be used with good results for pure water and for organic solvents. Static LALLS results for a series of polyacrylamides are given in Figure 21 (31). The linearity of the plots indicates the absence of the polyelectrolyte effect, implying that the chains are uncharged. The total volume of the loop employed for those measurements corresponds to about 2 mL. Such a configuration is expected to be quite valuable, as noted previously, for studies of biopolymers or those samples from which only small quantities can be devoted to scattering measurements.

Polyelectrolyte effects in SEC have been briefly reviewed by Kato and et al. (129). These workers studied the solution properties of a series of sodium

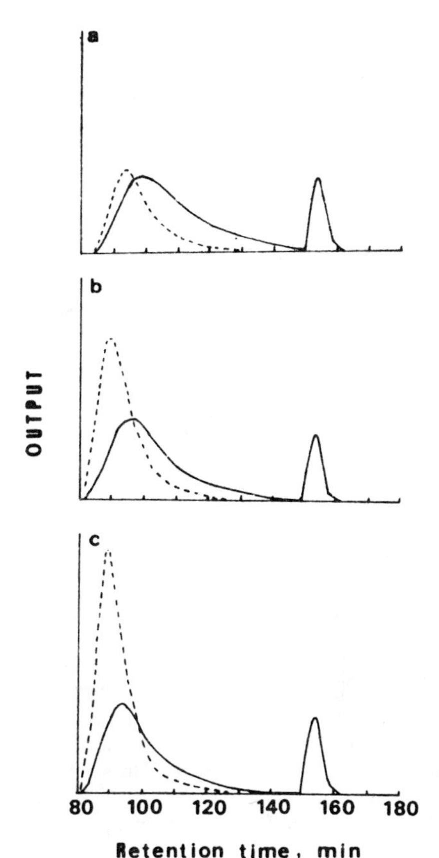

Figure 23. LALLS (dashed curve) and DRI (solid curve) tracings for ovalbumin following heat treatment at 80 °C and then immediate cooling to 25 °C. Panels a, b, and c correspond to times of 0, 6, and 22 h after cooling. Reprinted with permission A. Kato and T. Takagi, "Estimation of the Molecular Weight Distribution of Heat-Induced Ovalbumin Agregates by the Low-Angle Laser Light Scattering Technique Combined with High-Performance Gel Chromatography," *J. Agric. Food Chem.*, **35**, 633 (1987). Copyright 1987 American Chemical Society.

polyacrylate fractions in different NaCl solutions. The SEC–LALLS measurements were performed using the Tosoh LS-8000 instrument and $\bar{M}_w$ values were obtained by integration of the chromatograms. Calibration was accomplished by using dextran and ethylene oxide polymers. Table 5 gives $\bar{M}_w$ values determined by static scattering in 0.3 and 1.0 N NaCl and $\bar{M}_w$ from SEC–LALLS. Good agreement between the SEC–LALLS and static measurements was obtained indicating that the polyelectrolytes were properly chromatographed. Examination of the LALLS molecular weights as a function of elution volume, Figure 22, revealed deviations in linear response for the higher molecular weight fractions. The upward curvature was attributed to increasing branching as $\bar{M}_w$ increased. The extent of branching was estimated using the method outlined in Section 5.2, and the number of branch points per molecule was found to increase with increasing $\bar{M}_w$. This result was in agreement with the leveling off of the intrinsic viscosity at high molecular weights.

A number of different biopolymers have been studied by SEC–LALLS.

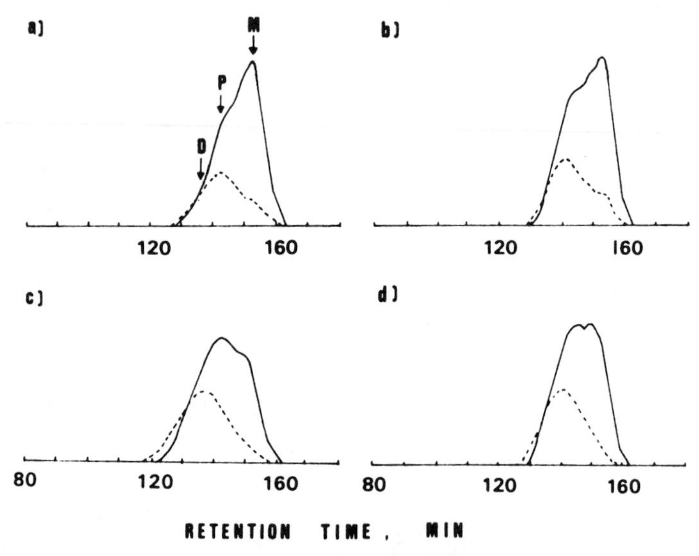

Figure 24. The effects of SDS and dithiothreitol on ovalbumin aggregates. Panels a and c correspond to addition of 0.5% SDS to ovalbumin following heat treatment at 80 and 90 °C, respectively. Panels b and d are for addition of 0.5% SDS and 0.02% dithiothreitol after treatments at 80 and 90 °C. M, P, and D denote monomer, pentamer and decamer. Reprinted with permission from A. Kato and T. Takagi, "Estimation of the Molecular Weight Distribution of Heat-Induced Ovalbumin Agregates by the Low-Angle Laser Light Scattering Technique Combined with High-Performance Gel Chromatography," *J. Agric. Food Chem.*, **35**, 633 (1987). Copyright © 1987 American Chemical Society.

Applications have been reviewed by Takagi (152) and more recently by Stuting et al. (153). Similar to synthetic polyelectrolytes, biopolymers can exhibit anomalous behavior in SEC. In addition, biopolymers are prone to aggregation, thus complex SEC–LALLS chromatograms can result. Figure 23 illustrates the LALLS and RI chromatograms for ovalbumin aggregates in the presence of sodium dodecyl sulfate (SDS) and dithiothreitol (148). The action of SDS on the aggregated protein can be observed by comparing Figure 23 with Figure 24. Note that the SDS effectively disrupted the high-molecular-weight component (early eluting peak) into oligomeric species. It should also be pointed out that the monomer or nonaggregated ovalbumin protein is the primary component of the chromatograms shown in Figure 24. Further details of SEC–LALLS applications to biopolymers are contained in the reviews cited above (152, 153).

7. CONCLUSIONS

As exemplified by the numerous applications (Tables 2 and 3), static LALLS is a preferred method for determination of $\bar{M}_w$ and A_2. The use of a LALLS detector following an SEC separation yields, without recourse to column calibration, the MWD and associated molecular weight averages. Structural and solution property determinations such as branching, the presence of high-molecular-weight components, and polymer association behavior have been studied by SEC–LALLS. In conjunction with SEC universal calibration, it is possible to estimate the prefactor and exponent of the Mark–Houwink–Sakurada equation.

ACKNOWLEDGMENTS

We acknowledge the support of the Diagnostic Instrumentation and Analysis Laboratory and Hercules Incorporated. We also thank the reviewers for some helpful comments and the editors for their patience. Finally, we thank Ms. C. Huntley for preparation assistance in the completion of this chapter.

INSTRUMENT COMPANIES

Light Scattering Photometer Companies

- LDC Analytical*, 3681 Industrial Park Rd. N., Riviera Beach, Florida 33404

- Outsuka Electronics (U.S.A.) Ltd.*, 2555 Midpoint Dr., Fort Collins, Colorado 80525
- Tosoh Corporation, 7-7, 1-Chome, Akasaka, Minato-ku, Tokyo, Japan
- Viscotek, 1032 Russell Dr., Porter, Texas 77365 (Supplies SEC-LALLS software.)
- Wyatt Technology Corporation*, 820 East Haley Street, Santa Barbara, California 93130

NOTATION

A	Virial coefficient or chromatogram area
B	$k_2 K(h_i)^{-1}$ (Eq. 35)
c	Solute concentration
D	Attenuator transmissions
D_{T}	Translational diffusion coefficient
$G_{\theta,0}$	Intensities KMX-6, CMX-100
h_i	Chromatogram height at elution volume i
I_0	Incident light intensity
i_s	Scattered light intensity
K	Optical constant
k	Proportionality constant or prefactor
l	Optical path length
M_i	Mass of scattering center
$\bar{M}_x$	x-Average molecular weight
m_{i}	Molality of component i
N	Number of scatterers
N_{A}	Avogadro's number
$\tilde{n}_0$	Refractive index
n_{i}	Moles of component i
$P(\theta)$	Particle scattering function
$\mathbf{q}$	Scatter wave vector
R_θ	Rayleigh ratio
R_T	Thermodynamic radius
r	Distance
$\langle S^2 \rangle$	Mean-square radius of gyration
u	Excluded volume
V	Unit volume
V_s	Effective spherical volume
W_x	Weight fraction of x

*These companies also supply differential refractometers.

X	Refractometer deflection
α	Local polarizability
β	Angle (Eq. 43)
γ	Preferential solvation coefficient
δ	Fluctuation
ε	Extenction coefficient
$[\eta]$	Intrinsic viscosity
θ	Scattering angle
λ	Effective wavelength
σ	Solid angle, KMX-6
v	Refractive index
ϕ	Phase angle

Superscripts

a	Exponent of $[\eta]$–M power law
b	Exponent of A_2–M_w power law
LS	Light scattering
RI	Refractive index

Subscripts

app, 2	Apparent quantity
b	Branched chain
l	Linear chain
n	Number average
w	Weight average
z	z average
θ	Scattering angle
μ	Constant chemical potential
v	Vertical polarization

REFERENCES

1. K. F. Freed, *Renormalization Group Theory of Macromolecules*, Wiley, New York, 1987, Chapter 10.
2. Y. Oono, "Statistical Physics of Polymer Solutions: Conformational Space Renormalization-group Approach," *Adv. Chem. Phys.*, **61**, 301 (1985).
3. H. Yamakawa, *Modern Theory of Polymer Solutions*, Harper & Row, New York, 1971.

4. J. S. Lindner, W. W. Wilson, and J. W. Mays, "Properties of Poly(α-methylstyrene) in Toluene: A Comparison of Experimental Results with Predictions of Renormalization Group Theory," *Macromolecules*, **21**, 3304 (1988).
5. H. Fujita, "Some Unsolved Problems in Dilute Polymer Solutions," *Macromolecules*, **21**, 171 (1988).
6. H. C. van de Hulst, *Light Scattering by Small Particles*, Wiley, New York, 1957.
7. M. Kerker, *The Scattering of Light and Other Electromagnetic Radiation*, Academic, New York, 1969.
8. C. Tanford, *Physical Chemistry of Macromolecules*, Wiley, New York, 1961, Chapter 5.
9. M. B. Huglin (Ed.), *Light Scattering From Polymer Solutions*, Academic, London, 1972.
10. K. A. Stacey, *Light Scattering in Physical Chemistry*, Butterworths, London, 1956.
11. P. Kratochvil, *Classical Light Scattering from Polymer Solutions*, Elsevier, Amsterdam, 1987.
12. M. B. Huglin, "Determination of Molecular Weights by Light Scattering," in *Topics in Current Chemistry*, No. 77, Inorganic and Physical Chemistry, Springer-Verlag, New York, 1978.
13. W. Burchard, "Static and Dynamic Light Scattering from Branched Polymers and Biopolymers," *Adv. Polym. Sci.*, **48**, 1 (1983).
14. H. Eisenberg, *Biological Macromolecules and Polyelectrolytes in Solution*, Clarendon, Oxford, 1976.
15. E. F. Casassa and G. C. Berry, "Light Scattering from Solutions of Macromolecules," in P. E. Slade (Ed.), *Molecular Weight Methods*, Part 1, Marcel Dekker, New York, 1975.
16. P. Debye, "A Photoelectric Instrument for Light Scattering Measurements and a Differential Refractometer," *J. Appl. Phys.*, **17**, 342 (1946).
17. A. Einstein, "Theorie der Opaleszenz von homogenen Flüssigheiten und Flüssigheitsgermischen in der Nähe des Kritischen Zustanles," *Ann. Physik*, **32**, 1275 (1910).
18. H. Yamakawa, *Modern Theory of Polymer Solutions*, Harper and Row, New York, 1971, Chapter V.
19. W. H. Stockmayer and E. F. Casassa, "The Third Virial Coefficient of Polymer Solutions," *J. Chem. Phys.*, **20**, 1560 (1952).
20. P. J. Flory, *Principles of Polymer Chemistry*, Cornell University, Ithaca, NY, 1953, Chapter XII, p. 532.
21. T. B. Grimley, "High Polymer Solutions," *Proc. R. Soc. London Ser. A.*, **212**, 339 (1952).
22. A. Ishihawka and R. Yoyama, "Theory of Dilute High Polymer Solutions (the Pearl Necklace Model)," *J. Chem. Phys.*, **25**, 712 (1956).
23. M. L. Huggins, "Some Properties of Solutions of Long-Chain Compounds," *J. Phys. Chem.*, **46**, 151 (1942); "Theory of Solutions of High Polymers," *J. Am. Chem. Soc.*, **64**, 1712 (1942).

24. The situation at the θ point is formally complex with three-body interactions becoming quite important, see, for example, J. F. Douglas and K. F. Freed, "Polymer Contraction Below the θ Point: A Renormalization Group Description," *Macromolecules* **18**, 2445 (1985); B. J. Cherayil, J. F. Douglas, and K. F. Freed, "Effect of Residual Interactions on Polymer Properties Near the Theta Point," *J. Chem. Phys.*, **83**, 5293 (1985).

25. P. O. O. Klames and H. A. Ende, "Second Virial Coefficients of Polymers in Solution," in J. Brandrup and E. H. Immergut (Eds.), *Polymer Handbook*, Wiley, New York, 1975, p. IV-9.

26. G. Berry and E. F. Casassa, "Thermodynamic and Hydrodynamic Behavior of Dilute Polymer Solutions," *J. Polym. Sci. Part D*, **4**, 1 (1970).

27. P. G. de Gennes, *Scaling Concepts in Polymer Physics*, Cornell University Press, Ithaca, NY, 1979, Chapter III, p. 78.

28. A. Gunier, "X-Ray Diffraction at Small Angles—Application to the Study of Ultramicroscopic Phenomenon, "*Ann. Physik*, **12**, 161 (1939).

29. P. Debye, "Molecular Weight Determination by Light Scattering," *J. Phys. Colloid Chem.*, **51**, 18 (1947).

30. B. H. Zimm, "Apparatus and Methods for Measurement and Interpretation of the Angular Variation of Light Scattering; Preliminary Results on Polystyrene Solutions," *J. Chem. Phys.*, **16**, 1099 (1948).

31. J. S. Lindner, "Solution Properties of Water-Soluble Macromolecules for Enhanced Oil Recovery Applications," *Ph.D. Thesis, Mississippi State University*, 1985, p. 244.

32. Z. Alexandrowicz, "Results of Osmotic and of Donnan Equilibria Measurements in Polymethacrylic Acid–Sodium Bromide Solutions, Part II," *J. Polym. Sci.*, **42**, 337 (1960).

33. E. F. Casassa and H. Eisenberg, "Thermodynamic Analysis of Multicomponent Solutions," *Adv. Protein Chem.*, **19**, 287 (1964).

34. J. Stejskel, M. J. Benes, P. Kratochvil, and J. Peska, "Solution Properties of Polymers of Pyridinium and Ammonium Methacrylate," *J. Polym. Sci., Polym. Phys. Ed.*, **12**, 1941 (1979).

35. M. Hert and C. Strazielle, "Influence of Molecular Weight, Polydispersity and Temperature on the Phenomena of Preferential Solvation of Linear Polystyrene in Different Solvent Mixtures," *Makromol. Chem.*, **175**, 2149 (1974).

36. P. Kratochvil, "Light Scattering as a Tool in the Study of Multicomponent Polymer Solutions," *J. Polym. Sci., Symp. Ed.*, **50**, 487 (1975).

37. M. Girolamo and J. R. Urwin, "Configuration of Block Copolymers of Polyisoprene and Polystyrene II. Molecular Dimensions," *Eur. Polym. J.*, **8**, 1159 (1972).

38. J. Stejskal, P. Kratochvil, and D. Straková, "Study of the Statistical Heterogeneity of Copolymers by Cross-Fractionation," *Macromolecules*, **14**, 150 (1981).

39. S. Teramachi, A. Hasegawa, and S. Yoshida, "Determination of the Correlation between Molecular Weight Distribution and Chemical Composition Distribution in a High-Conversion Copolymer," *Macromolecules*, **16**, 542 (1983).

40. H. Benoit and D. Froelich, "Application of Light Scattering to Copolymers," in M. Huglin (Ed.), *Light Scattering from Polymer Solutions*, Academic, New York, 1972, Chapter 11, p. 427.
41. J. Prud' homme and S. Bywater, "Light Scattering Studies on Polystyrene Polyisoprene Block Copolymer," *Macromolecules*, **4**, 543 (1972).
42. G. C. Berry, "Thermodynamic and Conformational Properties I; Light Scattering Studies on Dilute Solutions of Linear Polystyrenes," *J. Chem. Phys.*, **44**, 4350 (1966).
43. B. Chu, R. Xu, T. Maeda, and H. S. Dhadwal, "Prism Light Scattering Spectrometer," *Rev. Sci. Instr.*, **59**, 718 (1988).
44. H. R. Haller, C. Destor, and D. E. Cannel, "Photometer for Quasielastic and Classical Light Scattering," *Rev. Sci. Instr.*, **54**, 973 (1983).
45. J. D. Gassaway and R. W. Stapp, "A Laser-Based Instrument for Measurement of Particle-Size Distributions," Proceedings of the 24th Symposium on the Engineering Aspects of Magnetohydrodynamics, Butte, Montana 1986, p. 229.
46. W. D. Bachalo, "On-line Particle Diagnostic Systems for Application in Hostile Environments," in H. D. Thompson and W. H. Stevinson (Eds.), *Laser Velocimetry and Particle Sizing*, Hemisphere, Washington, DC, 1979, p. 506.
47. W. Kaye, A. J. Havlik, and J. B. McDaniel, "Light Scattering Measurements on Liquids at Small Angles," *J. Polym. Sci. Part B*, **9**, 695 (1971).
48. W. Kaye, "Low-Angle Laser Light Scattering," *Anal. Chem.*, **45**, 221A (1973).
49. R. J. Anderson, "A Photometer for Low-Angle Light Scattering Measurements," *Am. Lab.*, **8**(5), 95 (1976).
50. A. C. Duano and W. Kaye, "Gel-Permeation Chromatography I. Molecular Weight Detection by LALLS," *J. Polym. Sci., Polym. Chem. Ed.*, **12**, 1151 (1974).
51. P. J. Wyatt, C. Jackson, and G. K. Wyatt, "Absolute GPC Determinations of Molecular Weights and Sizes From Light Scattering," *Am. Lab.*, **20**(5), 86 (1988).
52. P. J. Wyatt, D. L. Hicks, C. Jackson, and G. K. Wyatt, "Part 2. Absolute GPC Determination of Molecular Weights and Sizes: Incorporation of Light Scattering Techniques into GPC–SEC Measurements," *Am. Lab.*, **20**(6), 108 (1988).
53. M. Mandel, F. A. Varkevisser, and C. J. Bloys Van Treslong, "Measurement of the Specific Refractive Index Increment of Polyelectrolytes in Aqueous Salt Solutions with the Chromatix KMX-16 Differential Refractometer," *Macromolecules*, **15**, 675 (1982).
54. J. M. Lorimer, "Refractive Index Increments of Polymers in Solution 3. Dependence on Concentration," *Polymer*, **13**, 274 (1972).
55. KMX-16 Application Note LS7, LDC Analytical.
56. M. B. Huglin, "Specific Refractive Index Increments of Polymers in Dilute Solutions," in J. Brandrupt and E. H. Immergut (Eds.), *Polymer Handbook*, Wiley, 1975, p. IV-267.
57. J. S. Lindner, W. W. Wilson, and J. W. Mays, unpublished data.
58. C. J. Stacy, "Molecular Weight and Structural Characteristics of Some Am-

ylopectins and their Beta-Amylase Limit Dextrins," Ph.D. Thesis, Purdue University, West Lafeyette, Indiana, 1956.

59. P. Katochvil, *Classical Light Scattering from Polymer Solutions*, Elsevier, Amsterdam, 1987, p. 75.

60. W. W. Wilson, "Light Beating Spectroscopy Applied to Biochemical Systems," Ph.D. Thesis, University of North Carolina, Chapel Hill, NC, 1973.

61. E. Pike, W. R. M. Pomeroy, and J. M. Vaughn, "Measurement of Rayleigh Ratios for Several Pure Liquids Using a Laser and Monitored Photon Counting," *J. Chem. Phys.*, **62**, 3188 (1975).

62. M. B. Huglin, S. J. O'Donohue, and M. A. Radwan, "Refractometric and Light Scattering Parameters at 633 nm for Polystyrene Solutions," *Eur. Polym. J.*, **25**, 543 (1989).

63. W. W. Wilson, M. S. Luzzana, J. T. Penniston, and C. S. Johnson, Jr., "Pregelation Aggregation of Sickel Cell Hemoglobin," *Proc. Natl. Acad. Sci.* USA, **71**, 1260 (1974).

64. M. L. Salin and W. W. Wilson, "Porcine Superoxide Dismutase," *Mol. Cell Biochem.*, **36**, 157 (1981).

65. J. S. Lindner, W. W. Wilson, J. W. Mays, L. J. Fetters, and N. Hadjichristidis, "Transport Properties of Poly(isobutylene)," to be submitted to *Macromolecules.*

66. M. B. Huglin, "Specific Refractive Index Inrements of Polymers in Solution, Part II, Scope and Applications," *J. Appl. Polym. Sci.*, **9**, 4003 (1965).

67. W. Lorimar and D. E. G. Jones, "Refractive Index Increments of Polymers in Solution, 2. Refractive Index Increments and Light Scattering in Polydisperse Systems of Low Molecular Weight," *Polymer*, **13**, 52 (1972).

68. J. P. Kratochvil, G. J. Dezelic, M. Kerker, and E. Matijivic, "Calibration of Light Scattering Instruments; A Critical Survey," *J. Polym. Sci.*, **57**, 59 (1962).

69. H. G. Barth and S. S. Huang, "The Use of Low-Angle Laser Light Scattering (LALLS) Photometry for the Measurement of Weight-Average Molecular Weights of Polymers," Conference Proceedings for the Society of Plastics Engineers, 44th Annual Technical Conferences (ANTEC), 473 (1986).

70. S. A. Berkowitz, "Rejection of Spike Noise from Size Exclusion Chromatography/Low-Angle Laser Light Scattering Experiments," *Anal. Chem.*, **58**, 2571 (1986).

71. O. Procházka and P. Kratochvil, "Analysis of the Accuracy of Determination of Molar Mass Distribution by GPC with an On-Line Light-Scattering Detector," *J. Appl. Polym. Sci.*, **31**, 919 (1986).

72. O. Procházka and P. Kratochvil, "An Analysis of the Accuracy of Determining Molar-Mass Averages of Polymers by GPC with an On-Line Light-Scattering Detector," *J. Appl. Polym. Sci.*, **34**, 2325 (1987).

73. M. Martin, "Polymer Analysis by Fractionation with On-Line Light Scattering Detector," *Chromatographia*, **15**, 426 (1982).

74. T. L. Welzen, "Second Virial Coefficients for Solutions of Polystyrene Mixtures Obtained with Low Angle Laser Scattering," *Br. Polym. J.*, **12**, 95 (1980).

75. T. Dumelow, S. R. Holdingant, and L. J. Maisey, "LALLS for Determination of Molecular Weights—an Inter-Laboratory Comparison," *Polym. Commun.* **24**, 307 (1983).

76. J. Roovers, N. Hadjichristidis, and L. J. Fetters, "Analysis and Dilute Solution Properties of 12- and 18- Arm Star Polystyrenes," *Macromolecules*, **16**, 214 (1983).

77. A. C. Ouano, "Solution Properties of Polymers by Low-Angle Laser Light Scattering Photometry," *J. Colloid Interface Sci.*, **63**, 275 (1978).

78. M. L. McConnell, "Polymer Molecular Weights and Molecular Weights Distributions by Low-Angle Laser Light Scattering," *Am. Lab.* **10**(5), 63 (1978).

79. M. Fukuda, M. Fukutomi, N. Bamba, Y. Kato, K. Arichika, M. Aiura, and T. Hashimoto, "Construction of a Low Angle Light Scattering Photometer and Evaluation of Its Performance," *Toyo Soda, Kenkya Hokoku*, **23**, 111 (1979).

80. R. C. Jordon, "Size Exclusion Chromatography with Low Angle Laser Light Scattering Detection," *J. Liq. Chromatog.*, **3**, 439 (1980).

81. C. Huber, K. H. LeOerer, "GPC of Polyamides," *J. Polym. Sci., Polym. Lett. Ed.*, **18**, 535 (1980).

82. R. C. Jordon and M. L. McConnell, "Characterization of Branched Polymers by Size Exclusion Chromatography with Light Scattering Detection," *Am. Chem. Soc. Symp. Ser.*, **138** [Size Exclusion Chromatography (GPC)], 107 (1980).

83. J. G. Rooney and G. Ver Strate, "On-Line Determination by Light Scattering of Mechanical Degradation in the GPC Process," *Chromatogr. Sci.*, **19** (Liq. Chromatogr. Polym-Related Material 3), 207 (1981).

84. M. Ye and L. Shi, "A Study of Mechanical Degradation of Polymer in High-Performance GPC," *J. Liq. Chromatogr.*, **5**, 1259 (1982).

85. T. B. MacRury and M. L. McConnell, "Measurement of the Absolute Molecular Weight and Molecular Weight Distribution of Polyolefins Using LALLS," *J. Appl. Polym. Sci.*, **24**, 651 (1979).

86. V. Grinshpun, K. F. O'Driscoll, and A. Rudin, "On the Accuracy of SEC Analysis of MWD of Polyethylenes", *J. Appl. Polym. Sci.*, **29**, 1071 (1984).

87. D. E. Axelson and W. C. Knapp, "SEC and LALLS, Application to the Study of Long Chain-Branched Polyethylene," *J. Appl. Polym. Sci.*, **25**, 119 (1980).

88. T. Hjiertberg, L. I. Kulin, and E. Soervik, "Laser Light Scattering as GPC Detector," *Polym. Test.*, **3**, 267 (1983).

89. A. Rudin, V. Grinshpun, and K. F. O'Driscoll, "Long-Chain Branching in Polyethylene," *J. Liq. Chromatogr.*, **7**, 1809 (1984).

90. V. Grinshpun and A. Rudin, "Measurement of Mark–Houwink Constants by SEC–LALLS," *Makromol. Chem. Rapid Commun.*, **6**, 219, (1985).

91. V. Grinshpun, A. Rudin, and D. Potter. "Comments on the Measurement of Long-Chain Branching by Size Exclusion Chromatography," *Polym. Bull.* (Berlin), **13**, 71 (1985).

92. V. Grinshpun and A. Rudin, "High Temperature GPC of Polypropylene," *J. Appl. Polym. Sci.* **30**, 2413 (1985).

93. S. Shiga, "Characterization of Polymers by GPC-LALLS, Branching Structure of LDPE." *Nippon Gomu Kyokaishi*, **59**, 162 (1986).
94. A. E. Hamielec, A. C. Ouano, and L. I. Nebenzahl, "Characterization of Branched Polyvinylacetate by GPC with LALLS Photometry," *J. Chromatogr.*, **1**, 527 (1978).
95. S. H. Agarwal, R. F. Jenkins, and R. S. Porter, "Mechanical Degradation and Effect of Shear on the Distribution of Long Branching the Polyvinylacetate," *J. Appl. Polym. Sci.*, **27**, 113 (1982).
96. H. Schorn, R. Kosfeld, and M. Hess, "Investigation of the Solution Crystallization Behavior of Polycarbonate," *J. Chromatogr.*, **353**, 273 (1986).
97. G. Pastuska, U. Just, and H. August, "Analysis of Polyamides, Determination of Molecular Weight and Its Distribution," *Angew. Makromol. Chem.*, **107**, 173 (1982).
98. D. J. Goedhart, J. B. Hussem, and B. P. M. Smeets, "GPC of Polyamides," *Chromatogr. Sci.*, **13** (Liq. Chromatogr. Polym.-Related Material 2), 203 (1980).
99. H. Schorn, R. Kosfeld, and M. Hess, "High-Performance Size Exclusion Chromatography of Polyamide 6," *J. Chromatogr.*, **282**, 579 (1983).
100. S. Berkowitz, "Viscosity-Molecular Weight Relationships for Poly(ethylene-terephthate) in Hexafluoroisopropanol-Pentafluorophenol Using SEC–LALLS," *J. Appl. Sci. Part 2*, **29**, 4353 (1984).
101. P. M. Cotts, "Characterization of Polyimides and Polyamic Acids in Dilute Solution," in K. L. Mittal (Ed.), *Polyimides: Synthesis, Characterization, and Applications* (Proceedings of the Technical Conference on Polyimides 1st, 1982), Plenum, New York, 1984, pp. 223–236.
102. L. E. Stillwagon, "Poly(vinylidine fluoride): Weight-Average Molecular Weight Measurements," *Org. Coat. Appl. Polym. Sci. Proc.*, **48**, 780 (1983).
103. J. D. Wellons and L. Gollob, "GPC and Light Scattering of Phenolic Resins-Problems in Determining Molecular Weights," *Wood Sci.*, **13**, 68 (1980).
104. N. Hadjichristidis, Z. Xu, L. J. Fetters, and J. Roovers, "The Characteristic Ratios of Stereoirregular Polybutadiene and Polysoprene," *J. Polym. Sci., Polym. Phys. Ed.*, **20**, 743 (1982).
105. S. Bo and R. Cheng, "Determination of the Monodisperse Mark–Houwink Equation of High-Vinyl Polybutadiene by Gel Chromatography and Light Scattering," *J. Liq. Chromatogr.*, **5**, 1404 (1982).
106. T. Hjertberg, L. I. Kulin, and E. Soervik, "Laser Light Scattering as GPC Detector," *Polym. Test*, **3**, 267 (1983).
107. D. J. Nagy, "Molecular Weight Determination of Poly(vinyl alcohol) using Aqueous Size Exclusion Chromatography/Low Angle Laser Light Scattering," *J. Polym. Sci., Polym. Lett. Ed.*, **24**, 87 (1986).
108. P. M. Cotts, "Solution Properties of Semirigid Polyquinolines I. Size Exclusion Chromatography with Light Scattering Detection of Poly(2,2′-*p*,*p*′-oxydi-*p*-phenylene) 6,6′-oxybis (4-phenylquinoline)," *J. Polym. Sci., Polym. Phys. Ed.*, **24**, 1493 (1986).

109. Q-C. Ying, G-Q. Chen, and L-R. Huang, "Determination of Molecular Weight of Polymer by Low Angle Laser Light Scattering," *Kao. Fen Tzu. Tung Hsun*, **1**, 24 (1981); *Chem. Abstr.*, **95**, 81693q (1981).

110. J. J. Cael, D. J. Cietek, and F. J. Kolpak, "Application of GPC/LALLS to Cellulose Research," *J. Appl. Polym. Sci., Appl. Polym. Sym.*, **37** (Proc. Cellul. Conf. 9th, 1982, Part 1), 509 (1982).

111. A. F. Cunningham, C. Heathcote, D. E. Hillman, and J. I. Paul, "Gel Permeation Chromatography of Nitrocellulose," *Chromatogr. Sci.*, **13** (Liq. Chromatogr. Polym. Related Materials 2), 175 (1980).

112. H. P. Hombach, "Virial Coefficients in Determination of Molecular Weights on Solutions of Coal Derivatives," *Fuel*, **60**, 663 (1981).

113. E. S. Olson and J. W. Diehl, "Size Exclusion Chromatography-Low-Angle Laser Light Scattering Photometry of Lignite Macromolecules," *J. Chromatogr.*, **349**, 337 (1985).

114. F. B. Malihi, C. Y. Kuo, and T. Provder, "Determination of the Absolute Molecular Weight of a Styrene–Butyl Acrylate Emulsion Copolymer by LALLS and GPC/LALLS," *J. Appl. Polym. Sci.*, **29**, 925 (1984).

115. R. C. Jordon, S. F. Silver, R. D. Sehon, and R. J. Rivard, "SEC with LALLS Detection. Application to Linear and Branched Block Copolymers," *Am. Chem. Soc., Symp. Ser.*, **245** (Size Exclusion Chromatogr.) 295 (1984).

116. R. C. Jordon, S. F. Silver, R. D. Sehon, and R. J. Rivard, "SEC with LALLS Detection: Application to Linear and Branched Block Copolymers," *Org. Coat. Appl. Polym. Sci. Proc.*, **48**, 755 (1983).

117. S. Djadoun, "Association Between Acidic and Basic Copolymers by LALLS," *Polym. Bull.*, **7**, 607 (1982).

118. S. Shiga and Y. Sato, "Characterization of Polymers by GPC–LALLS. II, Concentration Effect of Polydisperse Polymers in GPC," *Nippon Gomu Kyokaishi*, **57**, 229 (1984).

119. S. Shiga and Y. Sato, "Characterization of Polymers by GPC/LALLS, III, Structure of Branched EPDM," *Nippon Gomu Kyokaishi*, **57**, 811 (1984).

120. G. VerStrate, C. Cozewith, and W. Graessley, "Kinetics of Formation, Characterization and Consequences of Long Branching in Elastomers," *Polym. Prepr.*, **20**(2), 149 (1979).

121. D. B. Cotts, "Application of SEC–LALLS to the Characterization of Alternating, Random and Block Copolymers," *Org. Coat. Appl. Polym. Sci. Proc.*, **48**, 750 (1983).

122. B. L. Neff and J. R. Overton, "Characterization of Branched Copolyesters Using Gel Permeation Chromatography with On-Line LALLS," *Polym. Prepr.*, **23**(2), 130 (1982).

123. A. C. Ouano, B. L. Dawson, and D. E. Johnson, "Characterization of P(MMA) and P(MMA/MAA) by LALLS Photometry, GPC and Viscometry," *Chromatogr. Sci.*, **13**, (Liq. Chromatogr. Polym. Related Materials) 1 (1977).

124. Y. Saito and T. Sato, "Effects of Polyoxyethylene Chain Length on Micellar Structure," *J. Phys. Chem.*, **89**, 2110 (1985).

125. M. Fukutomi, M. Fukuda, and T. Hashimoto, "Evaluation of Molecular Weight of Water-Soluble Polymers by GPC-Light Scattering Instrument," *Tosoh, Kenkyu Ho Ko Ko*, **24**, 33 (1980).

126. A. Donald and M. Rinaudo, "Gel Permeation Chromatography of Polymer on Cationic Porous Silica Gels," *Polym. Commun.*, **25**, 55 (1984).

127. C. J. Kim, A. E. Hamielec, and A. Benedek, "Characterization of Nonionic Polyacrylamides by Aqueous Size Exclusion Chromatography Using a DRI/LALLS Detection System," *J. Liq. Chromatogr.*, **5**, 1277 (1982).

128. W. M. Kulicke and N. Boese, "Determination of the Molecular Weight Distribution and Stability Limits of Polyacrylamides by Using a Combined SEC Column and LALLS Photometer," *Colloid Polym. Sci.*, **262**, 197 (1984).

129. T. Kato, T. Tokuya, T. Nozaki, and A. Takeshi, "Molecular Characterization of Sodium Polyacrylate by an Aqueous GPC/LS Method," *Polymer*, **25**, 218 (1984).

130. T. Tagagi, J. Miyake, and T. Nashima, "Assesment Study on the Use of the LALLS Technique for the Estimation of Molecular Weight of Polypeptide Forming a Complex with SDS," *Biochim. Biophys. Acta.*, **626**, 5 (1980).

131. T. Takaki, "LALLS Photometer, Its Application to Molecular Weight Measurement by High Pressure Silica Gel Chromatography," in K. Osawa and Y. Tanaka (Eds.), *Kyodai Ryushi no Gerupamieishion Kuromatogurafi: Seitai Ryushi no Ryuekei Bunri*, Kitumi Shobo, Tokyo, 1980, p. 319; *Chem. Abstr.*, **95**, 14623f (1981).

132. K. Ledeer, C. Huber, M. Dunky, J. K. Fink, H. P. Ferber, and E. Nitsoh, "Studies on Hydroxyethyl Starch. Part I. Molecular Characterization by SEC Coupled with LALLS," *Arzneim Forsch.*, **35**, 610 (1985).

133. T. Takagi and S. Hizukuri, "Molecular Weight and Related Properties of Lily Amylose Determined by Monitoring of Elution From TSK-GEL PW HP GPC Columns by the LALLS Technique and Precision Differential Refractometry," *J. Biochem. (Tokyo)*, **95**, 1459 (1984).

134. S. Hizukuri and T. Takagi, "Estimation of the Distribution of Molecular Weight for Amylose by the LALLS Technique Combined with High Performance Gel Chromatography," *Carbohydr. Res.*, **134**, 1 (1984).

135. L. P. Yu and J. E. Rollings, "Low–Angle Laser Light Scattering-Aqueous Size Exclusion Chromatography of Polysaccharides: Molecular Weight Distribution and Polymer Branching Determination," *J. Appl. Polym. Sci.*, **33**, 1909 (1987).

136. F. Lambert, M. Miles, and M. Rinaudo, "Gel Permeation Chromatography of Xanthan Gum Using a Light Scattering Detector," *Polym. Bull.*, **7**, 185 (1982).

137. B. R. Vigayendran and T. Bone, "Absolute Molecular Weight and Molecular Weight Distribution of Guar by SEC and LALLS," *Carbohydr. Polym.*, **4**, 299 (1984).

138. M. Miya, R. Iwamoto, S. Yoshikawa, and S. Mima; "Determination of Molecular Weight Changes of Chitosan in Solution and Film by the GPL/LALLS Method," *Kobunshi Ronbunshu*, **43**, 83 (1986).

139. T. Kato, T. Tokuya, and A. Takahash, "Measurements of Molecular Weight and Molecular Weight Distribution for Water Soluble Cellulose Derivatives in the Film Coating of Tablets," *Kobunshi Ronhunshu*, **39**, 293 (1982).

140. W. D. Eigner, J. Billiani, and A. Huber, "Exclusion Chromatography Coupled with Small Angle Laser Light Scattering," *Papier (Darmstadt)*, **41**, 680 (1987).

141. S. Makino, S. Maezawa, R. Moriyama, and T. Takagi, "Determination of Polypeptide Chain Molecular Weights of Human and Bovine 3. Protein from Erythrocyte Membranes by Low Angle Laser Light Scattering Combined with High Performance Gel Chromatography in the Presence of Dodecyl Sulfate," *Biochim. Biophys. Acta*, **874**, 216 (1986).

142. T. Nakao, T. Ohno-Fujitani, and M. Nakao, "Sodium and Potassium Ion-Dependent Change in Oligomerization of Sodium-Potassium ATPase Detected by LALLS in Combination with High Performance Porous Silica Gel Chromatography," *J. Biochem.* (*Tokyo*), **94**, 689 (1983).

143. T. Takagi, S. Maezaua, and Y. Hayashi, "Determination of Subunit Molecular Weights of Canine Renal Sodium-Potassium ATPase by Low-Angle Laser Light Scattering Coupled with High Performance Gel Chromatography in the Presence of Sodium Dodecyl Sulfate," *J. Biochem.* (*Tokyo*), **101**, 805, (1987).

144. T. Takagi, "Confirmation of Molecular Weight of Aspergillus Oryzae β-amylase Using the LALLS Technique in Combination with High Pressure Silica Gel Chromatography," *J. Biochem.* (*Tokyo*), **89**, 363 (1981).

145. M. K. Gabriel and E. T. McGuinnes, "NAD^+ Kinase: Molecular Weight Determination by LALLS," *FEBS Lett.*, **175**, 419 (1984).

146. J. G. Blindels, B. M. DeMan, and H. J. Hoenders, "HPGPC of Bovine Eye Lens Proteins in Combination with LALLS: Superior Resolution of the Oligomeric β-Crystallines," *J. Chromatogr.*, **252**, 255 (1982).

147. A. Kato, Y. Nagase, N. Matsudomi, and K. Kobayash, "Determination of Molecular Weight of Soluble Ovalbumin Aggregates During Heat Denaturation Using LALLS Technique," *Agric. Biol. Chem.*, **47**, 1829 (1983).

148. A. Kato and T. Takagi, "Estimation of the Molecular Weight Distribution of Heat-Induced Ovalbumin Agregates by the Low Angle Laser Light Scattering Technique Combined with High-Performance Gel Chromatography," *J. Agric. Food Chem.*, **35**, 633 (1987).

149. W. Flapper, P. J. M. van den Oetelaar, C. P. M. Breed, J. Steenbergen, and H. J. Hoenders, "Determination of Serum Proteins by High Pressure Gel Permeation Chromatography with Low-Angle Laser Light Scattering, Compared with Analytical Ultracentrifugation," *Clin. Chem.*, (Winston-Salem, N.C.) **32**, 363 (1986).

150. T. Takagi, "Molecular Weight Determination of Proteins: Combination of Gel Chromatography and Light Scattering Photometer," *Tanpakushitsu Kakusan Koso*, **27**, 1526 (1982); *Chem. Abstr.* **97**, 158900f (1982).

151. T. Takagi, "Assessment Study on the Use of the LALLS Technique in Combination with the High-Performance Silica Gel Chromatography," *Protides Biol. Fluids*, **30**, 701 (1982).

152. T. Takagi, "Determination of Protein Molecular Weight by Gel Permeation Chromatography Equipped with Low Angle Laser Light Scattering Photometer," *Prog. HPLC*, **1**, 27 (1985).

153. H. H. Stuting, I. S. Krull, R. Mhatre, S. C. Krzysko, and H. G. Barth, "High Performance Liquid Chromatography of Biopolymers Using On-line Laser Light Scattering Photometry," LC-GC, 7, 402 (1989).

154. S. Maczawa and T. Takagi, "Monitoring of the Elution from a High Performance Gel Chromatography Column by a Spectrophotometer, a LALLS Photometer and a Precision Refractometer as a Versatile Way to Determine Protein Molecular Weight," *J. Chromatogr.* **280**, 124 (1983).

155. T. Takagi, "Molecular Weight and State of Assembly of Membrane Proteins: An Approach of LALLS Measurement," *Seikagaku*, **57**, 202 (1985).

156. M. N. Jones and P. Midgl, "LALLS from Surfactant-Soluble Biological Macromolecules," *Biochem. Soc. Trans.*, **12**, 625 (1984).

157. J. Miyake and T. Takagi, "A LALLS Study of the Association Behavior of a Major Membrane Protein of Rhodospirillum Rubrum Chromatophore," *Biochem, Biophys. Acta.*, **668**, 290 (1981).

158. K. Kameyama, T. Nakae, and T. Takagi, "Estimation of Molecular Weights of Membrane Proteins in the Presence of SDS by LALLS Combined with High Pressure Silica Gel Chromatography: Combination of the Trimer Structure of Porin of E. Coli Outer Membrane," *Biochem. Biophys. Acta*, **706**, 19 (1982).

159. W. E. Hennink, J. W. A. Vander Berg, and J. Feijrn, "Standardization of Heparins by Means of High Performance Liquid Chromatography with a Low Angle Laser Light Scattering Detector," *Thromb. Res.*, **45**, 463 (1987).

160. D. Lecacheux, Y. Mustiere, R. Panaras, and G. Brigand, "Molecular Weight of Scleroglucan and other Extracellular Microbial Polysaccharides by Size Exclusion Chromatography and Low Angle Laser Light Scattering," *Carbohydr. Polym.*, **6**, 477 (1986).

161. T. Nicolai, L. Van Dijk, J. A. P. P. Van Dijk, and J. A. Smit, "Molar Mass Characterization of DNA Fragments by Gel Permeation Chromatography Using a Low Angle Light Scattering Detector," *J. Chromatogr.*, **389**, 286, (1987).

162. A. Kazuhiko, A. Toshio, and M. L. McConnell, "Formation of Nuclei During Delay Time Prior to Aggregation of Deoxyhemoglobin S in Concentrated Phosphate Buffer," *Biochem. Biophys. Acta.*, **580**, 405 (1979).

CHAPTER

10

PHOTON CORRELATION SPECTROSCOPY

RICHARD B. FLIPPEN

E. I. duPont de Nemours & Company
Experimental Station
Wilmington, Delaware

1. INTRODUCTION

Light scattering by dispersions of particles and molecules has been a major technique for polymer analysis for many years; industrial analytical laboratories have used the techniques since at least 1944 (1). Absolute weight-average molecular weights, radii of gyration, virial coefficients, and molecular conformations in dilute solution are obtained from classical (static) light

Modern Methods of Polymer Characterization, Edited by Howard G. Barth and Jimmy W. Mays
ISBN 0-471-82814-9

scattering (SLS) (2) (see Chapter 9). Although the light scattering particles are in thermal motion during the measurement, the scattered light is collected in time periods much longer than the time scale of motion of the particles. The resultant scattered light is an average over the number of particles in the scattering volume. The measurement is equivalent to the steady state or static conditions.

Recently, instrumentation for frequency–time measurements and the development of fast multichannel analyzers has made possible the measurement of properties of scattered light that are the result of the motion of the light scattering particles. For reasons described below, this new technique is called quasielastic light scattering (QELS), photon correlation spectroscopy (PCS), or simply dynamic light scattering. This technique provides different information about the nature of light scattering particles than that obtained from classical light scattering and extends the analysis of polymer properties into new areas previously not readily accessible by other techniques.

This chapter will describe this light scattering technique and how it is applied for polymer analysis. Some background theory will first be given for understanding the concepts involved, and then the kinds of instrumentation necessary to do the measurements will be described. Some experimental results obtained with PCS will be compared with similar results obtained using other techniques to show the validity of application. A number of studies of PCS in the field of polymer analysis will be described to show the usefulness of this relatively new technique.

2. THEORY

Theories of light fluctuation spectroscopy will not be considered in detail here. Material sufficient for understanding the general principles and practical applications will be presented. Interested readers are referred to the works of Berne and Pecora (3), Chu (4), and the various conference proceedings on the subject (5–7) for details and more extensive discussions of dynamic light scattering. A recent review by Phillies (8) provides a good overview to the field.

In a solution of polymer molecules in a solvent or in a suspension of particles in a fluid, the molecules or particles are in irregular motion caused by the thermal energy imparted to them by collisions with the solvent or fluid molecules (Brownian motion). These collisions cause the particles to undergo both translational and rotational moion. The probability $P(\bar{\rho}, t)$ of finding a molecule at position $\bar{\rho}$ at time t if it is at the coordinate origin at time zero is given by

$$\frac{\partial P(\rho, t)}{\partial t} = D_t \bar{V}^2 P(\rho, t) \tag{1}$$

where D_t is the translational diffusion coefficient of the particle, and $\bar{V}$ is the spatial derivative. If a rod-shaped particle is similarly in Brownian motion, and its probability of having spherical coordinates θ, ϕ at a time t is $P(\theta\phi/\theta_0\theta_0, t)$, where θ_0 and ϕ_0 are the coordinates at a time zero, then P will be given by

$$\frac{\partial P}{\partial \tau} = \Omega \frac{1}{\sin^2\theta}\left[\left(\sin\theta\frac{\partial}{\partial\theta}\right)^2 + \left(\frac{\partial^2}{\partial\phi^2}\right)\right]P \tag{2}$$

where Ω is called the rotational diffusion coefficient. Thus, the diffusion coefficients of particles in solution or suspensions are related to the particle motion. Since moving particles will scatter light in a way that is quantitatively related to their motion, light scattering measurements provide a path to determine characteristic diffusion coefficients of particles and molecules. Diffusion coefficients in turn are related to such physical characteristics, as molecular weight, particle shapes and sizes, and solvent interactions.

When monochromatic light from a laser traverses a transparent medium, light is scattered by density fluctuations in the medium. If no shift in frequency of the scattered light occurs from that of the incident light, this scattering is called elastic. If these fluctuations do not change in time, or if the scattered light is measured over a long period as compared to the time of change of the fluctuations, then there is no change in the frequency of the scattered light. However, density fluctuations in solutions and suspensions are time dependent because of the Brownian motion of the particles. Thus, the frequency of light scattered by such material will show a frequency spectrum characteristic of the time dependence of the fluctuations and the motion of the particles (Doppler shift). Because the frequency difference of the scattered light with the incident light is small, this kind of scattering is called *quasielastic* light scattering.

The scattering particles in a sample of interest are usually in a relatively narrow size range; as a result, the particles move with similar velocities, although in random directions. Because of their similar motions, the light scattered from the particles is said to be correlated in time. Formally, a time correlation function between two signals A and B is given by

$$g(\tau) = \lim_{T\to\infty}\left(\frac{1}{T}\right)\int_{t_0}^{t_0+T} A(t)B(t-\tau)dt \tag{3}$$

where τ is a delay time, t_0 the starting time, and T the averaging time. If B is a delayed form of A, then the relation is called autocorrelation. The correlation function $g(\tau)$ is a real quantity that can be measured with appropriate apparatus as will be described.

A related quantity is the power spectrum $I(\omega)$ of a time correlation function defined as

$$I(\omega) = \frac{1}{2\pi} \int_{-\infty}^{\infty} e^{-i\omega t} \langle A(t)A(t-\tau) \rangle \, dt \tag{4}$$

$I(\omega)$ is the "amount" of A contained in the frequency ω, where $I(\omega)$ and $\langle A(t)A(t-\tau)\rangle$ are Fourier transforms of each other, so that an experimental determination of one quantity can be used to determine the other.

This form of light scattering is called photon correlation spectroscopy. In experiments used to determine the correlation function, the process is called homodyne or self-beating if only the light from the scattering volume is used in the detection system. If light from a local oscillator is mixed with the scattered light, the process is called heterodyne. Most experimental arrangements utilize the less complicated homodyne method (3).

In the simple case of an ensemble of hard, spherical, noninteracting identical particles moving in a fluid, the correlation function $g(\tau)$ is given by

$$g(\tau) = A_0 + A \exp(-\Gamma t) \tag{5}$$

where $\Gamma = D_t K^2$ and the wave vector $\mathbf{K} = 4\pi n \sin(\theta/2)/\lambda$. The term D_t is the translational diffusion coefficient, n is the index of refraction of the fluid, λ_0 is the wavelength of incident light, and θ is the angle of observation. Since Γ can be obtained by an appropriate fitting of experimental correlation data, and all of the parameters comprising K are known, the translational diffusion coefficient D_t can be calculated.

The term D_t can be used to determine some particulate and molecular properties of solutions and dispersions. In general

$$D_t^0 = kT/f \tag{6}$$

Where D_t^0 is the translational diffusion coefficient at infinite dilution obtained by extrapolation of PCS data for finite concentrations, k is Boltzmann's constant, and f is the molecular frictional coefficient (Stokes–Einstein relation). For spherical, noninteracting particles, $f = 3\pi\eta d$, where η is the fluid viscosity at temperature T, and d is the particle diameter. Thus, for this simple case PCS measurements will yield the size of the scattering particles. For nonspherical particles, the calculation for d will give an average dimension, which is adequate for many purposes. In general, $f = 6\pi\eta F(\bar{r}^2)^{1/2}$, where F is a form factor and $(\bar{r}^2)^{1/2}$ is the root-mean-square end-to-end distance for the particle or molecule.

In the case of molecular dispersions where molecules–solvent interactions are present, strong concentration effects may occur, so that

$$D_t = D_t^0(1 + k_D C + \cdots) \quad (7)$$

where

$$k_D = 2A_2M - k_f - (N_A V_1/M) \quad (8)$$

A_2 is the second viral coefficient, M the molecular weight, k_f the first-order frictional coefficient, N_A is Avogadro's number, and V_1 the molecular volume. Much work has been done to relate the value of k_D to molecular properties. For practical purposes it is often sufficient to calculate a hydrodynamic diameter d_H from

$$D_t = kT/3\pi\eta d_H \quad (9)$$

where d_H can be a measure of the effect of changing solvents, the effect of different solution ionic charge, temperature, and other factors.

The diffusion coefficient can be related to other molecular properties such as the sedimentation coefficient. Here, the Svedberg relation

$$M = sRT/D_0(1 - \bar{v}\rho) \quad (10)$$

relates the molecular weight of the solute M to the diffusion coefficient D_0 through the sedimentation coefficient s, gas constant R, solvent density ρ, and partial specific volume $\bar{v}$ of the solute molecule. For homologous polymers, the diffusion coefficient D_t is related to the molecular weight by

$$D_t^0 = k_A M^{-b} \quad (11)$$

which is similar to the Mark–Houwink relation between intrinsic viscosity and molecular weight. Measurements of diffusion coefficients and viscosities can be used together to determine the molecular weights of a series of homologous polymers. In the special case of hard, nonsolvated spheres, the volume of the particle $\frac{4}{3}\pi r^3$ is equal to $\bar{v}M/N_A$. In this simple case, if $\bar{v}$ is known, the PCS measurement will directly give a value for the molecular weight.

For rod-shaped molecules, rotational diffusion may also play a part in the light scattering. The total spectral density $S(\mathbf{K}, \omega)$ for such particles is given by

$$S(\mathbf{K}, \omega) = S_T(\mathbf{K}, \omega) + S_R(\mathbf{K}, \omega) + S_{TR}(\mathbf{K}, \omega) \quad (12)$$

where the three terms are, respectively, the translational, rotational, and mixed contributions. The term $\mathbf{K}$ is the wave vector and ω the frequency. Formulas for special cases can be developed (9) that relate the rotational diffusion coefficient D_r to a correlation function. However, D_r is typically much larger than D_t and is experimentally difficult to measure.

In most practical cases, solutions and dispersions of light-scattering particles will not be monodisperse. The correlation function in general will be given by a distribution of exponentials such that it can be expressed as a Laplace integral of the form

$$g(\tau) = \int_0^\infty G(\Gamma)e^{-\Gamma\tau}d\Gamma \tag{13}$$

and much theoretical effort has gone into the solution of this relation for the normalized distribution function of decay rates $G(\Gamma)$.

One of the most useful approaches to the application of PCS to polydisperse systems is the method of cumulants reported by Koppel (10). In this method, a moment-generating function $M(\tau, \mathbf{K}) = \overline{\exp(-\Gamma\tau)} = |g(\tau)|$ is used with a MacLauran series expansion of the logarithm of this function to average over all the light scattering particles. The expansion takes the form

$$A\frac{\ln g'(\tau)}{B} = \ln A - \Gamma(\tau) + \frac{\mu_2(\tau)}{2} + \cdots \tag{14}$$

where A is an optical constant and B the baseline determined by calculation or experiment. This method has been discussed in some detail recently by Phillies (11). The coefficients of the expansion successively give the average diffusion coefficient, the normalized distribution width, the normalized distribution skewness, and higher moments. Only the first two quantities are usually justified by the quality of the data. The average diffusion coefficient determined in this way is a z-average quantity as described by Chu (4). The attractiveness of this technique is that it can be carried out quickly by a small computer program. Data acquisition and processing can be done automatically to give rapid results.

Two other common approaches to the polydispersity problem, which are more time consuming and elaborate than the cumulant approach, are (a) to invert the Laplace integral with some form of transform to obtain $G(\Gamma)$, and (b) to assume a specific form for $G(\Gamma)$ to obtain a solution. Much progress has been made in recent years in the former approach using a histogram analysis by Gulari et al. (12) and an exponential sampling method by Provencher (13). The latter developed his method into an accessible general computer program CONTIN (13). In the histogram method, $|g(\tau)|$ is represented by a series of discrete steps

$$|g(\tau)| = \sum_{i=1}^{J} G(\Gamma_i) \int_{\Gamma_i - \Delta\Gamma/2}^{\Gamma_i + \Delta\Gamma/2} e^{-\Gamma/\Delta\tau}\, d\Gamma \tag{15}$$

$\Delta\Gamma$ is the width of each step. The step width and number of steps are varied until a best fit is obtained according to statistical rules. In the exponential sampling method, trial eigenvalues of the inverse Laplace transform are picked according to certain assumptions and then varied to obtain a best fit to the solution.

Other approaches to obtaining the size distribution from the correlation function data are various exponential function assumptions (14), cubic-B spline analysis (15), polynomial subdistribution methods (16), and others. Many assumptions are necessary to solve this problem, including the following: no unique solution of inverting Eq. 13 exists, a positive distribution is necessary, a smooth distribution with the fewest components work best, and so on. Solutions using this formalism can be obtained that have no relation to the actual experimental distribution (17).

Although, in principle, light scattering can be used to obtain molecular size distributions, the data usually are not clean enough of extraneous scatter to give a signal-to-noise (S/N) ratio sufficient to obtain more than the general outline of a particle size distribution. Binary or, in some cases, three or more major component distributions can be distinguished by PCS only if the dominant sizes are separated by a factor of two to three. In most cases even this resolution is difficult. For example, the base line B depends on the amount of extraneous matter present, like dust in the scattering volume; if B is large or uncertain, the accuracy of calculating Γ will be limited. Thus, size distribution analysis by PCS cannot be considered as normal competition to other techniques capable of finer resolution such as size exclusion chromatography and sedimentation field-flow fractionation. In the cases where it can be used, PCS does offer the advantages of small sample size and rapidity of results.

3. INSTRUMENTATION

A typical experimental arrangement for light scattering correlation measurements is shown in Figure 1. Detailed experimental setups are described in, for example, Han and McCracklin (14) and Chu (4). In common with classical (intensity) measurements, a light source, optics for defining the light path, a spectrometer with sample holder, and a sensitive light detector are needed. Additionally, for PCS measurements a correlator is necessary with either a dedicated microprocessor or a small computer for data analysis. Usually a

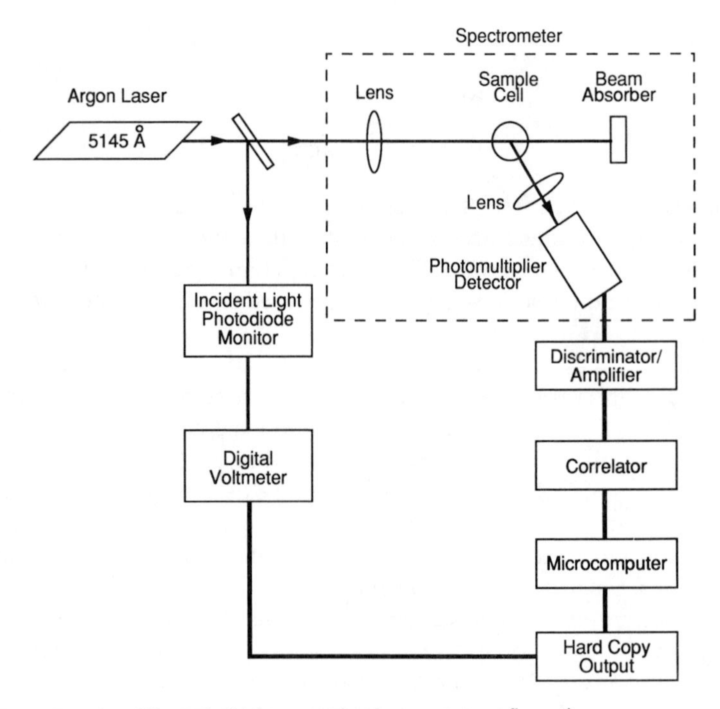

Figure 1. Light scattering instrument configuration.

monitor of the incident light intensity is included. Although this quantity is not needed in the calculations, it is useful in referring back to the experimental conditions of the measurement. A laser is used as the light source; for relatively large particles > 500 Å, which scatter light strongly, a 10–15-mW He–Ne laser can be used. For small particles < 500 Å, stronger sources such as a 1–2 W Ar-ion laser may be necessary for sufficient scattered light intensity from weakly scattering material. Lenses with pinhole guides–shields are used to focus the beam down to a 10–50-μm diameter in the center of the sample. Similar optics focus the scattered light from a 50–100-μL volume of the illuminated center of the sample onto the detector, typically a fast, low dark-current photomultiplier (PM) tube such as a ITT 130 or RCA C31034. These optics define the coherence area of the scattered light, which determines the S/N ratio of the measurement. Experimental considerations of coherence effects are covered by Chu (4).

An amplifier–discriminator circuit shapes the output of the PM tube to form a single voltage pulse per received photon, which is then fed into a special multichannel analyzer called a correlator. This device can have from 64 to 256 channels or more (with multiplexing). The channel width of the correlator can be adjusted typically from 0.1 μs to >1 s, appropriate for the decay-time associated with the scattering source. Data collection times from seconds to hours can be set for the strength of the scattered light from the sample and the level of S/N. The correlator, typically operated in an unclipped correlation mode at 1–10 counts per sample (4), measures the sample correlation function as an exponentially decreasing number of photon counts for increasing channel number. These data are read into a computer that is typically programmed to do a fit to the data to determine the decay constant Γ. As described by Eq. 5, the index of refraction of the sample fluid at the measurement temperature, the detector angle, and the laser wavelength are used with the correlation exponent to calculate the average particle diffusion coefficient. The correlation data also can be used with one of the methods discussed earlier to calculate an effective size distribution. An example of an experimental correlation function for a latex dispersion is shown in Figure 2, where the normalized logarithm of the correlation data output is plotted versus the channel number. Various fitting parameters are shown for both a second-order and a third-order fit to the data.

Before about 1980, experimenters had to assemble their own PCS apparatus using custom-built or commercially available components, such as lasers, spectrometers, correlators, and computers. After that date, several manufacturers began to market packaged devices containing a light source, sample holder, detector, and dedicated microprocessor, in addition to more sophisticated PCS apparatus components. These self-contained devices are usually pushbutton operated and menu-driven so that they can be used by nonexperts. Some of the currently available packaged PCS apparatuses are marketed by a number of companies listed at the end of this chapter. In

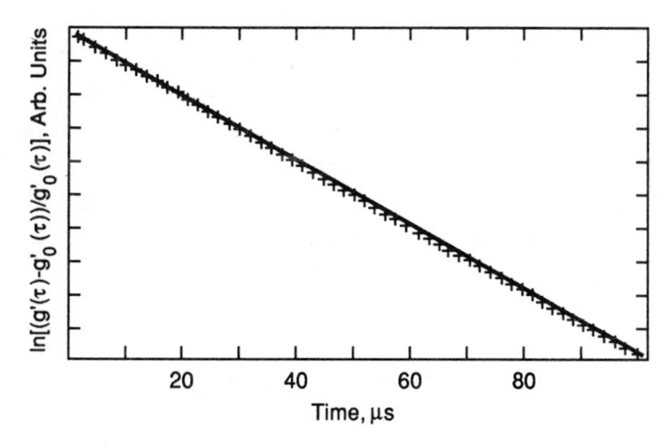

Figure 2. Typical normalized correlation function versus channel-number printout for a 754 Å latex dispersion using a third-order fit.

addition, Oros Instruments has recently introduced an online PCS detector for size exclusion chromatography of proteins, in which a 30-mW miniature solid-state laser at 780 nm is used.

Most commercial units work well for diffusion coefficient and effective particle size measurements of particles larger than about 200 Å. Below that size, scattered light intensities are often too low if a small laser is used to have enough sensitivity to do the measurement. Also, some of these devices have fixed-angle (90°) detectors, which can give misleading results with large, anisotropic or high aspect-ratio particles. These commercial units are programmed to determine diffusion coefficients, effective sizes, apparent molecular weights, and most have some form of size distribution analysis.

4. EXPERIMENTAL CONSIDERATIONS

The advantages of the PCS technique are rapid (often < 1 min) measurement, automatic data collection and analysis, small sample size (< 1 mL), almost any fluid can be used that is not strongly absorbing at the light wavelength used, and excellent reproducibility–accuracy (often 1–2%). The smallest sample particle sizes that can be measured is ~ 20 Å. The upper limit is dictated by the largest particle size that can be kept in suspension (several micrometer). As with all light scattering measurements, the sample solutions–suspensions must be homogeneous and free of foreign matter that would contribute to light scattering. In particular, dust particles are especially detrimental since they can act as local oscillators leading to unwanted heterodyne signals. Differences between the calculated and experimental baselines of greater than about 0.1% are often indicative of the presence of dust in the solutions. Types of samples that have been examined with PCS range from very dilute (< 0.001 g/mL) solutions to concentrated (~ 0.1 g/mL) solutions to solid polymers. The relevant analysis theories must progressively take into account single-particle scattering, multiple-particle scattering, and scattering from collective excitations. Examples of each of these sample types will be discussed in later sections. Factors that affect the molecular or particle diffusion, such as aggregation or disaggregation, comformational changes, chemical reactions, and solvent interactions can be studied with PCS. Examples of these effects also will be discussed.

5. COMPARATIVE ANALYSIS

Photon correlation spectroscopy analysis of particle diffusion coefficients, sizes, and size distributions can be tested against other methods for

Table 1. Analysis of Standard Latex Particles Using PCS

Sample	Nominal Diameter (Å)	Hydrodynamic Diameter (Å)
Polysciences 7599	7700	7800
Polysciences 7832	1700	1900
Dow	910	930
Dow	850	760

determining these quantities. For example, PCS results using a cumulants analysis for the size of standard latex dispersions used for electron microscope calibration are shown in Table 1 versus the manufacturer's results. Because it is difficult to avoid some aggregation in these dispersions, the average particle size by PCS will usually be slightly larger than the claimed particle size. In the case shown where the PCS result is smaller than the claimed size, it was found by other methods that the PCS size was accurate and that the "standard" value was incorrect. Chu and Gulari (17) found PCS results for the size of Dow 910 latex particles to be consistently 7% higher than the value for dry particles determined by electron microscopy. They did not speculate on reasons for the difference in results of the two methods.

Sedimentation field-flow fractionation (SFFF) has also been used to compare results from PCS (18). Sedimentation field-flow fractionation provides both weight- and number-average sizes, as well as size distributions. Such comparisons are shown in Table 2 for a series of experimental polymer lattices. The d_H value by PCS is a z-average value, so that this number will be close to the weight-average value of SFFF. The normalized second-cumulant coefficient by PCS should be proportional to the distribution width by SFFF. As shown in Table 2, the results are in excellent agreement.

Table 2. Comparison of PCS to Sedimentation Field-Flow Fractionation of Latices[a]

Sample	SFFF d_n (μm)	SFFF d_w (μm)	PCS d_H (μm)	SFFF DW (μm)	PCS (μ)	Remarks
101	0.201	0.244	0.223	0.119	0.318	Symmetrical
102	0.177	0.235	0.215	0.119	0.305	Skewed front
104	0.162	0.224	0.197	0.145	0.345	Bimodal
316	0.179	0.261	0.214	0.220	0.402	Biomodal
323	0.168	0.250	0.217	0.158	0.352	Skewed back
337	0.184	0.241	0.232	0.101	0.283	Symmetrical

[a] $\bar{d}_n$, $\bar{d}_w$, and $\bar{d}_H$ are number-, weight-, and hydrodynamic-average particle diameters, respectively. DW is the distribution width and μ is the second-cumulant coefficient.

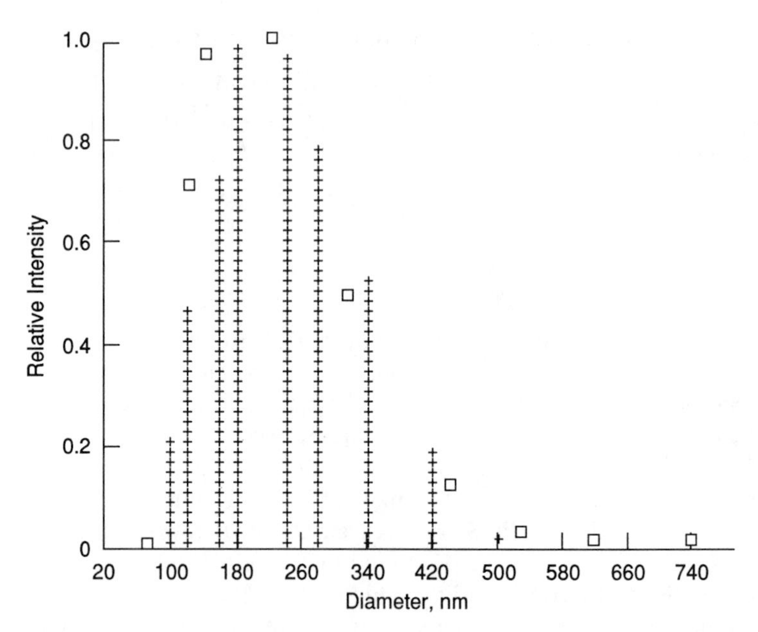

Figure 3. Size distribution of a latex suspension using PCS (+) and SFFF (□). The PCS results were obtained with a Brookhaven Instruments BI-2020 correlator using an inverse Laplace transform program. The SFFF results were obtained with a laboratory research apparatus.

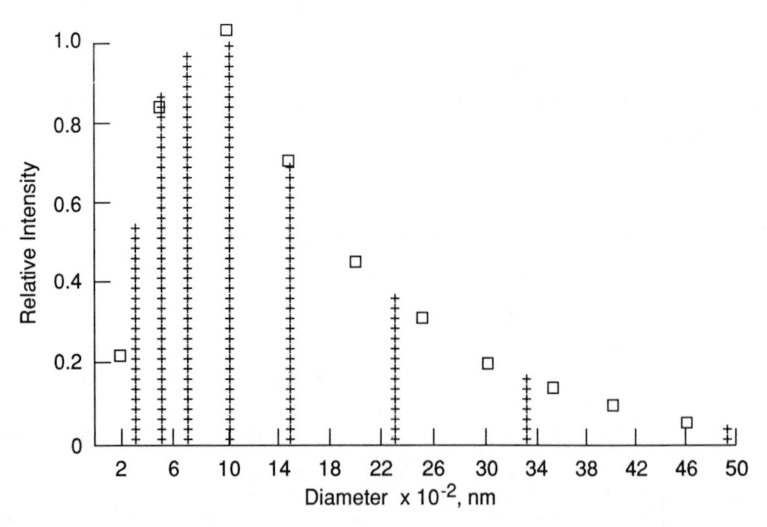

Figure 4. Size distribution of an inorganic dispersion using PCS (+) and sedimentation analysis (□). The latter was obtained with a Horiba instrument.

Sedimentation field-flow fractionation results for particle size distributions provide an additional check for the validity of PCS size distribution analysis, since the sample particle suspensions can be measured by both techniques and the results compared. The results for one such latex suspension are shown in Figure 3. where a commercial inverse Laplace transform program for a Brookhaven Instruments BI-2020 correlator has been used to determine a size distribution for the dispersion. The same sample was independently measured by SFFF, and the results for the two methods are shown plotted to scale. The agreement is very satisfactory. Another comparison is provided in Figure 4. Shown are results from a commercial sedimentation instrument for particle sizing (Horiba Instruments) for an inorganic dispersion plotted with PCS data for the same sample. Again, the agreement is good.

6. APPLICATIONS

In this section selected results are presented to give an overview of the types of applications that have employed PCS. No attempt has been made to give a complete review of this subject.

6.1. Diffusion Coefficients

The diffusion coefficients of particles suspended in a fluid are the fundamental measured quantities in a PCS experiment. Examples of the results of D_t for an experimental polymer poly(1,4-phenylene terephthalamide) dissolved in H_2SO_4 are shown as a function of concentration in Figure 5. The value of D_t for this sample is shown as a function of temperature in Figure 6. Most of the temperature dependence is caused by the change in viscosity of the fluid with temperature. Log D_t^0 for a number of low polydispersity poly(methyl methacrylate) samples dissolved in THF are shown plotted versus log— (molecular weight) in Figure 7. The inverse exponential dependence of D_t^0 with M is evident.

In the case of long, thin molecules, it becomes experimentally possible to determine the rotational as well as the translational diffusional coefficient. Several such measurements for tobacco mosiac virus, a particle 3000 Å in length, have been reported (19).

6.2. Particle Size Effects

Changes in diffusion coefficients of dissolved molecules are often the result of changes in properties of the carrier fluid, such as changes in viscosity with temperature shown in Figure 6. These effects can be separated from changes in

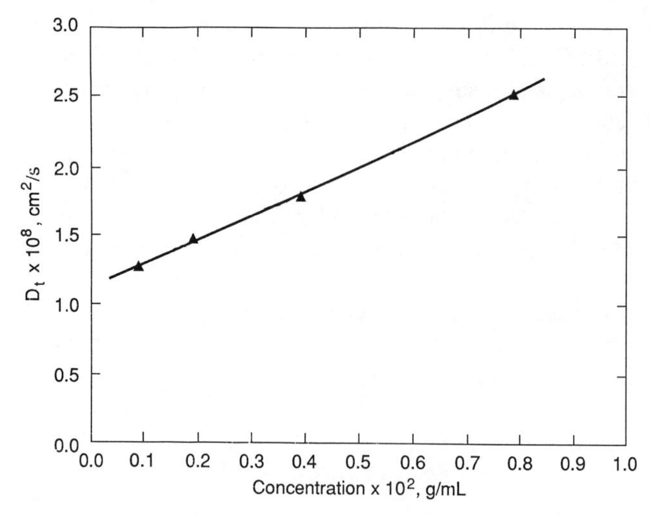

Figure 5. Diffusion coefficient versus concentration of poly(1,4-phenylene terephthalamide) in sulfuric acid.

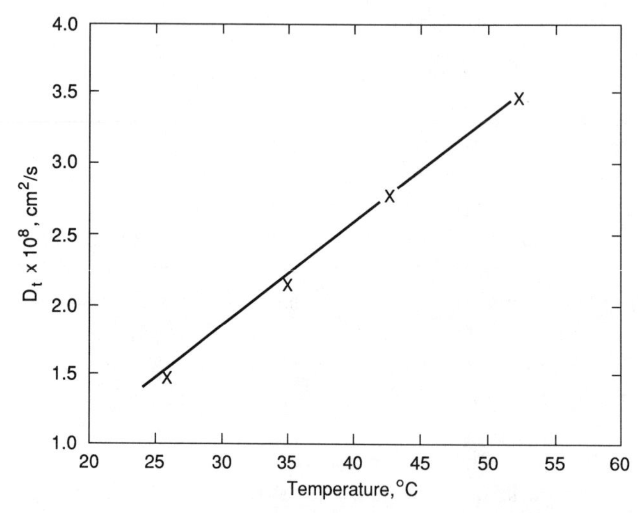

Figure 6. Diffusion coefficient versus temperature of poly(1,4-phenylene terephthalamide) in sulfuric acid. Concentration of the polymer is 2 mg/mL.

the molecules themselves by calculating the effective hydrodynamic diameter d_H discussed earlier from the Stokes–Einstein relation, thus removing the effects of temperature and viscosity of the carrier fluid (Eq. 9). Although d_H is not necessarily a measure of the scattering particle diameter, Phillies (20) has shown this D_t versus d_H relation has a wide range of validity.

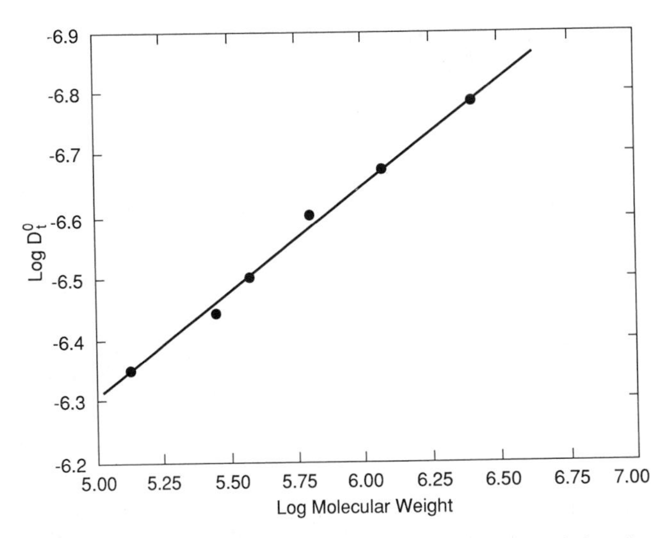

Figure 7. Diffusion coefficient versus molecular weight of poly(methyl methacrylate).

The hydrodynamic volume is effectively the diameter of the volume occupied by the light scattering particle as it moves about with kinetic energy. It reflects both the physical size of the particle and the range of solution interactions, which control the particle motion through frictional forces. If the particle conformation changes because of changes in a solution, this change will be reflected in d_H.

Changes in molecular conformation or particle size, which occur on a time scale of minutes or longer, can be followed with PCS. One such example is the formation of microemulsions. A mixture of H_2O–butyl alcohol–*n*-decane forms an emulsion that decreases from an initial cell size to an equilibrium size as time progresses. The cell size can be followed by PCS as a function of time with the results shown in Table 3.

Table 3. Size Changes of a Water/Butyl Alcohol/ *n*-Decane Microemulsion Measured by PCS

Time	Particle Diameter (Å)
1 min	7700
5 min	1300
2 h	690
24 h	380
14 days	78

Table 4. Effect of Solvents on the Hydrodynamic Diameter (Å) of Experimental Polymers

Sample	Amyl Methyl Ketone	Heptane	Butyl Acetate	Diacetone Alcohol	ρ-Xylene
110	2890	2290	3010	3240	2260
120	4360	2500	4630	3520	2540
135	3430	2610	3670	3810	2720
4340	3450	3980	2670	3390	3180
4341	3260	2890	3480	3260	2660

Solvation effects of polymer–solvent interactions can also be measured with PCS. Table 4 shows the infinite dilution hydrodynamic diameter of a number of different experimental polymers in six different solvents. Although some polymers (samples 4340 and 4341) show only small changes in d_H from solvent to solvent, others (sample 120) show almost a twofold change, and the effect varies from solvent to solvent for each polymer. Studies of this kind can be combined with viscosity and other measurements to provide an understanding of the interaction involved and guidance for practical applications.

6.3. Concentration Effects

The concentration dependence of D_t of a 233,000 molecular weight sample of polystyrene dissolved in ethyl methyl ketone is shown in Figure 8 along with a similar plot of D_t of a 265,000 molecular weight sample of poly(methyl

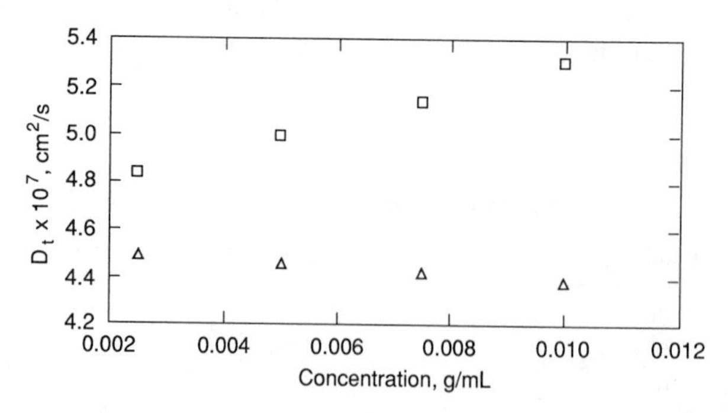

Figure 8. Diffusion coefficient versus concentration of poly(methyl methacrylate) (265,000 g/mol) (□) and polystyrene (233,000 g/mol) (△) in ethyl methyl ketone.

methacrylate) dissolved in the same solvent. The slopes have opposite signs, showing that the interactions involving these two polymers with the same solvent are different. The slope of D_t versus concentration in the dilute (< 1%) regime is comprised of a number of contributions given in Eq. 8. This relationship is simplified at the theta temperature T_θ where A_2 is zero. Pyun and Fixman (21) calculated that at this temperature, $k_f = 2.23\, N_A V_1/M$ as compared to Imai (22) who found $k_f = N_A V_1/M$. These relations can be tested for appropriate solutions using PCS. Gulari et al. (23) showed that for polystyrene in cyclohexane, the Pyun and Fixman theory agrees with their results.

As polymer solutions become more concentrated, the weak hydrodynamic interactions among molecules are overcome by stronger cooperative interactions as molecular entanglements become more prominent. For concentrations greater than an overlap concentration c^* (equal to $1/[\eta]$) but still less than a crossover to concentrated solutions, the scaling arguments of de Gennes (24) have provided most of the understanding of this solution region. In this theory, a key parameter is a correlation length ε, which defines a distance between two points of entanglement. The value of ε becomes shorter with increasing concentrations and is defined by $\varepsilon = a(c/c^*)^{-0.75}$, where a is a constant. Molecular diffusion is related to ε such that for $c > c^*$, the theroy predicts two diffusion modes: a fast mode given by $D_t \sim kT/6\,\pi\eta\varepsilon$ and a slow mode given by $D_s \sim (c/c^*)^{-1.75}$. The self-diffusion mode, D_s, is viewed as a motion of the molecular chain along its contour length.

Much work has been reported in the light scattering literature to confirm the predictions of this theory. Eisele and Burchard (25) confirmed both modes in poly(vinylpyrrolidone)/water above $c^* = 7.31 \times 10^{-3}$ g/mL. Statman and Chu (26) observe two translational motions in poly(n-butylisocyanate)–carbon tetrachloride. Chang and Yu (27) obtained similar results with self-diffusion in gelatin–water. Selser (28), however, claims PCS measurements of poly(α-methylstyrene)–toluene are inadequately explained by the de Gennes theory. Instead he interprets his results on the basis of a theory used by Ronca (29) who employs a system of linear and cross-connected bead and spring units to derive an expression for the dynamic structure factor $S(\mathbf{K}, \tau)$, where $S(\mathbf{K}, \tau)$ is the mean-square fluctuation of Fourier components of the density fluctuations. Here $S(\mathbf{K}, \tau)$ is a summation of the relative position vectors of all pairs of beads in the scattering volume. Measurements of polystyrene solutions by Patterson et al. (30) were found to be incapable of testing scaling law predictions and did not find the formation of a pseudogel network at c^*.

In recent years, PCS was used to study pure polymers in the bulk state. Patterson (31) reviewed work in this field. Light scattering in melts and solid polymers is caused by fluctuations in the dielectric tensor of the material. These fluctuations occur in the density and the optical anisotropy of the medium. Theories explaining the phenomena express the scattered light in terms of frequency-dependent moduli like specific heats C_v, modulus of compression k, shear modulus G, and thermal expansion coefficient K (32). Optical anisotropy occurs because of the inherent molecular structure and the anisotropic distribution of molecules relative to one another.

The total relaxation function for a bulk polymer is comprised of contributions from fast processes, like collision-induced anisotropic scattering, which with relaxation times $\sim 10^{-13}$s, cannot be observed with PCS, orientational fluctuations, and those due to the relaxing parts of k, G, and C_v. The latter two processes have been observed using PCS in a number of bulk polymers including polystyrene (33), poly(methyl methacrylate) (34), poly(ethyl methacrylate) (35), poly(n-butyl methacrylate) (36), and poly(propylene glycol) (37).

A Gaussian random process with a single relaxation time has a relaxation function $e^{-t/\tau}$. However, many of the experimental investigations of bulk polymers using PCS find the data are fitted well by a function $Ae^{-(t/\tau)\exp\gamma}$, with typically $0 < \gamma < 1$, where A and γ are empirical constants. This function can be interpreted as a distribution of relaxation processes with a distribution width parameter. The average relaxation $\langle\tau\rangle$ is given by $\langle\tau\rangle = (\tau/\gamma)\Gamma(1/\gamma)$, where $\Gamma(1/\gamma)$ is the PCS gamma function. The term $\langle\tau\rangle$ varies rapidly with temperature and pressure near the glass-transition temperature in polymers. The relaxation processes taking place in this polymer state have been examined with many other techniques involving dielectric and mechanical measurements. The PCS results agree well with the results of these other techniques and provide complimentary information about the distribution of relaxation times in these materials.

6.4. Macromolecular Conformation

The spatial volume occupied by a molecule as it moves about in solution is determined both by the molecular interactions present and by the molecule's shape. A measure of this volume is the hydrodynamic diameter d_H, which can be determined by PCS. In situations where translational and rotational diffusion coefficients are sufficiently separated in time that they can be separately measured by PCS, the molecular conformation can be estimated by fitting the data to various theoretical models. King et al. (19) determined the translational and rotational diffusion coefficients D_t and D_r of tobacco mosiac virus using PCS and then calculated its rodlike shape. Hwang and Cummings (38) calculated D_t and D_r from the measured correlation functions of collagen in solution and estimated a length of 2715 Å. Crosby et al. (39) calculated the persistence length of poly(p-phenylene benzbisthiazole) and characterized its conformation from measurements of the rotational relaxation time by PCS.

Zero and Pecora (40) similarly studied the properties of a long rigid-rod molecule, poly(γ-benzyl-L-glutamate), by measurements of translational and rotational diffusion coefficients by PCS. Kubota and Chu (41) analyzed PCS measurements of poly(hexyl isocyanate) on the basis of a semiflexible filament conformation.

Photon correlation spectroscopy is often used to measure changes in molecular shapes as reflected in changes in d_H. Lim et al. (42) studied a rod–coil conformational transition for 4-butoxycarbonylmethylurethane in chloroform–hexane using PCS. Ma and Wang (43) determined the conformational change in lysozyme when dimethyl sulfoxide (DMSO) is added to an aqueous solution of this protein. The DMSO molecules substitute for water molecules, and at concentrations of 60–70% DMSO, significant conformational changes are seen by PCS. Olson et al. (44) measured major changes in the conformation of transfer ribonucleic acid (tRNA) using PCS as the aqueous solution ionic strength is changed. Potkowski and Chu (45) studied tRNA–BSA association in buffered solutions using PCS and showed that both rate constants and aggregate size depend on the excess species.

Nishio et al. (46) examined a reversible coil–globule conformation change in polystyrene–cyclohexane. This change is the result of two forces acting on the polymer molecule; affinity among polymer units tends to compact the molecule while the thermal motion of individual units tends to expand it. At higher temperatures, the latter causes the molecule to be expanded, while the affinity of the segments at lower temperatures causes the molecule to collapse. However, translational diffusion takes place continuously at all temperatures so that PCS measurements will contain information about both translational and shape effects. The authors were able to separate the different contributions by first measuring the correlation function at low (forward) scattering angles where D_t dominates. This value was then used with data taken at higher angles to separate out internal mode correlations. With this procedure, the authors claim to show the coil–globule transition with temperature in this system. Park et al. (47) questioned the interpretation of a coil–globule transition made by Nishio et al. suggesting perhaps the latter's work was carried out in a metastable region.

6.5. Colloids, Microemulsions, and Micelles

Dispersions of small particles of one substance in a matrix of another substance are very common. These dispersions can be solid particles in a fluid (colloids), dispersions of small regions of one fluid in another nonmiscible fluid (microemulsions), or small regions of one fluid that have one or more boundary layers of other materials coating these regions dispersed in another fluid (micelles). If these mixtures were transparent to a light source and the particles differ in index of refraction from the matrix material, light will be scattered from the mixture, and the techniques of static and dynamic light scattering can be used to analyze these chemical–physical systems. In most cases, light scattering is the only method available to examine these systems because other techniques, such as electron microscopy, cannot be used for *in situ* measurements. Other techniques, like centrifugation and viscometry, impose shear forces which disturb the static distribution of the system.

Although both static and dynamic light scattering can be used to obtain information about particle dispersions, PCS is much faster and provides more size information than the former technique; many studies have appeared recently in the literature using PCS to analyze all three types of dispersions as the following examples show.

Colloidal suspensions are usually dispersions of insoluble particles in a carrier fluid. These dispersions may or may not show intraparticle force effects, and often in the dilute regime the hydrodynamic diameter derived from PCS measurements is independent of concentration because of the lack of particle–fluid interactions. Such behavior is shown in Figure 9, where d_H for organic dye molecules in isobutyl methyl ketone is plotted versus concentration.

Microemulsions, dispersions of one fluid in another, usually show concentration dependences because of the presence of interparticle forces. Oil–water microemulsions have been intensively studied because of the simplifying condition that electrical interactions are expected to be negligible in this system. Cazabat and Langevin (48) measured both the intensity and autocorrelation function of light scattered by oil-in-water emulsions. They were able to obtain information about droplet sizes and virial coefficients. In most cases, the droplets did not behave like hard spheres, showing that supplementary interactions were present. Gulari and Bedwell (49) similarly studied water–oil and oil–water microemulsions by determining hydrodynamic diameters and

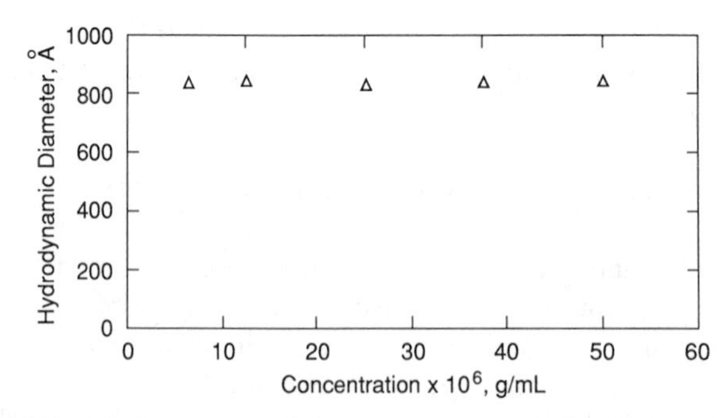

Figure 9. Hydrodynamic diameter d_H of an organic dye in isobutyl methyl ketone as a function of dye concentration.

polydispersities using PCS. They found the polydispersity and d_H increased with temperature and dispersed phase volume for water–oil with average droplet diameters ranging from 120 to 6000 Å. Some of the systems studied showed bimodal distribution functions. This finding was interpreted as showing the presence of surfactant micelles in equilibrium with the microemulsion droplets. Dorshow et al. (50) studied the static correlation length, osmotic compressibility, and diffusivity of an oil–water emulsion. The temperature dependence of these quantities near phase separation qualitatively resembled the critical behavior found for simple fluids and binary mixtures but with larger values of scaling exponents.

A number of micellar systems have been reported where mean size, shape, polydispersity, aggregation numbers, and change of phase were determined by static and dynamic light scattering as functions of temperature and constituent concentrations. Mazer et al. (51) studied sodium dodecyl sulfate in aqueous sodium chloride solutions. Schurtenberger et al. (52) measured bile salt solutions. Dorshow et al. (53) worked with cetyltrimethylammonium bromide micelles. Corti and Degiorgio (54) reported studies of water–*n*-dodecyl hexoxyethylene glycol monoether micelles.

6.6. Polyelectrolytes

The molecular conformation in solution of electrically charged polymer molecules is determined by the molecular charge distribution under the solution conditions. For example, the persistence length of a polymer is a measure of how far a polymer chain extends in a given direction and is a measure of the chain rigidity. The persistence length is made up of two contributions: a length given by the chain backbone rigidity and an electrostatic persistence length resulting from the repulsion between adjoining ionic sites. These sites are very dependent on the presence of compensating ionic species, such as added electrolyte, so that the charge repulsion, hence electrostatic persistence length, varies with electrolyte concentration. This behavior is exhibited by several aramid polymers, as discussed by Harwood and Fellers (55). The addition of an ionizing salt to the solution provides charge compensation, allowing the molecule to change its comformation. The effect can be seen with PCS as shown in Figure 10 where d_H is plotted versus the amount of added LiCl for poly(*m*-phenylene isophthalamide) in *N*,*N*-dimethylacetamide.

Nemoto et al. (56) made similar PCS studies of aqueous poly(L-lysine) hydrobromide solutions where sodium bromide was added. They found that the polymer chains assumed a considerably extended form at low concentrations of salt and polymer. Under some conditions internal cluster motions were observed, as shown by the presence of a slow correlation mode.

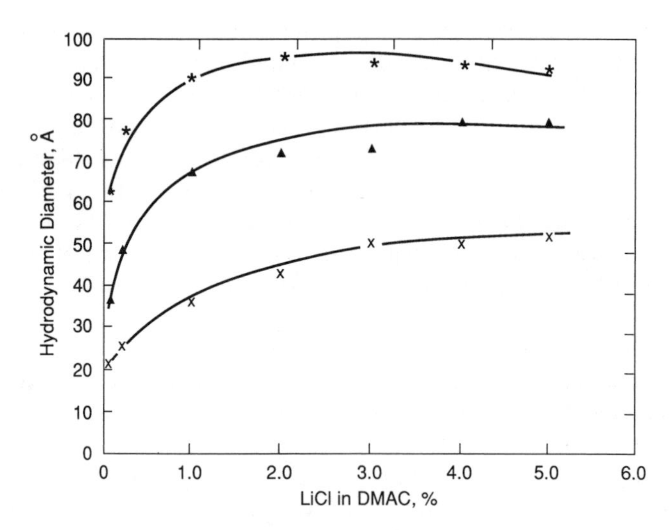

Figure 10. The influence of LiCl on the hydrodynamic diameter of poly(*m*-phenylene isophthalamide) in *N*,*N*-dimethylacetamide. Polymer concentration: ⋆ 2 mg/mL; ▲ 4 mg/mL; × 8 mg/mL.

Wunder et al. (57) examined a sodium salt of ethyl acrylate–acrylic acid copolymer using PCS. They concluded that for solutions containing 10^{-3} *M* or more of salt, D_t obtained by PCS could be analyzed according to hydrodynamic theory. Klooster et al. (58) in their study of partially neutralized poly(acrylic acid) in methanol, found that the behavior of this solution was different in the presence of Na^+ as compared with Li^+ counterions. The results were consistent with the occurrence of a conformational transition of the polymer at 20% neutralization by sodium methoxide.

Polyelectrolyte solutions can be studied by methods other than adding ionizing salts. Cotts and Berry (59) examined some rodlike polymers, like poly(1,4-phenylene benzobisoxazole), which are soluble only in strong protic acids. The solution properties were dependent on the ionic strength of the acid solvent. Koene and Mandel (60) observed aqueous solutions of sodium poly(styrene sulfonate), where by varying the polymer concentration they were able to affect the persistence length of the polymer molecules.

6.7. Copolymers and Multicomponent Systems

Solutions of copolymers, or two or more different polymers, present difficult analysis problems for light scattering because the different components each contribute to the scattered light. The amount of light scattered depends on the relative difference in index of refraction Δn between each solute and each solvent component. For a complete analysis, each Δn should be known, which may not be possible if each solute component is not soluble in each solvent component. To get around this problem, several experimenters have resorted to making some components of multicomponent solutions invisible by using solvents that exactly match the index of refraction of these components at a given temperature. Thus, in PCS measurements of such systems, the measured scattered light may be from only one part of a copolymer unit or one solute component, but the measured diffusion coefficient will reflect the whole particle motion. For example, Burchard et al. (61) studied a "star" molecule consisting of polyisoprene inner blocks and polystyrene outer blocks. They chose a solvent to match either the outer or the inner component index of refraction, thus making this component invisible by light scattering. In the former case, the results are as if the core had no outer components. In the latter case, the angular dependence of the first cumulant resembles that of a hollow sphere.

Several reports (62–64) used the same technique to study diffusion of a polymer test molecule within another polymer in solution. The intent in these works was to study polymer aggregation and entanglements.

Daivis et al. (65) used PCS to study ternary (polymer–polymer–solvent) solutions. Two distinct modes of decay are seen in autocorrelation functions of scattered light. They found that the diffusion coefficient of the fast mode corresponds to the mutual diffusion coefficient of the lower molecular weight polymer. The diffusion coefficient of the slow mode corresponds to the intradiffusion or self diffusion coefficient of the higher molecular weight polymer.

6.8. Miscellaneous Applications

Photon correlation spectroscopy can be used to study the nature of light scattering systems that show time dependent phenomena. For example, acidic solutions of tetraethoxysilane will gel after a period of time (66). The diffusion coefficients of the inorganic oxide species in solution can be measured by PCS. The diffusion coefficient decreases as the particles aggregate with time, as shown in Figure 11, and as the long-range structure of the gel begins to appear an additional correlation component is eventually seen, as shown in the figure. Similar time effects are found in polylactide–chloroform solutions shown in Figure 12, where the solute diffusion coefficient decreases with time, eventually disappearing as the solution gels. Keates and Hallett (67) used PCS to measure the time for reassemblage of microtubules after fragmentation by shearing.

Nystrom and Roots (68) made PCS measurements of polystyrene–toluene solutions at pressures up to 5000 atm for different concentrations. They found that at low concentrations both the diffusion coefficient and hydrodynamic

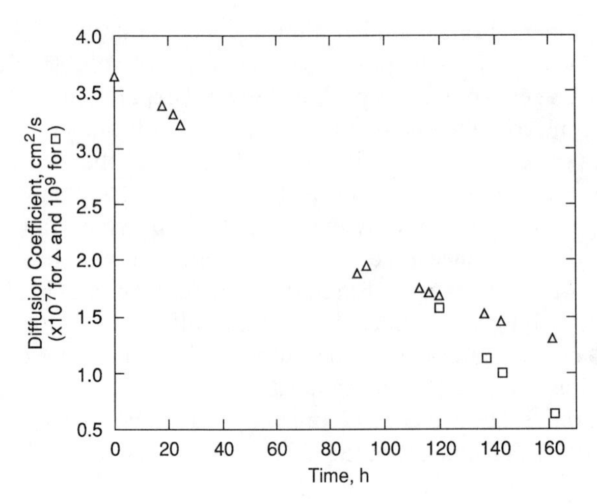

Figure 11. Diffusion coefficient versus time for two different tetraethoxysilanes.

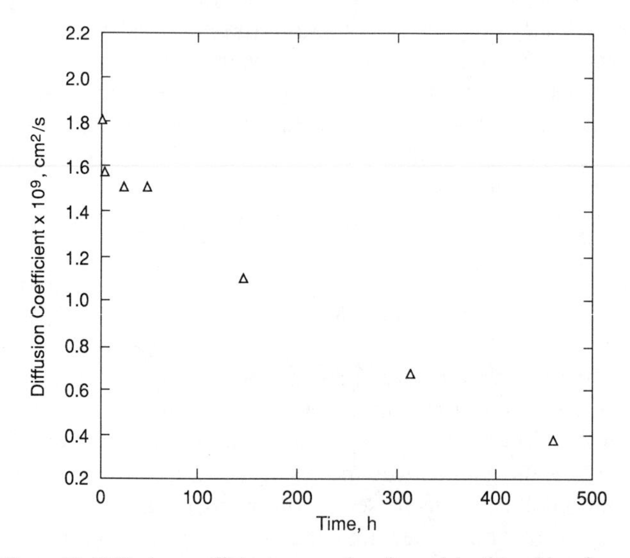

Figure 12. Diffusion coefficient versus time for polylactide–chloroform.

diameter were independent of pressure. At higher concentrations D_t was found to decrease. The decrease was related in all cases to the pressure dependence of the solvent viscosity. The same authors studied the effect of pressure on bovine serum albumin–water solutions (69). In this system the hydrodynamic diameter was seen to increase with pressure above 1000 atm caused by a denaturation of the protein. This effect was not reversible.

Since the Stokes–Einstein expression (Eq. 9) is a true relationship of D_t, d_H, and solvent η for hard, noninteracting spheres, the diffusion coefficient D_t can be measured by PCS for a known hydrodynamic diameter d_H of monodisperse latex spheres and this data used to calculate the viscosity of the suspending fluid. This technique is particularly advantageous for small volumes of fluids whose viscosity cannot be measured with usual techniques. Madonia et al. (70) used this procedure to measure fluid viscosity in macromolecular gelation processes involving the aggregation of sickle-cell hemoglobin. Measurements were made in a 1 × 1-mm cell where pH, temperature, and gelation time were varied.

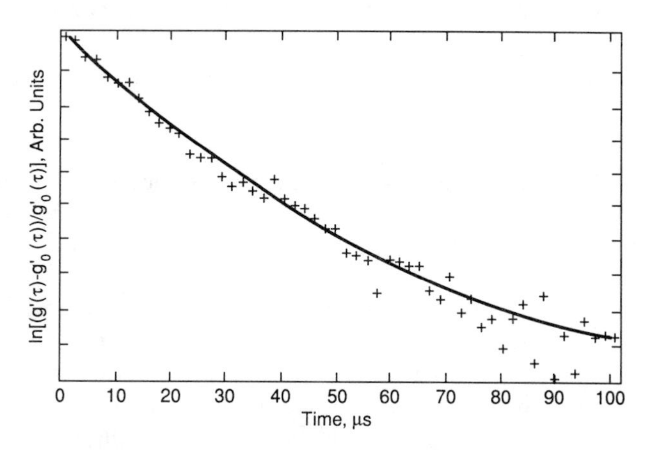

Figure 13. Correlation function of 20 ppm RuO_2 particles in water using a 500-mW light source.

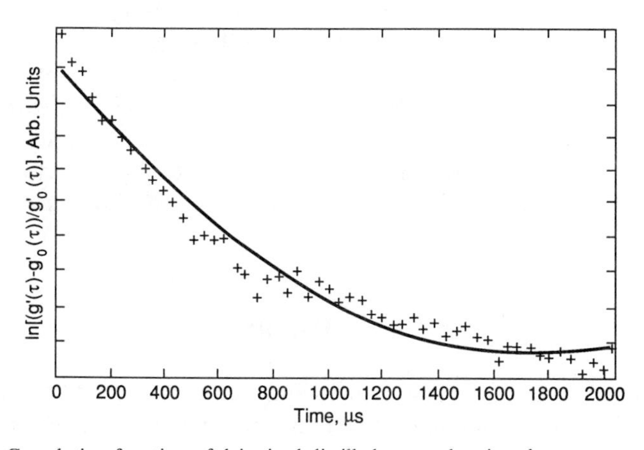

Figure 14. Correlation function of deionized distilled water showing the presence of particles.

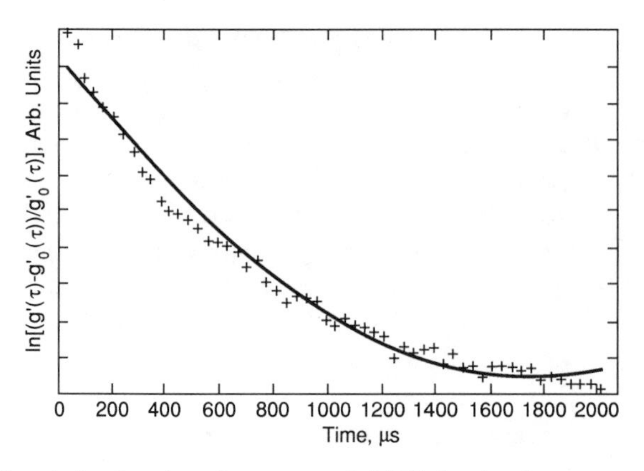

Figure 15. Correlation function of reagent grade THF showing the presence of particles.

Selser et al. (71) combined PCS with integrated optics to make possible the study of dynamic phenomena occurring at a solid–liquid interface. They were able to study the lateral diffusion of rhodopsin within one wavelength of a sample surface.

Photon correlation spectroscopy provides a rapid way of testing solvent cleanliness from particulate matter, since light scattering correlation can be seen for small concentrations of particles. Clean solutions will show only random scatter with no light correlation. For example, a measurable correlation shown in Figure 13 could be seen from a dispersion of RuO_2 particles in water at a concentration of 20 ppm using a laser light source of 500 mW; an apparent diameter of approximately 80 Å was obtained. Water is particularly prone to particulate contamination; a typical sample of deionized distilled water showed the correlation seen in Figure 14.

Although most reagent grade organic solvents are free of particulate matter, some materials, like THF, can undergo polymerization with aging or in the production process that create aggregates large enough to show correlation effects, as shown in Figure 15. The latter may interfere with the signal from a dissolved solute. Testing samples with PCS can often detect contaminated lots of this solvent.

7. SUMMARY

In this chapter some of the background and characteristics of PCS have been described. For dilute, nonlight absorbing solutions, PCS provides a convenient, rapid method of determining translational (and sometimes rotational) diffusion coefficients of particles from a few tens of angstroms to several micrometers in size under a variety of conditions. In many cases, the diffusion coefficient can be related to an effective particle size, so that the size of the light scattering particles can be related to its environment. Photon correlation spectroscopy is still a relatively new field, and there is room for progress both in experimental techniques and equipment and in theoretical interpretation of PCS data.

INSTRUMENT COMPANIES

- Brookhaven Instruments Corp., 750 Blue Point Road, Holtsville, New York, 11742
- Coulter Electronics, Inc., P. O. Box 2145, Hialeah, Florida, 33012
- Malvern Instruments, Inc., 10 Southville Rd., Southborough, Massachusetts, 01772
- Oros Instruments, Inc., 715 Banbury Ave., Slough SL1 4LJ, Berkshire, UK
- Particle Sizing System, 6780 Cortona Drive, Santa Barbara, California, 93617
- Otsuka Electronics Ltd., 2555 Midpoint Dr., Fort Collins, Colorado 80525
- Wyatt Technology Corporation, 820 East Haley Street, Santa Barbara, California, 93130

NOTATION

A_2	Second virial coefficient, mol cm^3/g^2
b	Exponent relating molecular weight to translational diffusion coefficient D_0 obtained from dynamic light scattering measurements
C	Total concentration, g/mL
D_s	Self-diffusion coefficient
D_t	Translational diffusion coefficient, cm^2/s
D_t^0	Diffusion coefficient at infinite dilution, cm^2/s
D_r	Rotational diffusion coefficient, cm^2/s
d_H	Hydrodynamic diameter, Å
F	Form factor
f	Molecular frictional coefficient
$g(\tau)$	Correlation function
$I(\omega)$	Scattered light power spectrum

$\mathbf{K}$	Magnitude of momentum transfer vector, $4\pi n_0 \sin(\theta/2)/\lambda_0$
k	Boltzmann constant
k_f	First-order functional coefficient
M	Molecular weight
N_A	Avogadro's number
n	Refractive index of solvent
P	Probability function
R	Gas constant, erg/°C/mol
R_H	Hydrodynamic radius, Å
$(\bar{r}^2)^{1/2}$	Root-mean-square end-to-end distance
s	Sedimentation constant
$S(\mathbf{K}, \omega)$	Spectral density
t	Time, s
T	Temperature, K; Averaging time
V	Molecular volume, mL
$\bar{v}$	Partial specific volume, mL/g
ε	Correlation length
Γ	Correlation function exponent
η	Solvent viscosity
λ	Wavelength of incident light
ϕ	Scattering angle
ρ	Density, g/mL
$\bar{\rho}$	Position vector, cm
τ	Delay time, s
θ	Scattering angle
ω	Frequency, s^{-1}

REFERENCES

1. P. P. Debye, "Analysis of G.R.S. Latex by Light-Scattering," Technical Report to the Rubber Reserve Compnay, Cornell University, Ithaca, NY, 1944.
2. P. W. Allen (Ed.), *Techniques of Polymer Characterization*, Butterworths, London, 1959.
3. B. J. Berne and R. Pecora, *Dynamic Light Scattering*, Wiley, New York, 1976.
4. B. Chu, *Laser Light Scattering*, Academic, New York, 1974.
5. S-H. Chen, B. Chu, and R. Nossal (Eds.), *Scattering Techniques Applied to Supramolecular and Nonequilibrium Systems*, Plenum, New York, 1981.
6. J. C. Earnshaw and M. W. Steer (Eds.), *The Application of Laser Light Scattering to the Study of Biological Motion*, Plenum, New York, 1983.
7. H. Z. Cummins and E. R. Pike (Eds.), *Photon Correlation Spectroscopy and Velocimetry*, Plenum, New York, 1977.

8. G. D. J. Phillies, "Applications of Quasi-elastic Light Scattering Spectroscopy in Polymer Characterization," *J. Appl. Polym. Sci., Appl. Polym. Symp.* **43**, 275 (1989).
9. C. Caroli, and O. Parodi, "Frequency Spectrum of the Depolarized Light Scattered by a Rigid Molecule in Solution," *J. Chem. Phys.*, **B2**, 1229 (1969).
10. D. E. Koppel, "Analysis of Macromolecular Polydispersity in Intensity Correlation Spectroscopy: The Method of Cumulants," *J. Chem. Phys.*, **57**, 4814 (1972).
11. G. D. J. Phillies, "Upon the Application of Cumulant Analysis to the Interpretation of Guasielastic Light Scattering Spectra," *J. Chem. Phys.*, **89**, 91 (1988).
12. E. Gulari, E. Gulari, and B. Chu, "A Second Comment on the Histogram Method in Photon Correlation Spectroscopy Applied in Dilute Polymer Solutions," *Polymer*, **23**, 649 (1982).
13. S. W. Provencher, "A Constrained Regularization Method for Inverting Data Represented by Linear Algebraic or Integral Equations," *Computer Phys. Commun.*, **27**, 213 (1982).
14. C. C. Han and F. L. McCracklin, "Molecular Weight and Polydispersity Measurements of Polystyrene by Quasielastic Light Scattering," *Polymer*, **20**, 427 (1979).
15. A. N. Lavery and J. C. Earnshaw, "Photon Correlation Spectroscopy of Particle Polydispersity: A Cubic B-Spline Analysis," *J. Chem. Phys.*, **80**, 5438 (1984).
16. C. Y. Cha and K. W. Min, "Photon Correlation Spectroscopy of Polymer Solutions," *J. Polym. Sci. Phys., Polym. Phys. Ed.*, **21**, 807 (1983).
17. B. Chu and E. Gulari, "Photon Correlation Measurements of Colloidal Size Distributions. II Details of Histogram Approach and Comparison of Methods of Data Analysis," *Phys. Scripta*, **19**, 476 (1979).
18. W. W. Yau and J. J. Kirkland, "Comparison of Sedimentation Field Flow Fractionation with Chromatographic Methods of Particlate and High Molecular Weight Macromolecular Characterizations," *J. Chromatogr.*, **218**, 217 (1981).
19. For example, T. A. King, A. Knox, and J. D. G. McAdam, "Translational and Rotational Diffusion of Tobacco Mosaic Virus from Polarized and Depolarized Light Scattering," *Biopolymers*, **12**, 1917 (1973).
20. G. D. J. Phillies, "Why Does the Generalized Stokes–Einstein Equation Work?," *Macromolecules*, **17**, 2050 (1984).
21. C. W. Pyun and M. Fixman, "Frictional Coefficient of Polymer Molecules in Solution," *J. Chem. Phys.*, **41**, 937 (1964).
22. S. Imai, "Concentration Dependency of the Sedimentation Constant," *J. Chem. Phys.*, **50**, 2116 (1969).
23. E. Gulari, E. Gulari, Y. Tsunashima, and B. Chu, "Polymer Diffusion in a Dilute Theta Solution. 1. Polystyrene in Cyclohexane," *Polymer*, **20**, 347 (1979).
24. P. G. de Gennes, "Dynamics of Entangled Polymer Solutions. I. The Rouse Model," *Macromolecules*, **9**, 587 (1976).
25. M. Eisele and W. Burchard, "Slow-mode Diffusion of Poly(vinyl pyrrolidone) in Semidilute Regime," *Macromolecules*, **17**, 1636 (1984).
26. D. Statman and B. Chu, "Dynamic Motions of Rodlike Polymers in Semi-

concentrated Solution: Poly(*n*-butyl isocyanate) in Carbon Tetrachloride," *Macromolecules*, **17**, 1537 (1984).

27. T. Chang and H. Yu, "Self-diffusion of Gelatin by Forced Rayleigh Scattering," *Macromolecules*, **17**, 115 (1984).
28. J. C. Selser, "A Photon Correlation Spectroscopy Study of the Dynamic Behavior of Poly(α-methylstyrene) in Moderately Concentrated Solution," *J. Chem. Phys.*, **79**, 1044 (1983).
29. G. Ronca, "Frequency Spectrum and Dynamic Correlations of Concentrated Polymer Liquids," *J. Chem. Phys.*, **79**, 1031 (1983).
30. G. D. Patterson, J. P. Jarry, and C. P. Lindsey, "Photon Correlation Spectroscopy of Polystyrene Solutions," *Macromolecules*, **13**, 228 (1980).
31. G. D. Patterson, "Photon Correlation Spectroscopy of Bulk Polymers," *Adv. Polym.*, **48**, 125 (1983).
32. G. D. Patterson and C. P. Lindsey, "Photon Correlation Spectroscopy Near the Glass Transition," *Macromolecules*, **14**, 83 (1981).
33. C. P. Lindsey, G. D. Patterson, and J. R. Steven, "Photon Correlation Spectroscopy of Polystyrene Near the Glass-Rubber Relaxation," *J. Polym. Sci., Polym. Phys. Ed.*, **17**, 1547 (1979).
34. G. D. Patterson, P. J. Carroll, and J. R. Stevens, "Photon Correlation Spectroscopy of Poly(methyl methacrylate) Near the Glass Transition," *J. Polym. Sci. Polym. Ed.*, **21**, 613 (1983).
35. G. Fytas, A. Patkowski, G. Meier, and T. Dorfmuller, "Separation of Two Relaxation Processes in Bulk Polymers Using Photon Correlation Spectroscopy at High Pressures," *Macromolecules*, **15**, 214 (1982).
36. G. Fytas, A. Patkowski, G. Meier, and T. Dorfmuller, "A High Pressure Photon Correlation Study of Bulk Poly(methyl acrylate). Comparison with Relaxation Processes in Poly(ethyl acrylate) and Related polymethacrylates," *J. Chem. Phys.*, **80**, 2214 (1284).
37. C. H. Wang, G. Fytas, D. Lilge, and T. Dorfmuller, "Laser Light Beating Spectroscopic Studies of Dynamics in Bulk Polymers: Poly(propylene glycol)," *Macromolecules*, **14**, 1363 (1982).
38. J. S. Hwang and H. Z. Cummings, "Dynamic Light Scattering Studies of Collagen," *J. Chem. Phys.*, **77**, 616 (1982).
39. C. R. Crosby, III, N. C. Ford, Jr., F. E. Karasz, and K. H. Langley, "Depolarized Dynamic Light Scattering of a Rigid Macromolecule Poly(p-phenylene benzbisthiazole)," *J. Chem. Phys.*, **75**, 4298 (1981).
40. K. M. Zero and R. Pecora, "Rotational and Translational Diffusion in Semidilute Solutions of Rigid-rod Macromolecules," *Macromolecules*, **15**, 87 (1982).
41. K. Kubota and B. Chu, "Quasi-elastic Light Scattering of Poly(hexyl isocyanate) in Hexane," *Macromolecules*, **16**, 105 (1983).
42. K. C. Lim, C. R. Fincher, Jr., and A. J. Heeger, "Rod-to-coil Transition of a Conjugated Polymer in Solution," *Phys. Rev. Lett.*, **50**, 1934 (1983).
43. R. J. Ma and C. H. Wang, "Studies of the Protein–Protein Interaction of

Lysozyme in Dimethyl Sulfoxide-Water Solutions by Quasielastic Light Scattering," *J. Phys. Chem.*, **87**, 679 (1983).

44. T. Olson, M. J. Fournier, K. H. Langley, and N. C. Ford, Jr., "Detection of a Major Conformational Change in Transfer Ribonucleic Acid by Laser Light Scattering," *J. Mol. Biol.*, **102**, 193 (1976).
45. A. Patkowski and B. Chu, "Studies of Protein-Nucleic Acid Interactions by Photon Correlation Spectroscopy," *J. Chem. Phys.*, **73**, 3082 (1980).
46. I. Nishio, G. Swislow, S-T Sun, and T. Tanaka, "Critical Density Fluctuations within a Single Polymer Chain," *Nature* (*London*), **300**, 243 (1982).
47. I. H. Park, Q.-W. Wang, and B. Chu, "Transition of Linear Polymer Dimensions from θ to Collapsed Regime. 1. Polystyrene/Cyclohexane System," *Macromolecules*, **20**, 1965 (1987).
48. A. M. Cazabat and D. Langevin, "Diffusion of Interacting Particles: Light Scattering Study of Microemulsions," *J. Chem. Phys.*, **74**, 3148 (1981).
49. E. Gulari and B. Bedwell, "Quasi-elastic Light Scattering Investigation of Microemulsions," *J. Colloid Interface Sci.*, **77**, 202 (1980).
50. R. Dorshow, F. de Buzzaccarini, C. A. Bunton, and D. F. Nicoli, "Critical-like Behavior Observed for a Five-component Microemulsion," *Phys. Rev. Lett.*, **47**, 1336 (1981).
51. N. A. Mazer, G. B. Benedek, and M. C. Carey, "An Investigation of the Micellar Phase of Sodium Dodecyl Sulfate in Aqueous Sodium Chloride Solutions using Quasielastic Light Scattering," *J. Chem. Phys.*, **80**, 1075 (1976).
52. P. Schurtenberger, N. Mazer, and W. Kanzig, "Static and Dynamic Light Scattering Studies of Micellar Growth and Interactions in Bile Salt Solutions," *J. Phys. Chem.*, **87**, 308 (1983).
53. R. Dorshow, J. Briggs, C. A. Bunton, and D. F. Nicoli, "Dynamic Light Scattering from Cetyltrimethylammonium Bromide Micelles. Intermicellar Interactions at Low Ionic Strengths," *J. Phys. Chem.*, **86**, 2388 (1982).
54. M. Corti and V. Degiorgio, "Critical Behavior of a Micellar Solution," *Phys. Rev. Lett.*, **45**, 1045 (1980).
55. D. D. Harwood and J. F. Fellers, "Imposed Polyelectrolyte Behavior of Poly(*m*-phenylene isophthalamide) in Lithium Chloride/Dimethylacetamide," *Macromolecules*, **12**, 693 (1979).
56. N. Nemoto, H. Matsuda, Y. Tsunashima, and M. Kurata, "Dynamic Light Scattering of Poly(L-lysine) Hydrobromide in Aqueous Solutions," *Macromolecules*, **17**, 1731 (1984).
57. S. Wunder, N. C. Ford, Jr., F. E. Karasz, and J. Tan, "Quasi-elastic Light Scattering from a Sodium Salt of Ethyl Acrylate–Acrylic Acid Copolymer," *J. Colloid Interface Sci.*, **63**, 290 (1978).
58. N. th. M. Klooster, F. van der Touw, and M. Mandel, "Solvent Effects in Polyelectrolyte Solutions. 2. Osmotic, Elastic Light Scattering, and Conductometric Measurements on (partially) Neutralized Poly(acrylic acid) in Methanol," *Macromolecules*, **17**, 2078 (1984).

59. P. M. Cotts and G. C. Berry, "Studies on Dilute Solutions of Rodlike Macroions. II. Electrostatic Effects," *J. Polym. Sci., Phys. Ed.*, **21**, 1255 (1983).

60. R. S. Koene and M. Mandel, M., "Quasi-elastic Light Scattering by Polyelectrolyte Solutions without Added Salt," *Macromolecules*, **16**, 973 (1983).

61. W. Burchard, K. Kajiwara, D. Nerger, and W. H. Stockmayer, "Regular Block-Copolymeric Star Molecules. Static Structure Factor and First Cumulant of the Dynamic Structure Factor," *Macromolecules*, **17**, 222 (1984).

62. B. Hanley, L. Wheeler, M. Tirrell, T. Lodge, C. C. Han, and B. Bauer, "Dynamic Light Scattering for Polystyrene in Isorefractive Poly(vinyl methyl ether) and *o*-Fluorotoluene Solvents," *Bull. Am. Phys. Soc.*, **30**, 247 (1985).

63. T. P. Lodge and L. Wheeler, "Diffusion of Branched Polymers in Concentrated Ternary Solutions by Dynamic Light Scattering," *Bull. Am. Phys. Soc.*, **30**, 247 (1985).

64. B. Hanley, M. Ramstad, and M. Tirrell, "Dynamic Light Scattering Studies on Poly(styrene-*co*-acrylonitrile) in Isorefractive Poly(methyl methacrylate) Solvents," *Bull. Am. Phys. Soc.*, **30**, 248 (1985).

65. P. Daivis, I. Snook, W. van Megen, B. N. Preston, and W. D. Comper, "Dynamic Light Scattering Measurements of Diffusion in Polymer–Polymer–Solvent Systems," *Macromolecules*, **17**, 2376 (1984).

66. G. W. Scherer, "Structural Evolution of Sol-Gel Glasses," *Yogyo-Kyokaishi*, **95**, 31 (1987).

67. R. A. B. Keates and R. Hallett, "Dynamic Instability of Sheared Microtubules Observed by Quasi-Elastic Light Scattering," *Science*, **241**, 1642 (1988).

68. J. Roots and B. Nystrom, "Photon Correlation Spectroscopy on Polystyrene Solutions under High Pressure," *Macromolecules*, **15**, 553 (1982).

69. B. Nystrom and J. Roots, "Dynamic Light Scattering Studies of Protein Solutions under High Pressure," *J. Chem. Phys.*, **78**, 2833 (1983).

70. F. Madonia, P. L. San Biagio, M. U. Palma, G. Schiliro, S. Musumeci, and G. Russo, "Photon Scattering as a Probe of Microviscosity and Channel Size in Gels such as Sickle Hemoglobin," *Nature* (*London*), **302**, 412 (1983).

71. J. C. Selser, K. J. Rothschild, J. D. Swalen, and F. Rondelez, "Study of Multilamellar Films of Photoreceptor Membrane by Photon-Correlation Spectroscopy Combined with Integrated Optics," *Phys. Rev. Lett.*, **48**, 1690 (1982).

CHAPTER

11

NMR CHARACTERIZATION OF POLYMERS

H. N. CHENG

Hercules Incorporated
Research Center
Wilmington, Delaware

Modern Methods of Polymer Characterization, Edited by Howard G. Barth and Jimmy W. Mays
ISBN 0-471-82814-9

1. INTRODUCTION

Nuclear magnetic resonance (NMR) spectroscopy is a well-established technique for the characterization of polymer structure. Beginning with ^{1}H NMR in the 1960s, numerous polymer systems have been studied and the technique has been firmly established (1). In the 1970s renewed interest arose as a result of the application of ^{13}C NMR (coupled with Fourier transform technology), which provided much more detailed and diagnostic information for homopolymers and copolymers (2–10).

Recent NMR studies of polymers have taken several directions. In conventional solution ^{1}H and ^{13}C NMR, more emphasis has been placed on in-depth studies involving polymerization mechanisms, copolymerization statistics, prediction of ^{13}C shifts, and applications software. Furthermore, advanced NMR techniques [e.g., two-dimensional (2D) NMR and spectral editing features] have been used to solve problems that were previously inaccessible. Yet, another area of investigation is multinuclear NMR by which ^{29}Si, ^{15}N, ^{19}F, ^{31}P, and ^{2}H have been successfully studied for numerous polymer systems.

In the last several years there has been a resurgence of interest in solid-state NMR, primarily as a result of recent theoretical and experimental developments that enable high-resolution liquidlike spectra to be obtained in the solid state. The combined techniques of cross-polarization and magic angle spinning are now standard practice, and the field is still growing with new techniques and approaches being reported.

This chapter provides a general review of the application of NMR to polymer systems. In view of the prolific literature on the subject, the examples chosen for this review are illustrative and not exhaustive, and frequently reflect the author's own research interests. Particular emphasis has been placed on the use of NMR as a problem-solving technique for the characterization of polymeric systems. Except for Section 9, this review will deal primarily with high-resolution solution NMR.

2. BASIC PRINCIPLES OF NMR INTERPRETATION

2.1. ^{1}H NMR

The ^{1}H NMR spectrum consists of a set of resonances (or spectral lines) corresponding to the different types of hydrogen atoms in the sample. There are three basic measurements that can be obtained from a set of resonances: (1) the *area* under the resonance, which is proportional to the amount of species present in the sample; (2) the position of the resonance or *chemical shift*, which is indicative of the identity of the species; and (3) the linewidth of the resonance ($1/\pi T_2^*$), which is related to the molecular environment of the particular ^{1}H (T_2^* being the spin–spin relaxation time). If scalar couplings are present, information related to the presence and the absence of neighboring ^{1}H nuclei can be determined.

The fact that the resonance area is proportional to the concentration of the species is the basis of quantitative NMR. By taking the ratios of different resonances corresponding to different species, the composition of multicomponent systems (e.g., copolymers) can be obtained. Nuclear magnetic resonance is a primary quantitative method that requires no calibration.

The spectral linewidth is a major parameter in solid-state NMR (see below), but it is less useful for solution NMR. In contrast, scalar coupling (spin–spin splitting) is very useful for structure determination and conformational analysis. Scalar coupling depends on the number of neighboring nuclei. In general, the number of lines obtained are given by $(2n_x I_x + 1)$ where n_x is the number of equivalent nuclei and I_x is the spin of nucleus x. For ^{1}H and ^{13}C ($I = \frac{1}{2}$), this expression reduces to $n + 1$. For example, in poly(diethylene glycol adipate) the methylenes (H_a, H_b) are nonequivalent and split each other.

$$\left[O-\underset{b}{CH_2}-\underset{a}{CH_2}-O-\underset{a}{CH_2}-\underset{b}{CH_2}-O-\overset{O}{\overset{\|}{C}}-(CH_2)_4\overset{O}{\overset{\|}{C}} \right]$$

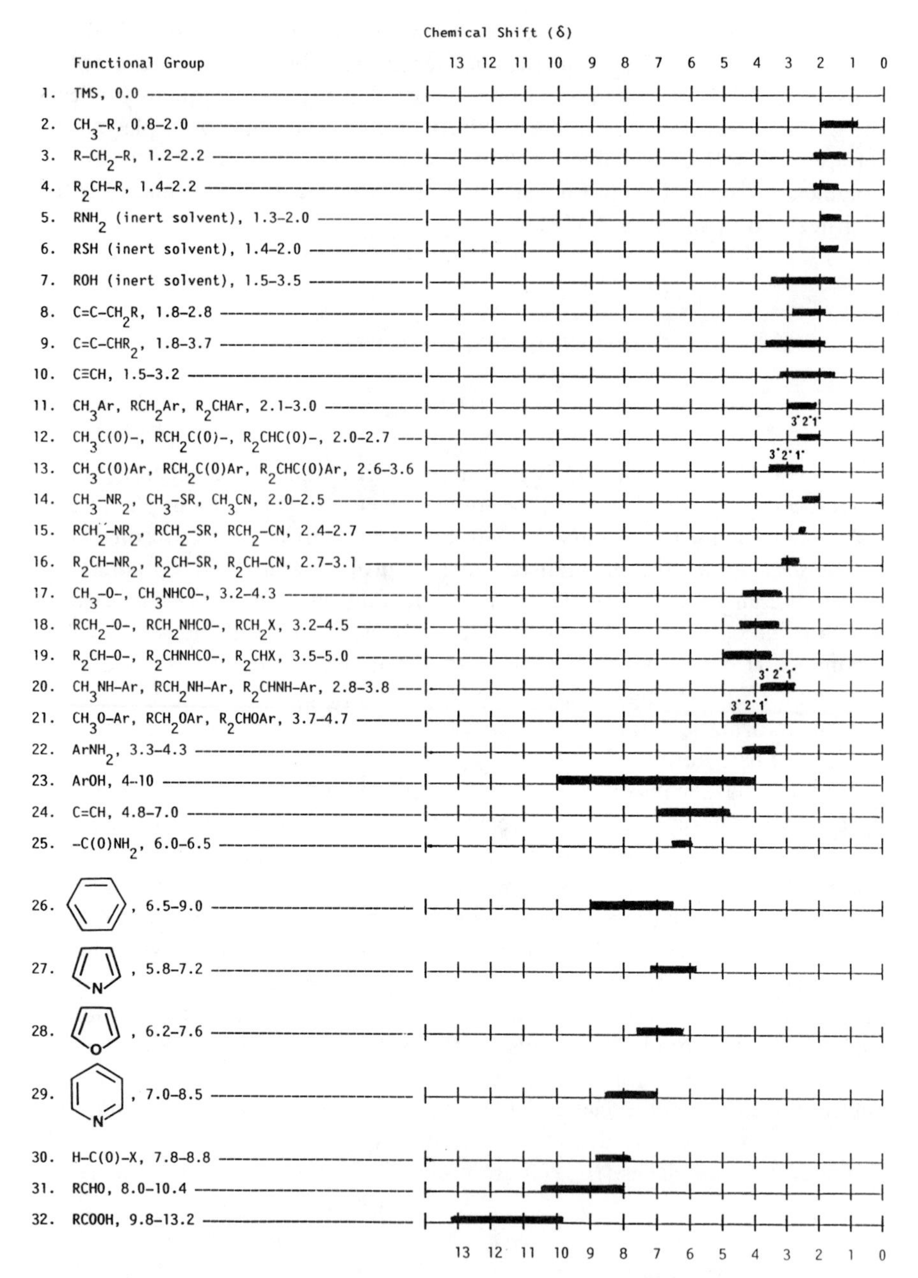

Figure 1. ¹H chemical shift positions of major functional groups pertaining to polymers; R = alkyl, Ar = aryl, X = Cl or Br, 1°, 2°, 3° = primary, secondary, and tertiary alkyl, respectively.

H_b has two neighboring H_a ($n = 2$); thus, it appears as a triplet (i.e., three lines). Similarly H_a has two neighboring H_b and also exists as a triplet.

The chemical shift δ is the major determinant in spectral interpretation. The ^{1}H NMR spectrum has a chemical shift range of 12 ppm. The chemical shift is diagnostic of different functional groups. An abbreviated shift scale is given in Figure 1.

As an example, Figure 2 shows the ^{1}H NMR spectrum of poly(ethylene glycol) adipate (11). Because both mono- and diethylene glycol are present in a ratio of 2.2:1.0, the assignments are given as follows:

$$\text{-}\!\!(\underset{4.27}{\text{OCH}_2\text{CH}_2}\text{O}\overset{\text{O}}{\overset{\|}{\text{C}}}\underset{2.38}{\text{CH}_2}\underset{1.68}{(\text{CH}_2)_2}\text{CH}_2\overset{\text{O}}{\overset{\|}{\text{C}}})\text{—}$$

$$(\underset{3.78}{\text{OCH}_2}\underset{4.20}{\text{CH}_2\text{O}}\underset{4.20}{\text{CH}_2}\underset{3.78}{\text{CH}_2}\text{O}\overset{\text{O}}{\overset{\|}{\text{C}}}\underset{2.38}{\text{CH}_2}\underset{1.68}{(\text{CH}_2)_2}\text{CH}_2\overset{\text{O}}{\overset{\|}{\text{C}}}\!\!)\text{-}$$

The different chemical shifts observed permit a very precise quantitative determination of the mono- and diethylene glycol content.

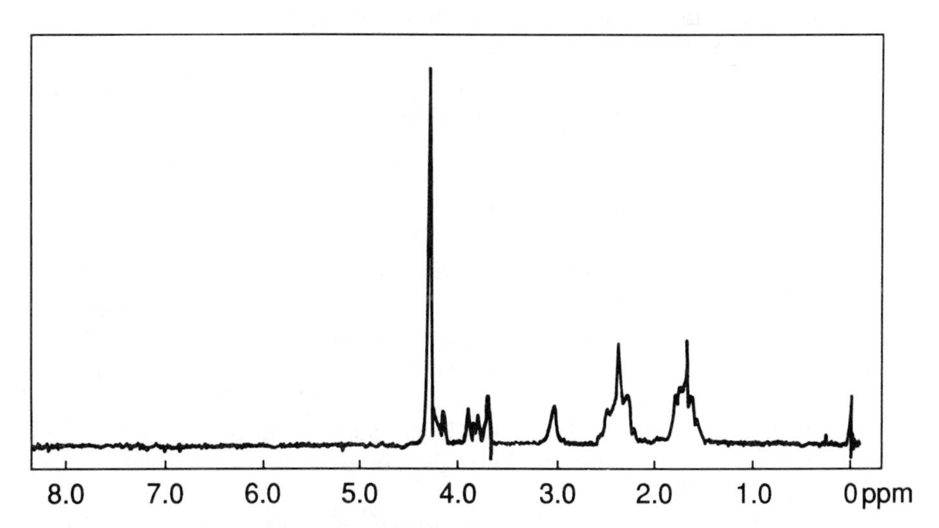

Figure 2. The ^{1}H NMR spectrum (60 MHz) of poly(ethylene glycol) adipate. Permission for publication herein of Sadtler Standard Spectra® has been granted, and the rights are reserved by Sadtler Research Laboratories, Division of Bio-Rad Laboratories, Inc.

2.2. ^{13}C NMR

Several features of the ^{13}C nucleus make it attractive for polymer analysis. First, ^{13}C is a rare nucleus (only 1.1% natural abundance); as a result, ^{13}C—^{13}C couplings do not usually show up in the spectrum. Through spin decoupling, the ^{13}C—^{1}H couplings can also be eliminated. In this way each resonance exists as a single line (Fig. 3). This simplifies the spectrum and makes assignments easier. Furthermore, the chemical shift range of ^{13}C is over 200 ppm. Thus, minor changes in structure cause shifts of carbon atoms that are four or five bonds away.

Previously, a major handicap in the use of ^{13}C was its low receptivity (1.76×10^{-4} that of ^{1}H). With the use of Fourier transform (FT) NMR and dedicated minicomputers, this difficulty has been largely overcome. Quantitation is also possible (see Section 4.1).

The interpretation of ^{13}C spectra is time consuming and requires a fair amount of experience. One common method involves looking up the chemical shifts in spectral libraries using either the compound in question or, if not available, compounds with similar structures. For this purpose, the spectral collections of Sadtler (12), Bremser et al. (13), Breitmaier et al. (14), and Stothers (15), among others, are very useful. In the last few years, many computer-assisted structure-determination methods have been developed (16, 17). The CNMR program (18) of Chemical Information Systems and the INKA program (19) have been generally accepted. Several other groups are also very active in advancing this important area (20–26).

Another commonly used method involves using empirical shift rules that were pioneered by Grant and Paul (27), who discovered that the ^{13}C shifts of alkanes can be approximated by the linear combination of additive terms related to the neighboring carbon atoms

$$\delta_{\text{obsd}} = -2.3 + n_\alpha S_\alpha + n_\beta S_\beta + n_\gamma S_\gamma + n_\delta S_\delta + n_\varepsilon S_\varepsilon + S_c \qquad (1)$$

where n_i refers to the number of i neighboring carbon atoms ($i = \alpha, \beta, \gamma, \delta$, and ε), S_i is a constant characteristic of the i carbon, and S_c represents steric corrective terms to be used for contiguous secondary, tertiary, and quaternary carbon atoms. Similar additive rules have also been reported for alcohols (28, 29), amines (30, 31), nitroalkanes (32), and others (33–35). Wehrli et al. (34), Pretsch and co-workers (35), and Cheng and Ellingsen (33) proposed additive rules for many functional groups. Computer approaches have been devised (33, 36, 37).

A recently updated set of additive parameters (33) is shown in Table 1. The term $n_\varepsilon S_\varepsilon$ is dropped because it is very small. The steric correction parameters S_c are given in Table 2.

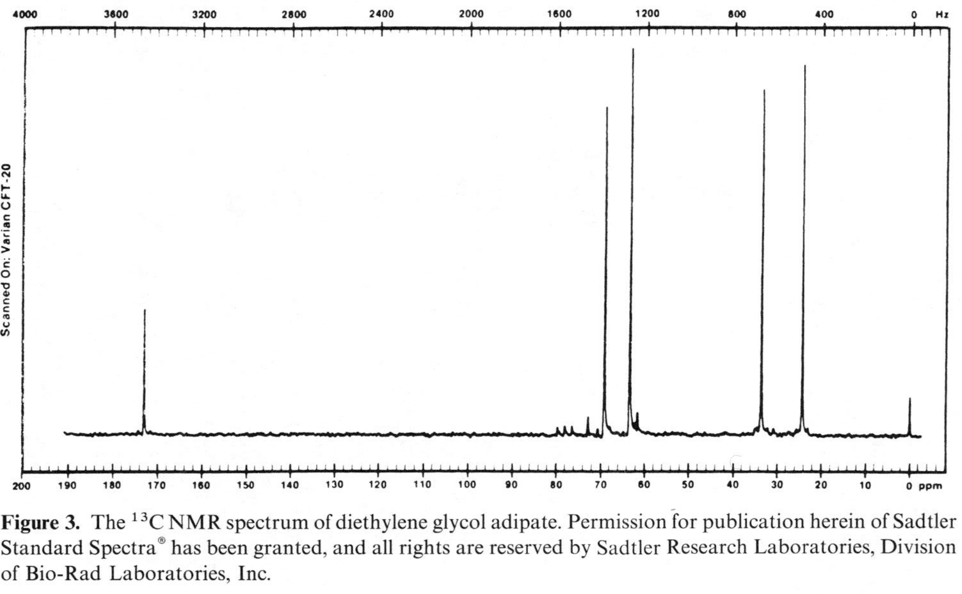

Figure 3. The ^{13}C NMR spectrum of diethylene glycol adipate. Permission for publication herein of Sadtler Standard Spectra® has been granted, and all rights are reserved by Sadtler Research Laboratories, Division of Bio-Rad Laboratories, Inc.

Table 1. Substituent Chemical Shift Parameters (in ppm) for ^{13}C NMR[a]

Substituent	Group	Code	α^b	β	γ	δ
Paraffin[c]	—C$\lessgtr$	C	9.1	9.4	−2.5	0.3
Ether, alcohol[c]	—O—	O	49.0	10.1	−6.0	0.3
Amine[c]	—N$\lessgtr$	N	28.3	11.3	−5.1	0.3
Ammonium[c]	—N—	N+	30.7	5.4	−7.2	−1.4
Thioether, Thiol[c]	—S—	S	11.0	12.0	−3.0	−0.5
Phenyl	—C_6H_5	PH	22.1	9.3	−2.6	0.3
Fluoro	—F	F	70.1(1,2) 69.0(3) 66.0(4)	7.8	−6.8	0.0
Chloro	—Cl	CL	31.1(1,2) 35.0(3) 43.0(4)	10.0	−5.1	−0.5
Bromo	—Br	BR	18.9(1,2) 27.9(3) 36.9(4)	11.0	−3.8	−0.7
Iodo	—I	I	−7.2(1,2) 3.8(3) 20.8(4)	10.9	−1.5	−0.9
Ammonium	—NH_3^+	NH3+	26.0	7.5	−4.6	−0.1
Nitrile	—CN	CN	3.1	2.4	−3.3	−0.5
Nitrate	—NO_2	NO2	62.0	4.4	−4.0	0
Peroxy, hydroperoxy	—OO—	OO	55.0	2.7	−4.0	0
Oxime, syn	—C=NOH	CNOS	11.7	0.6	−1.8	0.0
Oxime, anti	—C=NOH	CNOA	16.1	4.3	−1.5	0.0
Thiocyanate	—SCN	SCN	21.0	7.2	−4.0	0.3
Sulfoxide	—S(O)—	SO	31.1	9.0	−3.5	0.0
Sulfonate	—SO_3H	SO3H	38.9	0.5	−3.7	0.2
Aldehyde	—CHO	CHO	29.9	−0.6	−2.7	0.0
Ketone	—C(O)—	CO	22.5	3.0	−3.0	0.0
Acid	—COOH	COOH	20.1	2.0	−2.8	0.0
Carboxylate	—COO^-	COO−	24.5	3.5	−2.5	0.0
Acyl chloride	—COCl	COCL	33.1	2.3	−3.6	0.0
Ester	—C(O)O—'	COO	22.6	2.0	−2.8	0.0
Ester	—OC(O)—	OCO	54.5(1,2,3) 62.5(4)	6.5	−6.0	0.0
Amide	—CONH—	CON	22.0	2.6	−3.2	−0.4
Amide	—NHCO—	NCO	28.0	6.8	−5.1	0.0
Olefin[c]	C=C	C=C	21.5	6.9	−2.1	0.4
Acetylene[c]	C≡C	C3C	4.4	5.6	−3.4	−0.6

[a] $k_0 = -2.3$ for TMS reference.

[b] Number(s) in parentheses denotes the number of nonhydrogen substituents on the carbon in question.

[c] Steric correction parameters (Table 2) apply to these substituents.

Table 2. Steric Correction Parameters, $S_c(i, j)$[a]

i	$j=1$	$j=2$	$j=3$	$j=4$
Primary	0.0	0.0	−1.1	−3.4
Secondary	0.0	0.0	−2.5	−7.5
Tertiary	0.0	−3.7	−9.5	−15
Quarternary	−1.5	−8.4	−15	−25

[a]Designation i = the carbon in question and j = number of nonhydrogen substituents directly attached to the α substituent (applicable only to α substituents marked with footnote c in Table 1).

An example of the calculations needed is provided below:

$$CH_3\underset{}{\overset{\displaystyle OH}{\overset{|}{C}}}HCH_3$$

CH$_3$			CH
Base	−2.3	Base	−2.3
α-C	9.1	2α-C	18.2
β-C	9.4	α-OH	49.0
β-OH	10.1	$S(t, 1)$	0.0
$S(p, 3)$	−1.1		
Calcd	25.1	Calcd	64.9
Obsd	25.4	Obsd	63.7

A computer program (called CSHIFT) has been written (33) that provides convenient input of structure and rapid output of the predicted chemical shifts. The program has been adapted to several minicomputers and personal computers. A more refined program is also available (37).

In the case of poly(diethylene glycol) adipate spectrum shown in Fig. 3, these assignments can be readily made:

$$(\text{—O—}\underset{63}{CH_2}\text{—}\underset{69}{CH_2}\text{—O—}\underset{69}{CH_2}\text{—}\underset{63}{CH_2}\text{—O—}\underset{173}{\overset{\displaystyle O}{\overset{\|}{C}}}\text{—}\underset{34}{CH_2}\text{—}\underset{24}{CH_2}\text{—}CH_2\text{—}CH_2\text{—}\overset{\displaystyle O}{\overset{\|}{C}}\text{—})$$

3. THEORY AND PRACTICE OF POLYMER NMR

3.1. Copolymer Sequence and Reaction Probability Models

One of the unique features of NMR as applied to polymers is its ability to determine polymer microstructure and thereby polymerization statistics (1–9). In a copolymer, for example, the consecutive placement of the comonomers (say, A and B) upon polymerization can generate different comonomer sequence distributions. Three common types of copolymers are

Random copolymer	AABAABBAB
Block copolymer	AAAAABBBB
Alternating copolymer	ABABABABA

In the NMR experiment, the ^{1}H or ^{13}C nuclei in either comonomer A or B can "sense" that they have different neighbors. For vinyl comonomers, the different combinations of comonomers, taken two (diad) or three (triad) at a time, are shown below.

Diads (XX, XY, YY)

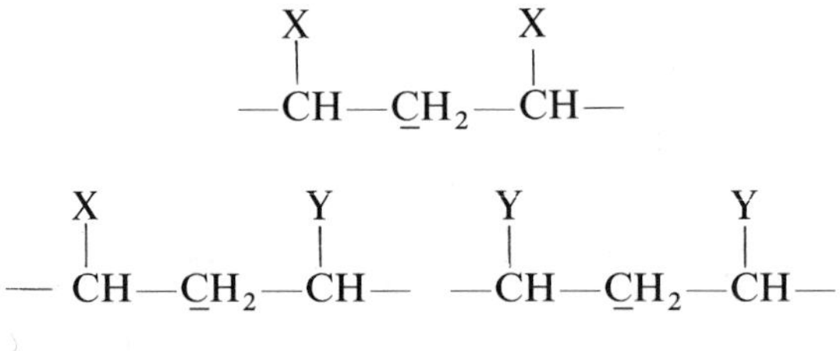

$$-\underset{\displaystyle |}{\overset{\displaystyle X}{CH}}-\underline{C}H_2-\underset{}{\overset{\displaystyle X}{CH}}-$$

$$-\overset{\displaystyle X}{CH}-\underline{C}H_2-\overset{\displaystyle Y}{CH}- \qquad -\overset{\displaystyle Y}{CH}-\underline{C}H_2-\overset{\displaystyle Y}{CH}-$$

Triads

$$\text{XXX:}\quad -\overset{\displaystyle X}{CH}-CH_2-\overset{\displaystyle X}{\underline{C}H}-CH_2-\overset{\displaystyle X}{CH}-$$

$$\text{XXY:}\quad -\overset{\displaystyle X}{CH}-CH_2-\overset{\displaystyle X}{\underline{C}H}-CH_2-\overset{\displaystyle Y}{CH}-$$

$$\text{YXY:}\quad -\overset{\displaystyle Y}{CH}-CH_2-\overset{\displaystyle X}{\underline{C}H}-CH_2-\overset{\displaystyle Y}{CH}-$$

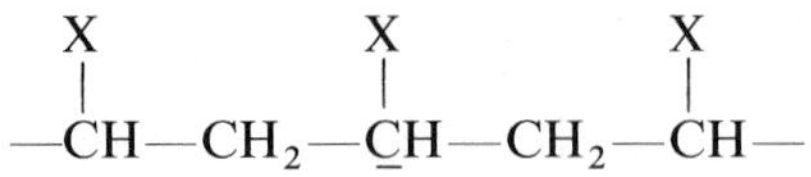

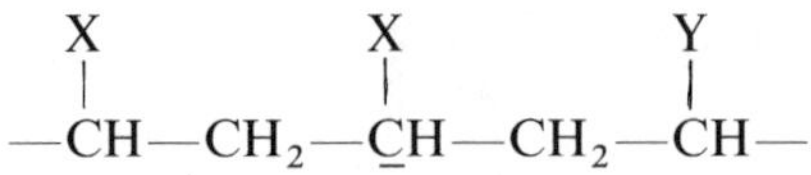

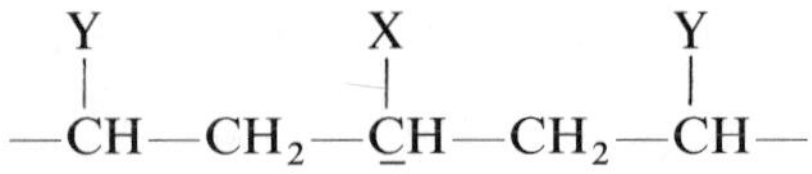

$$\text{YYY:}\quad \begin{array}{c}\mathrm{Y}\\|\\\mathrm{—CH}\end{array}\mathrm{—CH_2—}\begin{array}{c}\mathrm{Y}\\|\\\underline{\mathrm{C}}\mathrm{H}\end{array}\mathrm{—CH_2—}\begin{array}{c}\mathrm{Y}\\|\\\mathrm{CH—}\end{array}$$

$$\text{YYX:}\quad \begin{array}{c}\mathrm{Y}\\|\\\mathrm{—CH}\end{array}\mathrm{—CH_2—}\begin{array}{c}\mathrm{Y}\\|\\\underline{\mathrm{C}}\mathrm{H}\end{array}\mathrm{—CH_2—}\begin{array}{c}\mathrm{X}\\|\\\mathrm{CH—}\end{array}$$

$$\text{XYX:}\quad \begin{array}{c}\mathrm{X}\\|\\\mathrm{—CH}\end{array}\mathrm{—CH_2—}\begin{array}{c}\mathrm{Y}\\|\\\underline{\mathrm{C}}\mathrm{H}\end{array}\mathrm{—CH_2—}\begin{array}{c}\mathrm{X}\\|\\\mathrm{CH—}\end{array}$$

Thus, the methylene is sensitive as to whether the neighbors are X or Y substituents. In general, the methylenes with XX, XY, or YY neighbors (diads) will have different ^{1}H- or ^{13}C-chemical shifts in the NMR spectra. Similarly, the methines can be distinguished by their neighbors (triads), resulting in six resonances (or clusters of resonances) in the ^{13}C or ^{1}H NMR spectra.

In most copolymers even longer range sequence effects can be seen in the ^{13}C (and high-field ^{1}H) chemical shifts. Thus, for example, the methylenes may be sensitive not only to the nearest neighbors (diads), but to the next nearest neighbors as well. In other words, the XXXX, XXXY, and YXXY tetrads (below) may be wholly or partly resolved (i.e., having distinct resonances) in the NMR spectrum.

$$\begin{array}{c}\mathrm{X}\quad\;\;\mathrm{X}\\|\quad\;\;\;|\\\mathrm{CH\underline{C}H_2CH—}\\\text{— (XX diad)}\end{array}\longrightarrow\begin{array}{c}\mathrm{X}\qquad\mathrm{X}\qquad\mathrm{X}\qquad\mathrm{X}\\|\qquad\;|\qquad\;|\qquad\;|\\\mathrm{—CHCH_2CH\underline{C}H_2CHCH_2CH—}\\\text{(XXXX tetrad)}\end{array}$$

$$\begin{array}{c}\mathrm{X}\qquad\mathrm{X}\qquad\mathrm{X}\qquad\mathrm{Y}\\|\qquad\;|\qquad\;|\qquad\;|\\\mathrm{—CHCH_2CH\underline{C}H_2CHCH_2CH—}\\\text{(XXXY tetrad)}\end{array}$$

$$\begin{array}{c}\mathrm{Y}\qquad\mathrm{X}\qquad\mathrm{X}\qquad\mathrm{Y}\\|\qquad\;|\qquad\;|\qquad\;|\\\mathrm{—CHCH_2CH\underline{C}H_2CHCH_2CH—}\\\text{(YXXY tetrad)}\end{array}$$

A systematic list of n-ad sequences up to tetrads for vinyl polymers is given in Table 3.

A particularly useful feature of NMR is that if the experiment is carried out properly (see Section 4.1), the spectra are quantitative; that is, the areas under

Table 3. *n*-ad Sequences and Bernoullian Reaction Probabilities

n-ad	Sequence	Probability[a]	*n*-ad	Sequence	Probability[a]
Diad	AA	P_a^2	Triad	AAA	P_a^3
	AB	$2P_aP_b^2$		AAB	$2P_a^2P_b$
	BB	P_b^2		BAB	$P_aP_b^2$
				ABA	$P_a^2P_b$
Tetrad	AAAA	P_a^4		BBA	$2P_aP_b^2$
	AAAB	$2P_a^3P$		BBB	P_b^3
	BAAB	$P_a^2P_b^2$			
	AABA	$2P_a^3P_b$			
	AABB	$P_a^2P_b^2$			
	BABA	$2P_a^2P_b^2$			
	BABB	$2P_aP_b^3$			
	ABBA	$P_a^2P_b^2$			
	ABBB	$2P_aP_b^3$			
	BBBB	P_b^4			

[a] P_a,P_b = Bernoullian probabilities of A enchainment and B enchainment; $P_a + P_b = 1$.

the resonances are proportional to the molar concentrations of the nuclei in question. Thus, if one wishes to analyze an unknown copolymer, one can obtain the NMR spectrum, assign the resonances to their proper comonomer sequences (sometimes not a simple task), determine the areas under the resonances, and then directly obtain the copolymer composition, and the comonomer sequence distribution (diad, triad, tetrad, or even higher *n*-ads). This is the basis of structural studies by solution NMR.

To interpret the sequence data and to derive more information about the copolymerization, one needs to relate the sequence distribution to the statistics of copolymerization. A useful framework for such analyses is the *reaction probability models* (1, 2, 5, 38–40). The simplest of such models is the Bernoullian trial process whereby the addition of each monomer to a growing polymer chain is regarded as a random process, dictated by a Bernoullian probability, P_i.

In the first-order Markovian process, the addition of each monomer depends only on the last unit in the chain; this is also known as the terminal copolymerization model (41). Four propagating steps determine the copolymerization of two monomers, M_1 and M_2.

$$\begin{aligned}
&—M_1\cdot + \mathrm{M}_1 \xrightarrow{k_{11}} —\mathrm{M}_1\cdot \\
&—M_1\cdot + \mathrm{M}_2 \xrightarrow{k_{12}} —\mathrm{M}_2\cdot \\
&—M_2\cdot + \mathrm{M}_1 \xrightarrow{k_{21}} —\mathrm{M}_1\cdot \\
&—M_2\cdot + \mathrm{M}_2 \xrightarrow{k_{22}} —\mathrm{M}_2\cdot
\end{aligned} \qquad (2)$$

The reaction probabilities are given as follows:

$$P_{11} = \frac{k_{11}(\text{—M}_1\cdot)(\text{M}_1)}{k_{11}(\text{—M}_1\cdot)(\text{M}_1) + k_{12}(\text{—M}_1\cdot)(\text{M}_2)} = \frac{1}{1 + (\text{M}_2)/r_1(\text{M}_1)} \tag{3a}$$

$$P_{21} = \frac{k_{21}(\text{—M}_2\cdot)(\text{M}_1)}{k_{21}(\text{—M}_2\cdot)(\text{M}_1) + k_{22}(\text{—M}_2\cdot)(\text{M}_2)} = \frac{1}{1 + r_2(\text{M}_2)/(\text{M}_1)} \tag{3b}$$

where P_{11} is the probability of monomer M_1 adding to radical—$M_1\cdot$, P_{21} is the probability of monomer M_1 adding to radial—$M_2\cdot$, and r_1 and r_2 are the reactivity ratios; $r_1 = k_{11}/k_{12}$, $r_2 = k_{22}/k_{21}$. Two other equations can be written for P_{12} and P_{22}. The necessary conditions for the probabilities are $P_{11} + P_{21} = 1$ and $P_{21} + P_{22} = 1$; thus, there are only two independent parameters P_{12} and P_{21} that characterize the first-order Markovian process. If, in addition, $P_{12} + P_{21} = 1$, then the copolymerization is purely random, and the case is reduced to the Bernoullian model. Although Eqs. 2–3 have been written for free radical polymerization, identical equations can be used for ionic or Ziegler–Natta polymers.

The reaction probabilities P_{ij}, or alternatively the reactivity ratios (r_1 and r_2), determine how the monomers will arrange themselves on the polymer chain. Some limiting cases are given below.

Copolymer Types	Reactivity Ratios	Reaction Probabilities
Alternating $\leftarrow$AB(AB)$_n$AB$\rightarrow$	$r_1r_2 = 0$	$P_{12} = P_{21} = 1$
Random (ABBABAA)	$r_1r_2 = 1$	$P_{12} + P_{21} = 1$
Block $\leftarrow$A$_n$B$_n\rightarrow$	$r_1r_2 > 1$	$P_{12} = P_{21} = 0$

The second-order Markovian process takes into account the "penultimate effect." In this case, eight parameters determine the reaction probabilities, of which only four are independent (α, β, γ, δ)

$$\begin{array}{ll} \alpha = P_{111} & \bar{\alpha} = P_{112} \\ \beta = P_{121} & \bar{\beta} = P_{122} \\ \gamma = P_{211} & \bar{\gamma} = P_{212} \\ \delta = P_{221} & \bar{\delta} = P_{222} \end{array} \tag{4}$$

where P_{112} is the probability of monomer M_2 adding to radical —$M_1M_1\cdot$, and $\alpha + \bar{\alpha} = 1$, $\beta + \bar{\beta} = 1$, and so on. The equations relating the probabilities to monomer concentrations and reactivity ratios are given as follows:

$$P_{111} = \frac{r_1(\text{M}_1)}{r_1(\text{M}_1) + (\text{M}_2)} \qquad P_{211} = \frac{r'_1(\text{M}_1)}{r'_1(\text{M}_1) + (\text{M}_2)}$$

$$P_{121} = \frac{(\text{M}_1)}{(\text{M}_1) + r'_2(\text{M}_2)} \qquad P_{221} = \frac{(\text{M}_1)}{(\text{M}_1) + r_2(\text{M}_2)} \tag{5}$$

where $r_1 = k_{111}/k_{112}$, $r'_1 = k_{211}/k_{212}$, $r_2 = k_{222}/k_{221}$, and $r'_2 = k_{122}/k_{121}$.

In the case of terpolymerization, nine propagation reactions can be written corresponding to the reaction probabilities of three propagating radicals and three comonomers. Taking into account three normalization conditions, $P_{11} + P_{12} + P_{13} = 1$, and so on, one obtains six independent probabilities.

$$P_{12} = \frac{1}{1 + \frac{r_{12}(\text{M}_1)}{(\text{M}_2)} + \frac{r_{12}(\text{M}_3)}{r_{13}(\text{M}_2)}} \qquad P_{23} = \frac{1}{1 + \frac{r_{23}(\text{M}_1)}{r_{21}(\text{M}_3)} + \frac{r_{23}(\text{M}_2)}{(\text{M}_3)}}$$

$$P_{13} = \frac{1}{1 + \frac{r_{13}(\text{M}_1)}{(\text{M}_3)} + \frac{r_{13}(\text{M}_2)}{r_{12}(\text{M}_3)}} \qquad P_{31} = \frac{1}{1 + \frac{r_{31}(\text{M}_2)}{r_{32}(\text{M}_1)} + \frac{r_{31}(\text{M}_3)}{(\text{M}_1)}} \tag{6}$$

$$P_{21} = \frac{1}{1 + \frac{r_{21}(\text{M}_2)}{(\text{M}_1)} + \frac{r_{21}(\text{M}_3)}{r_{23}(\text{M}_1)}} \qquad P_{32} = \frac{1}{1 + \frac{r_{32}(\text{M}_1)}{r_{31}(\text{M}_2)} + \frac{r_{32}(\text{M}_3)}{(\text{M}_2)}}$$

Extension to the penultimate model (second-order Markovian process) is straightforward but laborious. Eighteen independent probabilities are needed. These take on the following forms

$$P_{111} = \frac{1}{1 + \frac{(\text{M}_2)}{r_{112}(\text{M}_1)} + \frac{(\text{M}_3)}{r_{113}(\text{M}_1)}}$$

$$P_{112} = \frac{1}{1 + \frac{r_{112}(\text{M}_1)}{(\text{M}_2)} + \frac{r_{112}(\text{M}_3)}{r_{113}(\text{M}_2)}} \tag{7}$$

$$P_{113} = \frac{1}{1 + \frac{r_{113}(\text{M}_1)}{(\text{M}_3)} + \frac{r_{113}(\text{M}_2)}{r_{112}(\text{M}_3)}}$$

and similarly for other probabilities. Here P_{113} is the probability of monomer M_3 adding to radical $—M_1M_1\cdot$, and $P_{111} + P_{112} + P_{113} = 1$. Also $r_{112} = k_{111}/k_{112}$ and $r_{113} = k_{111}/k_{113}$. Owing to the large number of adjustable parameters, this termonomer penultimate model is used infrequently.

The theoretical probabilities for diads, triads, tetrads, pentads (and in general n-ads) can be derived from the reaction probabilities (P_i, P_{ij}, or P_{ijk}). For the Bernoullian model, these n-ad probabilities follow the binomial distribution (1). Some lower n-ads are shown in Table 3. The probabilities for other models have been given elsewhere (1, 2, 5, 38–40).

3.2. Homopolymer Tacticity and Statistics

In vinyl polymers it is well known that homopolymer structures with meso (m) and racemic (r) configurations may exhibit different chemical shifts.

Diads

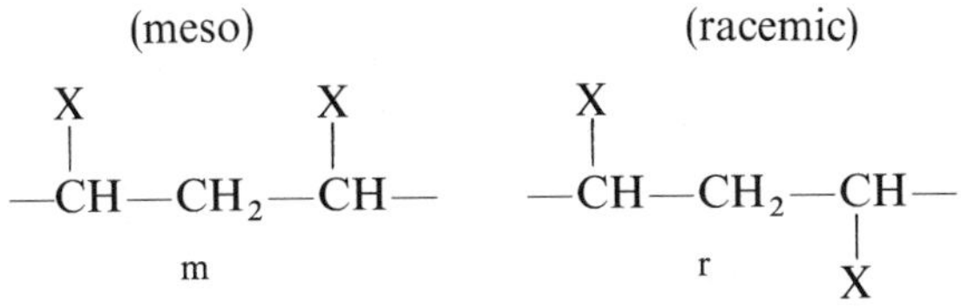

(meso): —CH(X)—CH_2—CH(X)— (m; both X above the chain)

(racemic): —CH(X)—CH_2—CH(X)— (r; first X above, second X below the chain)

Triads

—CH(X)—CH_2—CH(X)—CH_2—CH(X)— (m m; all X above the chain)

—CH(X)—CH_2—CH(X)—CH_2—CH(X)— (m r; first two X above, third X below the chain)

—CH(X)—CH_2—CH(X)—CH_2—CH(X)— (r r; first and third X above, middle X below the chain)

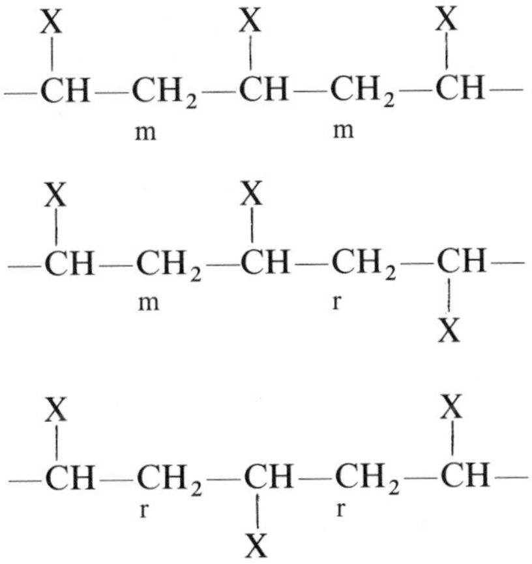

Thus, the methylene in the meso position generally resonates in a different position from the methylene in the racemic position. Similarly, the methine in mm, mr, rr triads usually have different ^{13}C- or ^{1}H-chemical shifts. In most polymers even longer configurational sequences can be distinguished. A systematic list of these longer tactic sequences is given in Table 4.

As in comonomer sequence distribution, homopolymer tacticity can be described by reaction probability models. In general, two formalisms have been used. In the more commonly used, Bovey formalism (1), tacticity is represented by the relative configuration of pairwise units.

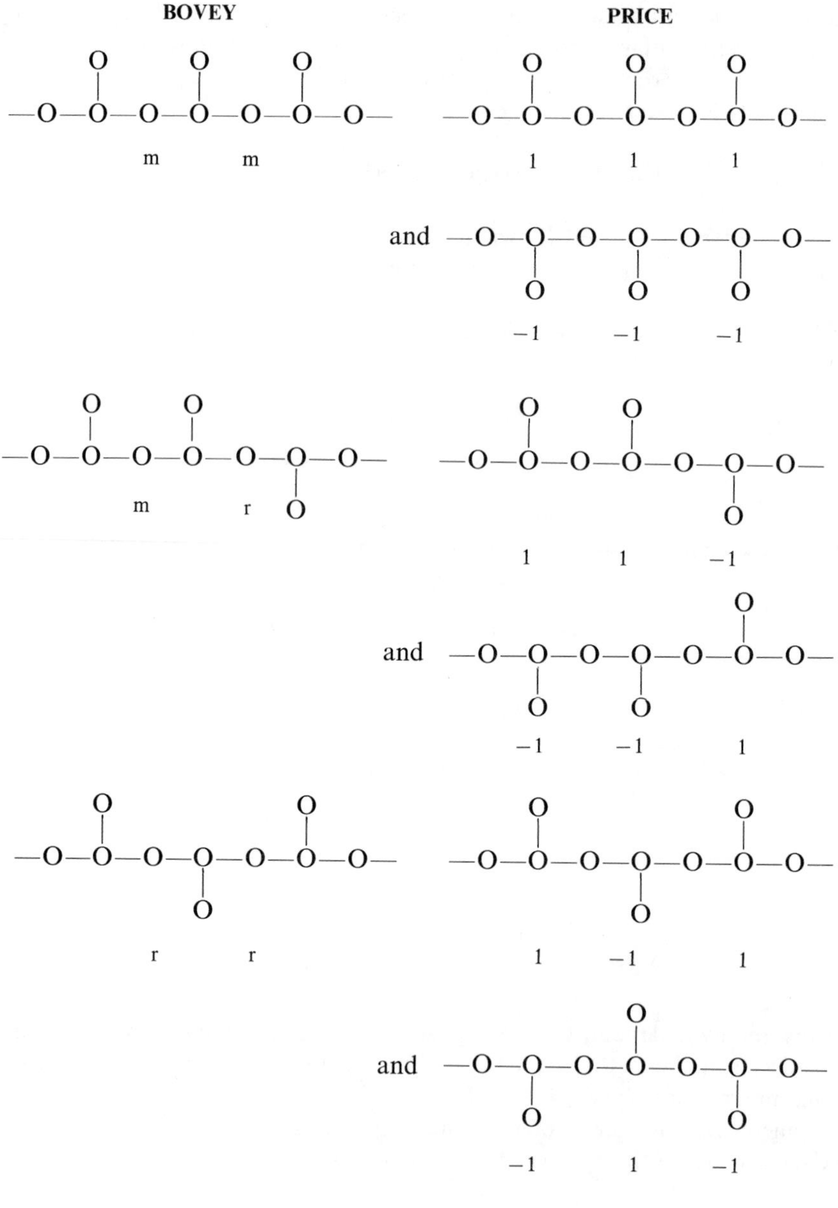

BOVEY **PRICE**

```
BOVEY                                   PRICE

    O     O     O                           O     O     O
    |     |     |                           |     |     |
—O—O—O—O—O—O—O—                         —O—O—O—O—O—O—O—
      m     m                               1     1     1

                                    and —O—O—O—O—O—O—O—
                                            |     |     |
                                            O     O     O
                                           −1    −1    −1

    O     O                                 O     O
    |     |                                 |     |
—O—O—O—O—O—O—O—                         —O—O—O—O—O—O—O—
      m     r O                                         |
                                                        O
                                            1     1    −1
                                                        O
                                                        |
                                    and —O—O—O—O—O—O—O—
                                            |     |
                                            O     O
                                           −1    −1     1

    O           O                           O           O
    |           |                           |           |
—O—O—O—O—O—O—O—                         —O—O—O—O—O—O—O—
          |                                       |
          O                                       O
      r     r                               1    −1     1
                                                  O
                                                  |
                                    and —O—O—O—O—O—O—O—
                                            |           |
                                            O           O
                                           −1     1    −1
```

Table 4. Second-Order Markovian[a] Expressions in Bovey's and Price's Formalisms for the *n*-ad sequences

n-ad	Price	Bovey
m	$cd + \bar{a}\bar{b}$	$(\bar{\alpha} + \gamma)\delta$
r	$2\bar{a}d$	$(\bar{\beta} + \delta)\bar{\alpha}$
mm	$acd + \bar{a}\bar{b}\bar{d}$	$\gamma\delta$
mr	$2\bar{a}cd + 2\bar{a}\bar{b}d$	$2\bar{\alpha}\delta$
rr	$\bar{a}bd + \bar{a}\bar{c}d$	$\bar{\alpha}\bar{\beta}$
mmm	$a^2cd + \bar{a}\bar{b}\bar{d}^2$	$\alpha\gamma\delta$
mmr	$2a\bar{a}cd + 2\bar{a}\bar{b}d\bar{d}$	$2\bar{\alpha}\gamma\delta$
rmr	$\bar{a}\bar{b}d^2 + \bar{a}^2cd$	$\bar{\alpha}\bar{\gamma}\delta$
mrm	$2\bar{a}\bar{b}cd$	$\bar{\alpha}\beta\delta$
rrm	$2\bar{a}bcd + 2\bar{a}\bar{b}\bar{c}d$	$2\bar{\alpha}\bar{\beta}\delta$
rrr	$2\bar{a}b\bar{c}d$	$\bar{\alpha}\bar{\beta}\bar{\delta}$
mmmm	$a^3cd + \bar{a}\bar{b}\bar{d}^3$	$\alpha^2\gamma\delta$
mmmr	$2\bar{a}a^2cd + 2\bar{a}\bar{b}d\bar{d}^2$	$2\alpha\bar{\alpha}\gamma\delta$
rmmr	$a\bar{a}^2cd + \bar{a}\bar{b}\bar{d}d^2$	$\bar{\alpha}^2\gamma\delta$
mmrr	$2a\bar{a}bcd + 2\bar{a}\bar{b}\bar{c}d\bar{d}$	$2\bar{\alpha}\bar{\beta}\gamma\delta$
mrmm	$2a\bar{a}\bar{b}cd + 2\bar{a}\bar{b}cd\bar{d}$	$2\bar{\alpha}\beta\gamma\delta$
rmrr	$2\bar{a}\bar{b}\bar{c}d^2 + 2\bar{a}^2bcd$	$2\bar{\alpha}\bar{\beta}\gamma\delta$
mrmr	$2\bar{a}\bar{b}cd^2 + 2\bar{a}^2\bar{b}cd$	$2\bar{\alpha}\beta\bar{\gamma}\delta$
rrrr	$\bar{a}b\bar{c}^2d + \bar{a}b^2\bar{c}d$	$\bar{\alpha}\bar{\beta}\bar{\delta}^2$
rrrm	$2\bar{a}bc\bar{c}d + 2\bar{a}b\bar{b}\bar{c}d$	$2\bar{\alpha}\bar{\beta}\delta\bar{\delta}$
mrrm	$\bar{a}bc^2d + \bar{a}\bar{b}^2\bar{c}d$	$\bar{\alpha}\bar{\beta}\delta^2$
k	$(\bar{a}\bar{b} + 2\bar{a}d + cd)^{-1}$	$(\bar{\alpha}\bar{\beta} + 2\bar{\alpha}\delta + \gamma\delta)^{-1}$

[a] Corresponding expressions for the first-order Markov can be obtained by substituting (for Price's formalism) $a = c = P_{11}, \bar{a} = \bar{c} = P_{1\bar{1}}$, $b = d = P_{\bar{1}1}$ and $\bar{b} = \bar{d} = P_{\bar{1}\bar{1}}$, and (for Bovey's formalism) $\alpha = \gamma = P_{mm}$. $\bar{\alpha} = \bar{\gamma} = \bar{P}_{mr}, \beta = \delta = P_{rm}$, $\bar{\beta} = \bar{\delta} = P_{rr}$. Similarly, Bernoullian expressions may be readily deduced: (for Price) $P_{11} = P_{\bar{1}1} = P_1$. $P_{1\bar{1}} = P_{\bar{1}\bar{1}} = P_{\bar{1}}$, and (for Bovey) $P_{mm} = P_{rm} = P_m$, $P_{mr} = P_{rr} = P_r$.

In Price's formalism (42), tacticity is depicted as a copolymer, with 1 for a configuration in the dextro direction, and -1 or $\bar{1}$ for the opposite levo direction (as above). For example, both (111) and ($\overline{111}$) sequence in Price's formalism are (mm) triads in Bovey's formalism.

An important distinction between the two formalisms is that Bovey's formalism gives the relative configuration, whereas Price's formalism gives the absolute configuration. The reaction probability models (Bernoullian and Markovian) hold for both formalisms, although they have different meanings. Bovey's first-order Markovian model, for example, means that the tacticity of incoming monomer *j* depends on the previous tacticity (between units *i* and

$i-1$ on the propagating chain). In the Price formalism, this means that the configuration of unit j depends on the configurations of *two* previous units (i and $i-1$). This represents the second-order Markovian model (in the Price formalism).

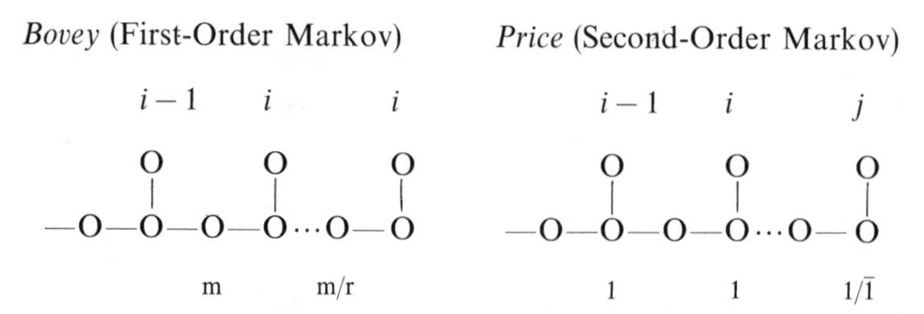

The reaction probability expressions can be readily derived (43). The derivation is identical to the comonomer sequence case delineated in the last section. In the Bovey case, the "comonomers" are (m) and (r) with first-order Markovian probabilities P_{mr} and P_{rm}. In the Price formalism, the "comonomers" are (1) and ($\bar{1}$), with corresponding second-order Markovian probabilities (a, b, c, d)

$$\text{Bovey} \quad (\text{m}) = kP_{rm} \tag{8a}$$

$$(\text{r}) = kP_{mr} \tag{8b}$$

$$\text{Price} \quad (\text{m}) = (11) + (\bar{1}\bar{1}) = k'(\text{cd} + \overline{\text{ab}}) \tag{9a}$$

$$(\text{r}) = (1\bar{1}) + (\bar{1}1) = k'(2\bar{\text{a}}\text{d}) \tag{9b}$$

where $k = (P_{mr} + P_{rm})^{-1}$ and $k' = (\overline{\text{ab}} + 2\,\bar{\text{a}}\text{d} + \text{cd})^{-1}$, and

$$\begin{array}{ll} \text{a} = P_{111} & \text{c} = P_{1\bar{1}1} \\ \text{b} = P_{1\bar{1}1} & \text{d} = P_{\bar{1}\bar{1}1} \end{array} \tag{10}$$

Thus, in Bovey's relative configurational scheme, two parameters are needed, whereas in Price's absolute configurational scheme, four parameters are used. Price's formalism reduces to Bovey's if the constraints of absolute configuration are lifted, that is,

$$\begin{array}{ll} \text{a} = \bar{\text{d}} & \bar{\text{a}} = \text{d} \\ \text{b} = \bar{\text{c}} & \bar{\text{b}} = \text{c} \end{array} \tag{11}$$

In this case

$$(\text{m}) = k'\text{c} = k'(1 - \text{b})$$

$$(\text{r}) = k'\text{d} = k'(1 - \text{a}) \tag{12}$$

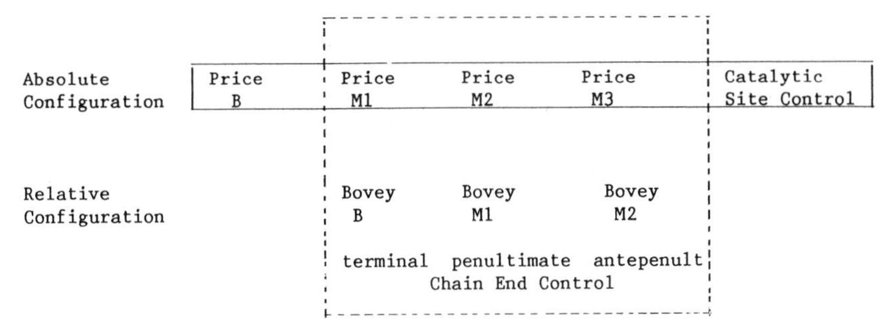

Figure 4. Relationships between the Bernoullian (B), first-order Markovian (M1), and second-order Markovian (M2) models in the Bovey and Price formalisms.

Thus, by using the absolute configurational scheme (Price), we gained two extra degrees of freedom (i.e., two additional parameters). If we set $a + d = 1$, and $b + c = 1$, then in effect we are using the relative configurational scheme of Bovey, where $c = P_{rm}$ and $d = P_{mr}$. Price's second-order Markov is therefore a generalization of Bovey's first-order Markov. This relationship is depicted schematically in Figure 4.

Analogous arguments can be used for Bovey's Bernoullian model. The corresponding model in the Price formalism is the first-order Markov (Fig. 4). The logical relationship between the two formalisms has been recently reviewed (43), and the reaction probabilities for both Price and Bovey formalisms are summarized in Table 4.

A useful feature of NMR studies of homopolymer tacticity is that information on polymerization mechanisms can frequently be obtained. For vinyl polymerization, two general kinds of mechanisms have been identified; one is chain-end controlling and the other is catalytic-site controlling. Usually free radical, anionic, and most cationic polymerizations are chain-end controlled, whereas many Ziegler–Natta polymerizations are considered to be catalytic-site controlled. The features of both chain-end and catalytic-site controlling mechanisms have recently been reviewed (44).

In a purely chain controlling mechanism, the absolute configuration is not needed, and Bovey's formalism works well. For polymerizations where the catalytic sites have a major influence, differentiation by the absolute configuration is necessary. Price's formalism should be used in this latter case.

A good example of catalytic-site controlled mechanism is demonstrated by the special case of Price's Bernoullian model. This model has no equivalence in the Bovey formalism. The catalytic site will have a reaction probability P_1 for d = configuration and $P_{\bar{1}}$ (or $1 - P_1$) for the $\underline{1}$-configuration. The various n-ad sequences are given as follows:

$$(\mathrm{m}) = (11) + (\bar{1}\bar{1}) = P_1^2 + (1 - P_1)^2 \tag{13a}$$

$$(\mathrm{r}) = (1\bar{1}) + (\bar{1}1) = 2P_1(1 - P_1) \tag{13b}$$

$$(\mathrm{mm}) = (111) + (\bar{1}\bar{1}\bar{1}) = P_1^3 + (1 - P_1)^3 \tag{13c}$$

$$\begin{aligned}(\mathrm{mr}) &= (1\bar{1}\bar{1}) + (\bar{1}\bar{1}1) + (\bar{1}11) + (11\bar{1}) \\ &= 2P_1(1 - P_1)^2 + 2P_1^2(1 - P_1)\end{aligned} \tag{13d}$$

$$(\mathrm{rr}) = (1\bar{1}1) + (\bar{1}1\bar{1}) = P_1(1 - P_1)^2 + P_1^2(1 - P_1) \tag{13e}$$

The above expressions are identical to the symmetrical enantiomorphic-site model (45, 46) found to be suitable for the highly isotactic portions of Ziegler–Natta polymers (47, 48).

A problem of considerable interest is the description of a polymerization that is controlled by *both* the catalyst sites and the chain ends. It is significant to note that Price's formalism with its emphasis on absolute configuration is a suitable description for such hybrid models (44). The various mechanisms of stereochemical control and the suitable models are summarized below.

Pure chain-end control	Bovey's B, M1, M2 Models
Pure catalyst-site control	Price's B Model
Both catalyst-site and chain-end control	Price's M1, M2 Models

3.3. Analytical versus Spectral Simulation Approaches

The analysis of polymer NMR spectra usually follows familiar patterns. One must first synthesize or otherwise acquire the polymer samples, obtain the spectra, and assign the observed resonance lines to specific sequences. The reaction probabilities and hence reactivity ratios are then deduced from the sequence distribution. This is known as the "analytical approach" and has been used for most polymer NMR studies; it has been highly successful in many cases.

Analytical Approach

NMR spectrum → spectral interpretation → sequence distribution → reaction probabilities

A more refined analytical approach is to approximate the copolymerization reaction with a statistical model (49–51). One can then associate every spectral intensity with a theoretical expression involving reaction probability parameters. The observed and the theoretical intensities for all the spectral lines are then compared and optimization is carried out to obtain the best-fit values of the reaction probability parameters. Depending on the goodness of fit, these probability parameters may then fully describe the structure of the polymer system in question. This model fitting approach has been successfully applied to a number of polymer systems (51–64). Two families of computer programs have been written for various polymer analysis: FITCO (59) and MIXCO (62, 63).

Computer-Assisted Analytical Approach

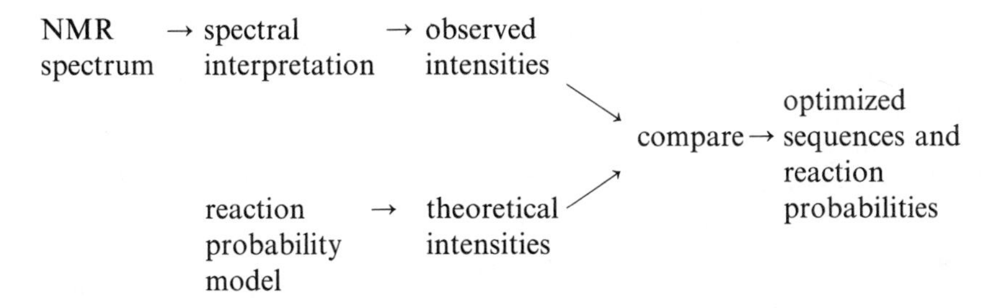

Both analytical approaches typically involve painstaking work, even for binary copolymers. For ternary copolymers, complete analyses are very difficult indeed and have been reported only rarely. An alternative method was recently proposed (40, 65) for vinyl and vinylidene copolymers, called the *spectral simulation* or "synthetic" approach. The idea is to take a given set of comonomer concentrations and the reactivity ratios and then through mathematical and empirical methods simulate a ^{13}C NMR spectrum for any vinyl copolymer. The resulting computer program is called PSPEC.

Spectral Simulation Approach

Reaction probability model → polymer chain generation → calculation of ^{13}C shifts → spectral simulation

An advantage of the PSPEC approach is that it is a predictive method; no laboratory work is needed to get a simulated spectrum. Thus, even for copolymers that cannot easily be prepared, one can obtain a ^{13}C NMR spectrum that bears a reasonable resemblance to the real spectrum.

The use of both analytical and spectral simulation approaches will be illustrated later in this review.

4. EXPERIMENTAL CONSIDERATIONS

4.1. Experimental Setup

For high-resolution solution NMR, the polymer samples are dissolved in suitable solvents, preferably with a chemical shift reference material. The common reference materials are tetramethylsilane (TMS), hexamethyldisiloxane (HMDS) for high-temperature work, sodium 3-(trimethylsilyl)propionate (TSP), sodium 2,2-dimethyl-2-silapentane-5-sulfonate (DSS), dioxane, and acetonitrile for aqueous solutions. The solvents must be deuterated to serve as field–frequency lock materials; otherwise, a deuterated cosolvent is needed. For ^{1}H NMR, a sample concentration of 1–5% (w/w) is usually satisfactory. For ^{13}C, higher concentrations are needed for improved sensitivity. The concentration is preferably larger than 10% (w/w).

The ^{1}H NMR spectroscopy is often carried out on continuous-wave (CW) instruments. The instrumental conditions are fairly standard except that for quantitative work care is needed to avoid saturation of the resonances. Because most polymer spectra have broader lines, the problem of saturation is not as severe as it is with organic compounds.

The ^{13}C NMR spectroscopy (and sometimes ^{1}H NMR) is carried out via the FT experiment. Depending on the problem, one may set up the experiment for either quantitative spectra, or for optimal sensitivity and resolution. These two experiments are frequently not compatible with each other as we shall see.

For spin $\frac{1}{2}$ nuclei like ^{13}C, the major factors affecting quantitation are spin–lattice relaxation time (T_1) and the nuclear overhauser effect (NOE). Indeed the spectral intensities depend critically on how fast the repetitive pulses are applied. If one waits long enough between pulses such that all the various ^{13}C spins are restored to thermal equilibrium, then all the peaks in the ^{13}C spectrum correspond exactly to their natural occurrence in the sample and one has a quantitative spectrum. Typically, this happens when the time between pulses (T_{aq}) $\geqslant 5T_1$. Such an experimental condition can become very time-consuming because some polymers have long T_1 values, as large as 1–2 s.

All the ^{13}C nuclei are normally coupled to neighboring protons, thereby giving rise to a multiplet structure for each carbon. To simplify the spectrum, a saturating field is applied at a proton resonance frequency, which effectively cuts off the coupling between ^{13}C and the proton and gives single lines for each carbon type on the ^{13}C spectrum. As a side effect, this saturating field also causes energy transfer between the proton and the carbon spins, causing an enhancement (or NOE) in peak intensities. Depending on the prevailing relaxation mechanisms, this NOE can be different for different ^{13}C spins.

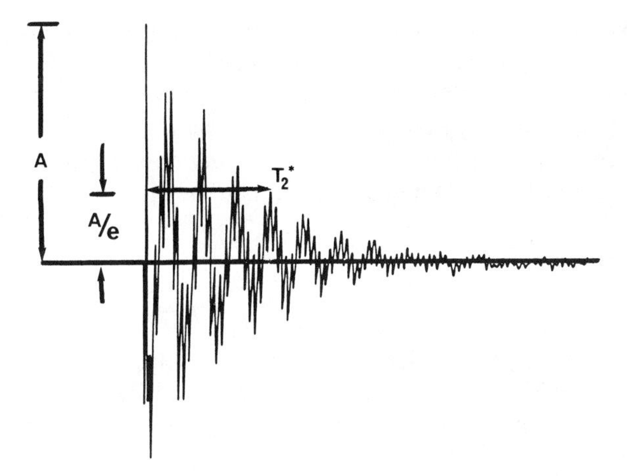

Figure 5. Free induction decay (fid) signal.

Methods to minimize NOE include gated decoupling and the use of relaxation reagents,* such as $Cr(acac)_2$, $Gd(fod)_3$, and $FeCl_3$.

The optimization of instrumental parameters is needed to improve sensitivity and resolution. These parameters include pulse angle, memory size, relaxation times, sweep width, and acquisition time. The complex interplay between these parameters has been described in the literature (66–69). The first task is to determine the acquisition time T_{aq}. There are two considerations. The first deals with the spectral resolution needed

$$T_{aq} = \frac{N}{2^* \mathrm{SW}} \tag{14}$$

where N = memory size and SW = sweep width. The values N and SW should be arranged such that there are at least two data points that define a peak. Further resolution enhancement may be obtained by zero filling the memory size after the free induction decay (fid) signal has been base line corrected. For illustration, with 16 K memory and 10,000 Hz sweep width, the acquisition time should be (Eq. 14) 16,000/(2·10,000) or 0.8 s.

The second consideration in acquisition time is sensitivity. It has been shown by Traficante and Kelly (69) that the maximum sensitivity is obtained when the acquisition time is set equal to the decay time constant T_2^* of an fid. In Figure 5, the T_2^* is defined as the time that the fid decays to $1/e$ of its initial

*Ligand abbreviations: fod = the anion of 6,6,7,7,8,8,8-heptafluoro-2,2-dimethyl-3,5-octanedione and acac = the anion of acetylacetone.

value. The T_2^* value may or may not be consistent with the requirement of resolution defined by Eq. 14. Where both sensitivity and resolution are crucial, a compromise is often necessary. One needs therefore, to adjust the memory size and the acquisition time to suit the problem.

When an optimal acquisition time is finally chosen, the Ernst equation relates the optimal pulse angle to the relaxation time T_1 (66):

$$\cos \alpha_E = \exp\left(-\frac{T_{aq}}{T_1}\right) \qquad (15)$$

where the optimal pulse angle α_E is known as the Ernst angle. For 16 K memory, 10,000-Hz sweep width, T_1 1.5 s, and acquisition time 0.8 s, the Ernst angle is 70°. Any changes in memory size, acquisition time, T_1, or T_2^* will alter the optimal pulse angle.

The competing requirements of quantitation and optimal sensitivity are clearly seen for polyolefin samples. In these cases $T_1 \sim 1.5$ s, then one needs to wait $5 \cdot T_1$ or 7.5 s between successive pulses. For maximum sensitivity, however, one should wait for only 0.8 s between the pulses (assuming a 10,000-Hz sweep width). In this case, the problem needs to be defined to decide what is more important. For polymers with large hindered groups (e.g., polystyrene), the T_1 values are short (200 ms), and it is possible to satisfy both requirements of quantitation and optimal sensitivity in one experiment.

4.2. Special Pulse Experiments

During the past decade, a number of special pulse experiments have been devised (70, 71). Perhaps the most useful experiments for polymer characterization are those that permit spectral editing to be obtained. These include:

SEFT (Spin–Echo Fourier Transform). This experiment (72, 73) distinguishes between carbon atoms of odd multiplicity (odd number of directly attached hydrogen atoms) and even multiplicity (even number of hydrogen atoms).

APT (Attached Proton Test). An improved version of SEFT, using double spin echoes. This experiment is widely used on account of its simplicity (74).

INEPT Sequence. This pulse sequence (75) uses the polarization transfer technique to provide sensitivity enhancement of a rare nucleus (e.g., ^{13}C), or to distinguish different proton multiplicities on carbon (similar to APT).

DEPT Experiment. This is a spectral editing pulse sequence (76) that can enable subspectra to be obtained consisting of the resonances of carbon atoms with proton multiplicities of 0, 1, 2, and 3, successively.

4.3. Two-Dimentional NMR

Two-dimensional (2D) NMR spectroscopy is a family of techniques that has been developed in the last few years (70, 71, 77–80). Two-dimensional experiments generally fall into two categories: J-resolved and shift correlated. J-resolved experiments include both homo- and heteronuclear types. Shift-correlated experiments constitute a broader class and can be brought about by homo- and heteronuclear scalar J coupling, dipolar coupling, chemical exchange, or multiple quantum coherence. The following list provides the more common 2D experiments in use. Most of these experiments have been used for polymer studies (54, 58, 61, 81–99). Some illustrations of these techniques are shown in later sections.

Acronym	Name	Function
2DJ	Homonuclear J-resolved 2D	H—H coupling patterns and precise J_{HH} measurements
	Heteronuclear J-resolved 2D	C—H coupling patterns and precise J_{CH} measurements
COSY	Homonuclear shift correlated 2D	Correlation of H signals through direct (vicinal) H—H coupling
SECSY	Spin–echo correlated spectroscopy	Same as COSY
NOESY	Nuclear Overhauser effect correlated spectroscopy	Correlation of H signals through NOE interactions
RELAY	Relayed coherence transfer –Homonuclear (H, H, H)–	Correlation of H signals through relayed coherence (long range)
	–Heteronuclear (H, H, C)–	Same as above, for ^{13}C—^{1}H correlation
TOCSY or HOHAHA	Total correlation spectroscopy	Correlation of very long range H signals

Acronym	Name	Function
CSCM or HETCOR	Heteronuclear shift-correlated 2D	Correlation of ^{13}C and ^{1}H signals through direct $^{1}J_{CH}$
COLOC	Correlation via long range coupling	Correlation of $^{13}C—^{1}H$ signals through long-range $^{13}C—^{1}H$ coupling
2D INADEQUATE	Multiple quantum 2D	Correlation of ^{13}C signals through direct C—C coupling

5. ^{1}H NMR ANALYSIS OF POLYMERS

Although ^{1}H NMR has been mostly supplanted by ^{13}C for polymer analysis, several features of ^{1}H NMR render it still attractive: (1) the instrumentation is less costly, (2) quantitation is easily accomplished, (3) analysis time is fast. These three factors make ^{1}H NMR highly suited for fast, routine analysis.

Relative to ^{13}C, ^{1}H NMR has other advantages: (1) ^{1}H NMR has very high sensitivity. This is the reason why ^{1}H is the primary NMR technique for biopolymers. (2) Information on ^{1}H—^{1}H scalar coupling is very useful for spectral interpretation and conformational analysis. (3) The OH and NH signals, which are undetectable by ^{13}C, can be directly observed. (4) The olefinic and the aromatic protons fall in different regions in the ^{1}H spectrum (olefinic, 4.8–6.8 ppm and aromatic, 6.8–8.0 ppm); they are overlapped in the ^{13}C spectrum. These advantages partially compensate for the relatively small ^{1}H-chemical shift range (12 ppm), and make ^{1}H NMR an indispensable technique in selected applications.

Because the ^{1}H spectra of many polymer systems are complex, several techniques have been developed to alleviate these problems. The techniques used prior to the 1970s include:

1. Use of computers to obtain chemical shifts and coupling constants, and to simulate the observed splitting patterns. Many computer programs have been written for this purpose (e.g., LAOCOON, NMRCAL, and ITRCAL).

Figure 6. Effect of shift reagents on the ^{1}H spectrum of EVAc: (*a*) $\rho = 0$; (*b*) Eu (fod)$_3$, $\rho = 0.00182$; (*c*) Eu (fod)$_3$, $\rho = 0.00364$; (*d*) Eu (fod)$_3$, $\rho = 0.00546$; (*e*) Eu (fod)$_3$, $\rho = 0.00728$; (*f*) Yb (fod)$_3$, $\rho \sim 0.0015$. Spectral line numbers 1–9, given in Table 5, have been indicated on the spectra. Reproduced with permission from Springer-Verlag, Heidelberg (96).

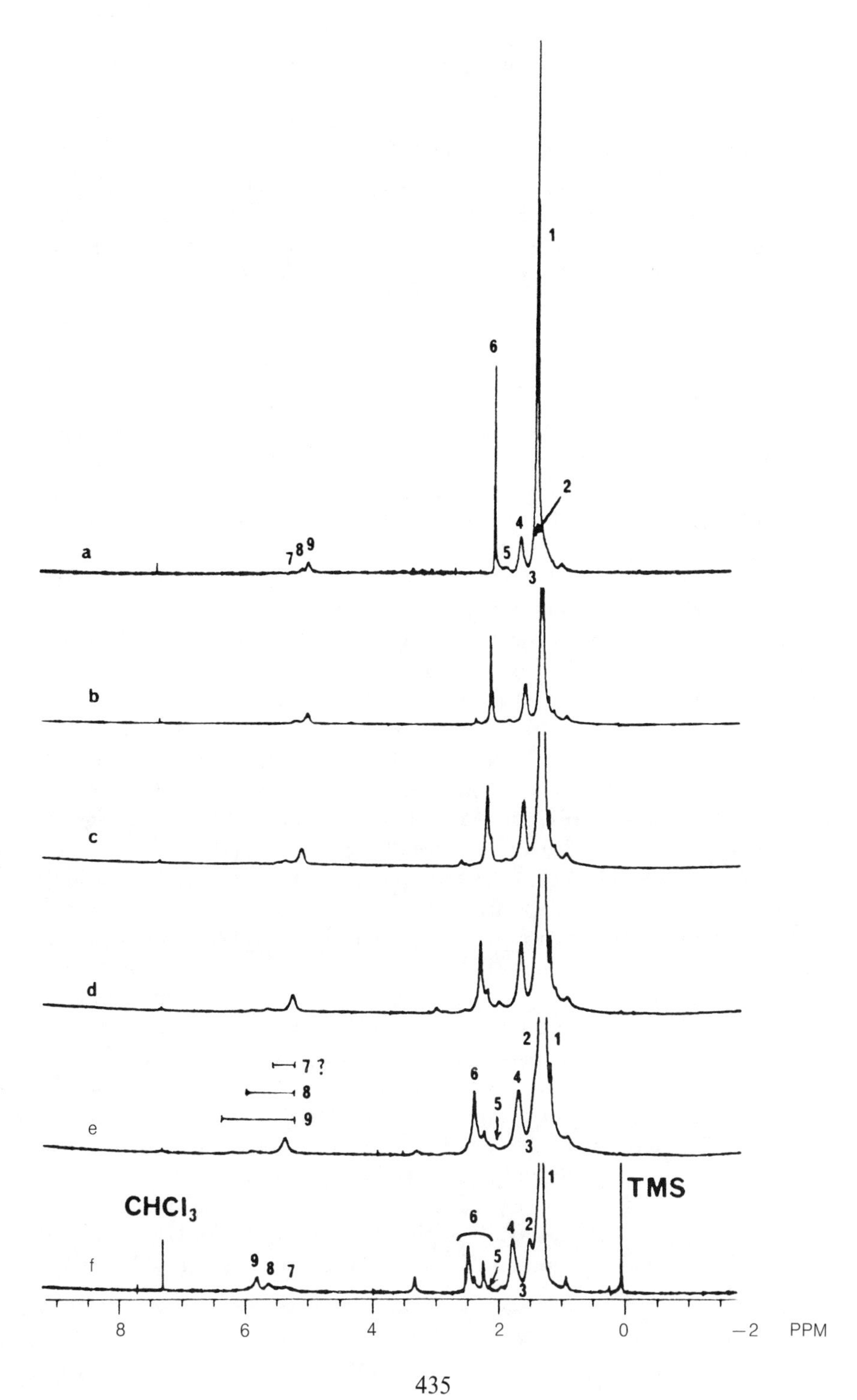
1
6
2
4
a
7
8
9
5
3
b
c
d
e
7 ?
8
9
6
5
4
2
1
3
CHCl3
TMS
f
9
8
7
6
4
2
5
1
3
8
6
4
2
0
−2
PPM

2. Use of lanthanide shift reagents (LSR) to "spread out" the chemical shifts of selected portions of the molecules.
3. Homonuclear decoupling to simplify the spectrum and to determine which spectral regions are coupled together.
4. Selective deuteration of polymers with (preferably) deuterium decoupling. This is superior to homonuclear decoupling but requires polymer synthesis.
5. Use of model compounds that provide information on chemical shifts and coupling constants in the polymer spectrum.

The application of these techniques has been amply documented and reviewed (1, 6, 9, 11).

With the increasing use of superconducting magnetic fields since the 1970s, the higher frequencies (up to 500–600 MHz) have given new opportunities for ^{1}H NMR spectroscopy. The high field increases the sensitivity, spreads out the chemical shifts, and simplifies the spectra of coupled spin systems. In addition, recent development of 2D NMR (70, 71, 77–80) opened the door to many applications. Two examples will be shown to illustrate the utility of COSY (^{1}H—^{1}H homonuclear shift correlation) and CSCM (^{1}H—^{13}C heteronuclear shift correlation) experiments.

Ethylene–vinyl acetate (EVAc) copolymers have been studied by both ^{1}H and ^{13}C NMR (100–110); however, several aspects of polymer structural and spectral assignments still attract attention (107–110). Recently, through the combined use of high field, lanthanide shift reagents, and 2D NMR techniques, the ^{1}H NMR spectra of EVAc have been fully analyzed (96).

Figure 6 shows the effect of the addition of ytterbium and europium shift reagents on the ^{1}H NMR spectra of EVAc. The addition of shift reagents causes a number of resonances to emerge and shift downfield from the large upfield resonances at 1.26 ppm. The ytterbium shift reagent, $Yb(fod)_3$,* gives larger lanthanide-induced shifts, and also provides better separation of overlapped resonances. At a shift reagent–substrate molar ratio (ρ) of about 0.0015, a number of distinct spectral features are observed (Fig. 6*f*).

To positively assign the ^{1}H resonances, one can use the ^{13}C—^{1}H heteronuclear shift correlated experiment (CSCM). By correlating the chemical shifts of directly bonded ^{13}C and ^{1}H, the information can be combined on spectral assignments for ^{1}H and ^{13}C to obtain a more complete interpretation of both spectra. A similar approach has been used for the analysis of several ethylene copolymers (54, 58, 61). It applies equally well to ^{1}H NMR spectra shifted by lanthanide shift reagents. The CSCM contour map for a $Yb(fod)_3$-

*See footnote on p. 431

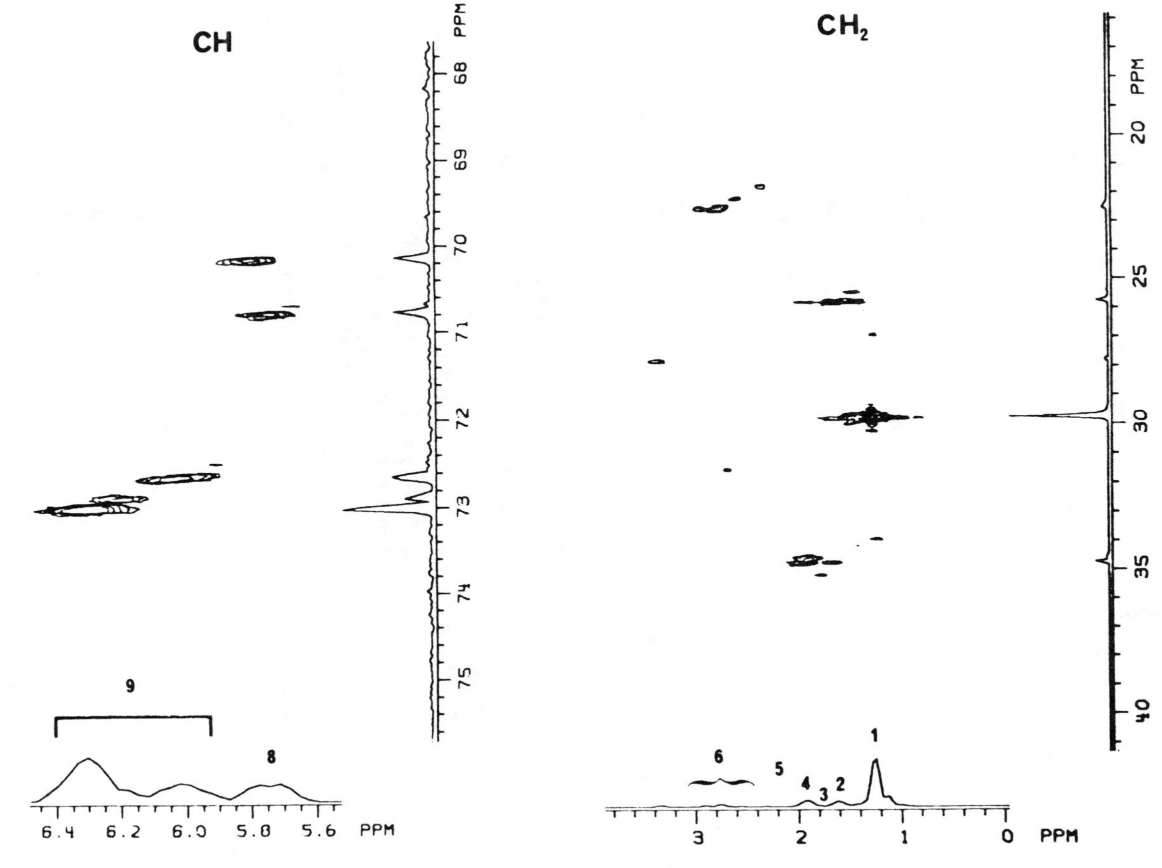

Figure 7. ^{13}C—^{1}H CSCM plots of the methylene and methine regions of EVAc with Yb (fod)$_3$ added ($\rho = 0.00235$). Reproduced with permission from Springer-Verlag, Heidelberg (96).

Table 5. ^{13}C and ^{1}H NMR Assignments in EVAc Copolymers

	^{1}H			^{13}C	
No.	Shift [$\rho = 0$]	Shift [Yb(fod)$_3$][a]	Sequence	Sequence	Shift
1	1.26	~1.26	$S_{\gamma\gamma} + S_{\gamma\delta} + S_{\delta\delta}$	Acetate	21.0
2	1.33	1.60	$S_{\beta\delta}$	$S_{\beta\beta}$	21.5
3	1.40	~1.70	$S_{\beta\beta}$	$S_{\beta\delta}$	25.7
4	1.50	1.93	$S_{\alpha\gamma} + S_{\alpha\delta}$	$S_{\gamma\gamma} + S_{\gamma\delta}$	29.8
5	1.73	2.08	$S_{\alpha\alpha}$	$S_{\delta\delta}$	30.0
6	1.92	2.58	Acetate	$S_{\alpha\delta}$	34.7
		2.74			
		2.90		$S_{\alpha\gamma}$	35.2
7	5.15	5.35	$T_{\beta\beta}$(VVV)	$S_{\alpha\alpha}$	39.6
8	5.03	5.75	$T_{\beta\delta}$(VVE)	$T_{\beta\beta}$	~68
				$T_{\beta\delta}$	71
9	4.90	5.90	$T_{\delta\delta}$(VEVEV)		
		6.00	$T_{\delta\delta}$(VEVEE)		
		6.30)	$T_{\delta\delta}$(EEVEE)	$T_{\delta\delta}$	73

[a] Shift observed with Yb(fod)$_3$ added, $\rho = 0.00235$.

Reproduced with permission from Springer-Verlag, Heidelberg (96).

shifted ^{1}H spectrum is given in Figure 7. On the basis of the 2D data, one can then make line-to-line correlations and provide assignments. For convenience, this is summarized in a graphical form in Table 5.

The terminology for comonomer sequences used in Table 5 has been borrowed from that used with ethylene–propylene copolymers (49). The backbone methine and methylene protons are represented by T and S, respectively. Greek subscripts are used to indicate distances the nearest methines are from the carbon in question. Some examples are shown below. Note that this terminology is different from those reported by others (105).

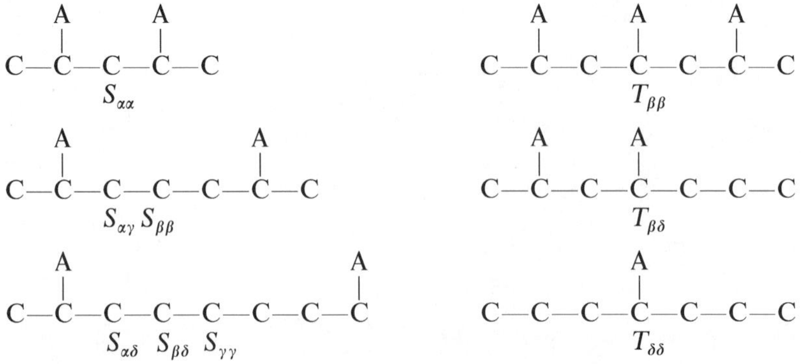

```
   A     A                         A     A     A
   |     |                         |     |     |
C—C—C—C—C                       C—C—C—C—C—C—C
      Sαα                                 Tββ

   A           A                   A     A
   |           |                   |     |
C—C—C—C—C—C—C                   C—C—C—C—C—C—C
      Sαγ Sββ                             Tβδ

   A                 A                   A
   |                 |                   |
C—C—C—C—C—C—C—C                 C—C—C—C—C—C—C
      Sαδ Sβδ Sγγ                         Tδδ
```

From the ^{13}C—^{1}H shift correlation, the LSR-shifted ^{1}H NMR spectrum can be fully interpreted. Rather detailed sequence information is available from this ^{1}H spectrum. Not surprisingly, the methine is shifted the most at about 6.0 ppm. The term $T_{\beta\beta}$ is the most upfield resonance, followed by $T_{\beta\delta}$ and $T_{\delta\delta}$. This is the same trend as ^{13}C NMR, but reversed from the normal (i.e., no LSR) ^{1}H spectrum. The acetate is now split into three resonances at about 2.74 ppm. Furthermore, $S_{\beta\beta}$ and $S_{\beta\delta}$ are all separately resolved.

Using the data from the 2D/shift reagent work, we can provide detailed spectral assignments of the normal (i.e., no LSR) one-dimensional ^{1}H and ^{13}C spectra of EVAc. For compactness, the ^{1}H assignments are included in the second column of Table 5. As expected, the ^{1}H resonances of many sequences

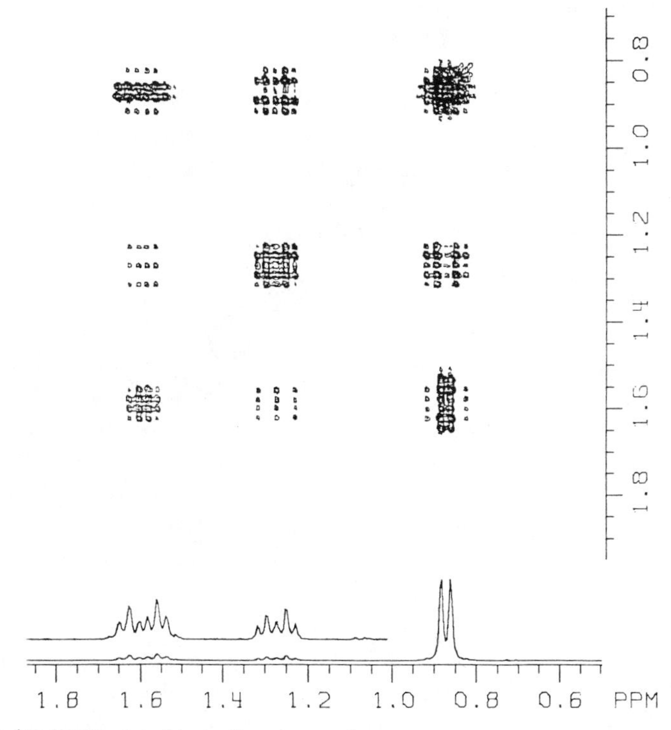

Figure 8. ^{1}H COSY plot of isotactic polypropylene.

$$\begin{array}{ccccc} & CH_3 & H_a & CH_3 & \\ & | & | & | & \\ — & C & — C — & C & — \\ & | & | & | & \\ & H_c & H_b & H_c & \end{array}$$

The assignments are: $H_a = 0.89$ ppm, $H_b = 1.29$ ppm, $H_c = 1.6$ ppm, $CH_3 = 0.87$ ppm.

are overlapped. For example, with this approach, the assignments for $S_{\beta\beta}$ and $S_{\beta\delta}$ (previously uncertain) have been made (96).

Quantitative analysis of the ^{1}H spectra of EVAc may be carried out through the computerized analytical approach. Interested readers may consult Ref. 61.

A second example of the utility of 2D NMR will be shown for polypropylene tacticity. Although the major features of both ^{1}H and ^{13}C spectra are well known (1, 2, 111–122), the assignments of a few specific tactic sequences have been uncertain until recently. Two-dimensional NMR played a major role in removing these residual uncertainties and provided definitive assignments for both ^{1}H- and ^{13}C NMR spectra (88).

For a predominantly isotactic polypropylene sample, the ^{1}H COSY plot (58) is shown in Figure 8. This gives the connectivities (of scalar J couplings) between the four distinct protons in the sample. In agreement with the

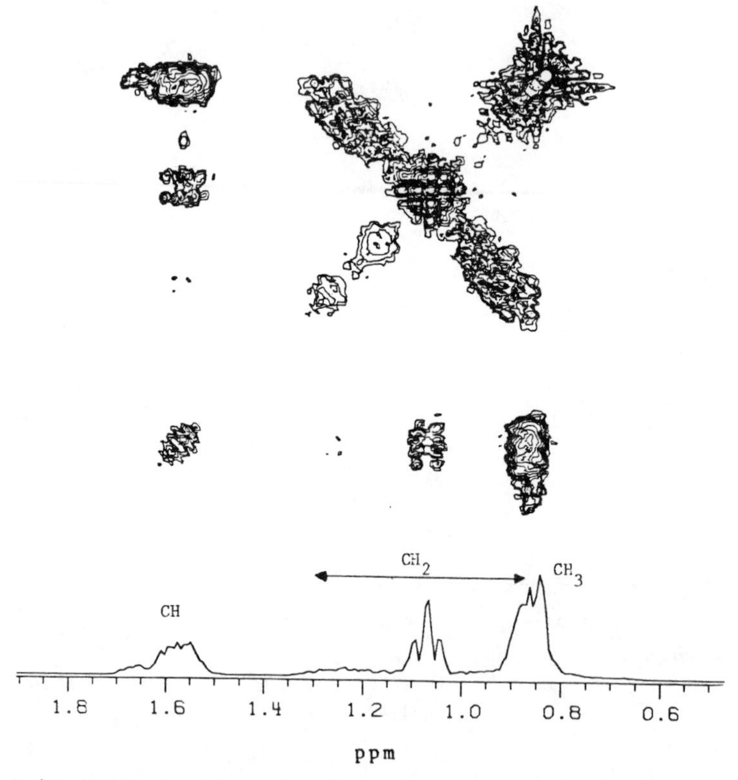

Figure 9. ^{1}H COSY plot of atactic polypropylene. Reproduced with permission from Springer-Verlag, Heidelberg (88). The detailed assignments of the methylenes are given in Table 7.

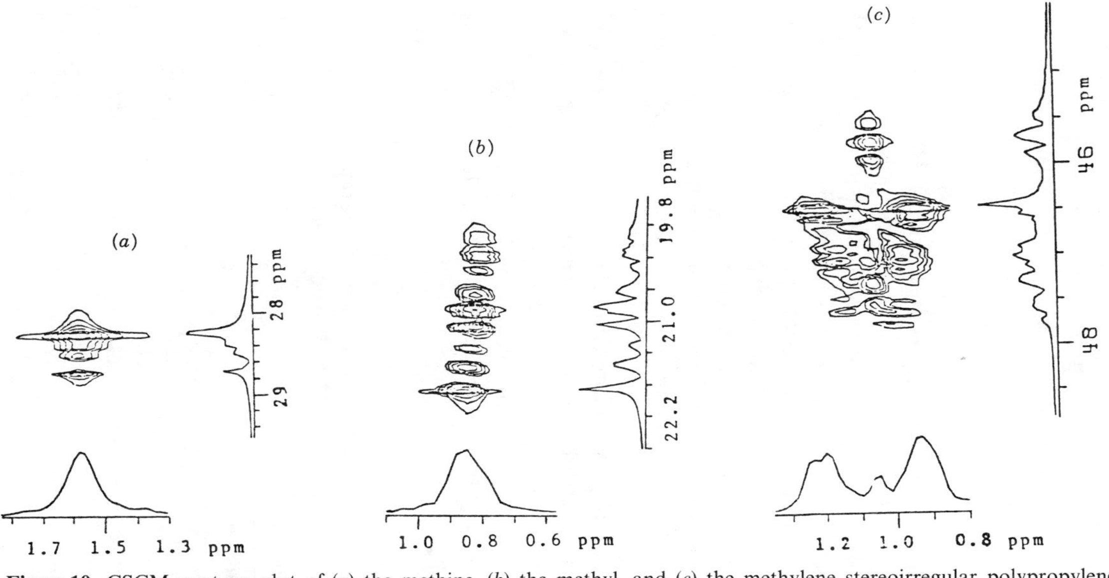

Figure 10. CSCM contour plot of (*a*) the methine, (*b*) the methyl, and (*c*) the methylene stereoirregular polypropylene. Reproduced with permission from Springer-Verlag, Heidelberg (88).

literature (116), the two methylene protons are inequivalent, the syn proton (H_a) being overlapped with the methyl groups.

The corresponding COSY plot of atactic polypropylene (88) is considerably more complex (Fig. 9). Here the methylene protons are shown to spread out over a large region (0.8–1.3 ppm). Cross-peaks indicate scalar couplings between methylenes (geminal, ^2J), between methylene and methine (^3J), and between methine and methyl (^3J). Detailed assignments of the methylene sequences are needed in order to proceed further with the ^{1}H-spectral analysis.

As in the case of EVAc copolymer, the use of the 2D ^{13}C—^{1}H shift correlation experiment (CSCM) permits complete ^{1}H assignments. The ^{13}C—^{1}H CSCM plots for selected regions of atactic polypropylene are shown in Figure 10. Of these, the least informative is the methine region (Fig. 10*a*). The proton lines are too squeezed together to provide separate assignments. The only distinctive feature is the isotactic (mmmm) sequence (at 28.7 ppm for ^{13}C).

As for the 2D spectrum of the methyl group (Fig. 10*b*), all the pentads in the ^{13}C NMR spectrum can be correlated to proton chemical shifts. The result is summarized in Table 6. It is interesting to note that the chemical shifts of the pentads occur in the same order in the ^{1}H spectrum as in the ^{13}C spectrum.

The methylene region is more complex. At the tetrad level, six sequences are possible. The tetrads (mmm), (mmr), (rmr), and (rrm) have two non-equivalent methylene hydrogen atoms, thereby producing eight distinct resonances. The methylene hydrogen atoms for the (rrr) and (mrm) tetrads are equivalent, thus giving two resonances. Thus, there are a total of 10 ^{1}H

Table 6. Stereoirregular Polypropylene: Methyl Assignments

Pentad	^{13}C	^{1}H[a]	^{1}H[b]
mmmm	21.8	0.871	0.871
mmmr	21.6		0.867
rmmr	21.4		0.859
mmrr	21.0		0.856
mrmm	20.8	0.849	0.852
rmrr	20.8		0.852
mrmr	20.6		0.846
rrrr	20.3		0.845
mrrr	20.2		0.837
mrrm	19.9	0.824	0.835

[a] See Ref. 118.
[b] See Ref. 54.

Table 7. NMR Assignments for the Methylene Regions of Stereoirregular Polypropylene

No.	^{13}C Shift	Hexad	Tetrad	^{1}H Shift
1	47.80	mrmrm		
2	47.68	mrmrr		
3	47.55	rrmrr		
4	47.54	mrrrm		
5	47.32	mrrrr		
6	47.18	mrmmr		
7	47.15	rrrrr	rrr	1.075
8	47.08	rmmrr	mrr	1.038 1.117
9	47.08	mrmmm	mrm	1.068
10	46.96	rrmmm	rmr	0.974 1.194
11	46.80	mrrmr	mmr	0.951 1.242
12	46.72	rmmmr	mmm	0.888 1.286
13	46.63	rrrmr		
14	46.58	mrrmm		
15	46.58	rmmmm		
16	46.50	mmmmm		
17	46.33	rrrmm		
18	45.88	rmrmr		
19	45.70	rmrmm		
20	45.52	mmrmm		

Reproduced with permission from Springer-Verlag, Heidelberg (88).

resonances. In contrast, the methylene ^{13}C resonances are resolved at the hexad level, and there are 20 hexads.

The ^{13}C—^{1}H CSCM plots for the methylenes are shown in Figure 10*c*. The ^{13}C spectral assignments have been most recently carried out by Zambelli et al. (120), Schilling and Tonelli (121), and Cheng and Lee (122). Using the ^{13}C assignments and the ^{13}C–^{1}H shift correlations, one can readily assign all ten ^{1}H resonances (88). The complete ^{1}H assignments are summarized on the right-hand side of Table 7.

6. ^{13}C NMR ANALYSIS OF POLYMERS

6.1. General Comments

Carbon-13 NMR is a very suitable method for the characterization of polymers. The range of chemical shifts is large, and the different *n*-ad sequences are frequently spread out to permit detailed assignments. The numerous applications of ^{13}C NMR to polymeric systems have been amply reviewed (2–9). The major applications of ^{13}C NMR can be categorized as follows.

6.1.1. *Polymer Identification by ^{13}C NMR*

The approaches used here are similar to those employed for organic compounds: (1) Look up the structure (or substructure) in polymer spectral libraries. The spectral collections of Sadtler (123) and Pham et al. (124) are especially useful. (2) Use empirical substituent chemical shift rules to predict the ^{13}C shifts.

6.1.2. *Homopolymer Tacticity*

The tacticity of all commercially important polymers have been studied by ^{13}C NMR. Not all the results are unambiguous. For example, it was only recently that a consensus was reached for the assignments of the tacticity of polystyrene (125–130).

6.1.3. *Copolymer Sequence Distribution*

The analytical approach (Section 3.3) has been generally successful in tackling sequence distribution of binary copolymers. For 21 common vinyl and vinylidene monomers, there could be a total of 420 binary copolymers and 1330 ternary copolymers. Only about 30 combinations have been characterized thus far by ^{13}C NMR (40). The traditional, "analytical" approach has been used in most reports. The synthetic approach (e.g., program PSPEC) can handle all vinyl and vinylidene copolymers (40). However, detailed ^{13}C-shift rules must be available.

6.1.4. *Determination of Chain Branching*

Most of the work done to date concentrates on polyethylene (56, 131–139) and poly(vinyl chloride) (140–143). Chain branching in irradiated polymers (144–148) has also been reported.

6.1.5. *Determination of Defect Structure*

A major defect structure of interest in vinyl polymers is the regiosequences due to head-to-head and tail-to-tail additions. Some examples have been given in Bovey (5). Another example (on irregular polypropylene) will be illustrated in Section 6.2.

6.1.6. *Substitution Patterns*

Cellulose may be substituted by nitro group, acetyl, carboxymethyl, hydroxyethyl, hydroxypropyl, and other functionalities. The substitution pattern determines the ultimate physical properties of the materials. The use of ^{13}C NMR spectroscopy turns out to be a powerful technique for this purpose (149–157).

6.1.7. *Determination of Molecular Weight*

In many samples, the polymer end groups can be directly observed. In some cases, they may be enriched with ^{13}C, ^{2}H, or ^{15}N. In still other cases, the initiator fragment (e.g., azobisisobutyronitrile) or a terminating reagent may be observed (158–163). The number-average molecular weight $\bar{M}_n$ may be calculated in these cases. Owing to spectral sensitivity, it is rare that $\bar{M}_n$ larger than 10,000 may be computed in this way. In special cases, for example, branched polyethylene (56), somewhat higher $\bar{M}_n$ may be determined. The identification of chain-end structures sometimes also enables the initiation or termination mechanism to be deducted (57, 138–163).

6.1.8. *Polymer Reactions*

Carbon-13 NMR can be used to follow the reactions of polymers. Many examples of these reactions occur in industry. Some outstanding examples are the chlorination of polyolefins (164, 165), poly(vinyl chloride) (166–168), and natural rubber (169), and the oxidation of polyethylene (170).

6.2. Examples of ^{13}C NMR Studies—Analytical Approaches

An example of polymer sequence analysis will be shown for *regioirregular polypropylenes* (55). These materials can be made through either vanadium-type catalysts or specially designed titanium catalysts. Apart from the lack of specific tacticity (i.e., stereoirregularity), they contain head-to-head and tail-to-tail structures for propylene (i.e., regioirregularity). The ^{13}C NMR spectra of these polymers are very complex.

The ^{13}C NMR spectrum for a sample of irregular polypropylene is shown in Figure 11. The spectral features are broad, indicating the presence of complex structures. Through the use of model polymers, APT experiment, and empirical additive shift rules, the spectrum has been fully assigned. The assignments are summarized in Table 8 and also indicated in Figure 11. It is important to note that the assignment scheme clearly shows the interplay of

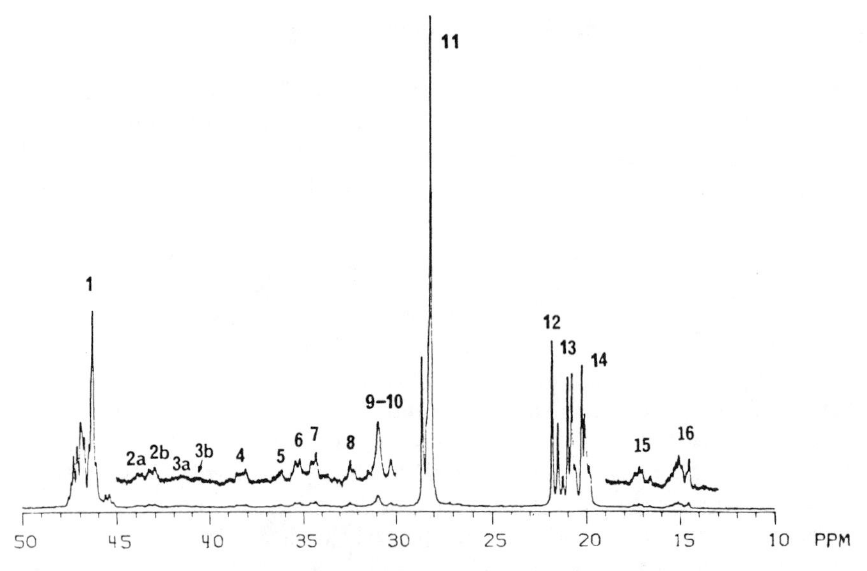

Figure 11. ^{13}C NMR spectrum at 90 MHz of a stereoirregular and regioirregular polypropylene. Reproduced with permission from Springer-Verlag, Heidelberg (55).

stereochemistry and regiochemistry, and as a result several sequence units are severely overlapped in the spectrum.

In Table 8, two sets of nomenclature have been used for the various microstructural units. The first is that of Carman et al. (49) as expanded by Smith (171) and Doi (172). In this system, S, T, P refer, respectively, to secondary (methylene), tertiary (methine), and primary (methyl) carbon atoms. Two Greek subscripts are used, referring to the distances the nearest methyl groups are placed from the carbon in question. Where four Greek subscripts appear, they refer to the distances of the two nearest methyl substituents on each side of the given carbon.

In the second nomenclature (55, 173), the polymer chain is represented by 0 and 1, with 0 for methylene and 1 for methyl–methine units. Furthermore, the absolute configuration is shown by 1 (for dextro placement) and $\bar{1}$ (for levo placement). Where an underlined $\underline{1}$ occurs, both dextro and levo placements apply. For convenience, tacticity is also represented by m (meso) or r (racemic) with a subscript corresponding to the number of methylene units between the methyls–methines, for example,

0110

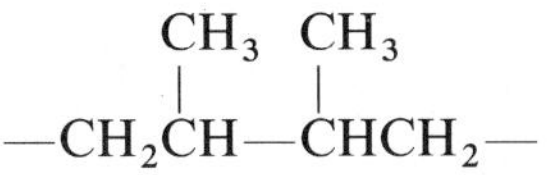

$$
\begin{array}{c}
\mathrm{CH_3}\quad\mathrm{CH_3}\\
|\qquad\;\;|\\
-\mathrm{CH_2CH}-\mathrm{CHCH_2}-
\end{array}
$$

m_0

$01\bar{1}0$

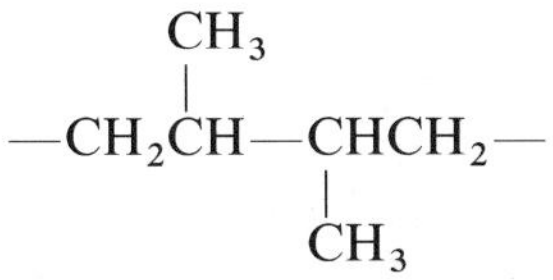

$$
\begin{array}{c}
\mathrm{CH_3}\\
|\\
-\mathrm{CH_2CH}-\mathrm{CHCH_2}-\\
|\\
\mathrm{CH_3}
\end{array}
$$

r_0

01010

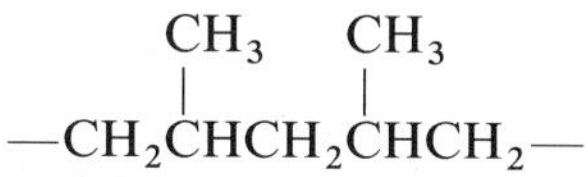

$$
\begin{array}{c}
\mathrm{CH_3}\qquad\mathrm{CH_3}\\
|\qquad\qquad|\\
-\mathrm{CH_2CHCH_2CHCH_2}-
\end{array}
$$

m_1

$010\bar{1}0$

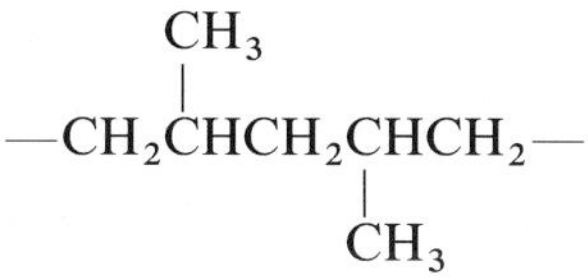

$$
\begin{array}{c}
\mathrm{CH_3}\\
|\\
-\mathrm{CH_2CHCH_2CHCH_2}-\\
|\\
\mathrm{CH_3}
\end{array}
$$

r_1

010010

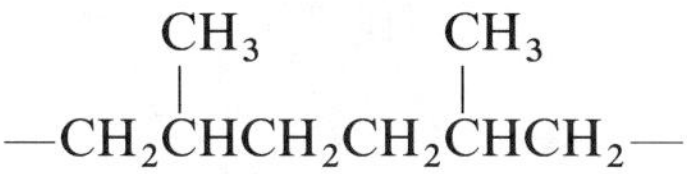

$$
\begin{array}{c}
\mathrm{CH_3}\qquad\qquad\mathrm{CH_3}\\
|\qquad\qquad\qquad|\\
-\mathrm{CH_2CHCH_2CH_2CHCH_2}-
\end{array}
$$

m_2

$0100\bar{1}0$

$$
\begin{array}{c}
\mathrm{CH_3}\\
|\\
-\mathrm{CH_2CHCH_2CH_2CHCH_2}-\\
|\\
\mathrm{CH_3}
\end{array}
$$

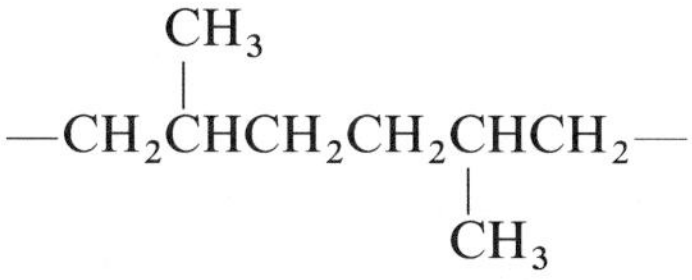

r_2

In NMR spectra with overlapping signals, the best analytical approach (in the view of this author) is to use the computer-assisted analytical approach (Section 3.3). In this case, the second-order Markovian probabilities can be used. Regioirregular polypropylene is regarded as the copolymer of head-first propylene (p) and tail-first propylene (q).

$$
\begin{array}{llll}
\alpha = P_{\mathrm{ppp}} & \beta = P_{\mathrm{pqp}} & \gamma = P_{\mathrm{qpp}} & \delta = P_{\mathrm{qqp}} \\
\bar{\alpha} = P_{\mathrm{ppq}} & \bar{\beta} = P_{\mathrm{pqq}} & \bar{\gamma} = P_{\mathrm{qpq}} & \bar{\delta} = P_{\mathrm{qqq}}
\end{array}
$$

where P_{ijk} is the probability of monomer k adding to a polymer chain end terminating in units i and j.

The theoretical expressions for the signal intensities can be derived (Table 9). The theoretical intensities are compared with the observed intensities through a simplex optimization process (59), thereby producing the best fit values for the probability parameters (α, β, γ, δ). Using these optimized α, β, γ, and δ, one can obtain the composition, and the diad and the triad comonomer sequences.

$$
\mathrm{p} = (01) = k(\bar{\alpha} + \gamma)\delta \tag{16a}
$$

Table 8. ^{1}C NMR Assignments for Irregular Polypropylene

No.	Shift (ppm)	Designation	Structure
1a	45.0–47.0	$S_{\gamma\alpha\alpha\gamma}$	$\underline{1}0\underline{1}\dot{0}\underline{1}0\underline{1}$
1b		$S_{\gamma\alpha\alpha\delta}$	$\underline{1}0\underline{1}\dot{0}\underline{1}00$
2a	~43.8	r_0–$S_{\beta\alpha\alpha\gamma}$	$0\bar{1}1\dot{0}\underline{1}0\underline{1}$
2b	~43.0	r_0–$S_{\beta\alpha\alpha\delta}$	$0\bar{1}1\dot{0}\underline{1}00$
3a	~41.8	m_0–$S_{\beta\alpha\alpha\gamma}$	$011\dot{0}\underline{1}0\underline{1}$
3b	~41.0	m_0–$S_{\beta\alpha\alpha\delta}$	$011\dot{0}\underline{1}00$
4a	39.0	m_0–$T_{\alpha\gamma}$	$01\dot{1}00\underline{1}$
4b	38.2	r_0–$T_{\alpha\gamma}$	$0\bar{1}\dot{1}00\underline{1}$
5a	36.8	m_0r_1–$S_{\gamma\alpha\beta\gamma}$	$\bar{1}01\dot{0}011 + \bar{1}01\dot{0}0\bar{1}\bar{1}$
5b	36.2	r_0r_1–$S_{\gamma\alpha\beta\gamma}$	$\bar{1}01\dot{0}01\bar{1} + \bar{1}01\dot{0}0\bar{1}1$
6	35.0–36.0	m_0–$T_{\alpha\beta}$	$01\dot{1}0\underline{1}$
		m_1–$S_{\gamma\alpha\beta\gamma}$	$101\dot{0}0\underline{1}\underline{1}$
		r_1–$S_{\gamma\alpha\beta\delta}$	$\bar{1}01\dot{0}0\underline{1}0\underline{1}$
7	33.7–35.0	r_0–$T_{\alpha\beta}$	$0\bar{1}\dot{1}0\underline{1}$
		m_1–$S_{\gamma\alpha\beta\delta}$	$101\dot{0}0\underline{1}0\underline{1}$
8	32.0–32.7	r_0–$S_{\beta\alpha\beta\gamma}$	$0\bar{1}1\dot{0}0\underline{1}\underline{1}0$
		r_0–$S_{\beta\alpha\beta\delta}$	$0\bar{1}1\dot{0}0\underline{1}0\underline{1}$
9	30.5–31.5	m_0–$S_{\beta\alpha\beta\gamma}$	$011\dot{0}0\underline{1}\underline{1}0$
		m_0–$S_{\beta\alpha\beta\delta}$	$011\dot{0}0\underline{1}0\underline{1}$
10a	31.5	$T_{\beta\gamma}$	$\underline{1}\underline{1}0\dot{1}00\underline{1}$
10b	31.0	$T_{\beta\gamma}$	$0\underline{1}0\dot{1}00\underline{1}$
11a	29.1	$T_{\beta\beta}$	$\underline{1}\underline{1}0\dot{1}0\underline{1}$
11b	29.0–28.1	$T_{\beta\beta}$	$0\underline{1}0\dot{1}0\underline{1}$
12	22.0–21.2	m_1m_1–$P_{\beta\beta}$	$10\bar{1}01$
13	21.2–20.5	m_1r_1–$P_{\beta\beta}$ and m_1–$P_{\beta\gamma}$	$10\dot{1}0\bar{1} + 10\dot{1}00\underline{1}$
14	20.5–19.5	r_1r_1–$P_{\beta\beta}$ and r_1–$P_{\beta\gamma}$	$\bar{1}0\dot{1}0\bar{1} + \bar{1}0\dot{1}00\underline{1}$
15	16.5–18.0	m_0–$P_{\alpha\beta}$ and m_0–$P_{\alpha\gamma}$	$01\dot{1}0\underline{1} + 01\dot{1}00\underline{1}$
16	14.5–15.5	r_0–$P_{\alpha\beta}$ and r_0–$P_{\alpha\gamma}$	$0\bar{1}\dot{1}0\underline{1} + 0\bar{1}\dot{1}00\underline{1}$

[a] See text for explanation of symbols.

Reproduced with permission from Springer-Verlag Heidelberg (55).

Table 9. Second-Order Markovian Reaction Probability Parameters

No.	Shift (ppm)	Assignments	Second-Order Markovian
1	45.0–47.0	$S_{\gamma\alpha\alpha\gamma} + S_{\gamma\alpha\alpha\delta}$	$\alpha\gamma\delta + \bar{\alpha}\bar{\beta}\bar{\delta}$
2–3	40.5–44.0	$S_{\beta\alpha\alpha\gamma} + S_{\beta\alpha\alpha\delta}$	$\bar{\alpha}\gamma\delta + \bar{\alpha}\bar{\beta}\delta$
4	38.0–39.5	$T_{\alpha\gamma}$	$\bar{\alpha}\beta\delta + \bar{\alpha}\bar{\gamma}\delta$
5–7	33.7–37.0	$T_{\alpha\beta} + S_{\gamma\alpha\beta\gamma} + S_{\gamma\alpha\beta\delta}$	$2\,\bar{\alpha}\gamma\delta + 2\,\bar{\alpha}\bar{\beta}\delta$
8–10	30.0–33.0	$T_{\beta\gamma} + S_{\beta\alpha\beta\gamma} + S_{\beta\alpha\beta\delta}$	$2\,\bar{\alpha}\delta$
11	28.0–29.1	$T_{\beta\beta}$	$\alpha\gamma\delta + \bar{\alpha}\bar{\beta}\bar{\delta}$
12–14	19.5–22.0	$P_{\beta\beta} + P_{\beta\gamma}$	$\gamma\delta + \bar{\alpha}\bar{\beta}$
15–16	14.5–18.0	$P_{\alpha\beta} + P_{\alpha\gamma}$	$2\,\bar{\alpha}\delta$
	Total		$3\,\bar{\alpha}\bar{\beta} + 6\,\bar{\alpha}\delta + 3\,\gamma\delta$

Reproduced with permission from Springer-Verlag Heidelberg (55).

$$\mathrm{q} = (10) = k(\bar{\beta} + \delta)\bar{\alpha} \tag{16b}$$

$$\mathrm{pp} + \mathrm{qq} = (0101) + (1010) = k(\overline{\alpha\beta} + \gamma\delta) \tag{16c}$$

$$\mathrm{pq} = (0110) = k(\bar{\alpha}\delta) \tag{16d}$$

$$\mathrm{qp} = (1001) = k(\bar{\alpha}\delta) \tag{16e}$$

$$\mathrm{ppp} + \mathrm{qqq} = (010101) + (101010) = k(\overline{\alpha\beta\delta} + \alpha\gamma\delta) \tag{16f}$$

$$\mathrm{ppq} + \mathrm{pqq} = (010110) + (011010) = k(\bar{\alpha}\gamma\delta + \bar{\alpha}\bar{\beta}\delta) \tag{16g}$$

$$\mathrm{qpp} + \mathrm{qqp} = (100101) + (101001) = k(\bar{\alpha}\gamma\delta + \bar{\alpha}\bar{\beta}\delta) \tag{16h}$$

$$\mathrm{qpq} + \mathrm{pqp} = (100110) + (011001) = k(\overline{\alpha\gamma}\delta + \bar{\alpha}\beta\delta) \tag{16i}$$

In the above equations, the normalization constant $k = (\bar{\alpha}\bar{\beta} + 2\,\bar{\alpha}\delta + \gamma\delta)^{-1}$. To reduce the second-order to the first-order Markovian statistics, one simply drops the first subscript, thus

$$\begin{aligned} P_{\mathrm{pp}} &= \alpha = \gamma \qquad & P_{\mathrm{qp}} &= \beta = \delta \\ P_{\mathrm{pq}} &= \bar{\alpha} = \bar{\gamma} \qquad & P_{\mathrm{qq}} &= \bar{\beta} = \bar{\delta} \end{aligned} \tag{17}$$

The product of reactivity ratios is given by

$$r_{\mathrm{p}} r_{\mathrm{q}} = \left(\frac{1}{P_{\mathrm{pq}}} - 1\right) \left(\frac{1}{P_{\mathrm{qp}}} - 1\right) \tag{18}$$

The entire computation has been computerized (Program FIT33). The user enters the spectral intensities and can choose either first- or second-order

Table 10. FIT33 Analysis of Irregular Polypropylene

Line	1	2–3	4	5–7	8–10	11	12–14	15–16
Area	100	3.2	1.5	7.0	10.7	97.7	103.4	6.5

$\alpha = 0.9695$,	$\beta = 0.8975$,	$\gamma = 0.9685$,	$\delta = 0.8590$
$p = 0.967$	$pp + qq = 0.941$		$ppp + qqq = 0.909$
$q = 0.033$	$pq = 0.295$		$ppq + qqq = 0.316$
	$qp = 0.295$		$qpp + qqp = 0.316$
$m_0/r_0 = 30/70$			$qpq + pqp = 0.274$

Reproduced with permission from Springer-Verlag, Heidelberg (55).

Markovian statistics. The program then calculates the optimal values of reaction probability parameters, from which the optimized composition (i.e., level of propylene inversion), as well as diad and triad sequence distribution, is computed. An example of the fitted results is shown in Table 10 for a sample with 3% propylene inversion (55). The data can be fitted equally well with a first-order Markovian model with $P_{pq} = 0.0307$, $P_{qp} = 0.9036$, and $r_1r_2 = 35$.

A major development is the realization that many polymers are mixtures of statistical polymers. As a result analysis of the NMR data through a single reaction probability model may not give the correct picture. Recently, several papers have appeared primarily dealing with ethylene–propylene copolymers and polypropylene using two-state models (174–178). Cheng recently showed that three-state models may be suitable in some cases (63, 179). A general methodology for the treatment of NMR data of *multistate models* was also introduced (62).

Table 11. Computerized Analytical Approach for Polymers Containing Multiple Components

Polymer	Problem	Program	Model	References
Polypropylene	Tacticity	MIXCO.C3	E/B 2-state	62
		MIXCO.C3X	E/E/B 3-state	62
Polybutylene	Tacticity	MIXCO.C4	E/B 2-state	62
		MIXCO.C4X	E/E/B 3-state	63
Ethylene-propylene	Sequence	MIXCO.TRIAD	B/B 2-state	62
			M1/M1 2-state	62
		MIXCO.TRIADX	B/B/B 3-state	63, 179
Ethylene-butylene	Sequence	MIXCO.TRIADX	B/B/B 3-state	180
Propylene-butylene	Sequence	MIXCO.TRIAD	B/B 2-state	62
			M1/M1 2-state	62
Acrylamide-acrylate	Sequence	MIXCO.TRIAD	M1/M1 2-state	62

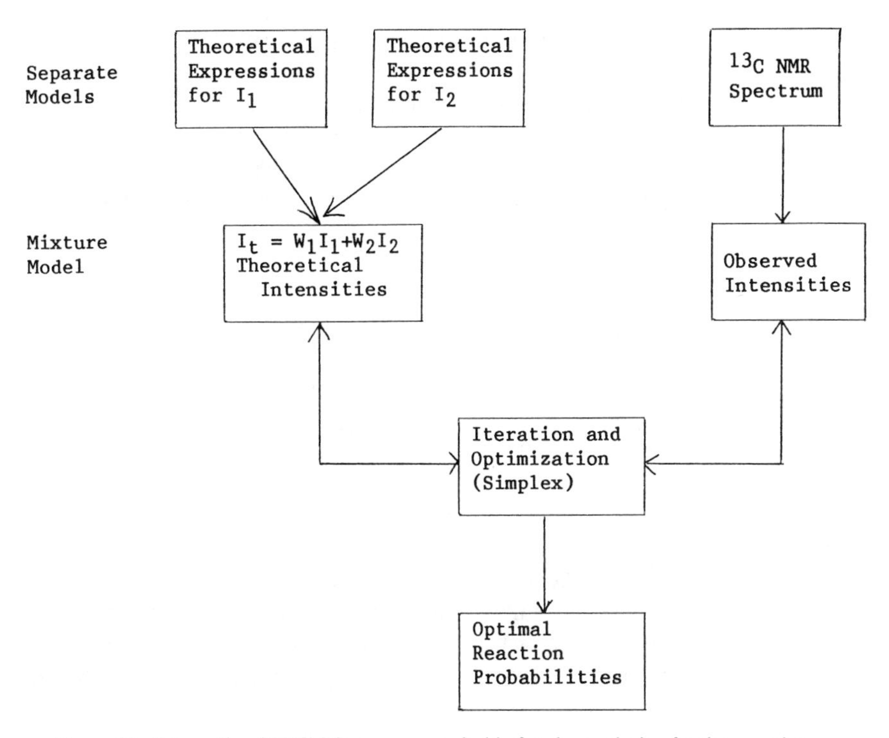

Figure 12. Schematic of MIXCO program, suitable for the analysis of polymer mixtures.

The methodology is based on the computerized analytical approach (59). A schematic is shown in Figure 12. Thus far, quite a number of polymeric systems (62, 63, 179, 180) have been treated by this method (Table 11). A family of computer programs (MIXCO) has been written.

Through this computer-assisted analytical method, the two-state models can easily be tested for either homopolymer tacticity or copolymer triad sequence. However, for three-state models, it is necessary to use data from fractionated polymers. In Table 11, the programs ending in X (MIXCO.C3X, MIXCO.C4X, and MIXCO.TRIADX) are used to treat the NMR data of a pair of polymer fractions at the same time. By this means, it was shown unambiguously that polypropylene (62) and poly(1-butene) (63) can be considered to consist of three components; two obeying the enantiomorphic-site (E) statistics and one obeying the Bernoullian (B) statistics. Similarly, for ethylene–propylene copolymer (179) and ethylene-1-butene copolymer (180), the discrete three-component model was found to be appropriate in the interpretation of the NMR data of polymer fractions.

6.3. Spectral Simulation Approaches

An alternative method of treating the ethylene and propylene polymers is to use the spectral simulation (synthetic) approach. In this case, detailed ^{13}C-shift rules have been devised (181, 182) taking into account stereosequences (i.e., tacticity) and regiosequence (i.e., head-to-head and tail-to-tail structures). The rules are different for methyl, methine, and methylene carbon atoms, and are summarized below.

$$\delta(CH_3) = 19.99 + \sum_i A_i + \sum_{i,j} B_{ij} \tag{19}$$

$$\delta(CH) = 33.26 + \sum_i A_i + \sum_{i,j} B_{ij} \tag{20}$$

$$\delta(CH_2) = 29.9 + \sum_i A_i + \sum_{i,j} B_{ij} + \sum_{i,j} C_{ij} + \sum_k D_k \tag{21}$$

where A_i corresponds to the substituent parameter for position i, B_{ij} is the corrective term for methyl groups at positions i and j on the same side of the methylene group, and C_{ij} is to be used for methyl groups at positions i and j on the opposite sides of the methylene group in question. The term D_k is specifically designed to take into account the effect of polypropylene tacticity on methylene carbon atoms. After a careful study of the published literature, consistent sets of values have been obtained (181).

The rules delineated above have been completely encoded in a computer program, CALMOD (181). As an example, a low-molecular-weight analog of a regioirregular structure can be shown below.

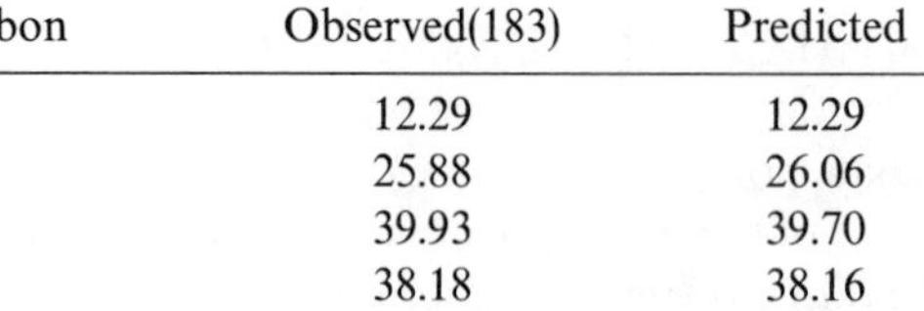

```
        6   7
        C   C           C   C
        |   |           |   |
C—C—C—C—C—C—C—C—C—C
1   2   3   4   5
```

Carbon	Observed(183)	Predicted
1	12.29	12.29
2	25.88	26.06
3	39.93	39.70
4	38.18	38.16
5	31.36	31.65
6	16.21	16.30
7	16.69	17.05

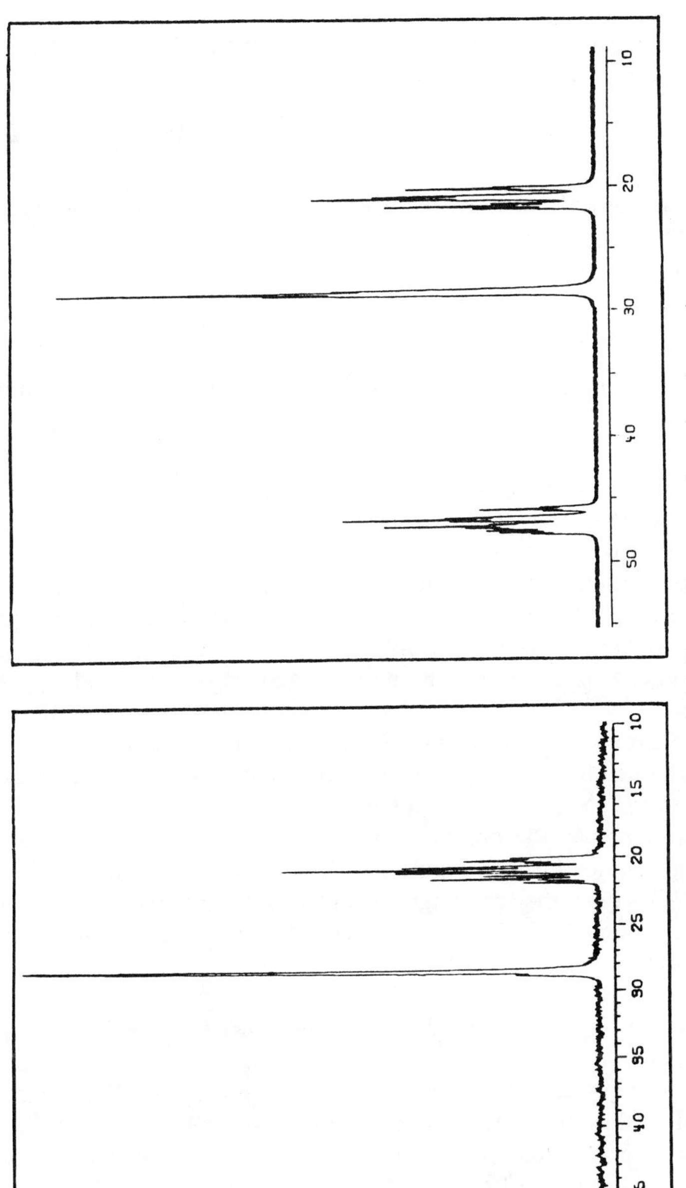

Figure 13. The observed (left) and the predicted (right) ^{13}C NMR spectra of atactic polypropylene.

In most cases examined, the rules give predicted shifts that are within 0.3 ppm of the observed values. More importantly, the rules give accurate relative shifts for related structures (e.g., stereosequences and regiosequences). The use of these rules therefore permits detailed assignments of ^{13}C NMR spectra.

These empirical shift rules have been applied successfully not only to low-molecular-weight compounds, but also to various polymers of ethylene and propylene (182). Polymer systems of significance include

$$C_3 \rightarrow \text{isotactic } C_3 \rightarrow \text{irregular } C_3 \rightarrow C_3 \text{ oil}$$

$$C_2/C_3 \rightarrow \text{stereoregular } C_2C_3 \rightarrow \text{ethylene–propylene rubber} \rightarrow \text{ethylene–propylene oil}$$

where polymer microstructure becomes increasingly complex as one moves from left to right. The ^{13}C NMR spectral features correspondingly increase in complexity from left to right.

To carry out spectral simulation, one needs a suitable reaction probability model. For ethylene–propylene copolymers, Bernoullian and first-order Markovian models can be easily applied. For homopolypropylene, both chain-end controlled models (Bernoullian and Markovian) and catalytic-site controlled models (enantiomorphic site) can be used (182). Recently, hybrid models (involving both chain-end and catalytic-site control at once) and mixture models (involving the sum of two or more polymer components) have been used in spectral simulation (44).

As an example, the observed and the simulated ^{13}C spectra of atactic polypropylene are given in Figure 13. The chemical shift values calculated by Eqs. 19–21 gave rather good agreement (44) with the observed shifts and the shifts calculated for the rotational isomeric state model of Tonelli and Schilling (121).

6.4. Miscellaneous Studies

One of the major developments in ^{13}C NMR of polymers is the application of the rotational isomeric state model (RIS) to predict relative ^{13}C shifts. For a given polymer, this requires the knowledge of the energy barriers to bond rotation and the proper RIS description (184). If this is known (e.g., from other measurements), then the relative shifts can be calculated. Tonelli and Schilling (185) have done most of the work in this area. Interested readers may consult Refs. 121, 128, 185, and 186.

The nature of the reaction mechanism for Ziegler–Natta polymerization has been vigorously pursued by Zambelli and co-workers (187–190) and by

Fink (191, 192) using ^{13}C labeling. For both groups ^{13}C labeling has provided very specific information on the catalyst–monomer complexes. Computer modeling and calculations have been done to rationalize the ^{13}C data (193, 194). The work is of considerable significance especially since the mechanism of the Ziegler–Natta polymerization has been in dispute for such a long time.

Two-dimensional NMR techniques have been used in abundance in ^{13}C NMR (98, 99), frequently in conjunction with ^{1}H NMR in CSCM-type experiments (see Section 4). Continued use of various 2D techniques is anticipated in the future.

7. NUCLEAR RELAXATION

7.1. Description of Polymer Relaxation

The use of NMR relaxation in polymers is well known and has been described in a number of reviews (10, 195–198). Carbon-13 is presently the most widely used technique. For ^{13}C, the dipolar relaxation is the predominant mechanism in polymeric systems. The relaxation parameters are given as follows:

$$\frac{1}{NT_1} = \frac{1}{10}\left(\frac{\gamma_H^2\gamma_C^2\hbar^2}{r_{CH}^6}\right)[J(\omega_H - \omega_C) + 3J(\omega_C) + 6J(\omega_H + \omega_C)] \tag{22}$$

$$\frac{1}{NT_2} = \frac{1}{20}\left(\frac{\gamma_H^2\gamma_C^2\hbar^2}{r_{CH}^6}\right)[4J(0) + J(\omega_H - \omega_C) + 3J(\omega_C) + 6J(\omega_H) + 6J(\omega_H + \omega_C)] \tag{23}$$

$$\eta = \frac{\gamma_H}{\gamma_C}\left[\frac{6J(\omega_H + \omega_C) - J(\omega_H - \omega_C)}{J(\omega_H - \omega_C) + 3J(\omega_C) + 6J(\omega_H + \omega_C)}\right] \tag{24}$$

where η is the NOE factor and is related to NOE (NOE $= \eta + 1$), $\hbar$ is Dirac's constant, γ_C and γ_H are the carbon and proton gyromagnetic ratios, respectively, r_{CH} is the internuclear C—H distance to the Nth proton, and ω_C and ω_H are the respective carbon and proton Larmor frequencies. Except for the internuclear distances, all N protons are considered equivalent.

The spectral density function $J(\omega)$ is defined by

$$J(\omega) = \int_{-\infty}^{\infty} G(\tau)\exp(-i\omega\tau)d\tau \tag{25}$$

where $G(\tau)$ is the autocorrelation function for the motion in question. In a highly simplified case, an isotropic motion characterized by a single correlation time τ_c can be applied in analogy to lower molecular weight analogs (199). The spectral density function is given by

$$J(\omega_i) = \frac{\tau_c}{1 + \omega_i^2 \tau_c^2} \tag{26}$$

An extension of Eq. 26 is to use a distribution of correlation times (200). More detailed mathematical descriptions of polymer chain motions have been given by Valeur et al. (201) (called the VJGM theory), Jones and Stockmayer (202), Bendler and Yaris (203), Hall and Helfand (204), Helfand (205) and recent updates (206–208). It is now generally accepted that motion along the polymer backbone is a diffusive process that decays with distance from the site of perturbation; thus,

perturbation → chain connectivities → motional correlations → diffusion

The mathematical description of motion varies with different treatments (201–205). The general form for the autocorrelation function is

$$G(\tau) = e^{-t/\tau_2} e^{t/\tau_1} I_0(t/\tau_1) \tag{27}$$

where the correlation time τ_1 characterizes diffusive motion along the polymer chain (e.g., three bond jumps), and the correlation time τ_2 characterizes all other processes (e.g., motional damping). The function I_0 can take on different forms depending on the detailed description of the models used.

For polymeric systems with double bonds in the backbone (e.g., diene polymers), local rotations that are nondiffusive in nature also occur. Description of these complex motions is more difficult to devise (206).

Another type of polymer motion is that due to side chains. In many cases the formalism initially derived by Woessner (209) applies:

$$J(\omega_i) = \frac{A\tau_R}{1 + \omega_i^2 \tau_R^2} + \frac{B\tau_B}{1 + \omega_i^2 \tau_B^2} + \frac{D\tau_D}{1 + \omega_i^2 \tau_D^2} \tag{28}$$

where

$$\tau_B^{-1} = \tau_R^{-1} + (6\tau_{\text{int}})^{-1} \tag{29}$$

$$\tau_D^{-1} = \tau_R^{-1} + 2(3\tau_{\text{int}})^{-1} \tag{30}$$

and A, B, and D are constants that depend on the geometry of the internal motion. The terms τ_R and τ_{int} are correlation times for isotropic segmental motion of the main chain and for internal rotation of side chains, respectively. A more detailed description of side-chain motion has been provided by Lipari and Szabo (210, 211). The combined effects of polymer main chain motion coupled with side-chain motion have been delineated by Pant et al. (208).

More recently, a different line of approach was developed by Bahar and Erman (212–214), using dynamic rotational isomeric state models. An improved version of this formalism has found good agreement with ^{1}H and ^{13}C T_1 data of poly(ethylene oxide) (215).

7.2. Applications

Studies of polymer relaxation can provide a significant amount of information. Thus far, most of the relaxation measurements have been used for three purposes.

1. To gain knowledge of polymer dynamics and to make structural correlations.
2. To serve as an aid in spectral assignments.
3. To probe polymer–polymer or polymer–solvent interactions.

Since the appearance of VJGM (201) and other refined theories (202–205), a number of studies have been carried out to test the theories and to derive more information on polymer motion. Most of the work has been reviewed by Heatley (196). In general, the agreement with the various theories has been good at temperatures much above T_g. The utility of Eq. 27 has since been established. However, NMR parameters (T_1, T_2, and NOE) are usually not sensitive enough to indicate which functional form is preferable for the term $I_0(t/\tau_1)$.

An important finding in polymer relaxation measurements was that resonances corresponding to different stereosequences sometimes have different T_1 values. In fact, frequently the isotactic sequence has the longest T_1, followed by heterotactic, and then syndiotactic. This trend was found for the ^{13}C relaxation of polypropylene (216, 217), poly(1-butene) (218), poly(methyl methacrylate) (219–222), and poly(vinyl chloride) (223), and for the ^{1}H relaxation of poly(methyl methacrylate) (224, 225). For polymer chains bearing rings, for example polystyrene (220, 226, 227), poly(α-methylstyrene) (228), and poly(vinyl pyrrolidone) (229), the isotactic forms have smaller ^{13}C T_1 values.

Proton magnetic resonance has been used very effectively to study polymer solution behavior. In an early work, for example, Anderson et al. (230) showed that for poly(ethylene oxide), polydimethylsiloxane, and polyisobutylene, T_1 is

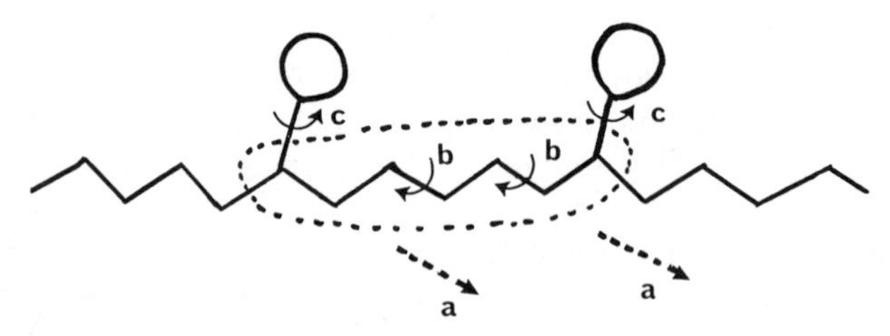

Figure 14. Schematic representation of three modes of motion that contribute to ^{13}C T_1 process. Motion a is the overall molecular tumbling. Motion b is segmental motion along the backbone bonds. Motion c is the side-chain rotation.

insensitive to the molecular weight of the polymer and to the concentration of the solution up to about 20% concentration. It is, however, dependent on solvent viscosity. In contrast, T_2 is a function of polymer concentration and molecular weight. This was interpreted as due to chain entanglements in polymer solutions. This T_2 dependence on chain entanglements was used effectively by Charlesby and co-workers (231, 232) to measure the degree of cross-linking in irradiated polymers.

The ^{13}C NMR relaxation data are easier to interpret. The observed T_1 can usually be attributed to three kinds of motion: overall tumbling, segmental motion, and side-chain motion. As in ^{1}H, the overall tumbling becomes insignificant at higher molecular weight (233). For most polymers, segmental motion dominates the relaxation of the backbone carbon atoms. Much work has been done using different relaxation models (199–209), as reviewed by Heatley (196). For side chains, additional motions around the side-chain axis is possible (Fig. 14). The observed T_1 in general will be the sum of all 3 motions.

$$\left(\frac{1}{T_1}\right)_{\text{obsd}} = \left(\frac{1}{T_1}\right)_{\text{tumbling}} + \left(\frac{1}{T_1}\right)_{\text{segmental}} + \left(\frac{1}{T_1}\right)_{\text{side chain}} \tag{31}$$

Polymer–solvent interactions have been studied by a number of workers. For example, Spavacek and Schneider (234) and Inoue et al. (220) have used ^{1}H T_1 to study the interaction of poly(methyl methacrylate) with several solvents. Cheng et al. (235) used ^{13}C NMR to monitor the interaction of poly(vinyl pyrrolidone) with water, tetrachloroethane, and benzene. Leyte and co-workers (236, 237) studied the dynamic effects of poly(ethylene oxide) and polyelectrolytes in water using ^{1}H, ^{13}C, and ^{17}O NMR relaxation.

In a few cases NMR relaxation has been used to probe polymer–polymer interactions. Bovey et al. (238) showed that there is no change in the ^{13}C T_1 values when poly(methyl methacrylate) and poly(vinylidene fluoride) are mixed together in solution. Since these two polymers form a compatible blend, the data mean either there is no complex formation between the two polymers, or alternatively, the time scale of interaction is much slower than those accessible to T_1 measurements (typically 10^{-9}–10^{-12} s).

7.3. Polymer Melt Relaxation

Different approaches are needed for slow molecular motions, for example polymers in the melt or in concentrated solutions, and polymers that are cross-linked or in the gel state. For such "entangled" polymers, two approaches have been proposed. Kimmich's model applies to T_1 and $T_{1\rho}$ (239, 240), and is based on a reptation model. Cohen–Addad's formalism is especially useful for T_2 processes (241, 242). These approaches are theoretically significant, and may be correlated to mechanical relaxation (243).

8. MULTINUCLEAR NMR

Although most of the polymer NMR work reported in the literature involves ^{1}H and ^{13}C, other nuclei may also provide valuable information in suitable cases (e.g., Table 12). In theory, any nucleus in the periodic table that possesses a nonzero spin can be potentially studied, and this includes 80% of the elements in the periodic table. Only three qualifications limit the general applicability of NMR: (1) NMR is not a trace-analytical tool and the concentration of the observed species should be at least in the range of 0.5%; (2) paramagnetic nuclei are observed with difficulty; and (3) for high-resolution liquid techniques, the sample must be soluble; otherwise solid-state techniques must be used.

The literature on multinuclear NMR has become very extensive (244–248). In polymer studies, however, there are relatively few papers. This is partly due to the newness of the field so that many researchers are not fully aware of the potentials of this technique. The multinuclear NMR studies of polymers have been recently reviewed (249) and for polymer characterization, multinuclear NMR can be applied to several areas as discussed below.

8.1. Studies of Polymer Structure

Studies of polymer structure by far constitute the largest use of multinuclear NMR. For this purpose, quadrupolar nuclei usually give linewidths too broad to provide microstructural information. Only spin $\frac{1}{2}$ nuclei are useful, particularly ^{15}N, ^{29}Si, ^{19}F, and ^{31}P.

Nitrogen-14 is 99.63% abundant but has a quadrupolar moment. Thus it is not difficult to see the ^{14}N signal, but the spectral line can be very broad. In

Table 12. Properties of Some Nuclei Used in Polymer NMR

	Spin	Natural Abundance (%)	NMR Frequence (MHz)	Gyromagnetic Ratio (γ) ($\times 10^7$ rad/T/s)	Quadruple Moment, Q ($\times 10^{28}$ m^2)	Relative Receptivity		Approximate Shift Range (ppm)
						D^P	D^C	
^1H	$\frac{1}{2}$	99.985	100.000	26.75		1.000	5.68×10^3	15
^2H	1	1.56×10^2	15.351	4.1064	2.73×10^{-3}	1.45×10^{-6}	8.21×10^3	15
^{13}C	$\frac{1}{2}$	1.108	25.145	6.7263		1.76×10^{-4}	1.00	220
^{14}N	1	99.634	7.224	1.9324	1.6×10^{-2}	1.00×10^{-3}	5.69	900
^{15}N	$\frac{1}{2}$	0.365	10.137	-2.7107		3.85×10^{-6}	2.19×10^{-2}	900
^{17}O	$\frac{5}{2}$	3.7×10^{-2}	13.557	-3.6266	-2.6×10^{-2}	1.08×10^{-5}	6.11×10^{-2}	800
^{19}F	$\frac{1}{2}$	100	94.094	25.17		0.8328	4.73×10^{-3}	800
^{23}Na	$\frac{3}{2}$	100	26.451	7.08	0.12	9.25×10^{-2}	5.25×10^{-2}	70
^{29}Si	$\frac{1}{2}$	4.70	19.867	-5.3141		3.69×10^{-4}	2.09	250
^{31}P	$\frac{1}{2}$	100	40.481	10.8290		0.0663	3.77×10^2	700

contrast, ^{15}N has spin $\frac{1}{2}$ but is only 0.31% abundant. It is a difficult nucleus to observe without isotopic enrichment, especially since the NOE is negative. Nevertheless, owing to immense chemical interest, a large body of data has been accumulated over the years. Prior to 1982 Kricheldorf and his co-workers had done most of the ^{15}N-polymer NMR work (249). Since then, a number of polymers have been studied:

Polyamides

Alternating and random copolyamides (250, 251).
Nylon 66 (252)

Polyamines

Poly(vinyl amine) and its pH dependence (253)
Polyethyleneimine-based polymers (254)

Thermosets

Urea–formaldehyde and melamine–formaldehyde resins (255, 256)
Bisphenol-A diglycidyl ether (DGEBA)–amine polymers (257, 258)

Silicon-29 has spin $\frac{1}{2}$, 4.7% natural abundance, and a negative NOE. It is a relatively easy nucleus to study. The most commonly studied silicon-containing organic polymers are the polysiloxanes (silicones) and polysilanes. The structures found in silicones and silicates can be depicted as follows (259):

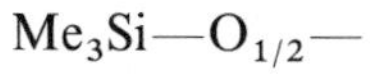

$$\underset{\text{M}}{Me_3Si{-}O_{1/2}{-}} \qquad \underset{\text{D}}{{-}O_{1/2}{-}\overset{Me}{\underset{Me}{Si}}{-}O_{1/2}{-}} \qquad \underset{\text{D}'}{{-}O_{1/2}{-}\overset{Me}{\underset{H}{Si}}{-}O_{1/2}{-}}$$

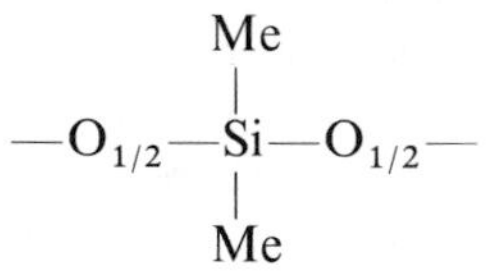

$$\underset{\text{T}}{{-}O_{1/2}{-}\overset{O_{1/2}}{\underset{Me}{Si}}{-}O_{1/2}{-}} \qquad \underset{\text{Q}}{{-}O_{1/2}{-}\overset{O_{1/2}}{\underset{O_{1/2}}{Si}}{-}O_{1/2}{-}}$$

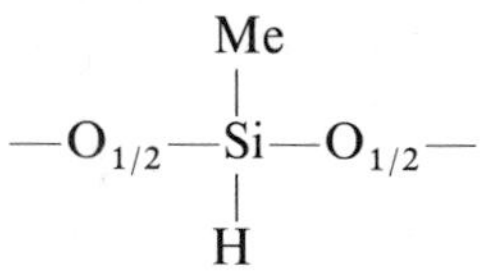

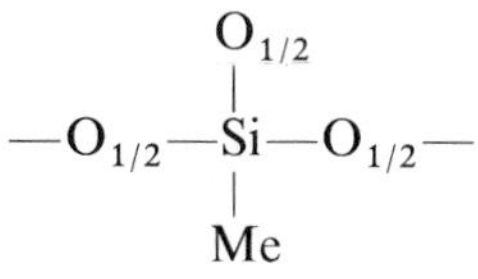

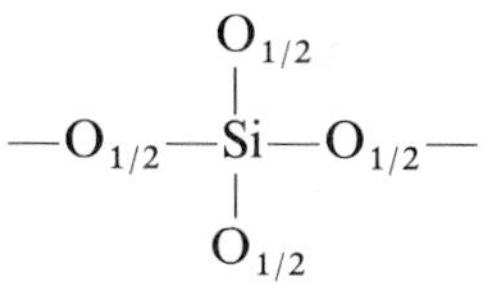

A sample ^{29}Si spectrum is given in Figure 15. The range of ^{29}Si-chemical shift from M to D′ is 30 ppm; in contrast the ^{1}H range is 0.2 ppm and the ^{13}C range

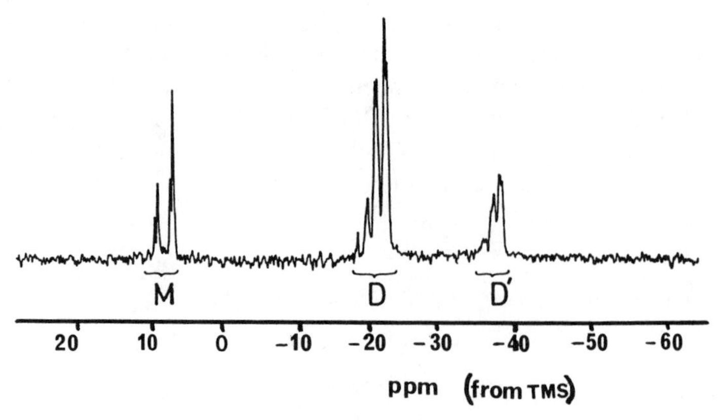

Figure 15. ^{29}Si (^{1}H) spectrum of a copolymer mixture of overall average composition $M_2D_{5.5}D'_{2.5}$. The assignments are indicated in the figure. Reprinted from Ref. 259 by courtesy of Marcel Dekker, Inc.

is 2.0 ppm. A comparison of the end groups (M) versus the internal groups (D, D′) gives the number-average molecular weight. The fine structures in M and D correspond to various sequence lengths. The fine structures in D′ possess information on both sequence lengths as well as tacticity (259).

Recently, a number of additional papers have appeared using ^{29}Si for the characterization of polysiloxanes. Laude and Wilkins (260, 261) used a recycled-flow method to sidestep the problem of long T_1 values and determined the molecular weights of several polysiloxanes. Jancke and Engelhardt (262, 263) analyzed the ^{29}Si spectral region corresponding to the M, D, and T structures; their work is especially relevant for cross-linked silicone resins. A report also appeared on the ^{29}Si NMR characterization of polysiloxane oligomers terminated with amine functional groups (264).

In recent years, the soluble polysilanes, $-\!\!(\text{SiRR}'_n)\!\!-$, have attracted some attention. Several dialkylsilanes were studied by ^{13}C, ^{1}H, and ^{29}Si NMR through the 2D COSY method (265). Further studies by ^{29}Si NMR of a number of polysilanes were also reported (266, 267).

With a receptivity of 0.83 relative to ^{1}H, ^{19}F is extremely easy to observe. The ^{19}F NMR papers on fluorine-containing polymers prior to 1982 have been previously cited (249). Particularly significant is the work of Murasheva et al. (268) attempting to use empirical additive rules for ^{19}F-chemical shifts, and Tonelli et al. (269) using the theoretically rigorous rotational isomeric state (RIS) models with γ-gauche parameters. Since 1982, many more papers have appeared. Some representative research activities are summarized below.

Vinyl and Vinylidene Polymers

Poly(vinyl fluoride) (268–271)
Poly(fluoromethylene) (269, 270)
Poly(vinylidene fluoride) (268–270, 272, 273)
Poly(trifluoroethylene) (269, 270, 274)
Copolymers and terpolymer of vinyl fluoride, hexafluoropropylene, and tetrafluoroethylene (275)

Other Fluorine-Containing Polymers

Poly(*p*-fluorostyrene) (276)
Fluorinated polyurethanes (277)

End-Group Analysis with ^{19}F

Polymers made with *p*-fluorobenzoyl peroxide (278)

It may be noted that the paper by Tonelli et al. (270) gives a good overview of their successful efforts to interpret the ^{19}F NMR spectra with γ-gauche RIS models. Several papers cited above use 2D methods to assist spectral interpretation (271, 273, 276). The paper by Gerig (276), in fact, first reported the use of 2D NMR for polymers.

Phosphorus-31 with a spin of $\frac{1}{2}$ and a receptivity of 0.66 (relative to proton) can be readily observed. In polyphosphazenes, for example, ^{31}P chemical shift data were routinely reported (279–282).

8.2. Interaction of Polymers with Solvents and Additives

These are usually studied by monitoring the nuclear relaxation times or spectral linewidths. The bulk of the studies were done by ^{1}H and ^{13}C. In multinuclear NMR, occasional reports appear, using ^{2}H, ^{17}O, or ^{23}Na. Some examples have been previously reported (249).

Two recent reports dealt with ^{23}Na in two different polymeric systems. In alkali cellulose, sodium hydroxide is employed to induce structural modifications as a prelude to mercerization or esterification. Sodium-23 NMR was used (283) to understand this process. In a separate report, the ion binding of sodium cation to polyelectrolytes was studied by ^{23}Na NMR (284). The ^{23}Na-relaxation data were consistent with Poisson–Boltzmann equation.

8.3. Polymer Dynamics

The study of polymer-chain motion is readily amenable to NMR relaxation experiments. Most of the work was done with ^{1}H and ^{13}C (see Section 7).

Multinuclear NMR studies are relatively rare because most common polymers only contain C, H, N, and O. For nuclei with low receptivity, such as ^{2}H and ^{15}N, isotopic enrichment may be needed to decrease data acquisition time. A few cases where ^{19}F, ^{2}H, ^{15}N, and ^{29}Si were involved have been previously reported (249). In recent years, most of the multinuclear relaxation studies were carried out on bulk polymers, which is beyond the scope of this chapter.

9. SOLID-STATE NMR

Solid-state NMR is an old discipline that is undergoing a renaissance. Much of the renewed interest is due to recent experimental and instrumental developments such as high-powered decoupling, cross-polarization, magic angle spinning, and multipulse sequences. The large (and growing) body of recent literature on this subject is witness to its current popularity and its problem solving potential (285). In this section, an attempt is made to review briefly solid-state NMR spectroscopy as applied to synthetic polymers (286, 287). The literature is classified into three areas: (1) wide-line NMR, (2) pulse techniques, and (3) high-resolution solid-state NMR.

9.1. Wide-Line NMR

Wide-line NMR is the exact analog of continuous-wave NMR of liquids and solutions. As in liquid NMR the sample is placed in a magnetic field and a secondary field of varying frequency is applied to give the resonance signals at the appropriate frequencies. In liquids, since the molecules move very fast (Brownian motion), most of the intramolecular forces are averaged out. As a result, narrow spectral lines (0.2–5.0 Hz) are observed. In the solid state, however, the molecules are relatively immobile; thus each molecule interacts much more strongly with the neighboring molecules. These strong interactions produce a very broad line shape in the NMR spectrum, giving rise to the name "wide-line NMR."

Wide-line NMR has been successfully used for many polymers. Both ^{1}H and ^{19}F are the nuclei usually being studied because of their favorable natural abundance (100%). In recent years, with computer techniques, studies have been made with deuterium (^{2}H) and other less abundant nuclei. Since only broad patterns are usually obtained as the NMR signals, most studies were made by carefully analyzing the spectral line shapes. Several examples are listed below.

1. Determination of crystallinity of semicrystalline polymers (288–290).
2. Motional transition in polymers (through measurements of line widths and second moments) (291).
3. Studies of oriented polymer fibers and films (292–302).
4. Absorption of water in cellulose (303–305).
5. Interaction of polymers with plasticizers (306).
6. Studies of X-ray induced polymerization of several vinyl monomers (307, 308).

Further examples can be found in a review article (309).

9.2. Simple Pulse Experiments

Nuclear relaxation times (T_1, T_2, and $T_{1\rho}$) have been determined and reported for most common polymers. The results have been successfully correlated with relaxation measurements by dynamic mechanical relaxation, dielectric relaxation, and Brillouin and Rayleigh light scattering (310). The principal benefits of NMR are its experimental accuracy and its sound theoretical foundation. For these reasons, there are many reports on the use of NMR pulse techniques in polymer research (311). An example is poly(vinyl acetate). The various relaxation times (T_1, T_2, and $T_{1\rho}$) as a function of temperature are given in Figure 16. It is clear that all three relaxation times are sensitive to molecular motions, but to different extents. At the glass transition ($\sim 100\,°C$), all three relaxation times register abrupt changes, T_1 and $T_{1\rho}$ showing minima, and T_2 showing a steep rise. At about $-230\,°C$, both T_1 and $T_{1\rho}$ show minima, but T_2 barely manifests any change. This is γ-relaxation, and has been attributed to methyl group rotation. Considering the low activation energy (~ 1 kcal/mol), the rotation probably occurs by quantum mechanical tunneling. At $-20\,°C$, there is a hint of β relaxation (probably corresponding to ester rotation) in the $T_{1\rho}$ curve only.

Other examples of pulsed NMR include:

1. Measurement of the degree of cross-linking and entanglement (229, 312–315).
2. Studies of curing process of epoxy resins (316).
3. Relaxation processes in polymer fibers, for example, polypropylene (317).
4. Effect of fillers on polymers (317–319).
5. Studies of polyblends (286a, 320–323).

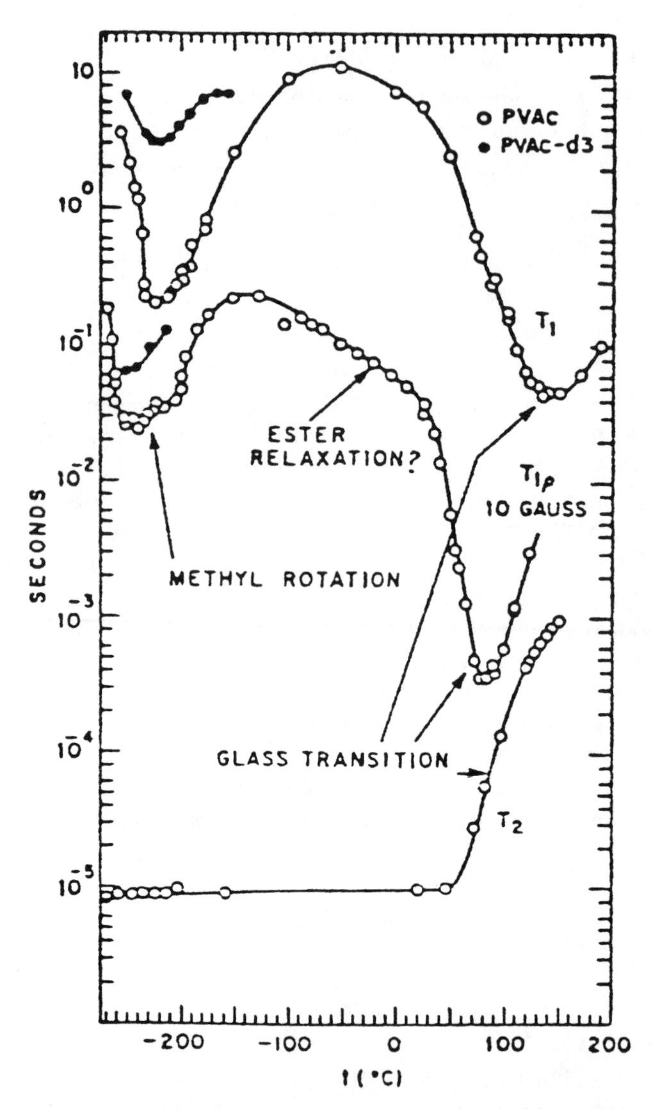

Figure 16. NMR relaxation data for poly(vinyl acetate). Reprinted with permission from D. W. McCall, "Nuclear Magnetic Resonance Studies of Molecular Relaxation Mechanisms in Polymers," *Acc. Chem. Res.*, **4**, 223 (1971). Copyright © 1971 American Chemical Society.

9.3. High-Resolution Solid-State NMR

The last few years saw intense research activities directed towards obtaining solutionlike spectral lines from solid samples. Many techniques have been well established, and high-resolution solid-state NMR has rapidly become a routine problem-solving tool.

A brief description of the background of high-resolution solid-state NMR is given. Previously, ^{13}C NMR studies of solids were hampered by three major problems.

1. $^{13}C—^{1}H$ dipole–dipole interaction, a primary reason for broad lines in solid-state spectra.
2. Chemical-shift anisotropy, which is a result of the different orientations the molecules can assume in a magnetic field.
3. Long spin–lattice relaxation times (T_1) for ^{13}C in solids (on the order of 1 min). Since ^{13}C is a rare spin (natural abundance, 1%), a spectrum with good S/N necessarily requires time averaging. The long T_1 means that each experiment must be done over a long period of time. This restricts sample turnaround.

(It may be noted that in liquids, the molecules are free to tumble, thus averaging out the dipole–dipole interaction and the chemical shift anisotropy, and shortening the T_1.)

To achieve high resolution (narrow lines) for solids, these problems must be overcome. Problem 1 is solved mostly by high-powered decoupling techniques. This is achieved by irradiating the proton nuclei with a strong radiofrequency field at the proton Larmor resonance frequency, so as to induce rapid transitions in the proton energy levels and effectively average out most of the dipole–dipole interactions. Problem 2 and residual dipole–dipole interactions are taken care of by a technique known as magic angle spinning (MAS). The sample is aligned at an angle of 54.7° relative to the external magnetic field and spun at a rate comparable to the range of the chemical shift anisotropy. For an anisotropy of 5000 Hz (200 ppm for 25 MHz ^{13}C instrument), the sample must be spun almost at 5000 cycles/s (30,000 rpm) to remove the anisotropy broadening effectively. Finally, problem 3 can be removed in many cases by using the cross-polarization (CP) technique. This method utilizes the shorter T_1 and the larger magnetic polarization of the protons. By irradiating ^{13}C and ^{1}H in a certain way, magnetic polarization can be transferred from the protons to the carbons. The ^{13}C signals are sampled the same way as before, but now the shorter proton T_1 determines the sampling rate. This reduces the time it takes to obtain a spectrum.

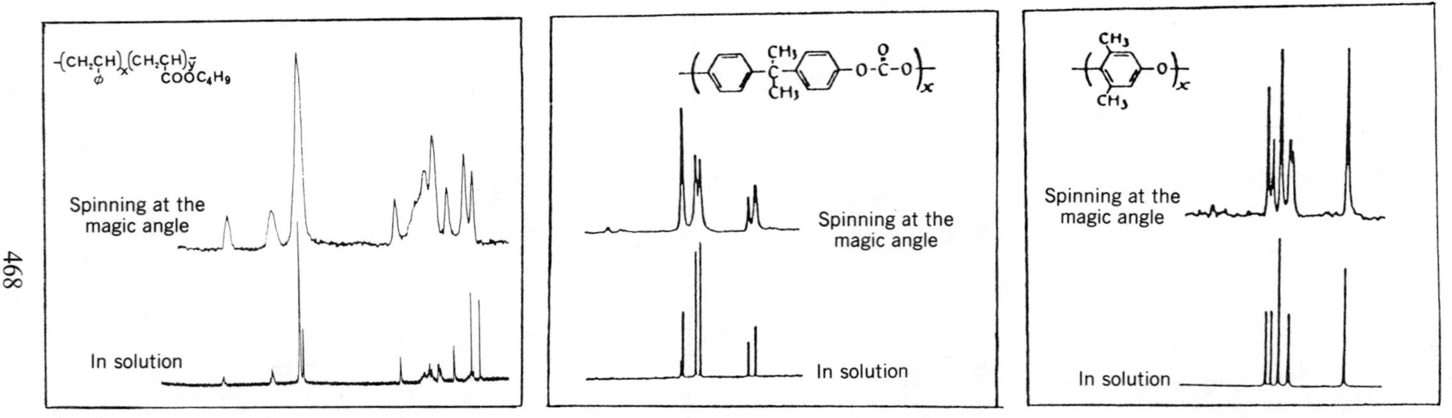

Figure 17. Comparison of solid-state NMR spectra (by CP–MAS method) with solution NMR: poly(styrene-*co*-butyl acrylate), polycarbonate, and poly(phenylene oxide). Reprinted with permission from J. Schaefer, E. O. Stejskal, and R. Buchadahl, “Magic-Angle Carbon-13 NMR Analysis of Motion in Solid Glassy Polymers,” *Macromolecules,* **10**, 384 (1977). Copyright © 1977 American Chemical Society.

Most of the papers published on polymers use all three techniques: strong decoupling, MAS, and CP. Several examples of interest are given in the following sections.

9.3.1. *Structural Analysis*

The high-resolution solid-state NMR spectra of poly(styrene-*co*-butyl acrylate), polycarbonate, and poly(phenylene oxide) are given in Figure 17. For comparison, the liquid-state (solution) NMR of these polymers are also given (324, 325). Although the solid-state spectra possess slightly broader line widths, it is clear that a considerable amount of structural information is present, and the solid-state ^{13}C spectra can be interpreted just like the solution spectra. One use of this technique is in the routine analysis and identification of polymers that cannot be studied in solution. This includes materials such as thermosetting polymers, cross-linked gels, vulcanized elastomers, filled plastics, and polymer blends.

For example, Koenig and co-workers (326) used CP–MAS technique to characterize polybutadiene cross-linked with dicumyl peroxide. Cis–trans conversion and double-bond migration were found, together with increased line widths with increased cross-linking. Similar polymers were studied by Curran and Padwa (327). Komoroski et al. (328) studied vulcanized natural rubber and directly observed sulfur cross-link sites. In addition, some isomerization of the double bond was found. This problem was also studied by Koenig and co-workers (329).

As a cautionary note, not all polymers can have narrow-line spectra as in Figure 17. In some atactic polymers, for example, poly(vinyl chloride), the stereochemical configurations cause residual broadening not averaged out by CP–MAS. In such systems, the information content necessarily decreases.

9.3.2. *Measurement of Relaxation Times by High-Resolution Solid-State NMR*

With CP–MAS experiments, two relaxation times can be easily obtained, $T_{1\rho}$ and T_{CH}. The $T_{1\rho}$ is the rotating-frame spin–lattice relaxation time. This is sensitive to motions with frequencies in the 10–200-kHz range. The T_{CH} is the characteristic time it takes to transfer the proton polarization to the ^{13}C polarization. This is sensitive to near static interactions.

One use of $T_{1\rho}$ is illustrated for polyoxymethylene. The $T_{1\rho}$ data obtained by CP–MAS is given in Figure 18. Apparently, two components are present, the large $T_{1\rho}$ about 17 ms and the shorter $T_{1\rho}$ at 1.5 ms. These correspond to the amorphous and crystalline materials. For this polymer, it

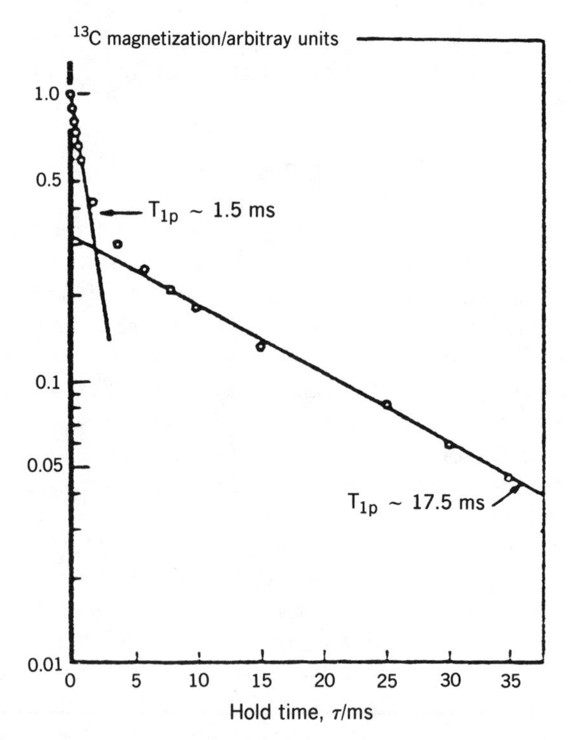

Figure 18. $T_{1\rho}$ measurement for a sample spinning at the magic angle with a frequency of 2.1 kHz. The carbon magnetization is prepared via cross-polarization. Reprinted with permission from W. S. Veeman, E. M. Menger, W. Ritchey, and E. deBoer, "High-Resolution Carbon-13 Nuclear Magnetic Resonance of Solid Poly(oxymethylene)," *Macromolecules*, **12**, 924 (1979). Copyright © 1979 American Chemical Society.

was shown that the percent crystallinity calculated from $T_{1\rho}$ data agrees very well with the results of X-ray diffraction (330).

9.3.3. Correlation of Relaxation Parameters with Physical Properties

One potential use of this technique is to correlate the NMR parameters with the end-use physical properties. We have seen earlier how NMR can determine T_g and the degree of crystallinity. Attempts have also been made to correlate various NMR parameters to flexural modulus and impact strength. The work done by Schaefer et al. (325), for example, showed great promise. For seven polymers, $T_{1\rho}$ and T_{CH} were measured together with the polymer impact strength (notched Izod). The data are plotted in Figure 19. It is apparent that a clear correlation exists between $(T_{CH}/T_{1\rho})$ and impact strength.

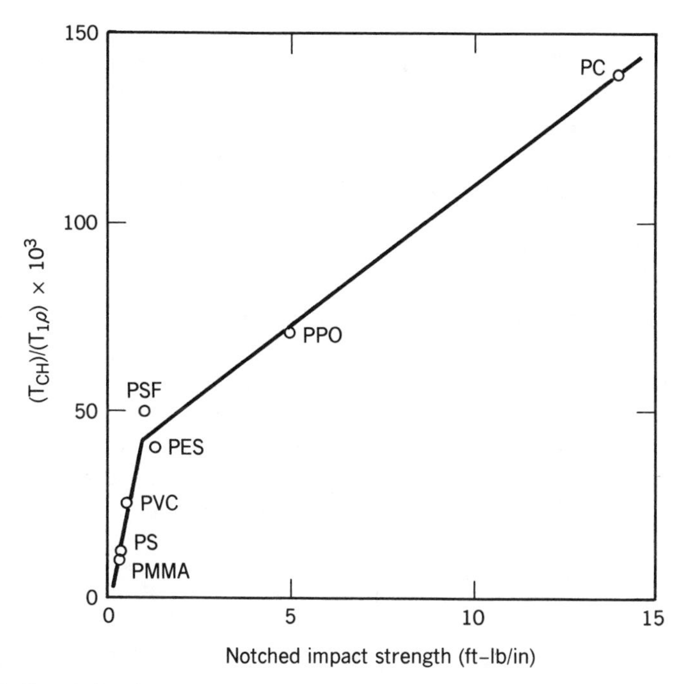

Figure 19. Correlation between notched impact strength (in ft-lb/in. of notch radius) and a ratio of ^{13}C NMR parameters. PC = polycarbonate, PPO = poly(phenylene oxide), PSF = polysulfone, PES = poly(ether sulfone), PVC = poly(vinyl chloride), PS = polystyrene, PMMA = poly(methyl methacrylate). Reprinted with permission from J. Schaefer, E. O. Stejskal, and R. Buchdahl, "Magic-Angle Carbon-13 NMR Analysis of Motion in Solid Glassy Polymers," *Macromolecules*, **10**, 384 (1977). Copyright © 1977 American Chemical Society.

Although a detailed understanding of this phenomenon is lacking, the existence of such a correlation is of considerable practical interest.

9.4. Other Solid State Studies

Apart from CP–MAS techniques, there is also an alternative method to average out dipolar interactions and obtain narrow lines for solid samples. This is the multipulse technique whereby pulse cycles of specific phase and timing relationships are used. One can conceive this method heuristically as a rapid reorientation of the nuclear spins in many directions such that the overall dipole–dipole interaction is averaged out. The original multipulse method was proposed by Waugh et al. (331) (called WAHUHA) and was later pursued by others (332, 333). Thus far, this approach has been applied extensively to ^{1}H NMR (334, 335). Attempts have also been made to combine this technique with cross-polarization–magic angle spinning (CRAMPS) (336, 337). A good example was shown by Gerstein and co-workers (338) for poly(ethylene terephthalate).

Another important development is the use of high-speed MAS, up to 18 kHz (339) to obtain CP–MAS ^{19}F spectra of fluorocarbon polymers. Sufficient resolution was obtained to determine the relative intensities of pentad sequences in bulk copolymers.

Spiess and Sillescu (340, 341) pioneered the use of ^{2}H-quadrupolar spin–echo technique in polymers. The molecular fragment of interest is deuterated, and the ^{2}H spectrum is obtained by the spin–echo technique. Through line shape analysis, the detailed motional behavior of the polymer can be studied. Two elegant illustrations of this technique were reported for polyethylene (342) and poly(butylene terephthalate) (343).

Recently, 2D NMR techniques have been used for solid polymers. This area has been reviewed by Blumich and Spiess (286d).

10. SUMMARY

Nuclear magnetic resonance spectroscopy, as applied to synthetic polymers, is in the advanced stage of development. A wide array of experimental techniques have been developed, and all common polymeric systems have been studied. Many of these techniques have been reviewed in this chapter and illustrations made of selected applications.

In general, the major types of information available from NMR studies of polymers are (1) polymer microstructure, (2) polymerization mechanisms, (3) polymer dynamics and interactions, and (4) morphology and structure–property correlations. The various techniques described in this chapter, be they solution or solid-state, 1D or 2D, computer-assisted or manual, ^{1}H, ^{13}C, or heteronuclear, all contribute in different ways to the information base. Given the diversity of polymer chemistry and the ingenuity of NMR spectroscopists, one can anticipate continued vitality in this field of research for quite some time to come.

REFERENCES

1. F. A. Bovey, *High Resolution NMR of Macromolecules*, Academic, New York, 1972.
2. J. C. Randall, *Polymer Sequence Determination*, Academic, New York, 1977.
3. W. M. Pasika (Ed.), *Carbon-13-NMR in Polymer Science*, *Am. Chem. Soc. Symp. Ser.*, **103** (1979).

4. A. E. Woodward and F. A. Bovey (Ed.), *Polymer Characterization by ESR and NMR, Am. Chem. Soc. Symp. Ser.*, **142** (1980).
5. F. A. Bovey, *Chain Structure and Conformation of Macromolecules*, Academic, New York, 1982.
6. Y. E. Shapiro, "Analysis of Chain Microstructure by ^{1}H and ^{13}C NMR Spectroscopy," *Bull. Magn. Reson.*, **7**, 27 (1985).
7. J. R. Ebdon, "Copolymer Characterization by ^{13}C NMR," in J. V. Dawkins (Ed.), *Developments in Polymer Characterization*, Vol. 2, Applied Science, New York, 1982, pp. 1–29.
8. J. C. Randall (Ed.), *NMR and Macromolecules, Am. Chem. Soc. Symp. Ser.*, **247** (1984).
9. F. A. Bovey, "Structure of Chains by Solution NMR Spectroscopy," in G. Allen and J. C. Bevington (Ed.), *Comprehensive Polymer Science*, Vol. 1, Pergamon, Oxford, 1989, pp. 339–375.
10. F. Heatley, "Dynamics of Chains in Solutions by NMR Spectroscopy," in G. Allen and J. C. Bevington (Ed.), *Comprehensive Polymer Science*, Vol. 1, Pergamon, Oxford, 1989, pp. 377–396.
11. W. W. Simons and M. Zanger, *The Sadtler Guide to the NMR Spectra of Polymers*, Sadtler Research Labratories, Philadelphia, PA, 1973.
12. *Sadtler Standard Carbon-13 NMR Spectra*, Sadtler Research Laboratories Philadelphia, PA.
13. W. Bremser, L. Ernst, B. Francke, R. Gerhards, and A. Hardt, *Carbon-13 NMR Spectral Data*, VCH, Weinheim, Federal Republic of Germany, 1986.
14. E. Breitmaier, G. Haas, and W. Volter, *Atlas of Carbon-13 NMR Data*, Heyden, London, 1979.
15. J. B. Stothers, *Carbon-13 NMR Spectroscopy*, Academic, New York, 1972.
16. N. A. B. Gray, "Computer Assisted Analysis of Carbon-13 NMR Spectral Data," *Prog. NMR Spectros.*, **15**, 201 (1982).
17. Several recent articles are given in *J. Chem. Inf. Computer Sci.*, **25**, 224, 231, 252, 326, and 344 (1985).
18. D. L. Dalrymple, C. L. Wilkins, G. W. A. Milne, and S. R. Heller, "A Carbon-13 Nuclear Magentic Resonance Spectral Data Base and Search System," *Org. Magn. Reson.*, **11**, 535 (1978).
19. INKA Database, Scientific Information Service Inc., 7 Woodland Avenue, Larchmont, NY 10538.
20. C. Djerassi, D. H. Smith, C. W. Crandell, N. A. B. Gray, J. G. Nourse, and M. R. Lindley, "Applications of Artificial Intelligence for Chemical Inference. XLII. The DENDRAL Project: Computational Aids to Natural Products Structure Elucidation," *Pure Appl. Chem.*, **54**, 2425 (1982).
21. M. R. Lindley, N. A. B. Gray, D. H. Smith, and C. Djerassi, "Applications of Artifical Intelligence for Chemical Inference. 40. Computerized Approach to the Verification of Carbon-13 NMR Spectral Assignments," *J. Org. Chem.*, **47**, 1027 (1982).

22. Y. Takahashi, S. Maeda, and S. Sasaki, "Automated Recognition of Common Geometrical Patterns Among a Variety of Three-Dimensional Molecular Structures," *Anal. Chim. Acta*, **200**, 363 (1987).

23. S. Sasaki and Y. Kudo, "Structure Elucidation System Using Structural Information from Multisources:CHEMICS," *J. Chem. Inf. Computer Sci.*, **25**, 252 (1985).

24. M. Carabedian, I. Dagane, and J. E. Dubois, "Elucidation by Progressive Intersection of Ordered Substructures from Carbon-13 NMR," *Anal. Chem.*, **60**, 2186 (1988).

25. M. E. Munk and B. D. Christie, "The Characterization of Structure by Computer," *Anal. Chim. Acta*, **216**, 57 (1989).

26. W. Bremser and W. Fachinger, "Multidimensional Spectroscopy," *Magn. Res. Chem.*, **23**, 1056 (1985).

27. D. M. Grant and E. G. Paul, "Carbon-13 Magnetic Resonance. II. Chemical Shift Data for the Alkanes," *J. Am. Chem. Soc.*, **86**, 2984 (1964).

28. J. D. Roberts, F. J. Weigert, J. I. Kroschwitz, and H. J. Reich, "NMR Spectroscopy. Carbon-13 Chemical Shifts in Acyclic and Alicyclic Alcohols," *J. Am. Chem. Soc.*, **92**, 1338 (1970).

29. A. Ejchart, "^{13}C NMR Chemical Shifts in Aliphatic Alcohols and the γ-shift Caused by Hydroxyl and Methyl Groups," *Org. Magn. Reson.*, **9**, 351 (1977).

30. H. Eggert and C. Djerassi, "Carbon-13 NMR Spectra of Acyclic Aliphatic Amines," *J. Am. Chem. Soc.*, **95**, 3710 (1973).

31. J. E. Sarneski, H. L. Suprenant, F. K. Molen, and C. N. Reilley, "Chemical Shifts and Protonation Shifts in Carbon-13 NMR Studies of Aqueous Amines," *Anal. Chem.*, **47**, 2116 (1975).

32. A. Ejchart, "Substituent Effect of the Nitro Group on Alkanes in Carbon-13 NMR Spectroscopy," *Org. Magn. Reson.*, **10**, 263 (1977).

33. H. N. Cheng and S. J. Ellingsen, "Carbon-13 NMR Spectral Interpretation by a Computerized Substituent Chemical Shift Method," *J. Chem. Inf. Computer Sci.*, **23**, 197 (1983), and references cited therein.

34. F. W. Wehrli, A. P. Marchand, and S. Wehrli, *Interpretation of Carbon-13 NMR Spectra*, 2nd ed., Wiley, Chichester, England, 1988.

35. E. Pretsch, J. T. Clerc, J. Seibl, and W. Simon, *Tables of Spectral Data for Structure Determination of Organic Compounds*, Springer-Verlag, Berlin, 1983.

36. J. T. Clerc and H. Sommerauer, "A Minicomputer Program Based on Additivity Rules for the Estimation of ^{13}C-NMR Chemical Shifts," *Anal. Chim. Acta*, **95**, 33 (1977).

37. H. N. Cheng and M. A. Bennett, "Trends in ^{13}C-NMR Shifts and Computer-Aided Shift Prediction," *Anal. Chim. Acta*, **242**, 43 (1991).

38. G. G. Lowry, *Markov Chains and Monte Carlo Calculations in Polymer Science*, Marcel Dekker, New York, 1970.

39. J. L. Koenig, *Chemical Microstructure of Polymer Chains*, Wiley-Interscience, New York, 1980.

40. H. N. Cheng and M. A. Bennett, "General Analysis of the Carbon-13 NMR Spectra of Vinyl Copolymers by the Spectral Simulation Approach," *Anal. Chem.*, **56**, 2320 (1984).
41. G. E. Ham (Ed.), *Copolymerization*, Wiley-Interscience, New York, 1964.
42. F. P. Price, *Markov Chains and Monte Carlo Calculations in Polymer Science*, Marcel Dekker, New York, 1970, Chapter 7.
43. H. N. Cheng, "Stereochemistry of Vinyl Polymers and NMR Characterization," *J. Appl. Polym. Sci.*, **36**, 229 (1988).
44. H. N. Cheng, "Statistical Modelling and NMR Analysis of Polyolefins," in R. P. Quirk (Ed.), *Transition Metal Catalyzed Polymerizations*, Cambridge University, Cambridge, 1988, pp. 599–623.
45. R. A. Shelden, T. Fueno, T. Tsunetsuga, and J. Furukawa, "A One Parameter Model for Isotactic Polymerization Based on Enantiomorphic Catalyst Sites," *J. Polym. Sci.*, **B3**, 23 (1965).
46. T. Fueno, R. A. Shelden, and J. Furukawa, "Probabilistic Considerations of the Tacticity of Optically Active Polymers," *J. Polym. Sci.*, **A3**, 1279 (1965).
47. Y. Doi, "Structure and Stereochemistry of Atactic Polypropylenes. Statistical Model of Chain Propagation," *Makromol. Chem. Rapid Commun.*, **3**, 635 (1982).
48. A. Zambelli, P. Locatelli, M. C. Sacchi, and I. Tritto, "Isotactic Polymerization of Propene: Stereoregularity of the Insertion of the First Monomer Unit as a Fingerprint of the Catalytic Active Site," *Macromolecules*, **15**, 831 (1982).
49. C. J. Carman, R. A. Harrington, and C. E. Wilkes, "Monomer Sequence Distribution in Ethylene–Propylene Rubber Measured by Carbon-13 NMR. 3. Use of Reaction Probability Model," *Macromolecules*, **10**, 536 (1977).
50. T. Moritani and H. Iwasaki, "Carbon-13 NMR Study on Sequence Distribution and Anomalous Linkage in Ethylene–Vinyl Alcohol Copolymers," *Macromolecules*, **11**, 1251 (1978).
51. H. N. Cheng, "Markovian Statistics and Simplex Algorithm for Carbon-13 Nuclear Magnetic Resonance Spectra of Ethylene–Propylene Copolymers," *Anal. Chem.*, **54**, 1828 (1982).
52. H. N. Cheng, "^{13}C NMR Analysis of Propylene–Butylene Copolymers by a Reaction Probability Model," *J. Polym. Sci., Polym. Phys. Ed.*, **21**, 573 (1983).
53. H. N. Cheng, "Characterization of 1-Octene Copolymers by ^{13}C-NMR," *Polym. Commun.*, **25**, 99 (1984).
54. H. N. Cheng and G. H. Lee, "Two Dimensional NMR Spectroscopy and Automated ^{1}H NMR Analysis of Ethylene–Propylene Copolymers," *Polym. Bull.*, **12**, 463 (1984).
55. H. N. Cheng, "^{13}C-NMR Determination of Inverted Propylene Units in Polypropylene," *Polym. Bull.*, **14**, 347 (1985)
56. H. N. Cheng, "Determination of Polyethylene Branching Through Computerized ^{13}C-NMR Analysis," *Polym. Bull.*, **16**, 445 (1986).
57. H. N. Cheng and D. A. Smith, "^{13}C-NMR Studies of Low-Molecular Weight

Ethylene–Propylene Copolymers and Characterization of Polymer Chain Ends," *Macromolecules*, **19**, 2065 (1986).

58. H. N. Cheng and G. H. Lee, "Two-Dimensional NMR Characterization of Propylene Copolymers," *J. Polym. Sci., Polym. Phys. Ed.*, **25**, 2355 (1987).
59. H. N. Cheng, "Computerized Model Fitting Approach for the NMR Analysis of Polymers," *J. Chem. Inf. Computer Sci.*, **27**, 8 (1987).
60. H. N. Cheng and T. S. Dziemianowicz, "^{13}C-NMR Characterization of Copolymers of Ethylene and Perdeuteroethylene," *Makromol. Chem.*, **189**, 845 (1988).
61. H. N. Cheng and G. H. Lee, "Characterization of Ethylene Copolymers with ^{1}H NMR Techniques and Reaction Probability Models," *Macromolecules*, **21**, 3164 (1988).
62. H. N. Cheng, "^{13}C-NMR Sequence Determination for Multicomponent Polymer Mixtures," *J. Appl. Polym. Sci.*, **35**, 1639 (1988).
63. H. N. Cheng, "^{13}C-NMR Analysis of Multicomponent Polymer Systems," in T. Provder (Ed.), *Computer Applications in Applied Polymer Science, Am. Chem. Soc. Symp. Ser.*, **404**, 174 (1989).
64. H. N. Cheng, "Analytical and Synthetic Approaches for the NMR Characterization of Polymers," *J. Appl. Polym. Sci., Appl. Polym. Symp.*, **43**, 129 (1989).
65. H. N. Cheng, "Automated Approaches for the ^{13}C NMR Characterization of Polyolefins," in R. P. Quirk (Ed.), *Transition-Metal Catalyzed Polymerizations: Alkenes and Dienes*, MMI/Harwood Academic, New York, 1983.
66. R. R. Ernst and W. A. Anderson, "Application of Fourier Transform Spectroscopy to Magnetic Resonance," *Rev. Sci. Instr.*, **37**, 93 (1966).
67. R. Freeman and H. D. W. Hill, "Phase and Intensity Anomalies in Fourier Transform NMR," *J. Magn. Reson.*, **4**, 366 (1971).
68. D. E. Jones and H. Sternlicht, "Fourier Transform Nuclear Magnetic Resonance. I. Repetitive Pulse," *J. Magn. Reson.*, **6**, 167 (1972).
69. D. D. Traficante and M. M. Kelly, Paper presented at Experimental NMR Conference, Asilomar, CA, 1979.
70. R. Benn and H. Gunther, "Modern Pulse Methods in High-Resolution NMR Spectroscopy," *Angew. Chem. Int. Ed. Engl.*, **22**, 350 (1983).
71. G. A. Morris, "Modern NMR Techniques for Structure Elucidation," *Magn. Res. Chem.*, **24**, 371 (1986).
72. D. W. Brown, T. T. Nakashima, and D. L. Rabenstein, "Simplification and Assignment of Carbon-13 NMR Spectra with Spin-Echo Fourier Transform Techniques," *J. Magn. Reson.*, **45**, 302 (1981).
73. D. L. Rabenstein and T. T. Nakashima, "Spin Echo Fourier Transform NMR Spectroscopy," *Anal. Chem.*, **51**, 1465A (1979).
74. S. L. Patt and J. N. Shoolery, "Attached Proton Test for Carbon-13 NMR," *J. Magn. Reson.*, **46**, 535 (1982).
75. G. A. Morris and R. Freeman, "Enhancement of Nuclear Magnetic Resonance Signals by Polarization Transfer," *J. Am. Chem. Soc.*, **101**, 760 (1979).

76. D. M. Doddrell, D. T. Pegg, and M. R. Bendall, "Distortionless Enhancement of NMR Signals by Polarization Transfer," *J. Magn. Reson.*, **48**, 323 (1982).

77. A. Bax, *Two-Dimensional Nuclear Magnetic Resonance in Liquid*, Delft University, Dordrecht, 1982.

78. R. R. Ernst, G. Bodenhausen, and A. Wokaun, *Principles of Nuclear Magnetic Resonance in One and Two Dimensions*, Clarendon, Oxford, 1987.

79. J. Schraml and J. M. Bellama, *Two-Dimensional NMR Spectroscopy*, Wiley, New York, 1988.

80. G. E. Martin and A. S. Zektzer, *Two-Dimensional NMR Methods for Establishing Molecular Connectivity*, VCH, New York, 1988.

81. S. Macura and L. R. Brown, "Improved Sensitivity and Resolution in Two-Dimensional Homonuclear *J*-Resolved NMR Spectroscopy of Macromolecules," *J. Magn. Reson.*, **53**, 529 (1983).

82. M. D. Bruch and F. A. Bovey, "Proton Resonance Assignments in Copolymer Spectra by Two-Dimensional NMR," *Macromolecules*, **17**, 978 (1984).

83. G. P. Gippert and L. R. Brown, "Absolute Configurational Assignments for the ^{1}H NMR Spectrum of Poly(vinyl alcohol) by Use of Two-Dimensional NMR Methods," *Polym. Bull.*, **11**, 585 (1984).

84. L. R. Brown, "Differential Scaling Along ω_1 in COSY Experiments," *J. Magn. Reson.*, **57**, 513 (1984).

85. M. D. Bruch, F. A. Bovey, and R. E. Cais, "Microstructure Analysis of Poly(vinyl fluoride) by Fluorine-19 Two-Dimensional *J*-Correlated NMR Spectroscopy," *Macromolecules*, **17**, 2547 (1984).

86. M. D. Bruch, F. A. Bovey, R. E. Cais, and J. H. Noggle, "Elucidation of the Proton NMR Spectrum of Poly(propylene oxide) by Two-Dimensional *J*-Resolved Spectroscopy," *Macromolecules*, **18**, 1253 (1985).

87. F. C. Schilling, F. A. Bovey, M. D. Bruch, and S. A. Kozlowski, "Observation of the Stereochemical Configuration of Poly(methyl methacrylate) by Proton Two-Dimensional *J*-Correlated and NOE-Correlated NMR Spectroscopy," *Macromolecules*, **18**, 1418 (1985).

88. H. N. Cheng and G. H. Lee, "Two-Dimensional NMR Studies of Polypropylene Tacticity," *Polym. Bull.*, **13**, 549 (1985).

89. C. Chang, D. D. Muccio, and T. St. Pierre, "Heteronuclear Correlated Two-Dimensional NMR Spectroscopy Applied to the Diad Assignment of Vinyl Polymers," *Macromolecules*, **18**, 2334 (1985).

90. G. J. Ray, R. E. Pauls, J. J. Lewis, and L. B. Rogers, "Structure Determination by Two-Dimensional NMR of α-Hydro-ω-Butyl Oligostyrenes Fractionated by Liquid Chromatography," *Makromol. Chem.*, **186**, 1135 (1985).

91. P. A. Mirau and F. A. Bovey, "Two-Dimensional Nuclear Magnetic Resonance Analysis of Poly(vinyl chloride) Microstructure," *Macromolecules*, **19**, 210 (1986).

92. M. W. Crowther, N. M. Szeverenyi, and G. C. Levy, "Absolute Tacticity Assignments of Poly(vinyl chloride) via the Two-Dimensional NMR Spin-Lock RELAY Experiment," *Macromolecules*, **19**, 1333 (1986).

93. M. D. Bruch and J. K. Bonestill, "Interpretation of the Proton NMR Spectrum of Poly(vinyl butyral) by Two-Dimensional NMR," *Macromolecules*, **19**, 1622 (1986).
94. S. A. Heffner, F. A. Bovey, L. A. Verge, P. A. Mirau, and A. E. Tonelli, "Two-Dimensional ^{1}H and ^{13}C NMR Spectroscopy of Styrene–Methyl Methacrylate Copolymers," *Macromolecules*, **19**, 1628 (1986).
95. M. D. Bruch and W. G. Payne, "Assignment of Monomer Sequences in the ^{13}C and ^{1}H NMR Spectra of Several Ethylene-Containing Co- and Terpolymers by Two-Dimensional NMR Spectroscopy," *Macromolecules*, **19**, 2712 (1986).
96. H. N. Cheng and G. H. Lee, "The Combined Use of Two-Dimensional NMR and Lanthanide Shift Reagents for the Characterization of Ethylene–Vinyl Acetate Copolymers," *Polym. Bull.*, **19**, 89 (1988).
97. M. D. Bruch, "Microstructure Analysis of Poly(ethylene-*co*-vinyl alcohol) by Two-Dimensional NMR Spectroscopy," *Macromolecules*, **21**, 2707 (1988).
98. F. A. Bovey and P. A. Mirau, "The Two-Dimensional Nuclear Magnetic Resonance Spectroscopy of Macromolecules," *Acc. Chem. Res.*, **21**, 37 (1988).
99. H. N. Cheng and G. H. Lee, "Two-Dimensional NMR Analysis of Polymers in Solution," *Trends Anal. Chem.*, **9**, 285 (1990).
100. J. Schaefer, "Random Monomer Distributions in Copolymers. Copolymerizations of Ethylene–Vinyl Chloride and Ethylene–Vinyl Acetate," *J. Phys. Chem.*, **70**, 1975 (1966).
101. T. K. Wu, "NMR Studies on Microstructure of Ethylene Copolymers. II. Sequence Distribution in Ethylene–Vinyl Acetate Copolymers," *J. Polym. Sci., Part A2*, **8**, 167 (1970).
102. Y. Yawaka, T. Tsuchihara, N. Tanaka, K. Kosaka, Y. Hirakida, and M. Ogawa, "Analysis of Vinyl Acetate–Ethylene Copolymers by High Resolution NMR," *Kobunshi Kagaku*, **28**, 459 (1971).
103. T. Okada and T. Ikushige, "^{1}H-NMR Spectra of Ethylene–Vinyl Acetate Copolymers in the Presence of Shift Reagent," *Polym. J.*, **9**, 121 (1977).
104. M. Delfini, A. L. Segre and F. Conti, "Sequence Distributions in Ethylene–Vinyl Acetate Copolymers. I. Carbon-13 NMR Studies," *Macromolecules*, **6**, 456 (1973).
105. T. K. Wu, D. W. Ovenall, and G. S. Reddy, "NMR Studies on Microstructure of Ethylene Copolymers. VI. Carbon-13 Spectra of Ethylene–Vinyl Acetate Copolymers," *J. Polym. Sci., Poly. Phys. Ed.*, **12**, 901 (1974).
106. B. Ibrahim, A. R. Katritzky, A. Smith, and D. E. Weiss, "Carbon-13 NMR Spectroscopy of Polymers. Part 1. High Resolution Carbon-13 Nuclear Magnetic Resonance Spectroscopy: Tacticity Studies on Poly(vinyl acetate) and Monomer Distribution Analysis in Ethylene–Vinyl Acetate Copolymers," *J. Chem. Soc., Perkin Trans.* **2**, 1537 (1974).
107. J. C. Randall, *Polymer Sequence Determination, Carbon-13 NMR Method*, Academic, New York, 1977, pp. 138–144.
108. H. N. Sung and J. H. Noggle, "Carbon-13 NMR of Poly(vinyl acetate) and

Ethylene–Vinyl Acetate Copolymers," *J. Polym. Sci., Polym. Phys. Ed.*, **19**, 1593 (1981).

109. M. F. Grenier-Loustalot, "Microstructure et Determination des Ramifications Alkyle dars Une Serio de Copolymeres Ethylene Acetate de Vinyle," *Eur. Polym. J.*, **21**, 361 (1985).

110. V. N. Viswanadhan and W. L. Mattice, "Short-Branch Formation via the Roedel Mechanism in Copolymers of Ethylene and Vinyl Acetate," *J. Polym. Sci., Polym. Phys. Ed.*, **23**, 1957 (1985).

111. A. Zambelli and A. Segre, "Detection of the Tetrads in Amorphous Polypropylenes by NMR," *J. Polymer Sci., Part B*, **6**, 473 (1968).

112. F. Heatley and A. Zambelli, "Polymer NMR Spectroscopy. XVII. Tetrad Resonances in Polypropylene," *Macromolecules*, **2**, 618 (1969).

113. F. Heatley, R. Salovey, and F. A. Bovey, "Polymer NMR Resonance Spectroscopy. XVIII. NMR Spectrum, Dimensions, and Steric Interactions of Isotactic Polypropylene," *Macromolecules*, **2**, 619 (1969).

114. P. J. Flory and Y. Fujiwara, "Conformations of Tetrads in Vinyl Polymers and NMR Spectra of the Methylenic Protons," *Macromolecules*, **2**, 327 (1969).

115. P. J. Flory, "Stereochemical Constitution and NMR Spectra of Polypropylenes," *Macromolecules*, **3**, 613 (1970).

116. R. C. Ferguson, "220-MHz Proton Magnetic Resonance Spectra of Polymers. II. Polypropylene and Ethylene–Propylene Copolymers," *Macromolecules*, **4**, 324 (1971).

117. A. Zambelli, L. Zetta, C. Sacchi, and C. Wolfsgruber, "Methylene Proton Magnetic Resonance of Some Partially Deuterated Polypropylenes," *Macromolecules*, **5**, 440 (1972).

118. F. C. Stehling and J. R. Knox, "Stereochemical Configuration of Polypropylene by Hydrogen Nuclear Magnetic Resonance," *Macromolecules*, **8**, 595 (1975).

119. A. Zambelli, P. Locatelli, G. Bajo, and F. A. Bovey, "Model Compounds and ^{13}C-NMR Observation of Stereosequences of Polypropylene," *Macromolecules*, **8**, 687 (1975).

120. A. Zambelli, P. Locatelli, A. Provasoli, and D. R. Ferro, "Correlation between Carbon-13 NMR Chemical Shifts and Conformation of Polymers. 3. Hexad Sequence Assignments of Methylene Spectra of Polypropylene," *Macromolecules*, **13**, 267 (1980).

121. F. C. Schilling and A. E. Tonelli, "Carbon-13 Nuclear Magnetic Resonance of Atactic Polypropylene," *Macromolecules*, **13**, 270 (1980).

122. H. N. Cheng and G. H. Lee, "^{13}C NMR Assignments of the Methylene Carbons in Polypropylene," *Macromolecules*, **20**, 436 (1987).

123. *Sadtler Carbon-13 NMR of Monomers and Polymers*, Sadtler Research Laboratories, Philadelphia, PA, 1985.

124. Q. T. Pham, R. Petiaud, and H. Waton, *Proton and Carbon NMR Spectra of Polymers*, Wiley, Chichester and New York, 1984.

125. Y. Inone, A. Nishioka, and R. Chujo, "Carbon-13 NMR Spectroscopy of Polystyrene and Poly-α-methylstyrene," *Makromol. Chem.*, **156**, 207 (1972).

126. K. Matsuzaki, T. Uryu, T. Seki, K. Osada, and T. Kawamura, "Stereoregularity of Polystyrene and Mechanism of Polymerization," *Makromol. Chem.*, **176**, 3051 (1975).

127. J. C. Randall, "The Distribution of Stereochemical Configurations in Polystyrene as Observed with ^{13}C-NMR," *J. Polym. Sci., Polym. Phys. Ed.*, **13**, 889 (1975).

128. A. E. Tonelli, "Stereosequence-Dependent ^{13}C-NMR Chemical Shifts in Polystyrene," *Macromolecules*, **16**, 604 (1983).

129. H. Sato and Y. Tanaka, "NMR Spectra of Styrene Oligomers and Polymers," *Am. Chem. Soc. Symp. Ser.*, **247**, 181 (1984), and references cited therein.

130. H. J. Harwood, T. Chen, and F. Lin, "75 MHz Carbon-13 NMR Studies on Polystyrene and Epimerized Isotactic Polystyrenes," *Am. Chem. Soc. Symp. Ser.*, **247**, 197 (1984), and references cited therein.

131. D. E. Dorman, E. P. Otocka, and F. A. Bovey, "Carbon-13 Observations of the Nature of the Short-Chain Branches in Low-Density Polyethylene," *Macromolecules*, **5**, 574 (1972).

132. J. C. Randall, "Carbon-13 NMR of Ethylene-1-Olefin Copolymers: Extension to the Short-Chain Branch Distribution in Low-Density Polyethylene," *J. Polym. Sci., Polym. Phys. Ed.*, **11**, 275 (1973).

133. M. E. A. Cudby and A. Bunn, "Determination of Chain Branching in Low-Density Polyethylene by ^{13}C NMR and IR Spectroscopy," *Polymer*, **17**, 345 (1976).

134. F. A. Bovey, F. C. Schilling, F. L. McCracken, and H. L. Wagner, "Short-Chain and Long-Chain Branching in Low-Density Polyethytlene," *Macromolecules*, **9**, 76 (1976).

135. D. E. Axelson, L. Mandelkern, and G. C. Levy, "Carbon-13 Spin Relaxation Parameters of Branched Polyethylenes. Ramifications for Quantitative Analysis," *Macromolecules*, **10**, 557 (1977).

136. J. Spavacek, "Short-Chain Branching in High Density Polyethylene: ^{13}C NMR Study," *Polymer*, **19**, 1149 (1978).

137. D. E. Axelson, G. C. Levy, and L. Mandelkern, "A Quantitative Analysis of Low-Density (branched) Polyethylenes by Carbon-13 Fourier Transform NMR at 67.9 MHz," *Macromolecules*, **12**, 41 (1979).

138. J. C. Randall, "Characterization of Long-Chain Branching in Polyethylenes Using High-Field Carbon-13 NMR," *Am. Chem. Soc. Symp. Ser.*, **142**, 93 (1980).

139. T. Usami and S. Takayama, "Fine-Branching Structure in High-Pressure, Low-Density Polyethylenes by 50.10 MHz ^{13}C-NMR Analysis," *Macromolecules*, **17**, 1756 (1984), and references cited therein.

140. W. H. Starnes, Jr., F. C. Schilling, K. B. Abbas, I. M. Plitz, R. C. Hartless, and F. A. Bovey, "Structural Selectivities in the Reduction of Poly(vinyl chloride) with Lithium Aluminum Hydride and Tri-*n*-butyltin hydride," *Macromolecules*, **12**, 13 (1979).

141. W. H. Starnes, Jr., F. C. Schilling, K. B. Abbas, R. E. Cais, and F. A. Bovey, "Mechanism for the Formation of Chloromethyl Branches in Poly(vinyl chloride)," *Macromolecules*, **12**, 556 (1979).

142. W. H. Starnes, Jr., F. C. Schilling, I. M. Plitz, R. E. Cais, and F. A. Bovey, "Detailed Microstructure and Concentration of the Chlorinated *n*-Butyl Branches in Poly(vinyl chloride)," *Polym. Bull.*, **4**, 555 (1981).

143. W. H. Starnes, Jr., G. M. Villacorta, and F. C. Schilling, "Formation of 1-Ethyl-2-(Long Alkyl) Cyclopentane End-Groups During Organotin Hydride Reductions of Poly(vinyl chloride)," *Polym. Preprints*, **22** (2), 307 (1981).

144. F. A. Bovey, F. C. Schilling, and H. N. Cheng, "^{13}C NMR Observation of the Effects of High Energy Radiation and Oxidation on Polyethylene and Model Paraffins," *Adv. Chem. Ser.*, **169**, 133 (1978).

145. J. C. Randall, F. J. Zoepfl, and J. Silverman, "High-Resolutin Solution Carbon-13 NMR Measurements of Irradiated Polyethylene," *Radiat. Phys. Chem.*, **22**, 183 (1983).

146. W. K. Busfield, J. V. Hanna, J. H. O'Donnell, and A. K. Whittaker, "A ^{13}C NMR Study of Radiation-Induced Racemization in Isotactic Polypropylene," *Br. Polym. J.*, **19**, 223 (1987).

147. M. Tabata, J. Sohma, W. Yang, K. Yokota, H. Yamaoka, and T. Matsuyama, "Magnetic Resonance Studies on Crosslinks in Polyethylene and its Model Compounds," *Radiat. Phys. Chem.*, **30**, 147 (1987).

148. P. F. Barron, W. K. Busfield, and J. V. Hanna, "Extensive Steroregularity Changes Induced in Molten Isotactic Polypropylene by High Energy Radiation and Other Morphology Effects," *J. Polym. Sci., Polym. Lett.*, **26**, 225 (1988).

149. T. K. Wu, "Carbon-13 and Proton NMR Studies of Cellulose Nitrates," *Macromolecules*, **13**, 74 (1980).

150. D. S. Lee and A. S. Perlin, "^{13}C-NMR-Spectral and Related Studies on the Distribution of Substituents in O-(2-Hydroxypropyl) Cellulose," *Carbohydr. Res.*, **106**, 1 (1982).

151. D. S. Lee and A. S. Perlin, "Use of Methanolysis in the Characterization of *O*-(2-hydroxypropyl) Cellulose by ^{13}C-NMR Spectroscopy," *Carbohydr. Res.*, **124**, 172 (1983).

152. J. Reuben and H. T. Conner, "Analysis of the Carbon-13 NMR Spectrum of Hydrolyzed *O*-(carboxymethyl) Cellulose: Monomer Composition and Substitution Patterns," *Carbohydr. Res.*, **115**, 1 (1983).

153. J. Reuben, "Description and Analysis of Hydroxyethyl Cellulose," *Macromolecules*, **17**, 156 (1984).

154. K. Kimura, T. Shigemura, M. Kubo, and Y. Maru, "^{13}C-NMR Study of *O*-(2-hydroxypropyl) Cellulose," *Makromol. Chem.*, **186**, 61 (1985).

155. J. Reuben, "Analysis of the ^{13}C-NMR Spectra of Hydrolyzed and Methanolyzed *O*-Methylcelluloses: Monomer Compositions and Models for their Description," *Carbohydr. Res.*, **157**, 201 (1986).

156. J. Reuben and T. E. Casti, "Distribution of Substituents in *O*-(2-Hydroxyethyl) Cellulose: A ^{13}C-NMR Approach," *Carbohydr. Res.*, **163**, 91 (1987).
157. J. Reuben, in M. Yalpani (Ed.), *Industrial Polysaccharides: Genetic Engineering, Structure/Property Relations, and Applications*, Elsevier, Amsterdam, 1987, pp. 305–309.
158. J. C. Bevington, J. R. Ebdon, and T. N. Huckerby, "An Appraisal of NMR Methods from Study of End-Groups Derived from Initiators in Radical Polymerizations," *Eur. Polym. J.*, **21**, 685 (1985).
159. D. E. Axelson and K. E. Russell, "Characterization of Polymers by Means of ^{13}C NMR Spectroscopy," *Prog. Polym. Sci.*, **11**, 221 (1985).
160. K. Hatada, K. Ute, K. Tanaka, M. Imanari, and N. Fujii, "Two-Dimensional NMR Spectra of Isotactic Poly(methyl methacrylate) Prepared with *t*-C_4H_9MgBr and Detailed Examination of Tacticity," *Polym. J.*, **19**, 425 (1987).
161. J. C. Bevington, S. W. Breuer, N. J. Heseltine, T. N. Huckerby, and S. C. Varma, "Reactions of the 2-Cyano-2-Propyl Radical with Vinyl Acetate and Other Monomers: A Study of Head Addition," *J. Polym. Sci., Polym. Chem. Ed.*, **25**, 1085 (1987).
162. K. Kodaira, K. Ito, and S. Iyoda, "Study of End groups in Oligo- and Polystyrene Prepared using Azobisisobutyronitrile," *Polym. Commun.*, **29**, 83 (1988).
163. K. Hatada and T. Kitayama, "Detailed Studies of Polymer Structure by NMR Spectroscopy—Instrumental Developments and New Techniques," *Polym. News*, **13**, 244 (1988), and references cited therein.
164. F. Keller and C. Mügge, "Determination of Parameter Sets of Substituent Increments from Carbon-13 NMR Spectra of Copolymers and Application to the Assignment of Chlorine-Containing Polymers," *Z. Polymerforsch.*, **27**, 347 (1976).
165. P. Pinther, F. Keller, and M. Hartmann, "High Resolution Nuclear Magnetic Resonance Studies of the Microstructure of Chlorine-Containing Polymers. VIII./1. Line Assignments for the Carbon-13 NMR Spectra of Chlorinated Polypropylene," *Acta Polym.*, **31**, 299 (1980).
166. F. Keller, "Completion of the Increment System for Calculating the Chemical Shift of Carbon-13 NMR of Chlorine-Containing Polymers," *Z. Polymerforsch.*, **28**, 515 (1977).
167. R. A. Komoroski, R. G. Parker, and M. H. Lehr, "50.3 MHz Carbon-13 NMR Study of Chlorinated Poly(vinyl chloride) Microstructure and the Mechanism of Chlorination," *Macromolecules*, **15**, 844 (1982).
168. R. A. Komoroski, R. G. Parker, and J. P. Shockcor, "^{13}C-NMR Microstructural Analysis of Chlorinated Poly(vinyl chloride) in Terms of Three-Carbon Sequences," *Macromolecules*, **18**, 1257 (1985).
169. M. V. Eskina, A. S. Khachaturov, L. B. Krentsel, K. K. Yutudzhyan, and A. D. Litmanovitch, "A Study of the Structure of Natural Rubber by the ^{13}C-NMR Method," *Vysokomol. Soedin.*, **A30**, 142 (1988), and references cited therein.
170. H. N. Cheng, F. C. Schilling, and F. A. Bovey, "^{13}C-NMR Studies of the Oxidation of Polyethylene," *Macromolecules*, **9**, 363 (1976).

171. W. V. Smith, "Sequence Distribution in Ethylene Propylene Copolymers. I. Relations Between Multads and Between Multads and the ^{13}C-NMR Spectrum," *J. Polym. Sci., Polym. Phys. Ed.*, **18**, 1573 (1980).

172. Y. Doi, "Sequence Distributions of Inverted Propylene Units in Polypropylenes Measured by ^{13}C-NMR," *Macromolecules*, **12**, 248 (1979).

173. H. N. Cheng, "^{13}C-NMR Analysis of Ethylene–Propylene Rubbers," *Macromolecules*, **17**, 1950 (1984).

174. Y. Inoue, Y. Itabashi, R. Chujo, and Y. Doi, "Studies of the Stereospecific Polymerization Mechanism of Propylene by a Modified Ziegler–Natta Catalyst Based on 125 MHz ^{13}C-NMR," *Polymer*, **25**, 1640 (1984), and references cited therein.

175. Y. Doi, "Structure and Stereochemistry of Atactic Polypropylenes. Statistical Model of Chain Propagation," *Makromol. Chem., Rapid Commun.*, **3**, 635 (1982).

176. J. F. Ross, "Statistical Interpretation of Non-Homogeneous Copolymer Data," in R. P. Quirk (Ed.), *Transition Metal Catalyzed Polymerizations*, Cambridge University Press, Cambridge, 1988, pp. 799–814.

177. S. Floyd, "A Theoretical Interpretation of Reactivity Ratio Products in Copolymers Formed from Two Fractions Differing in Composition," *J. Appl. Polym. Sci.*, **34**, 2559 (1987).

178. C. Cozewith, "Interpretation of ^{13}C-NMR Sequence Distribution for Ethylene–Propylene Copolymers Made with Heterogeneous Catalysts," *Macromolecules*, **20**, 1237 (1987).

179. H. N. Cheng and M. Kakugo, to be published.

180. H. N. Cheng, manuscript in preparation.

181. H. N. Cheng and M. A. Bennett, "Additive Rules for the ^{13}C-NMR Shifts of Methyl-Substituted Alkanes and Ethylene–Propylene Copolymers," *Makromol. Chem.*, **188**, 135 (1987).

182. H. N. Cheng and M. A. Bennett, "Spectral Simulation and Characterization of Polymers from Ethene and Propene by ^{13}C-NMR," *Makromol. Chem.*, **188**, 2665 (1987).

183. M. Moeller and H. J. Cantow, "^{13}C-NMR Studies on Ditactic Poly(α-olefins). 1. Poly(1,2-dimethyltetramethylenes) and Their Tetrad Models," *Macromolecules*, **17**, 733 (1984).

184. P. J. Flory, *Statisical Mechanics of Polymer Chains*, Wiley-Interscience, New York, 1969.

185. A. E. Tonelli and F. C' Schilling, "Carbon-13 NMR Chemical Shifts and the Microstructure of Polymers," *Acc. Chem. Res.*, **14**, 233 (1981).

186. A. E. Tonelli, "^{13}C-NMR Chemical Shifts and the Conformations of Rigid Polypeptides," *Biopolymers*, **23**, 819 (1984), and references cited therein.

187. A. Zambelli, P. Ammendola, M. C. Sacchi, P. Locatelli, and G. Zannoni, "Isotactic Polymerization of 3-Methyl-1-pentene: Enantioselectivity and Diastereoselectivity," *Macromolecules*, **16**, 341 (1983).

188. A. Zambelli, P. Ammendola, M. C. Sacchi, and P. Locatelli, "Stereospecific Polymerization of α-Olefins: End-Groups and Reaction Mechanism," *Am. Chem. Soc. Symp. Ser.*, **247**, 223 (1984).

189. P. Ammendola, A. Vitagliano, L. Oliva, and A. Zambelli, "Ethylene–Propene Copolymerization in the Presence of a ^{13}C-Enriched Catalyst: End Group Analysis and Monomer Reactivities in the First Insertion Steps," *Makromol. Chem.*, **185**, 2421 (1984).

190. P. Locatelli, M. C. Sacchi, I. Tritto, G. Zannoni, A. Zambelli, and V. Piscitelli, "Isotactic Polymerization of Propene: Initiation at Titanium-Phenyl Bonds," *Macromolecules*, **18**, 627 (1985).

191. G. Fink, "Ethylene Insertion with Soluble Ziegler Catalysts: Direct Insight by Use of Reacting $^{13}C_2H_4$ by Means of ^{13}C-NMR Spectroscopy," in R. P. Quirk, (Ed.), *Transition Metal Catalyzed Polymerizations, Alkenes and Dienes*, MMI Press/Harwood Academic, New York, 1983, pp. 495–508.

192. G. Fink, in M. Fontanille and A. Guyot (Ed.), *Recent Advances in Mechanistic and Synthetic Aspects of Polymerization*, Reidel, Dordrecht, 1987, pp. 515–533.

193. P. Corradini, V. Barone, and G. Guerra, "Steric Control in the First Step of the Isospecific Ziegler–Natta Polymerization of Propene," *Macromolecules*, **15**, 1242 (1982).

194. P. Corradini, G. Guerra, and V. Barone, "Conformational Analysis of Polypropylene Chains Bound to Model Catalytic Sites," *Eur. Polym. J.*, **20**, 1177 (1984).

195. J. Schaefer, in G. C. Levy (Ed.), *Topics in Carbon-13 NMR Spectroscopy*, Vol. 1, Wiley-Interscience, New York, 1974.

196. F. Heatley, "Nuclear Magnetic Relaxation of Synthetic Polymers in Dilute Solution," *Prog. NMR Spectrosc.*, **13**, 47 (1979).

197. F. A. Bovey and L. W. Jelinski, "The Observation of Chain Motion in Macromolecules by ^{13}C- and ^{2}H-NMR Spectroscopy," *J. Phys. Chem.*, **89**, 571 (1985).

198. A. V. Cunliffe, "Synthetic Macromolecules," in G. A. Webb (Ed.), *Nuclear Magnetic Resonance*, Royal Society Chemistry, London, 1987, pp. 248–254.

199. N. Bloembergen, E. M. Purcell, and R. V. Pound, "Relaxation Effects in NMR Absorption," *Phys. Rev.*, **73**, 679 (1948).

200. J. Schaefer, "Distributions of Correlation Times and the Carbon-13 NMR Spectra of Polymers," *Macromolecules*, **6**, 882 (1973).

201. B. Valeur, J. P. Jarry, F. Geny, and L. Monnerie, "Dynamics of Macromolecular Chains. I. Theory of Motions on a Tetrahedral Lattice," "II. Orientation Relaxation Generated by Elementary Three-Bond Motions and Notion of an Independent Kinetic Segment," "Long-Time Orientation Relaxation for an Internal Bond of a Chain," *J. Polym. Sci., Polym. Phys. Ed.*, **13**, 667, 675, 2251 (1975).

202. A. A. Jones and W. H. Stockmayer, "Models for Spin Relaxation in Dilute Solutions of Randomly Coiled Polymers," *J. Polym. Sci., Polym. Phys. Ed.*, **15**, 847 (1977).

203. J. T. Bendler and R. Yaris, "A Solvable Model of Polymer Main-Chain Dynamics with Applications to Spin Relaxation," *Macromolecules*, **11**, 650 (1978).

204. C. K. Hall and E. Helfand, "Conformational State Relaxation in Polymers: Time-Correlation Functions," *J. Chem. Phys.*, **77**, 3275 (1982).

205. E. Helfand, "Dynamics of Conformational Transitions in Polymers," *J. Polym. Sci., Polym. Symp.*, **73**, 39 (1985).

206. L. Monnerie and F. Laupretre, "Spectroscopic Investigation of Local Molecular Motions in Polymers," in R. Daudel et al. (Ed.), *Structure and Dynamics of Molecular Systems*, Reidel, Dordrecht, 1986, pp. 129–154.

207. D. Perchak, J. Skolnick, and R. Yaris, "Computer Simulations of Simple Models of the Ring-Flip Process in Polycarbonate," *Macromolecules*, **20**, 121 (1987).

208. B. B. Pant, J. Skolnick, and R. Yaris, "Damped Orientational Diffusion Model of Polymer Local, Main Chain Motion. 4. Effects of Probes and Side Chains," *Macromolecules*, **18**, 253 (1985) and references cited therein.

209. D. E. Woessner, "Nuclear Spin Lattice Relaxation in Axially Symmetric Ellipsoids with Internal Motion," *J. Chem. Phys.*, **50**, 719 (1969).

210. G. Lipari and A. Szabo, "NMR Relaxation in Nucleic Acid Fragments: Models for Internal Motion," *Biochemistry*, **20**, 6250 (1981).

211. G. Lipari and A. Szabo, "Pade' Approximants to Correlation Functions for Restricted Rotational Diffusion," *J. Chem. Phys.*, **75**, 2971 (1981).

212. I. Bahar and B. Erman, "Investigation of Local Motions in Polymers by the Dynamic Rotational Isomeric State Model," *Macromolecules*, **20**, 1369 (1987).

213. I. Bahar and B. Erman, "Activation Energies of Local Conformational Transitions in Polymer Chains," *Macromolecules*, **20**, 2310 (1987).

214. I. Bahar, B. Erman, and L. Monnerie, "Comparison of Dynamic Rotational Isomeric State Results with Previous Expressions for Local Chain Motion," *Macromolecules*, **22**, 431 (1989).

215. I. Bahar, B. Erman, and L. Monnerie, "Application of the Dynamic Rotational Isomeric State Model to Poly(ethylene oxide) and Comparison with Nuclear Magnetic Relaxation Data," *Macromolecules*, **22**, 2396 (1989).

216. J. C. Randall, "Carbon-13-NMR Spin–Lattice Relaxation Times of Isotactic and Syndiotactic Sequences in Amorphous Polypropylene," *J. Polym. Sci., Polym. Phys. Ed.*, **14**, 1693 (1976).

217. T. Asakura and Y. Doi, "Carbon-13-NMR Spin–Lattice Relaxation Times of Inverted Monomeric Units in Polypropylene," *Macromolecules*, **14**, 72 (1981).

218. T. Asakura and Y. Doi, "Carbon-13-NMR Study of the Chain Dynamics of Polypropylene and Poly(1-butene) and the Stereochemical Dependence of the Segmental Mobility," *Macromolecules*, **16**, 786 (1983).

219. J. R. Lyerla, T. T. Horikawa, and D. E. Johnson, "Carbon-13 Relaxation Study of Stereoregular Poly(methyl methacrylate) in Solution," *J. Am. Chem. Soc.*, **99**, 2463 (1977).

220. Y. Inoue, T. Konno, R. Chujo, and A. Nishioka, "Carbon-13 Spin–Lattice Relaxation Times of Poly(methyl methacrylate) and Polystyrene in Solution," *Makromol. Chem.*, **178**, 2131 (1977).

221. F. Heatley and A. Begum, "Molecular Motion of Poly(methyl methacrylate), Polystyrene and Poly(propylene oxide) in Solution Studied by ^{13}C-NMR Spin–Lattice Relaxation Measurements: Effects Due to Distribution of Correlation Times," *Polymer*, **17**, 399 (1976).

222. Y. Inoue and T. Konno, "NMR Studies of Conformation and Molecular Motion of Poly(methyl methacrylate) in Solution," *Makromol. Chem.*, **179**, 1311 (1978).

223. F. C. Schilling, "Effect of Stereosequence on Carbon-13-Spin–Lattice Relaxation Times for Poly(vinyl chloride)," *Macromolecules*, **11**, 1290 (1978).

224. K. Hatada, J. Okamoto, K. Ohta, and H. Yuki, "Proton Spin–Lattice Relaxation Times of Polymers of Various Tacticities in Solution," *J. Polym. Sci., Polym. Lett. Ed.*, **14**, 51 (1976).

225. K. Hatada, H. Ishikawa, T. Kitayama, and H. Yuki, "Mechanism of Proton Spin–Lattice Relaxation in Poly(methyl methacrylate)," *Makromol. Chem.*, **178**, 2753 (1977).

226. Y. Inoue and T. Konno, "A Carbon-13 NMR Study of Molecular Motion of Polystyrene in Solution," *Polym. J.*, **8**, 457 (1976).

227. W. Gronski and N. Murayama, "^{13}C-Relaxationsuntersuchungen zum EinfluB des Lösungsmittels auf die Kettenbewegung Von Polystyrol," *Makromol. Chem.*, **179**, 1509, 1521 (1978).

228. Y. Inoue and Y. Kawamura, "Configuration Dependence of Carbon-13 Spin–Lattice Relaxation Times of Poly(α-methylstyrene)," *Polymer*, **23**, 1997 (1982).

229. H. N. Cheng, to be published.

230. J. E. Anderson, K. J. Liu, and R. Ullman, "Nuclear Magnetic Relaxation in Polymer Solutions," *Discuss. Faraday Soc.*, **49**, 257 (1970).

231. R. Folland, J. H. Steven, and A. Charlesby, "Proton Spin Relaxation in Liquid Poly-dimethylsiloxane: Molecular Motion and Network Formation," *J. Polym. Sci., Polym. Phys. Ed.*, **16**, 1041 (1978).

232. A. Charlesby, "The Use of Pulsed NMR Techniques in the Measurement of Radiation Effects in Polymer," *Radiat. Phys. Chem.*, **14**, 919 (1979).

233. A. Allerhand and R. C. Hailstone, "Carbon-13 Fourier Transform NMR. X. Effect of Molecular Weight on ^{13}C Spin–Lattice Relaxation Times of Polystyrene in Solution," *J. Chem. Phys.*, **56**, 3718 (1972).

234. J. Spevacek and B. Schneider, "Effect of Stereoregularity and Solvent Upon Molecular Motion and Structure of Stereoregular Poly(methyl methacrylates) in Solution. ^{13}C- and ^{1}H-NMR. Relaxation Study," *Polymer*, **19**, 63 (1978).

235. H. N. Cheng, T. E. Smith, and D. M. Vitus, "Studies of Solution Dynamics of Poly(*N*-Vinyl Pyrrolidone) and its Iodine Adduct," *J. Polym. Sci., Polym. Phys. Ed.*, **23**, 461 (1985).

236. J. Breen, D. Van Duijn, J. de Bleijser, and J. C. Leyte, "Poly(ethylene oxide) Dynamics in Aqueous Solutions Studied by Nuclear Magnetic Relaxation," *Ber. Bunsenges. Phys. Chem.*, **90**, 1112 (1986).

237. J. R. C. vander Maarel, D. Lankhorst, J. de Bleijsen, and J. C. Leyte, "Water

Dynamics in Polyelectrolyte Solutions from Deuterium and Oxygen-17 Nuclear Magnetic Relaxation," *Macromolecules*, **20**, 2390 (1987).

238. F. A. Bovey, F. C. Schilling, T. K. Kwei, and H. L. Frisch, "Dynamic Carbon-13 NMR Measurements on Poly(vinylidene fluoride) and Poly(methyl methacrylate) and their Mixed Solutions," *Macromolecules*, **10**, 559 (1977).
239. R. Kimmich, "Molecular Motion in Polymer Melts: 1. Description by Components and NMR Relaxation Behaviour," *Polymer*, **18**, 233 (1977).
240. R. Kimmich, "Characteristic Molecular Weights in the Dynamics of Polymer Melts: NMR and Zero-Shear Viscosity," *Polymer*, **25**, 187 (1984).
241. J. P. Cohen-Addad, "Nuclear Magnetic Resonance Investigations of Properties of the Terminal-Chain Diffusional Spectrum of Molten Poly(dimethylsiloxane)," *Macromolecules*, **18**, 1101 (1985) and references cited therein.
242. J. P. Cohen-Addad, "NMR and Macromolecular Migration in a Melt or in Concentrated Solutions," *Polymer*, **26**, 97 (1985).
243. J. P. Cohen-Addad, "Entangled Linear Polymer Chains in Melts: NMR and Rouse or Reptation Models; Stress Relaxation," *Polymer*, **24**, 1128 (1983).
244. R. K. Harris and B. E. Mann (Ed.), *NMR and the Periodic Table*, Academic, London, 1978.
245. C. Brevard and P. Granger, *Handbook of High Resolution Multinuclear NMR*, Wiley, New York, 1981.
246. J. B. Lambert and F. G. Riddell (Ed.), *The Multinuclear Approach to NMR Spectroscopy*, NATO ASI Series, Reidel, Dordrecht, 1983.
247. P. Laszlo (Ed.), *NMR of Newly Accessible Nuclei: Chemical and Biochemical Applications*, Academic, New York, 1983.
248. J. Mason (Ed.), *Multinuclear NMR*, Plenum, New York, 1987.
249. H. N. Cheng, "Multinuclear NMR of Polymers," *Proc. 41st ANTEC Soc. Plastic Eng.*, **29**, 500 (1983).
250. H. R. Kricheldorf, S. V. Joshi, and W. E. Hull, "^{15}N-NMR Spectroscopy. XXXV. Sequence Analysis of Alternating and Random Copolyamides Made Up of Aliphatic and Aromatic ω-Amino Acids," *J. Polym. Sci., Polym. Chem. Ed.*, **20**, 2791 (1982).
251. H. R. Kricheldorf, "Nitrogen-15 NMR Spectroscopic Characterization of Copolyamides and Polypeptides," *Pure Appl. Chem.*, **54**, 467 (1982).
252. B. S. Holmes, W. B. Moniz, and R. C. Ferguson, "NMR Study of Nylon 66 in Solution (proton, carbon-13 and nitrogen-15 NMR using adiabatic *J* cross polarization)," *Macromolecules*, **15**, 129 (1982).
253. C. Chang, F. Fish, D. D. Muccio, and T. St. Pierre, "^{13}C- and ^{15}N-NMR pH Titration of Poly(vinylamine): A Two-Stage Process Sensitive to Polymer Tacticity," *Macromolecules*, **20**, 621 (1987).
254. D. E. Axelson and S. L. Blake, "Multinuclear (^{13}C, ^{15}N) NMR Study of Polyethyleneimine-Based Polymers," *J. Polym. Sci., Polym. Chem. Ed.*, **23**, 2507 (1985).

255. J. R. Ebdon, P. E. Heaton, T. N. Huckerby, W. T. S. O'Rourke, and J. Parkin, "Characterization of Urea–Formaldehyde and Melamine-Formaldehyde Adducts and Resins by ^{15}N-NMR Spectroscopy," *Polymer*, **25**, 821 (1984).

256. J. R. Ebdon, B. J. Hunt, W. T. S. O'Rourke, and J. Parkin, "Characterization of Some Melamine–Formaldehyde Condensates and Some Cured Resins by ^{1}H-, ^{13}C- and ^{15}N-NMR Spectroscopy," *Br. Polym. J.*, **20**, 327 (1988).

257. H. H. Hoerhold, J. Klee, H. Schuetz, and R. Radeglia, "Uncrosslinked Epoxide–Amine Addition Polymers, 15. ^{15}N-NMR Spectroscopy of DGEBA–Aniline–Addition Polymers," *Angew. Makromol. Chem.*, **144**, 1 (1986).

258. M. F. Grenier-Loustalot and P. Grenier, "Reaction Epoxy–Amine: Suivi du Mecanisme Reactionnel et de la Cinetique par RMN ^{13}C, ^{15}N et HPLC," *J. Polym. Sci., Polym. Chem. Ed.*, **22**, 4011 (1984).

259. R. K. Harris and B. J. Kimber, "^{29}Si-NMR as a Tool for Studying Silicones," *Appl. Spectrosc. Rev.*, **10**, 117 (1975).

260. D. A. Laude and C. L. Wilkins, "Applications of a Recycled-Flow Fourier Transform Nuclear Magnetic Resonance System: Molecular Weight Determination of Siloxane Polymers by ^{29}Si-NMR," *Macromolecules*, **19**, 2295 (1986).

261. D. A. Laude and C. L. Wilkins, "Analytical Applications of a Recycled Flow NMR System: Quantitative Analysis of Slowly Relaxing Nuclei," *Anal. Chem.*, **57**, 1286 (1985).

262. G. Engelhardt and H. Jancke, "Structure Investigation of Organosilicon Polymers by Silicon-29 NMR," *Polym. Bull.*, **5**, 577 (1981).

263. H. Jancke and G. Engelhardt, "Structural Dependence of Silicon-29 NMR Chemical Shifts of $CH_3Si(O_{0.5})_3$ "T" Groups in Oligomeric and Polymeric Siloxanes," *Z. Chem.*, **23**, 253 (1983).

264. P. J. A. Brandt, R. Subramanian, P. M. Sormani, T. C. Ward, and J. E. McGrath, "Silicon-29 NMR of Functional Polysiloxane Oligomers," *Polym. Preprint*, **26** (2), 213 (1985).

265. F. C. Schilling, F. A. Bovey, and J. M. Zeigler, "Characterization of Polysilanes by Carbon-13, Silicon-29, and Proton NMR," *Macromolecules*, **19**, 2309 (1986).

266. A. R. Wolff, I. Nozue, J. Maxka, and R. West, "^{29}Si-NMR of Dimethyl and Phenylmethyl Containing Polysilanes," *J. Polym. Sci., Polym. Chem. Ed.*, **26**, 701 (1988).

267. A. R. Wolff, J. Maxka, and R. West, "^{29}Si-NMR of Dialkylpolysilanes," *J. Polym. Sci., Polym. Chem. Ed.*, **26**, 713 (1988).

268. E. M. Marasheva, A. S. Shashkov, and A. A. Dontsov, "Analysis of the ^{19}F-NMR Spectra of Copolymers of Vinylidene Fluoride with Tetrafluoroethylene, and of Vinylidene Fluoride with Tetrafluoroethylene and Hexafluoropropylene. The Use of an Empirical Additive Scheme and of the Principle of Alternation," *Vysokomol. Soedin.*, **A23**, 632 (1981), and references cited therein.

269. A. E. Tonelli, F. C. Schilling, and R. E. Cais, "Carbon-13 NMR Chemical Shifts and the Microstructures of the Fluoropolymers Poly(vinylidene fluoride), Poly(fluoromethylene), Poly(vinyl fluoride) and Poly(trifluoroethylene)," *Polym. Preprint*, **22** (1), 271 (1981).

270. A. E. Tonelli, F. C. Schilling, and R. E. Cais, "Fluorine-19 NMR Chemical Shifts and the Microstructure of Fluoro Polymers," *Macromolecules*, **15**, 849 (1982).

271. M. D. Bruch, F. A. Bovey, and R. E. Cais, "Microstructure Analysis of Poly(vinyl fluoride) by Fluorine-19-Two-Dimensional *J*-Correlated NMR Spectroscopy," *Macromolecules*, **17**, 2547 (1984).

272. R. C. Ferguson and D. W. Ovenall, "Microstructure of Poly(vinylidene Fluoride) by High Field Fluorine-19 NMR Spectroscopy," *Polym. Preprint*, **25** (1), 340 (1984).

273. R. E. Cais and J. M. Kometani, "Synthesis and Two-Dimensional ^{19}F-NMR of Highly Aregic Poly(vinylidene fluoride)," *Macromolecules*, **18**, 1354 (1985).

274. R. E. Cais and J. M. Kometani, "Synthesis of Pure Head-to-Tail Poly(trifluoroethylenes) and Their Characterization by 470-MHz Fluorine-19 NMR," *Macromolecules*, **17**, 1932 (1984).

275. M. Pianca, P. Bonardelli, M. Tato, G. Cirillo, and G. Moggi, "Composition and Sequence Distribution of Vinylidene Fluoride Copolymer and Terpolymer Fluoroelastomers. Determination by ^{19}F NMR and Correlation with some Properties," *Polymer*, **28**, 224 (1987).

276. J. T. Gerig, "Fluorine Proton Chemical Shift Correlation in Poly(*p*-fluorostyrene)," *Macromolecules*, **16**, 1797 (1983).

277. B. S. Holmes and T. M. Keller, "NMR Characterization of Fluorinated Urethanes," *Polym. Preprint*, **25** (1), 338 (1984).

278. J. C. Bevington, T. N. Huckerby, and N. Vickerstaff, "Multinuclear NMR Studies of End-groups in Polymers. 4. Application of F-NMR to Polymers Prepared Using *p*-Fluorobenzoyl Peroxide," *Makromol. Chem., Rapid Commun.*, **4**, 349 (1983).

279. H. R. Allcock and S. Kwon, "An Ionically Cross-Linkable Polyphosphazene: Poly[bis(carboxylatophenoxy)phosphazene] and its Hydrogels and Membranes," *Macromolecules*, **22**, 75 (1989).

280. H. R. Allcock and S. Kwon, "Glyceryl Polyphosphazenes: Synthesis, Properties, and Hydrolysis," *Macromolecules*, **21**, 1980 (1988), and references cited therein.

281. H. R. Allcock, P. E. Austin, T. X. Neenan, J. T. Sisko, P. M. Blonsky, and D. F. Shriver, "Polyphosphazenes with Etheric Side Groups: Prospective Biomedical and Solid Electrolyte Polymers," *Macromolecules*, **19**, 1508 (1986).

282. P. Wisian-Neilson and R. R. Ford, "Alcohol Derivatives of Poly(methylphenylphosphazene)," *Macromolecules*, **22**, 72 (1989).

283. J. Kunze, A. Ebert, B. Schroeter, K. Frigge, and B. Phillipp, "^{13}C- and ^{23}Na-NMR Investigations on Alkali Cellulose," *Polym. Bull.*, **5**, 399 (1981).

284. G. Gunnarsson and H. Gustavsson, "Ion Binding to Polyelectrolytes as Described by the Poisson–Boltzmann Equation," *J. Chem. Soc., Faraday Trans. 1*, **78**, 2901 (1982).

285. Some recent books on solid-state NMR include: (a) M. Mehring, *Principles of High Resolution NMR in Solids*, 2nd ed., Springer-Verlag, Berlin, 1983. (b) C. A. Fyfe, *Solid State NMR for Chemists*, CFC, Guelph, Canada, 1983. (c) B. C.

Gerstein and C. R. Dybowski, *Transient Techniques in the NMR of Solids*, Academic, New York, 1985.

286. Some recent reviews on polymer applications include: (a) V. J. McBriety, in G. Allen, and J. C. Bevington (Ed.), "NMR Spectroscopy of Polymers in the Solid-State,"*Comprehensive Polymer Science*, Vol. 1, Pergamon, Oxford, 1989, pp. 397–428. (b) P. Meier, "NMR of Solid Polymers," *Kunstoffe*, **79**, 63 (1989). (c) R. Voelkel, "High-Resolution Solid-State ^{13}C-NMR Spectroscopy of Polymers," *Angew. Chem. Int. Ed. Engl.*, **27**, 1468 (1988). (d) B. Blumich and H. W. Spiess, "Two-Dimensional Solid-State NMR Spectroscopy: New Possibilities for the Investigation of the Structure and Dynamics of Solid Polymers," *Angew. Chem. Int. Ed. Engl.*, **27**, 1655 (1988). (e) F. Laupretre, "Application of High Resolution Solid State Carbon-13 NMR to Polymers," *Prog. Polym. Sci.*, **15**, 425 (1990).
287. R. A. Komoroski (Ed.), *High Resolution NMR Spectroscopy of Synthetic Polymers in Bulk*, VCH, Deerfield Beach, FL, 1986.
288. C. W. Wilson and G. E. Pake, "NMR Determination of Degree of Crystallinity in Two Polymers," *J. Polym. Sci.*, **10**, 503 (1953).
289. W. P. Slichter and D. W. McCall, "Note on the Degree of Crystallinity in Polymers as Found by NMR," *J. Polym. Sci.*, **25**, 230 (1957).
290. M. J. Richardson, "Crystallinity Determination in Polymers and a Quantitative Comparison for Polyethylene," *Br. Polym. J.*, **1**, 132 (1969) and references cited therein.
291. V. J. McBriety and D. C. Douglass, "Recent Advances in the NMR of Solid Polymers," *J. Polym. Sci., Macromol. Rev.*, **16**, 295 (1981).
292. V. J. McBriety, I. R. McDonald, and I. M. Ward, "Investigation of the Molecular Structure in Drawn Polyethylene Based Upon NMR Fourth Moment Measurements," *J. Phys. D*, **4**, 88 (1971).
293. V. J. McBriety and I. M. Ward, "Investigation of the Orientation Distribution Functions in Drawn Polyethylene by Broad-Line NMR," *J. Phys. D*, **1**, 1529 (1968).
294. O. Phaovibul, J. Loboda-Cackovic, H. Cackovic, and R. Hosemann, "Chain Conformation in Linear Polyethylene and Paraffins Defined by Four NMR Components," *Makromol. Chem.*, **175**, 2991 (1974).
295. D. Hyndman and G. F. Origlio, "Nuclear Magnetic Resonance in Polyethylene and Polypropylene Fibers," *J. Polym. Sci.*, **39**, 556 (1959).
296. R. P. Wool, M. I. Lohse, and T. J. Rowland, "Broad-Line NMR Studies of Deformation and Recovery in Hard Elastic Polypropylene," *J. Polym. Sci., Polym. Lett.*, **17**, 385 (1979).
297. M. Kashiwagi and I. M. Ward, "The Measurement of Molecular Orientation in Drawn Poly(vinyl chloride) by Broad-Line NMR," *Polymer*, **13**, 145 (1972).
298. M. Kashiwagi, M. J. Folkes, and I. M. Ward, "The Measurement of Molecular Orientation in Drawn Polymer(methyl Methacrylate) by Broad-Line NMR," *Polymer*, **12**, 697 (1971).
299. M. Kashiwagi, A. Cunningham, A. J. Manuel, and I. M. Ward, "An Investigation

of Molecular Orientation in Oriented Poly(ethylene terephthalate) Films," *Polymer*, **14**, 111 (1973).

300. V. J. McBriety and I. R. McDonald, "NMR of Oriented Polymers: the Effects of Molecular Motion on the Second and Fourth Moments of Drawn Polyoxymethylene," *J. Phys. D*, **6**, 131 (1973).
301. V. J. McBriety and I. R. McDonald, "Nuclear Magnetic Relaxation in Linear and Branched Polyethylene," *Polymer*, **16**, 125 (1975).
302. A. J. Brandolini, K. J. Rocco, and C. Dybowski, "Solid-State ^{19}F-NMR Investigation of Annealed Poly(tetra fluoroethylene)," *Macromolecules*, **17**, 1455 (1984), and references cited therein.
303. W. W. Fleming, "Evidence of Distinct Water Species in Cellulosic Environments from Broad-Line NMR," *J. Polym. Sci., Polym. Phys. Ed.*, **17**, 199 (1979), and references cited therein.
304. T. F. Child and D. W. Jones, "Broad-Line NMR Measurement of Water Accessibility in Cotton and Wood Pulp Celluloses," *Cellulose Chem. Technol.*, **7**, 525 (1973).
305. M. Kimura, H. Hatakeyama, M. Usuda, and J. Nakano, "Studies on Absorbed Water in Cellulose by Broad-Line NMR," *J. Appl. Polym. Sci.*, **16**, 1749 (1972).
306. N. A. Novikov, A. S. Shashkov, and F. A. Galil-Ogly, "The NMR Study of the Interaction Between Polymer and a Plasticizer," *Vysokomol. Soedin.*, **A15**, 1068 (1973).
307. J. P. Quaegebeur et al., "ESR and NMR Study of the γ-Ray-Induced Postpolymerization of Vinyl Monomers Absorbed on Zeolite," *J. Polym. Sci., Polym. Chem. Ed.*, **14**, 2703 (1976).
308. C. Chachaty, M. Latimer, and A. Forchioni, "NMR and ESR Studies of the Solid-State Polymerization of Vinyl Monomers. III. Methacrylic Acid," *J. Polym. Sci., Polym. Chem. Ed.*, **13**, 189 (1975).
309. J. D. Memory and D. Lawing, "Broadline NMR of Polymers," *Magn. Reson. Rev.*, **5**, 69 (1979).
310. For good reviews of pulse NMR Studies of polymers, see (a) V. J. McBriety, "NMR of Solid Polymers: a Review," *Polymer*, **15**, 503 (1974). (b) D. W. McCall, "Dielectric, Mechanical and NMR Relaxation," *NBS Special Publication*, **310**, 475 (1976). (c) W. P. Slichter, "NMR Studies of Multiple Relaxations in Polymers," *J. Polym. Sci., Part C*, **14**, 33 (1966).
311. D. W. McCall, "Nuclear Magnetic Resonance Studies of Molecular Relaxation Mechanisms in Polymers," *Acc. Chem. Res.*, **4**, 223 (1971).
312. A. Charlesby and R. Folland, "The Use of Pulsed NMR to Follow Radiation Effects in Long-Chain Polymers," *Radiat. Phys. Chem.*, **15**, 393 (1980).
313. A. Charlesby, P. Käfer, and R. Folland, "Study of Very High Molecular Weight Polyethylene Using Pulsed NMR Techniques," *Radiat. Phys. Chem.*, **11**, 83 (1978).
314. R. Folland, J. H. Steven, and A. Charlesby, "Pulsed NMR Studies of Crosslinking and Entanglements in High Molecular Weight Linear Poly(dimethyl-siloxanes)," *Radiat. Phys. Chem.*, **10**, 61 (1977) and references cited therein.

315. R. Folland and A. Charlesby, "Pulsed NMR of *cis*-Polyisoprene: 1 & 2," *Polymer*, **20**, 207, and 211 (1979).
316. D. W. Larsen and J. H. Strange, "Pulsed NMR Study of Molecular Motion in the Uncured Diglycidyl Ether of Bisphenol-A," and "Pulsed NMR Study of Molecular Motion in the Diglycidyl Ether of Bisphenol-A Cured with 4,4′-Methylenedianiline," *J. Polym. Sci., Polym. Phys. Ed.*, **11**, 65 and 449 (1973).
317. V. J. McBriety, D. C. Douglass, and D. R. Falcone, "Nuclear Magnetic Relaxation in Polypropylene," *J. Chem. Soc., Faraday Trans. 2*, **68**, 1051 (1972) and references cited therein.
318. S. Kaufmann, W. P. Slichter, and D. D. Davis, "NMR Study of Rubber-Carbon Black Interactions," *J. Polym. Sci., Part A2*, **9**, 829 (1971).
319. D. H. Droste, A. T. DiBenedetto, and E. O. Stejskal, "Multiple Phases in Filled Polymers Detected by Nuclear Spin Relaxation Studies," *J. Polym. Sci., Part A2*, **9**, 187 (1971).
320. T. K. Kwei, T. Nishi, and R. F. Roberts, "Compatible Polymer Mixtures," *Macromolecules*, **7**, 667 (1974).
321. T. Nishi, T. T. Wang, and T. K. Kwei, "Thermally Induced Phase Separation Behavior of Compatible Polymer Mixtures," *Macromolecules*, **8**, 227 (1975).
322. T. Nishi, T. K. Kwei, and T. T. Wang, "Physical Properties of Poly(vinyl chloride) Copolyester Thermoplastic Elastomer Mixtures," *J. Appl. Phys.*, **46**, 4157 (1975).
323. V. J. McBriety, D. C. Douglass, and P. J. Barham, "Oriented Polymers from Solution. III. NMR of Polyethylene/Polypropylene Blends," *J. Polym. Sci., Polym. Phys.*, **18**, 1561 (1980).
324. M. J. Sullivan, unpublished data.
325. J. Schaefer, E. O. Stejskal, and R. Buchdahl, "Magic-Angle Carbon-1' NMR Analysis of Motion in Solid Glassy Polymers," *Macromolecules*, **10**, 384 (1977).
326. D. J. Patterson, J. L. Koenig, and J. R. Shelton, "Vulcanization Studies of Elastomers Using Solid-State Carbon-13 NMR," *Rubber Chem. Technol.*, **56**, 971 (1983).
327. S. A. Curran and A. R. Padwa, "^{13}C-NMR Analysis of Polybutadiene via Cross Polarization and Magic-Angle Spinning," *Macromolecules*, **20**, 625 (1987).
328. R. A. Komoroski, J. P. Shockcor, E. C. Gregg, and J. L. Savoca, "Characterization of Elastomers and Rubber Chemicals by Modern NMR Techniques," *Rubber Chem. Technol.*, **59**, 328 (1986).
329. M. Andreis, J. Lin, and J. L. Koenig, "Solid-State ^{13}C-NMR Studies of Vulcanized Elastomers. V. Observation of New Structures in Sulfur Vulcanized Natural Rubber," *J. Polym. Sci., Polym. Phys. Ed.*, **27**, 1389 (1989)
330. W. S. Veeman, E. M. Menger, W. Ritchey, and E. deBoer, "High-Resolution Carbon-13 Nuclear Magnetic Resonance of Solid Poly(oxymethylene)," *Macromolecules*, **12**, 924 (1979).
331. J. S. Waugh, L. M. Huber, and U. Haeberlen, "Approach to High-Resolution NMR in Solids," *Phys. Rev. Lett.*, **20**, 180 (1968).
332. For example, W. K. Rhim, D. D. Elleman, and R. W. Vaughan, "Enhanced Resolution for Solid-State NMR," *J. Chem. Phys.*, **58**, 1772 (1973).

333. D. P. Burum and W. K. Rhim, "Analysis of Multiple Pulse NMR in Solids," *J. Chem. Phys.*, **71**, 944 (1979).

334. B. C. Gerstein, "High Resolution NMR Spectrometry of Solids Parts I and II," *Anal. Chem.*, **55**, 781A and 899A (1983).

335. C. Dybowski, "Solid-State Investigations via NMR," *Chem Tech*, **1985**, 186.

336. B. C. Gerstein, R. G. Pembleton, R. C. Wilson, and L. M. Ryan, "High-Resolution NMR in Randomly Oriented Solids with Homonuclear Dipolar Broadening: Combined Multiple Pulse NMR and Magic-Angle Spinning," *J. Chem. Phys.*, **66**, 361 (1977).

337. C. E. Bronnimann, B. L. Hawkins, M. Zhang, and G. E. Maciel, "Combined Rotation and Multiple Pulse Spectroscopy as an Analytical Proton NMR Technique for Solids," *Anal. Chem.*, **60**, 1743 (1988).

338. T. T. P. Cheung, B. C. Gerstein, L. M. Ryan, R. E. Taylor, and C. R. Dybowski, "High-Resolution ^{1}H-Solid-State NMR Studies of Poly(ethylene terephathalate)," *J. Chem. Phys.*, **73**, 6059 (1980).

339. S. F. Dec, R. A. Wind, and G. E. Maciel, "Solid-State Fluorine-19 NMR Study of Fluorocarbon Polymers," *Macromolecules*, **20**, 2754 (1987).

340. H. Sillescu, "Recent Advances of ^{2}H-NMR for Studying Molecular Motion in Solid Polymers," *Pure Appl. Chem.*, **54**, 619 (1982).

341. H. W. Spiess, "Molecular Dynamics of Solid Polymers as Revealed by Deuteron NMR," *Colloid Polym. Sci.*, **261**, 193 (1983).

342. D. Hentschel, H. Sillescu, and H. W. Spiess, "Molecular Motion in Solid Polyethylene as Studied by 2D Wide Line NMR Spectroscopy," *Makromol. Chem.*, **180**, 241 (1979).

343. L. W. Jelinski, J. J. Dumais, and A. K. Engel, "Multitechnique Solid-State NMR Approach to Assessing Molecular Motion: Poly(butylene terephthalate) and Poly(butylene terephthalate)-Containing Segmented Copolymers," *Macromolecules*, **16**, 403 (1983).

CHAPTER

12

POLYMER CHARACTERIZATION USING MASS SPECTROMETRY

PAUL VOUROS and JOHN W. WRONKA

Department of Chemistry and Barnett Institute of Chemical Analysis
Northeastern University
Boston, Massachusetts

Modern Methods of Polymer Characterization, Edited by Howard G. Barth and Jimmy W. Mays
ISBN 0-471-82814-9

1. INTRODUCTION

During the past three decades mass spectrometry (MS) has evolved into an extremely important technique in chemical analysis. Its high sensitivity and dynamic range, coupled with both specificity and selectivity, provide a powerful method for the structural characterization of organic molecules. For most of this time, mass spectrometry has required vaporization of analytes in their intact form to acquire a representative mass spectrum. This has posed an obvious limitation for the structural characterization of polymers that degrade thermally before vaporization. Despite these limitations and the complexity of the resulting spectra, MS has continued to play a significant role in the field of polymer analysis even when employing conventional vapor phase ionization methods.

As of the late 1970s, MS has entered a renaissance period. New ionization methods have been developed that no longer require the classical thermal vaporization of the sample, while novel techniques of mass analysis and molecular ion fragmentation are beginning to provide a new data base for structural determination of organic molecules. These developments, along with the construction of instruments with expanded mass range and the emergence of combined high-performance liquid chromatography (HPLC)–MS, are gradually introducing new opportunities in the field of polymer analysis. This chapter will review some highlights of MS as it pertains to polymer analysis. These include the relationship of structural features to the spectral pattern as a function of the ionization process, using simple examples drawn primarily from studies of industrial polymers. In certain instances, data from work on biopolymers will be included, especially for the newer ionization methods, in view of the major concentration of research effort in that area. The implications of these results to industrial polymer analysis are obvious. The emphasis will be mainly on fundamental considerations associated with the formation and interpretation of the mass spectra of polymers. The objective is to present a somewhat basic and introductory, rather than exhaustive, picture of the field of polymer MS. For a more detailed treatment of the subject the reader is referred to an excellent recent review by Schulten and Lattimer (1).

When considering the application of MS to the analysis of polymers there is the potential for a wealth of information. Given the success that MS has enjoyed in the structural characterization and trace analysis of organic compounds, it is not unreasonable to expect the acquisition of data on the following:

1. Molecular weight distribution and average molecular weight.
2. Fingerprint pattern for polymer identification.
3. The sequence of monomeric units.
4. Branching, cross-linking, or other side-chain substitution.
5. Copolymer structures or variations in the polymeric system.
6. Additives or impurities present.

In practice, no single mass spectrometric technique can provide an answer to all these questions for a typical polymeric mixture. Frequently, a number of ancillary separation methods such as column chromatography, gas chromatography (GC), HPLC and/or chemical modification steps have to precede the MS analysis. These can be carried out offline but, in many cases, also in tandem. Equally important, however, the extent to which MS will be applicable to the solution of a given problem will depend on the method of ionization employed. In addition, procedures used for mass analysis, especially when coupled with collision-induced dissociation (CID), are expected to play an increasingly greater role in the field of polymer analysis. Accordingly, we will examine the different types of ionization methods, and the types of spectral patterns that they yield with typical oligomers. A discussion of recent advances in ion separation techniques will follow and selected examples will illustrate the significance of these principles.

2. IONIZATION TECHNIQUES

2.1. Vapor Phase Methods

The most conventional mode of spectrum acquisition in MS relies on the vaporization of the sample followed by ionization of the vaporized neutral molecules. This mode is sufficient to obtain the spectra of small, thermally stable, polymers, but larger polymers normally break down into pyrolytic products.

2.1.1. Electron Impact Ionization

The electron bombardment process produces a radical molecular ion typically depicted with the symbol $M^{+\bullet}$. For a hypothetical molecule ABC the sequence of events is illustrated by Reactions 1–5.

$$ABC + e^- \rightarrow ABC^{+\bullet}(M^{+\bullet}) + 2e^- \quad (1)$$

$$ABC^{+\bullet} \rightarrow AB^+ + C^\bullet \quad (2)$$

$$\rightarrow AB^{\bullet} + C^{+} \text{ and so on} \tag{3}$$

$$\rightarrow AC^{+} + B^{\bullet} \tag{4}$$

$$ABC^{+\bullet} + ABC \rightarrow ABCA^{+} + BC^{\bullet} \tag{5}$$

Reactions 1–3 yield a series of ions that can be used to "piece" the original structure together, effectively as in a jigsaw puzzle. Electron energies of 70 eV are typical of electron impact (EI) MS, although lower energies (10–20 eV) may enhance the relative intensity of molecular ion peaks and reduce fragmentation or ionically induced rearrangements. Rearrangements (Reac-

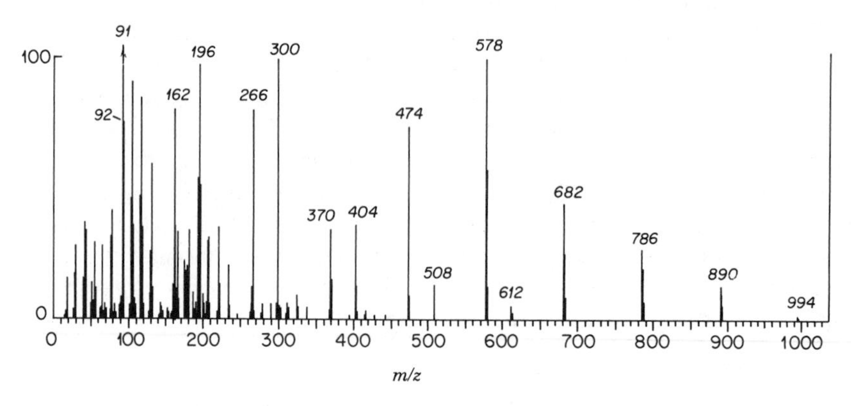

Figure 1. Electron impact (70 eV) mass spectrum of polystyrene 600. Maximum probe temperature: 225 °C; Ion source temperature: 225 °C. Reprinted with permission from F. Beckwitz and H. Heusinger, "Mass Spectrometric Characterization of Oligostyrol," *Agnew. Makromol. Chem.*, **46**, 143 (1975).

Table 1. Principal Ion Series (*m/z* Values) from Mass Spectrum (EI) of Polystyrene 600[a]

Series A	Series B
162	92
266	196
370	300
474	404
578	508
682	612
786	
890	
994	

[a]See Figure 1.

tion 4) and ion–molecule collisions (Reaction 5) can introduce considerable complexity to the spectral interpretation process especially in the presence of thermal rearrangements and/or polymeric degradation. In that case, it becomes extremely difficult to distinguish between ions produced from ionic fragmentation of larger species, and those generated from the ionization of small molecular moieties produced from pyrolysis. Under the simplest of conditions, however, that is, no thermal breakdown, EI ionization can provide unequivocal information about the structure of an oligomer.

The utility of EI ionization is illustrated by consideration of the EI mass spectrum of polystyrene 600 shown in Figure 1 (2). Structure **a**, the bracketed portion bearing a mass of 104 g/mol, represents the repeating unit of the polymer. Examination of the spectrum permits the recognition of two ion series exhibiting peaks every 104 g/mol, as outlined in Table 1.

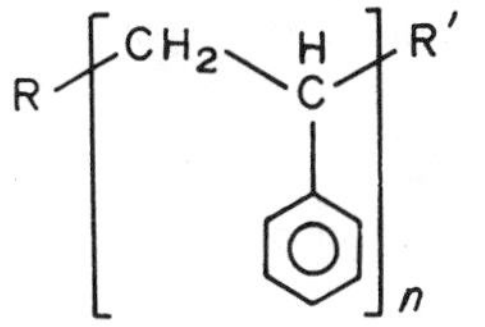

a

Pairs of related peaks in series A and B differ by 70 amu corresponding to the mass of a pentene moiety. Any given mass of series A can be viewed in terms of a molecular–ion containing n styrene moieties (n 104) + 58 (butane). This can help explain the formation of the series B peaks via a McLafferty rearrangement (Scheme 1). Thus the mass spectral data help determine that the polymer contains a butyl terminal group. Metastable peaks, which decompose into smaller fragment ions on the time scale of the experiment, support the process indicated in Scheme 1.

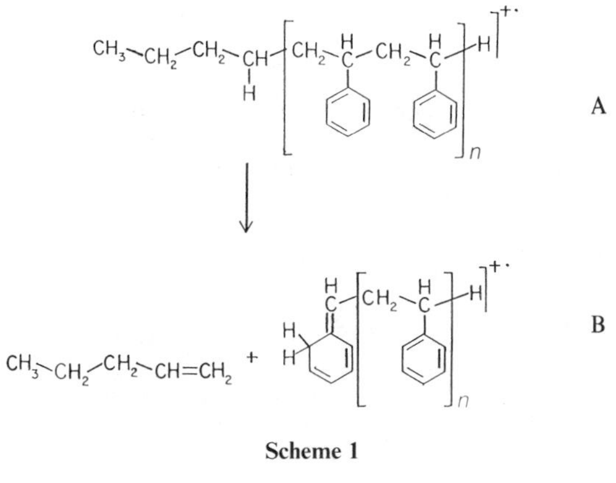

A

B

Scheme 1

The example discussed in Figure 1 represents in many ways an idealized system. Aromatic and/or conjugated systems are known to give few fragment ions upon ionization. This simplifies the spectral pattern considerably. Moreover, most common polymers pyrolyze into a multitude of products whose EI spectra are often superimposed onto each other. Specific pyrolytic bond cleavages are favored, depending on the temperature and the substrate, and this leads into different types of products. Understanding of the thermal degradation behavior of polymers facilitates the interpretation of these complex EI spectra and the acquisition of useful structural information.

As outlined by Luderwald (3), for synthetic polymers the typical thermal degradation behavior involves one or more of the following processes:

1. Statistical cleavage of polymer chains.
2. Degradation from the end group or some activated center to produce a monomer or to retropolymerize.
3. Thermal degradation and/or intramolecular reactions of side chains as a prereaction of main chain cleavage.
4. Thermal degradation via cyclic polymers.

The following examples reflect some of these features, the complexities associated with the interpretation of the spectra of polymeric systems, as well as the reasoning that enters into the rationalization of the spectral data.

The degree of substitution at reactive centers affects the thermal breakdown process and in turn, the mass spectral profile. Comparison of the pyrolysis–MS spectra (EI, 16 eV) of polymers **I**, **II**, and **III** illustrates this point.

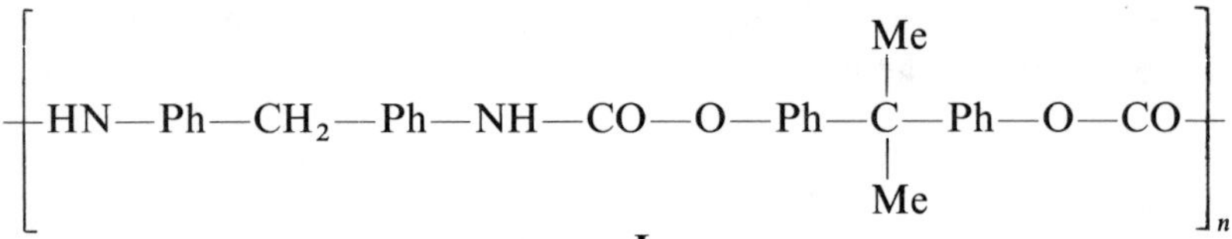

$$\left[\text{—HN—Ph—CH}_2\text{—Ph—NH—CO—O—Ph—}\overset{\text{Me}}{\underset{\text{Me}}{\text{C}}}\text{—Ph—O—CO—}\right]_n$$

I

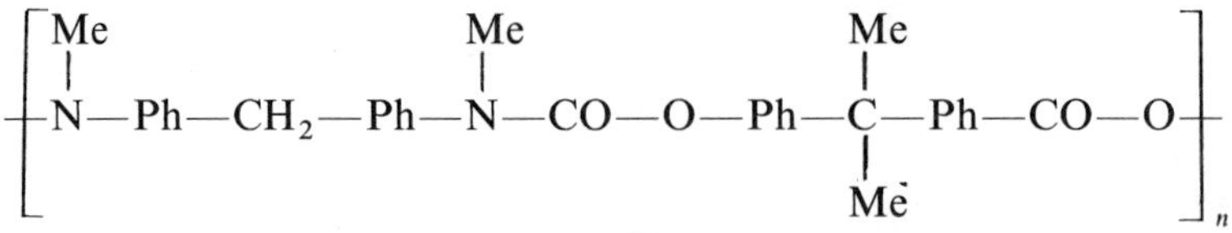

$$\left[\text{—}\overset{\text{Me}}{\text{N}}\text{—Ph—CH}_2\text{—Ph—}\overset{\text{Me}}{\text{N}}\text{—CO—O—Ph—}\overset{\text{Me}}{\underset{\text{Me}}{\text{C}}}\text{—Ph—CO—O—}\right]_n$$

II

$$\left[\text{—}\overset{\text{Me}}{\text{N}}\text{—Ph—CH}_2\text{—Ph—}\overset{\text{Me}}{\text{N}}\text{—CO—O—CH}_2\text{—}\overset{\text{Me}}{\underset{\text{Me}}{\text{C}}}\text{—CH}_2\text{—CO—O—}\right]_n$$

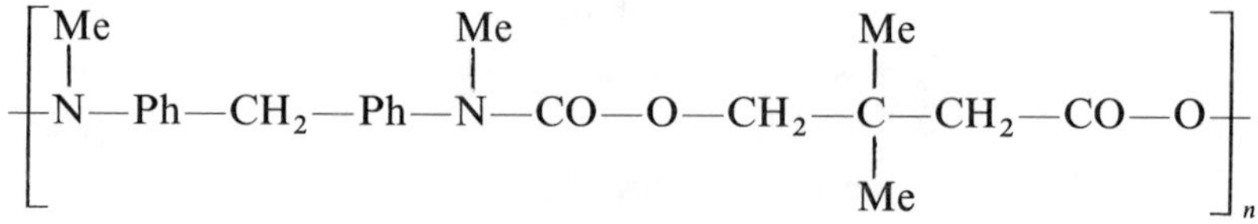

III

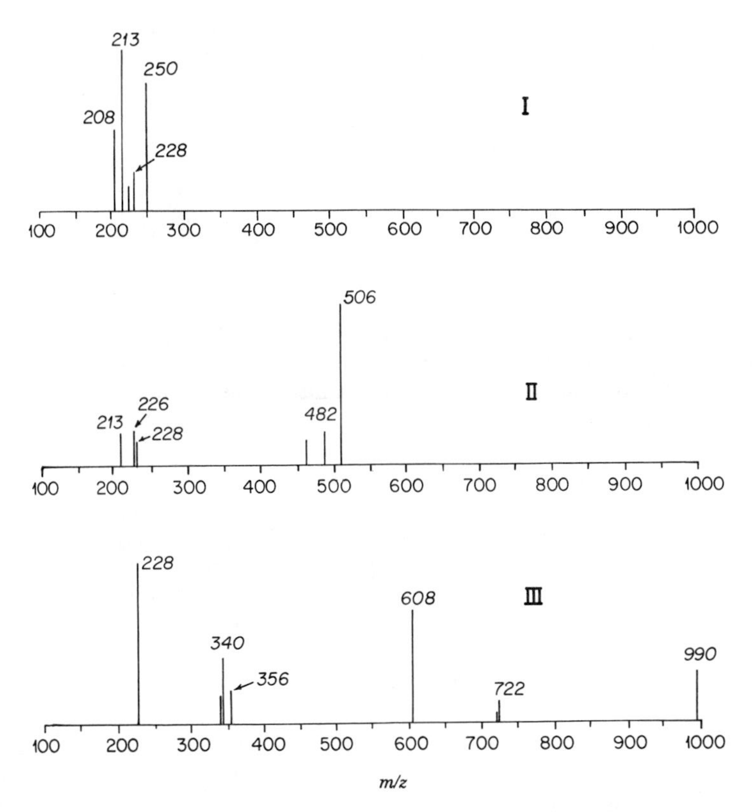

Figure 2. Pyrolysis mass spectra of polyurethane polymer **I** (190 °C, 16 eV), **II** (350 °C, 16 eV), and **III** (350 °C, 16 eV). (Reprinted from Ref. 4 with permission from the publisher).

Polymers **I** and **II** are rich in aromatic groups but differ in terms of the N substitution (H vs methyl). Polymer **III** has an aliphatic segment instead of the bisphenol in the repeating unit. Pyrolysis of **I** in the MS ion source gave an EI mass spectrum consisting of essentially six or seven major peaks (Fig. 2). These can be explained according to ions of structures **b**–**e**. No ions were found beyond m/z 250.

$$\left[O{=}C{=}N{-}Ph{-}\overset{\displaystyle Me}{\underset{\displaystyle Me}{C}}{-}Ph{-}NCO \right]^{+\cdot}$$

b
(m/z 250)

$$\left[HO{-}Ph{-}\overset{\displaystyle Me}{\underset{\displaystyle Me}{C}}{-}Ph{-}OH \right]^{+\cdot}$$

c
(m/z 228)

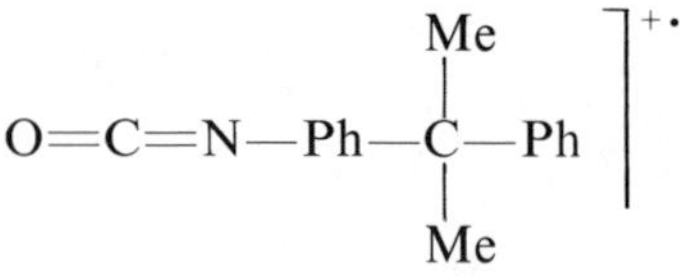

$$\left[O{=}C{=}N{-}Ph{-}\underset{Me}{\overset{Me}{\overset{|}{\underset{|}{C}}}}{-}Ph \right]^{+\bullet}$$

d
(m/z 208)

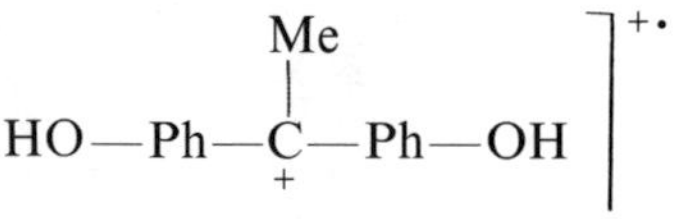

$$\left[HO{-}Ph{-}\underset{+}{\overset{Me}{\overset{|}{C}}}{-}Ph{-}OH \right]^{+\bullet}$$

e
(m/z 213)

By comparison, in polymer **II** the mass spectrum of the pyrolysis products (16 eV, 350 °C) is dominated by an ion of m/z 506, which corresponds to one with the cyclic structure **f**. No ions of higher mass were found in the spectrum of the pyrolyzate. The neutral compound of molecular weight 506 was in fact isolated from the pyrolysis products and its structure further confirmed by NMR.

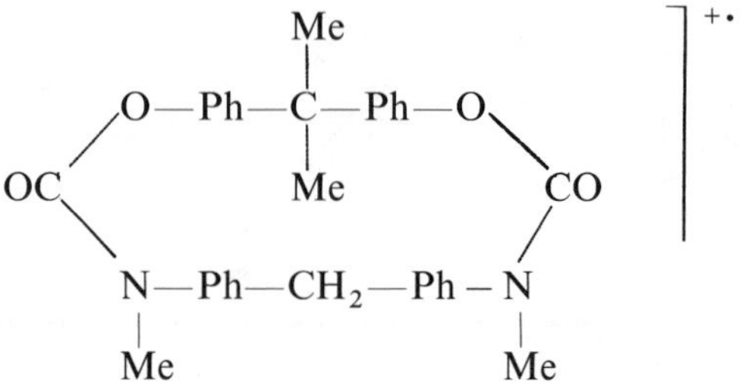

$$\left[\begin{array}{c} O{-}Ph{-}\underset{Me}{\overset{Me}{\overset{|}{\underset{|}{C}}}}{-}Ph{-}O \\ OC \qquad\qquad\qquad\qquad CO \\ \underset{Me}{\underset{|}{N}}{-}Ph{-}CH_2{-}Ph{-}\underset{Me}{\underset{|}{N}} \end{array} \right]^{+\bullet}$$

f
(m/z 506)

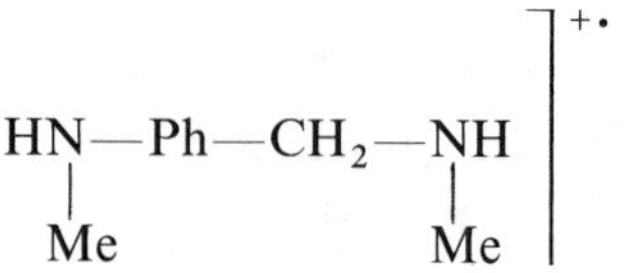

$$\left[\underset{Me}{\underset{|}{HN}}{-}Ph{-}CH_2{-}\underset{Me}{\underset{|}{NH}} \right]^{+\bullet}$$

g
(m/z 226)

$$\left[\underset{Me}{\underset{|}{HN}}{-}Ph{-}CH_2{-}Ph{-}\underset{Me}{\underset{|}{N}}{-}CO{-}O{-}Ph{-}\underset{Me}{\overset{Me}{\overset{|}{\underset{|}{C}}}}{-}Ph{-}OH \right]^{+\bullet}$$

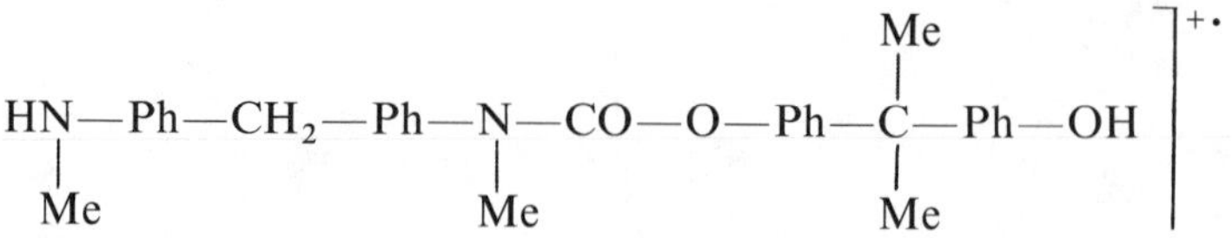

h
(m/z 480)

In contrast, the partially aliphatic character of polymer **III** helped generate a series of open-chain fragments extending as high as m/z 990 (16 eV, 350 °C). It has been rationalized (4) that intramolecular rearrangements, which involve methyl and hydrogen shifts, form many of these ions. Spectral data from these and other related polymers can help establish criteria for the recognition of functional group substitution in polyurethanes using pyrolysis–MS. The complexity of the spectral pattern generated from direct pyrolysis–MS–EI is further illustrated by a consideration of the polysiloxane polymers IV and **V** (5,6). Their mass spectra (18 eV) obtained at a pyrolysis temperature of 470 °C are compared in Figure 3.

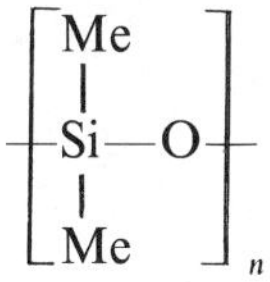

$$\left[-\mathrm{Si(Me)_2-O}-\right]_n \qquad\qquad \left[-\mathrm{Si(Me)_2-Ph-Si(Me)_2-O-Si(Me)_2-O}-\right]_n$$

IV 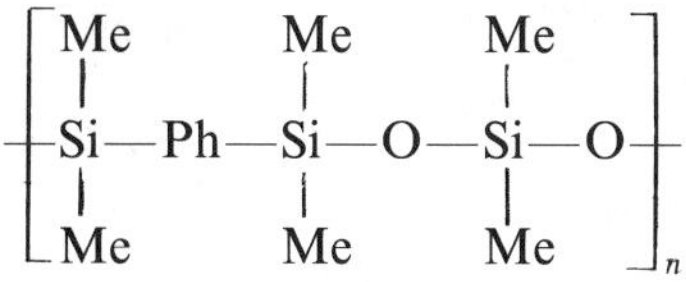 **V**

The first step in the pyrolysis of **IV** is thermal degradation to form a series of cyclic polymers via interaction of two activated centers

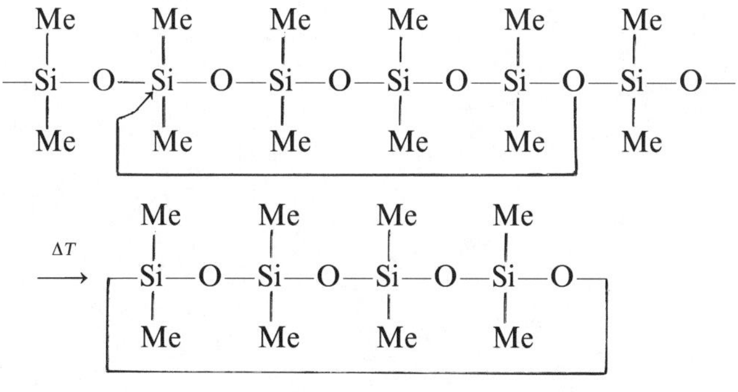

$$-\mathrm{Si(Me)_2-O-Si(Me)_2-O-Si(Me)_2-O-Si(Me)_2-O-Si(Me)_2-O-Si(Me)_2-O}-$$

$$\xrightarrow{\Delta T} \mathrm{cyclo}\left[\mathrm{Si(Me)_2-O-Si(Me)_2-O-Si(Me)_2-O-Si(Me)_2-O}\right]$$

or in general

$$\mathrm{cyclo}\left[-\mathrm{Si(Me)_2-O}-\right]_n$$

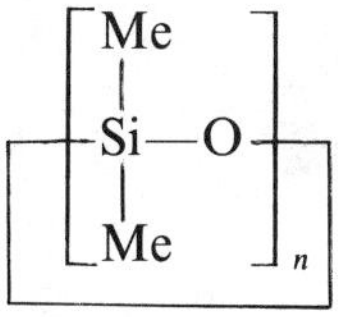

Ionization of these pyrolysis products by EI (18 eV) yields two types of key fragments.

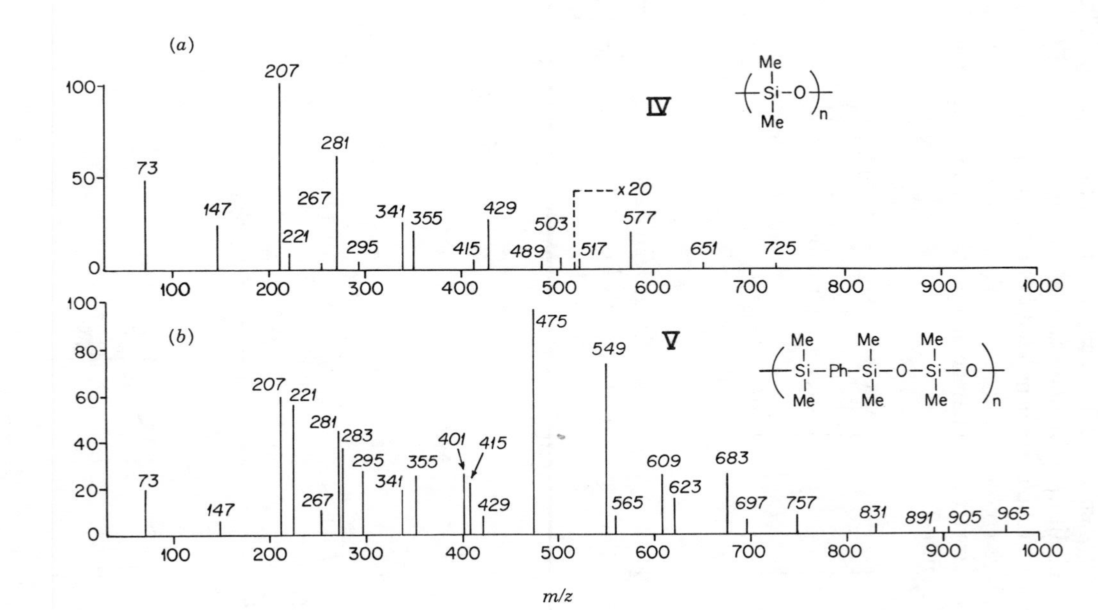

Figure 3. Pyrolysis–EI mass spectra of polymers IV (*a* 18 eV, 350 °C) and **V** (*b* 18 eV, 350 °C). Reprinted with permission from A. Ballistreri, D. Garozzo, and G. Mantaudo, "Mass Spectral Characterization and Thermal Decomposition Mechanism of Poly(dimethyl siloxane)," *Macromolecules,* **17**, 1312 (1984). Copyright © 1984 American Chemical Society.

1. Loss of a methyl group from a given cyclic polymer molecular–ion via direct cleavage to give $[M - 15]^+$ ions according to, for example, the reaction

$$\left[\overline{-\mathrm{Si(Me)_2}-\mathrm{O}-\mathrm{Si(Me)_2}-\mathrm{O}-\mathrm{Si(Me)_2}-\mathrm{O}-\mathrm{Si(Me)_2}-\mathrm{O}-}\right]^{+\cdot} \xrightarrow[-\mathrm{CH_3}]{} \overline{-\mathrm{Si(Me)_2}-\mathrm{O}-\mathrm{Si(Me)_2}-\mathrm{O}-\overset{+}{\mathrm{Si}}(\mathrm{Me})-\mathrm{O}-\mathrm{Si(Me)_2}-\mathrm{O}-}$$

These ions can be depicted with the general structure

$$\overline{\left[-\mathrm{Si(Me)_2}-\mathrm{O}-\right]_{n-1}\text{------}\overset{+}{\mathrm{Si}}(\mathrm{Me})-\mathrm{O}-}$$

i

and are responsible for the series of peaks at m/z 207 ($n = 3$), m/z 281 ($n = 4$), m/z 355 ($n = 5$)$\cdots m/z$ 725 ($n = 10$). For any given cyclic oligomer, these are the highest mass peaks since no molecular–ions are observed.

2. Alternatively, EI of the cyclic oligomer can also induce a fragmentation accompanied by a methyl shift. Using again the example of the tetramer, this can be rationalized according to the reaction

$$\left[\overline{-\mathrm{Si(Me)_2}-\mathrm{O}-\mathrm{Si(Me)_2}-\mathrm{O}-\mathrm{Si(Me)_2}-\mathrm{O}-\mathrm{Si(Me)_2}-\mathrm{O}-}\right]^{+\cdot} \xrightarrow[-\mathrm{CH_3}]{} \mathrm{Me}-\mathrm{Si(Me)_2}-\mathrm{O}-\mathrm{Si(Me)_2}^{+}$$

In general, the fragment ions of this type can be depicted with the structure **j**,

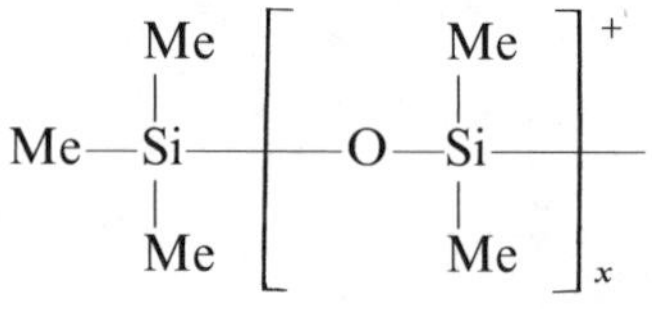

$$\mathrm{Me{-}Si(Me)_2}\left[\mathrm{{-}O{-}Si(Me)_2{-}}\right]_x^{+\cdot}$$

j

and are responsible for the series of peaks at m/z 73 ($x = 0$), m/z 147 ($x = 1$), m/z 221 ($x = 2$)$\cdots m/z$ 517 ($x = 6$).

Four ions (m/z 267, m/z 341, m/z 415, and m/z 489) in the spectrum of polymer **IV** cannot be accounted for by the above fragmentation processes. Metastable transitions indicate that they are formed by elimination of tetramethylsilane (TMS) (loss of 88 g/mol) from m/z 355, 429, 503, and 577, respectively. This would involve some complex rearrangement process from ions of the type **i** beginning with $n = 5$ to $n = 8$ for the four ions indicated.

The same general considerations that entered into the interpretation of the spectrum of the polydimethylsiloxane (**IV**), can also be used to explain the spectral patterns of the silarylene–siloxane polymer (**V**). As a first step, it is important to consider here the equivalency of the Si atoms in each repeating unit. It is observed that only one Si atom in **V** is attached to two oxygen atoms; both of the other Si atoms are linked to an oxygen and to a phenyl ring in the silphenylene portion of the unit. As a result, the thermal degradation produces a variety of products that contain pure siloxane, pure phenyldisilanol units, or combinations of siloxane and phenyldisilanol units.

In the thermal degradation of polymer **V**, intramolecular silicon–oxygen interactions yield a series of cyclic polymers as in the case of **IV**. Simple cyclic polymers with structures **k**, **l**, and **m** can be postulated to explain the indicated series of ions. It is interesting that the identity of the primary repeating unit is, to a large extent, lost; it is only found in the cyclic polymer **k** and, effectively reflected in only two major spectral ions at m/z 549 and 831.

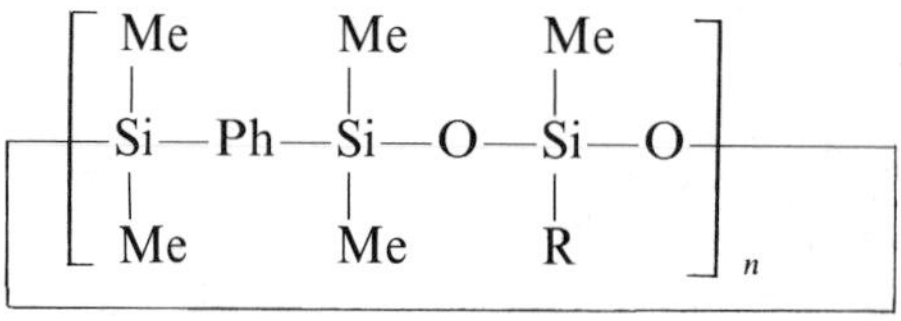

$$\mathrm{cyclo}\left[\mathrm{{-}Si(Me)_2{-}Ph{-}Si(Me)_2{-}O{-}Si(Me)(R){-}O{-}}\right]_n$$

k (R = Me)
k′ (R = Ph)

k	$M^{+\cdot}$	$[M-CH_3]^+$
$n = 2$	564	549
$n = 3$	846	831

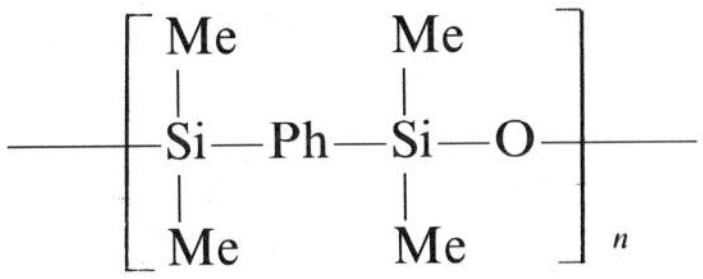

$$\left[-\overset{\text{Me}}{\underset{\text{Me}}{\text{Si}}}-\text{Ph}-\overset{\text{Me}}{\underset{\text{Me}}{\text{Si}}}-\text{O}- \right]_n$$

l

l	$M^{+\cdot}$	$[M-CH_3]^+$
$n=2$	416	401
$n=3$	624	609
$n=4$	832	817

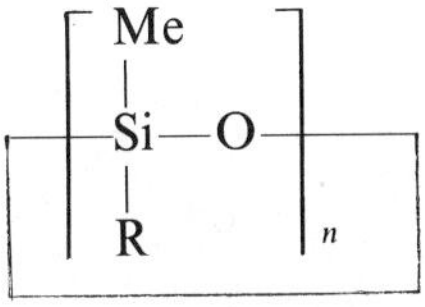

$$\left[-\overset{\text{Me}}{\underset{\text{R}}{\text{Si}}}-\text{O}- \right]_n \text{ (cyclic)}$$

m (R = Me)
m′ (R = Ph)

m	$M^{+\cdot}$	$[M-CH_3]^+$
$n=3$	222	207
$n=4$	296	281
$n=5$	370	355
$n=6$	444	429
$n=7$	518	503

In addition to these "simple" cyclic polymers, combination polymers of the type **n** and **o** can account for the bulk of the remaining peaks in the spectrum of **V**. These bear no similarity to the original structure in so far as the primary repeating unit is concerned.

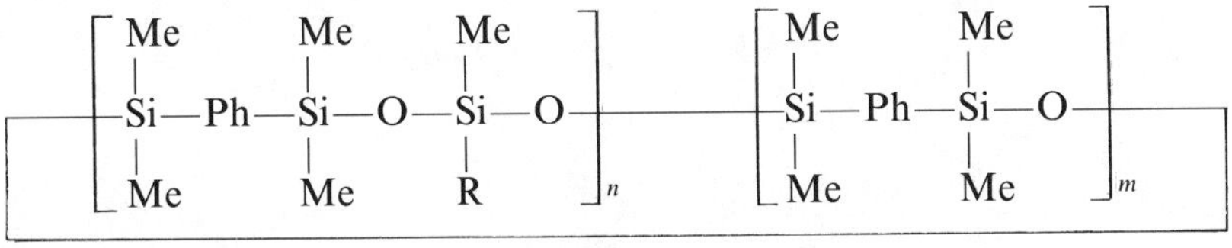

$$\left[-\overset{\text{Me}}{\underset{\text{Me}}{\text{Si}}}-\text{Ph}-\overset{\text{Me}}{\underset{\text{Me}}{\text{Si}}}-\text{O}-\overset{\text{Me}}{\underset{\text{R}}{\text{Si}}}-\text{O}- \right]_n \left[-\overset{\text{Me}}{\underset{\text{Me}}{\text{Si}}}-\text{Ph}-\overset{\text{Me}}{\underset{\text{Me}}{\text{Si}}}-\text{O}- \right]_m \text{ (cyclic)}$$

n (R = Me)
n′ (R = Ph)

n	$M^{+\cdot}$	$[M-CH_3]^+$
$n=1, m=1$	430	415
$n=1, m=2$	638	623
$n=1, m=3$	906	891
⋮	⋮	⋮
n = 2, $m=2$	980	965

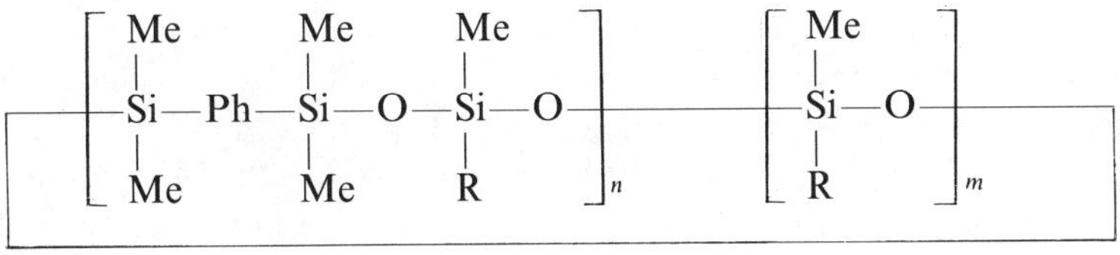

$$\left[-\overset{\text{Me}}{\underset{\text{Me}}{\text{Si}}}-\text{Ph}-\overset{\text{Me}}{\underset{\text{Me}}{\text{Si}}}-\text{O}-\overset{\text{Me}}{\underset{\text{R}}{\text{Si}}}-\text{O}- \right]_n \left[-\overset{\text{Me}}{\underset{\text{R}}{\text{Si}}}-\text{O}- \right]_m \text{ (cyclic)}$$

o (R = Me)
o′ (R = Ph)

o	$M^{+\cdot}$	$[M—CH_3]^+$
$n=2, m=2$	430	415
⋮	⋮	⋮
$n=2, m=1$	638	623
$n=2, m=2$	712	697
$n=3, m=1$	920	905

2.1.2. Chemical Ionization

Chemical ionization (CI) provides an energetically softer ionization process than EI, which results in reduced fragmentation of the analyte (7). Unlike in electron impact, ionization of a sample molecule results from the interaction of charged reagent gas ions produced by EI at pressures typically ranging from 0.5 to 1.0 torr. Methane, isobutane, and ammonia are the most common reagent gases used in CI. The principal events occurring under methane CI conditions are summarized in Reactions 6–11:

$$CH_4 + e^- \rightarrow CH_4^{+\cdot} + 2e^- \quad (6)$$

$$CH_4^{+\cdot} \rightarrow CH_3^+ + H^\cdot \quad (7)$$

$$CH_4^{+\cdot} + CH_4 \rightarrow CH_5^+ + H_2 \quad (8)$$

$$CH_3^+ + CH_4^+ \rightarrow C_2H_5^+ + H_2 \quad (9)$$

$$CH_5^+ + M \rightarrow [M+H]^+ + CH_4 \quad (10)$$

$$C_2H_5^+ + M \rightarrow [M+C_2H_5]^+ \quad (11)$$

While CH_5^+, a strong Brønsted acid, provides the main pathway for ionization of a sample molecule M via protonation, formation of adduct ions such as $[M+C_2H_5]^+$ and others also occurs to an appreciable extent. The difference in proton affinity between M and the reagent gas conjugate base (CH_4) controls the degree of fragmentation of $[M+H]^+$. When the reagent gas is ammonia, the principal reacting ion, NH_4^+, is a weaker protonating agent than CH_5^+ and analyte spectra with reduced fragmentation are produced. If the proton affinity of M is lower than that of NH_3, adduct ions $[M+NH_4]^+$ frequently result. Similarly, with isobutane which is a softer reagent gas than CH_4, both protonated $[M+H]^+$ and $[M+C_4H_9]^+$ adduct ions are typical. For oligomer mixtures introduced into the MS via the direct insertion probe, the CI spectra provide a more simplified picture of the mixture.

The combination of EI and CI can be used to provide structural detail about the sample. However, as pointed out by Shimizu and Munson (8), when oligomers are pyrolyzed off the direct insertion probe into the MS, it is important to beware of the possibility of changes in the consistency of the pyrolysis products. These may come about as a result of neutral–neutral reactions between radical species produced during the pyrolysis and the reagent gas. Such artifacts can be ultimately recognized by varying the reagent gas.

Figure 4 compares the EI and CI (*i*-butane) mass spectra up to 400 g/mol of a polystyrene sample pyrolyzed at 390 °C. The protonated monomer, dimer, and trimer are observed at *m*/*z* 105, 209, and 313 in the CI spectrum. Corresponding adducts $C_4H_9^+$ appear 56 g/mol higher at *m*/*z* 161, 265, and 369. While the multitude of these adduct ions may seem confusing, these 56 g/mol mass differences help confirm the presence of the related molecular species produced from the thermal degradation of the polymer. Consider, for example, the molecular–ion of the dimer (*m*/*z* 194) in the EI spectrum. Its recognition may be somewhat obscured by the cluster of peaks in that mass

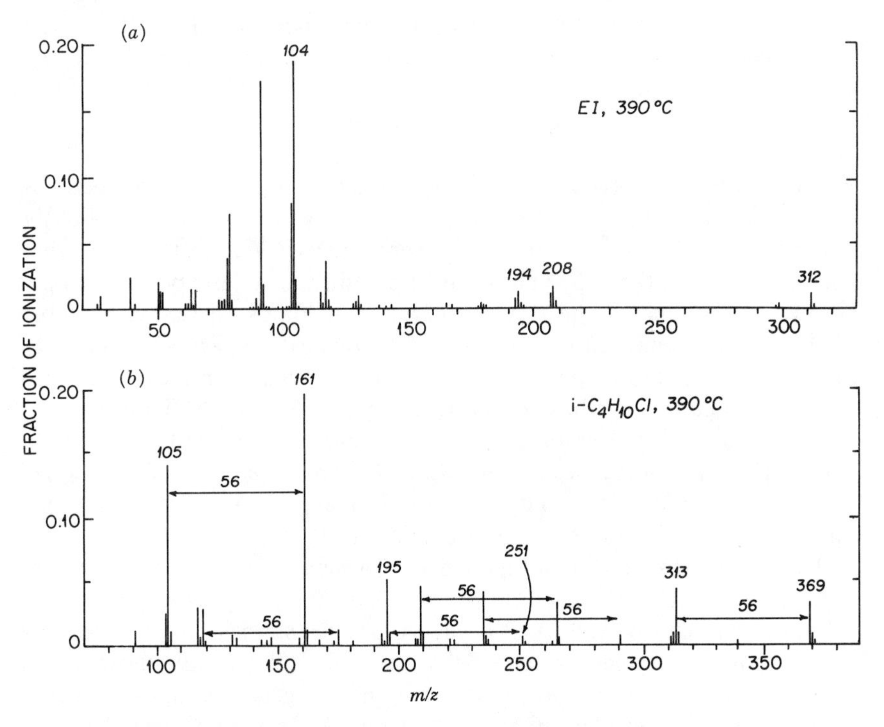

Figure 4. Comparison of EI (*a*) and CI–*i*-butane (*b*) mass spectra of pyrolysis products of polystyrene at 390 °C. Reprinted from Ref. 8 with permission from the publisher.

region. However, in view of the pairs of peaks at *m/z* 195 and *m/z* 251 in the CI spectrum, it is reasonable to propose the occurrence of a molecular species, possibly of composition $C_{15}H_{14}$, mass 194. Similar conclusions may be drawn about other peaks in the spectrum.

2.1.3. Direct Chemical Ionization

It is clear that, with increasing molecular weight, the ions observed in a mass spectrum are more likely to appear because of the ionization of pyrolytic degradation products rather than due to ionically induced fragmentations of vaporized molecular species. The technique of direct chemical ionization (DCI) mass spectrometry, in which a wire coated with a nonvolatile sample is introduced into the CI reagent gas plasma via a modified insertion probe (9), can provide molecular weight information on larger and/or more thermally labile molecules than conventional CI. Even highly labile molecules can be vaporized in their intact form using heated wires coated with inert materials such as polyimides as shown by Reinhold and Carr (10). The technique has been employed primarily in the study of biomolecules. The mass spectra of permethylated polysaccharides provide molecular ions as well as fragmentation patterns useful for sequencing of the oligomer.

2.1.4. Field Ionization

When a molecule is brought near a metal anode at a high positive potential, conditions may develop for the removal of an electron at field values of the order of 10^7–10^8 V/cm. The process is referred to as field ionization (FI) when the molecule is in the vapor phase (as opposed to field desorption discussed below), and is one of the softest ionization methods currently available. Fragmentation is minimized, and molecular–ion peaks typically dominate the spectra. Figure 5 shows the FI spectrum of a polystyrene sample pyrolyzed at 670 °C (11) and this can be contrasted to the CI spectrum of Figure 4, which was obtained with polystyrene pyrolyzed at 390 °C. The simplicity of the FI spectra allow a more facile determination of the relative ratios of a monomer (*m/z* 104), dimer (*m/z* 208), and a trimer (*m/z* 312) in the pyrolyzate and their variation as a function of pyrolysis conditions.

Evident in the spectrum of Figure 5 is the absence of oligomers beyond the trimer. In general, the detection of molecular–ions of primary, high-mass thermal degradation products is desirable, as such species can provide more information about the polymer. Schulten et al. (12) recently showed that improvements in heating procedures can, indeed, generate high-molecular-weight thermal degradation products. Coupled with improved detection methods using post acceleration of the ions, the procedure has provided as

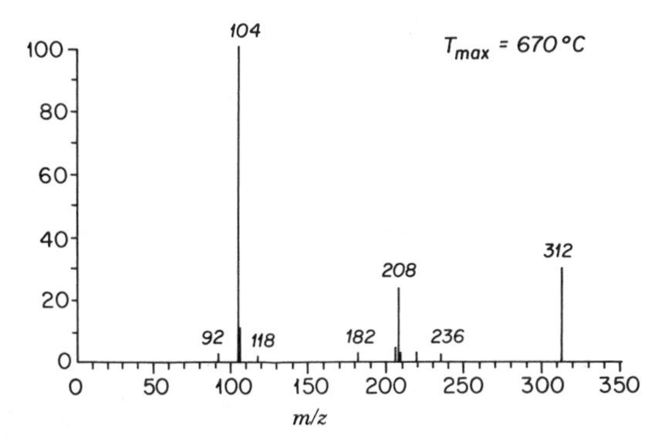

Figure 5. Field ionization mass spectrum of polystyrene pyrolyzed at 670 °C. Reprinted from Ref. 11 with permission from Hüthig & Wepf Verlag, Basel.

much as a 50-fold improvement in sensitivity at higher mass. For polyethylene, the authors have shown examples of a homologous series of pyrolyzates with a mass as high as 2000 g/mol using pyrolysis–FI–MS.

2.2. Condensed Phase Techniques

Direct desorption of ionized species from the condensed phase, effectively a concurrent ionization and vaporization, is a promising alternative for the analysis of high-molecular-weight nonvolatile substances. Initially the most popular approach involved an electric field-assisted surface desorption process (field desorption, FD), while most recently particle bombardment methods and direct ionization from ionic solutions have started to become more attractive. There are inherent advantages and disadvantages associated with these ionization methods and many of their special features will be elaborated in the following sections. These ionization methods are classified somewhat broadly into surface ionization methods and into techniques that involve ionization from solutions.

2.2.1. Surface Ionization Methods

2.2.1.1. Field Desorption. An extension of field ionization, field desorption mass spectrometry (FDMS) gained prominence in the early 1970s following the discovery of Beckey that wire anodes activated with benzonitrile and coated with a sample, could produce mass spectra of thermally labile compounds when placed in an FI source (13). Since FD can bypass the thermal vaporization step, the mass spectra usually show even less fragmentation than those obtained under FI conditions, where the molecules are of a higher initial

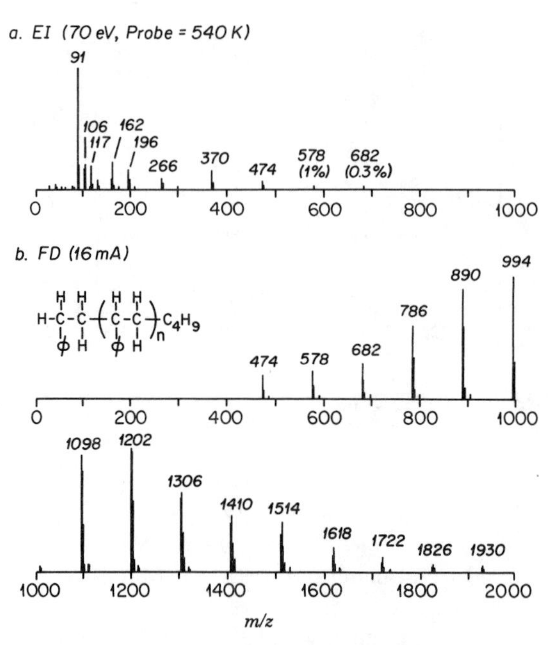

Figure 6. Electron impact (70 eV) (*a*) and field desorption (*b*) mass spectra of a mixture of polystyrene oligomers. Emitter current for FD: 16 mA.

internal energy. For nonvolatile compounds, heating of the emitter–anode is usually necessary. This heating may also promote fragmentation.

The value of FDMS is illustrated in Figure 6, which compares the EI (70 eV) and FD mass spectra of a sample containing polystyrene oligomers up to 18 monomer units (MW 1930) (14). The sample had a peak molecular weight of 1020 g/mol as determined by intrinsic viscosity. In the EI spectra one observes only weak peaks up to 682 g/mol corresponding to the hexamer. However, in the FD spectrum, where the emitter wire was heated to 16 mA, oligomer molecular–ions up to 1930 g/mol (octadecamer) are observed. The molecular–ions are in multiples of $(58 + n*104)$, indicating a polymer terminating in a butyl group. With the use of the molecular–ion masses from FDMS and the peak heights from an LC chromatogram, the calculated number-average molecular weight

$$\bar{M}_n = \frac{\sum N_i M_i}{\sum N_i} \tag{1}$$

where M_i = molecular weight of a given oligomer
N_i = number of molecules with specified molecular weight

was found to be 853 g/mol compared to 811 ± 5% g/mol by vapor pressure osmometry (VPO). The same extinction coefficient value was assumed for all polystyrene oligomers. No monomer, dimer, or trimer appear in the FDMS spectrum because these molecules were completely vaporized from the emitter at 16 mA. This can account for the higher $\bar{M}_n$ value obtained in the FDMS as opposed to the VPO measurement.

2.2.1.2. Secondary Ion Mass Spectrometry. The use of a primary ion beam to bombard a target surface and "sputter-off" secondary ions, which are then resolved mass spectrometrically, is well known for the analysis of inorganic materials (15). Argon or xenon ions generated by a discharge process are a typical source for the primary ion beam. This ion beam can be focused to provide surface analysis with a definition approaching 1 μm in-depth analysis of monolayers with 80–100-Å resolution (16). Mass spectra of organic compounds result if they are dissolved in a liquid matrix such as glycerol. This is analogous to the fast atom bombardment (FAB) ionization process discussed below. Under those conditions, secondary ion mass spectroscopy (SIMS) and FAB spectra are substantially similar and charging effects previously thought to occur in SIMS due to the use of ion as opposed to atom bombardment appear to be inconsequential (17). Hercules and co-workers (18) used SIMS effectively for the analysis of polymers and, in particular, the detection of high-mass ions using a time-of-flight (TOF) analyzer (see Section 3). The SIMS spectra of a series of polyamides (nylons), showed fragmentation of the backbone of the polymer and low-mass-fragment ions. Protonation and cationization of segments of the polymer chain with Ag^+, Na^+, or K^+, produced high-mass ions characteristic of the polymer and the repeat unit sequence in the polymer chain. The same group has also showed that charging of the polymer surface during SIMS can be prevented by placing a conducting grid on the polymer surface and keeping the primary ion beam flux at a low level. Thick polymer films (e.g., ~ 1 mm of poly(dimethyl siloxane) in the mass range of 4500 g/mol, were analyzed in this fashion. The mass of the polymer unit was ascertained from the m/z values and the spacings between peaks (19).

2.2.1.3. Fast Atom Bombardment. Barber et al. (20) introduced the soft ionization technique of fast atom bombardment (FAB)–MS in 1981. In FAB ionization, a highly energetic neutral atom beam (6–10 keV), typically Ar or Xe, bombards a glycerol film containing an analyte. A schematic diagram of the relative disposition of the atom beam, and the target surface is shown in Figure 7. An incident angle of approximately 20° is necessary for optimum ionization efficiency and sample ion recovery; if the beam is directed perpendicular to the target surface (angle = 0°) destruction of the sample and

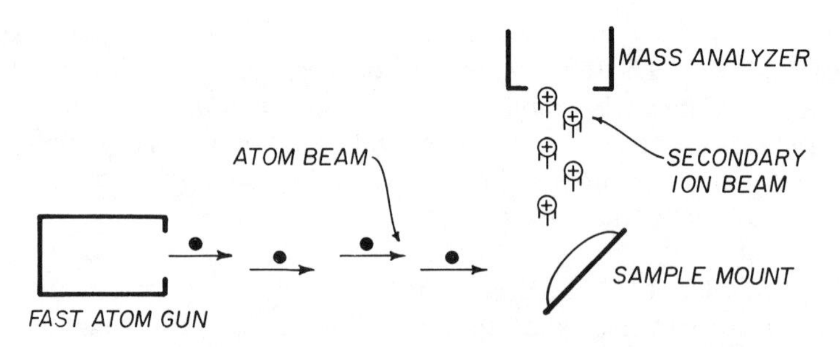

Figure 7. Schematic diagram showing relative disposition of atom beam, target surface, and mass analyzer in fast atom bombardment. (Reproduced from J. T. Watson, *Introduction to Mass Spectrometry*, Raven Press, New York, 1984, with permission from the publisher).

the matrix may result, whereas, a 90° angle of atom beam incidence simply skims the surface with minimal energy transfer and ion production. The glycerol or other suitable viscous matrix provides a continuously renewable surface for the incoming atom beam, thus maintaining a continuous spectrum for several minutes (21, 22).

The fundamental ionization process in FAB–MS is a proton or other ion (e.g., Na^+ or K^+) transfer from the matrix to the dissolved analyte. To a certain extent, this may be visualized as a chemical ionization in the condensed phase, or at the immediate interface between liquid and vapor, with a portion of the resulting ions from each species being transferred to the vapor phase. It appears that the ion peak intensities in FAB spectra correlate closely with ionic content in the matrix (23), although evidence for alternative mechanisms also exists (24). Sample ionization can be promoted with increased ionic strength of the solvent matrix. This is especially true for analytes of lower polarity. High-molecular-weight ionized molecules can be effectively sublimed into the vapor phase with minimal thermal degradation and, depending on the type of molecular species involved, form structurally informative fragment ions. Cochran (25) has addressed the applicability of FAB–MS to the analysis of synthetic polymers and polymer additives. The possible pitfalls associated with the occurrence of reactions between the analyte and the matrix have been cited. For example, reduction processes may result in the incorporation of two or more hydrogen atoms from the matrix and affect molecular weight determination.

Figure 8 shows the FAB spectrum of a low MW poly(ethersulfone) polymer, $—O—Ph—SO_2—Ph—O—Ph—SO_2—Ph—O—$, obtained from a glycerol solution doped with NH_4Cl and dilute HCl and containing 500 ng of the sample (26). The backbone structure of the polymer has a mass of 248 g/mol and a signal corresponding to as high as the pentamer, m/z 1240, is

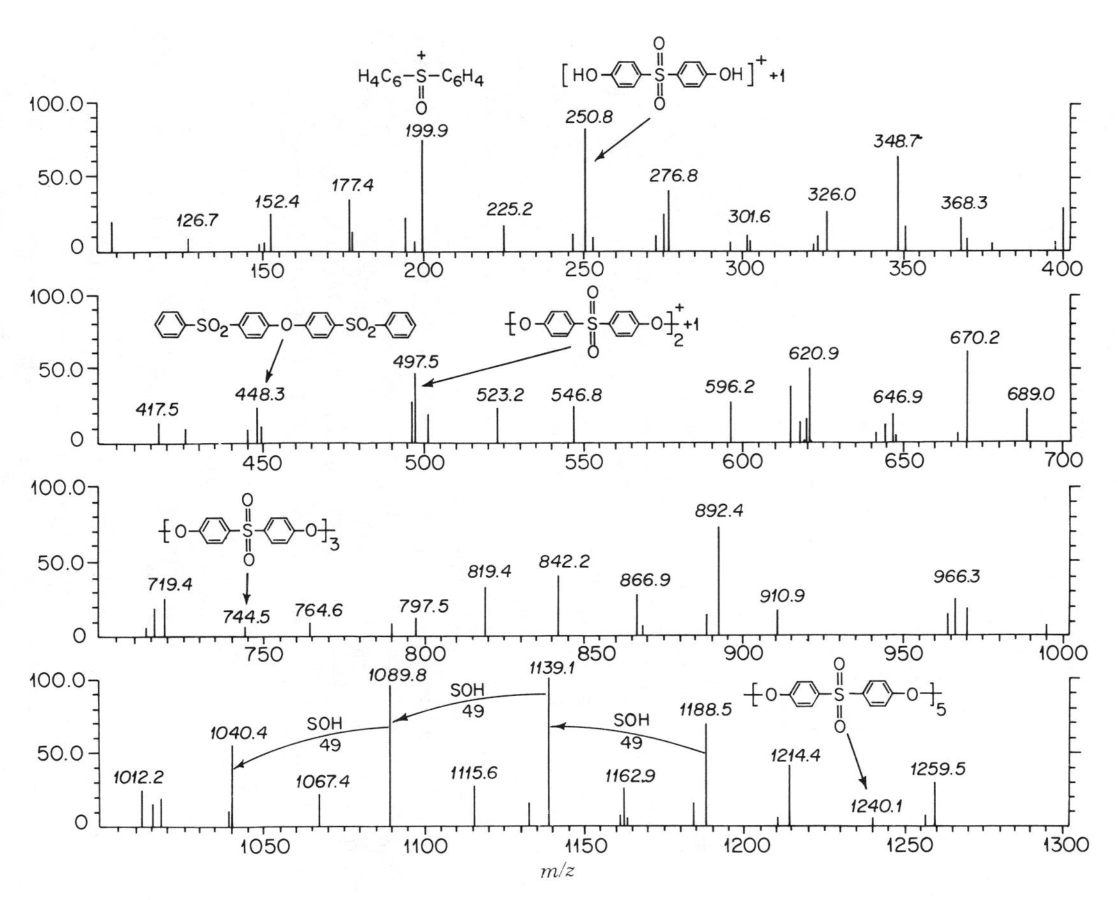

Figure 8. FAB mass spectrum of a commercial poly(ether sulfone). Reprinted from Ref. 26 with permission from the author.

observed. In addition, considerable fragmentation is indicated suggesting that, despite the obvious complexity of the spectra, FAB ionization can be used to provide a fingerprint of a polymer system. In principle, the various peaks shown in the mass spectrum can be explained in terms of fragmentation of the system. A series of ions separated by 49 mass units (S=O + H) is observed and the *m/z* 251 ion, *p*, is characteristic of the polymer backbone.

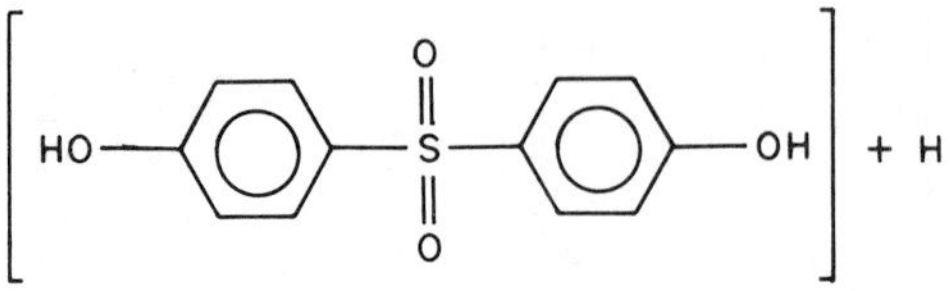

$$\left[HO-C_6H_4-S(=O)_2-C_6H_4-OH \right] + H$$

p
(*m/z* 251)

Ionization by FAB may not always result in fragmentation of the backbone of large polymers. In fact, as shown by Montaudo and co-workers (27), caution has to be exercised in the FAB analysis of polymers and in the interpretation of mass spectral data. For example, two polyesters and two polyamides were subjected to FAB analysis by dissolving in a glycerol matrix. Peaks were observed in the spectra of the crude polymers, which could be, at least in part, interpreted in terms of the fragmentation of the polymer chain. However, upon purification of the polymers by removal of lower molecular weight compounds, no such peaks were observed in the spectra of the samples. The FAB spectra of the oligomers extracted from the polymers, however, were substantially similar to those of the crude polymer. This is illustrated in Figure 9, which shows the spectra of the crude polymer (Fig. 9*a*) and of the extract (Fig. 9*b*), respectively, of poly(ε-caprolactone). The spectral peaks can be readily assigned to protonated molecular ions of cyclic oligomers and open linear oligomers:

$$\text{cyclo-}[-O-(CH_2)_5-CO-]_n \qquad m/z\ 115 + n114$$

$$H-[-O-(CH_2)_5-CO-]_n-OH \qquad m/z\ 247 + n114$$

Selective fragmentation of the polymeric backbone does not seem to occur to any significant extent in the spectra of these polymers.

Different types of chemical additives in polymer formulations prolong the useful lifetimes of these formulations or improve physical properties such as optical characteristics or mechanical strength. It is important to be able to identify these additives, monitor them for any chemical transformations, and quantify them. Sensitive techniques are needed for these analyses since the

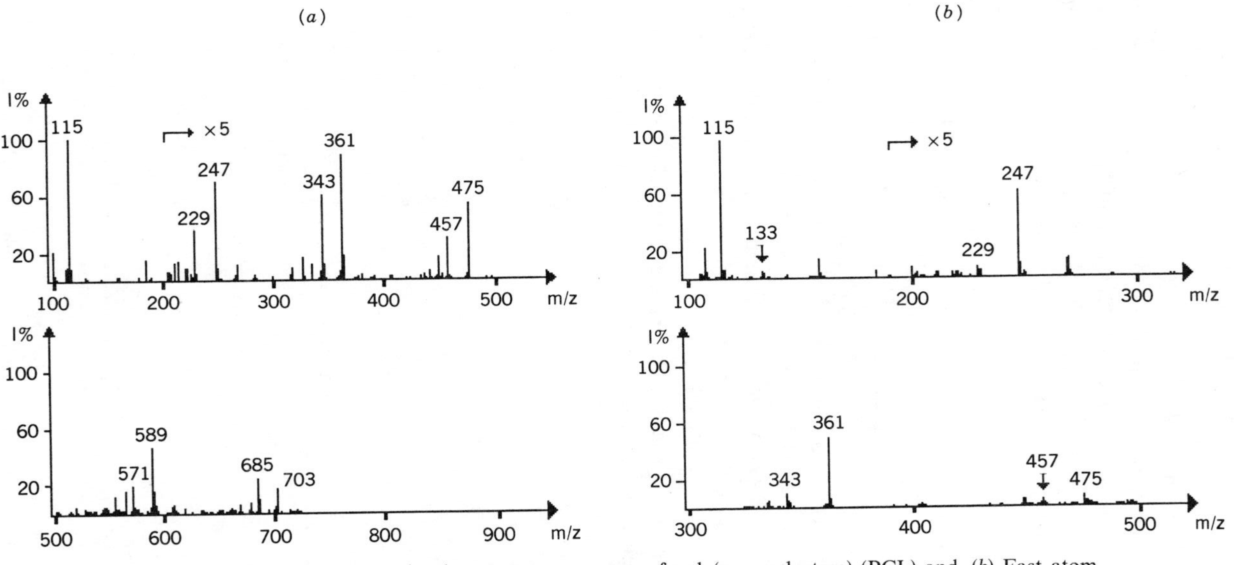

Figure 9. (*a*) Fast atom bombardment mass spectrum of poly(ε-caprolactam) (PCL) and, (*b*) Fast atom bombardment mass spectrum of mixture extracted from PCL. Reprinted with permission from A. Ballistreri, D. Garozzo, M. Giuffrida, and G. Montaudo, "Identification of the Ions Produced by FAS–MS in Some Polyesters and Polyamides," *Anal. Chem.*, **59**, 2024 (1987). Copyright © 1987 American Chemical Society.

additives are usually present in concentrations below 1%. Moreover, many additives are not sufficiently volatile for analysis by GC.

Youssefi (28) recently proposed the use of FAB for the detection and characterization of polymer additives. As shown in an earlier example (Fig. 8), to obtain a FAB spectrum of a neutral polymer such as poly(ether sulfone), it was necessary to dope the matrix with NH_4Cl and HCl to enhance its ionic properties. A simple extract of the polymer, that is, a mixture of oligomers and co-extracted additives, was subjected to FAB–MS analysis. In the absence of any ionic additive, it was significant that the oligomers did not respond to ionization by FAB. However, the antioxidant tris(2,4-di-*t*-butylphenyl) phosphite present in the extract yielded a well-defined FAB mass spectrum as shown in Figure 10. The protonated molecular–ion appears at *m/z* 647 while the base peak at *m/z* 441 is caused by the loss of HO-Ph-(t-Bu)$_2$ group from $[M + H]^+$. The cluster ions at *m/z* 661 and 662 are caused by the phosphate form of the additive (Fig. 10). These results indicate that FAB ionization can be used selectively for the detection of polymer additives. Many advantages can be realized when this approach is combined with MS–MS techniques and collision-induced dissociation methods discussed below.

2.2.1.4. Californium-252 Plasma Desorption. The technique of ^{252}Cf desorption MS falls under the general category of particle-induced bombardment ion

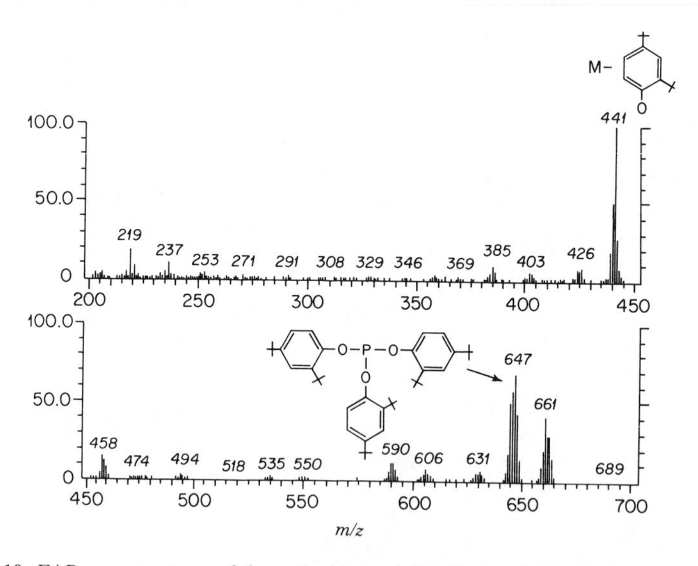

Figure 10. FAB mass spectrum of the antioxidant tris(2,4-di-*t*-butylphenyl) phosphite obtained from a polymer extract in the presence of oligomers of the polymer. Reprinted from Ref. 28 with permission from the author.

desorption, such as SIMS and FAB–MS. The desorption process is induced by the interaction with the sample with two ionic particles, 110-MeV $^{100}Sr^{+}_{20}$ and 75-MeV $^{150}Ba^{+}_{20}$, emitted during the nuclear fission of ^{252}Cf. The two atomic ion fragments from a single fission event are hurled in opposite directions: one activating a fission initiation detector and the other accelerated towards a sample foil on the reverse side (i.e., facing the MS analyzer), which is the sample being analyzed. Molecular and fragment ions from the sample are mass analyzed based on their different arrival times at an electron multiplier. The arrival of ions at the desorbed ion detector sends a signal pulse whose timing is referenced against the signal pulse received from the fission-fragment-detector, which monitors the initiation of the bombardment process. The cycle repeats itself approximately every 50 μs corresponding to the next fission process (29, 30). A TOF or Fourier tranform (FT) ion cyclotron resonance (ICR) mass analyzer (see discussion below) is necessary to conduct the mass analysis, thus placing a number of limitations on the optimum mass resolution that can be achieved. Mass spectra of large biomolecules in excess of 10,000 g/mol have been obtained via this method, but, with some exceptions (30), its application to synthetic polymers has been limited.

2.2.1.5. Laser Desorption. The basic principle of laser desorption ionization is similar to that of SIMS, FAB, or ^{252}Cf desorption, except that an intense photon beam is used for target bombardment instead of a particle beam. A well-focused pulsed laser is normally used, preferably with short duration pulses (in the nanosecond range), because the ions produced from longer pulses or large diameter beams have a large spread in kinetic energy and spatial distribution that results in poor mass resolution (31). As a consequence, laser desorption MS is compatible with pulsed nonscanning mass analyzers (TOF or FT–ICR) rather than scanning mass analyzers. In that respect LD–MS is more similar to ^{252}Cf desorption than either SIMS or FAB. The processes taking place on the target surface during laser bombardment have been discussed by Hercules et al. (32) and a model has been proposed. Five basic processes have been suggested, which may involve direct ionization of the solid at the exact area of impact of the laser beam, ionization and desorption of the solid in the immediate vicinity of impact, surface ionization, gas phase ionization from ion–molecule collisions, and emission of neutrals. Interestingly, organic compounds may produce spectra containing radical molecular ions (as in EI) and/or adducts of the type $[M + H]^+$, $[M + Na]^+$, $[M + K]^+$, and so on.

Laser desorption offers specific advantages over other surface ionization methods. In contrast to FAB or SIMS it requires no matrix, and large biomolecules or polymers can be directly ionized and vaporized with minimal fragmentation. Laser desorption was employed to produce the mass spectrum of a biopolymer with a molecular weight approaching 200,000 g/mol (33). Thus molecular–ion peaks in spectra obtained by LD–MS are typically of higher relative abundance than in those obtained by FAB. Moreover, it can be highly specific and selective. For example, Hau and Marshall (34) used LD–MS to detect and identify dyes in solid poly(methyl methacrylate) commercial plastics at concentrations of 0.1% by weight. Its general use for the characterization of additives in polymers is indicated.

Laser desorption may also be used to desorb neutral molecules, which are then ionized by another method, such as a second laser (35). The advantage of using a pulsed laser to thermally desorb nonvolatile samples is that thermal decomposition is greatly reduced, since the sample is only heated where the laser is focused. Schlag and co-workers (36) further reduced the thermal decomposition by immediately cooling the desorbed sample with a supersonic jet. This allows the desorption of intact neutral molecules, which are then selectively ionized with resonant laser ionization. The complete desorption–ionization method is shown in Figure 11. Since the cooling narrows the

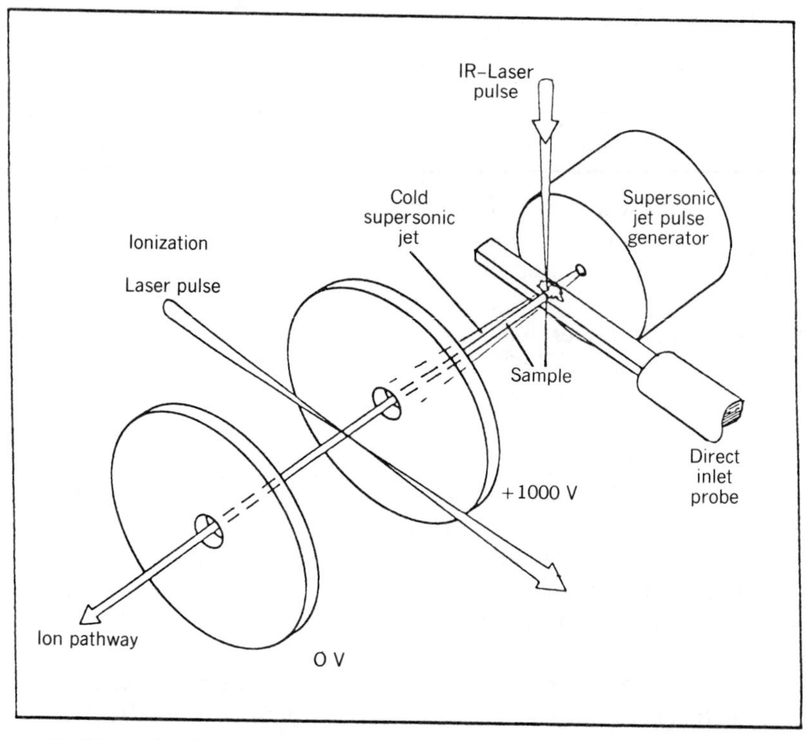

Figure 11. Schematic of the multistage ion source including laser desorption, supersonic jet cooling, and resonance ionization.

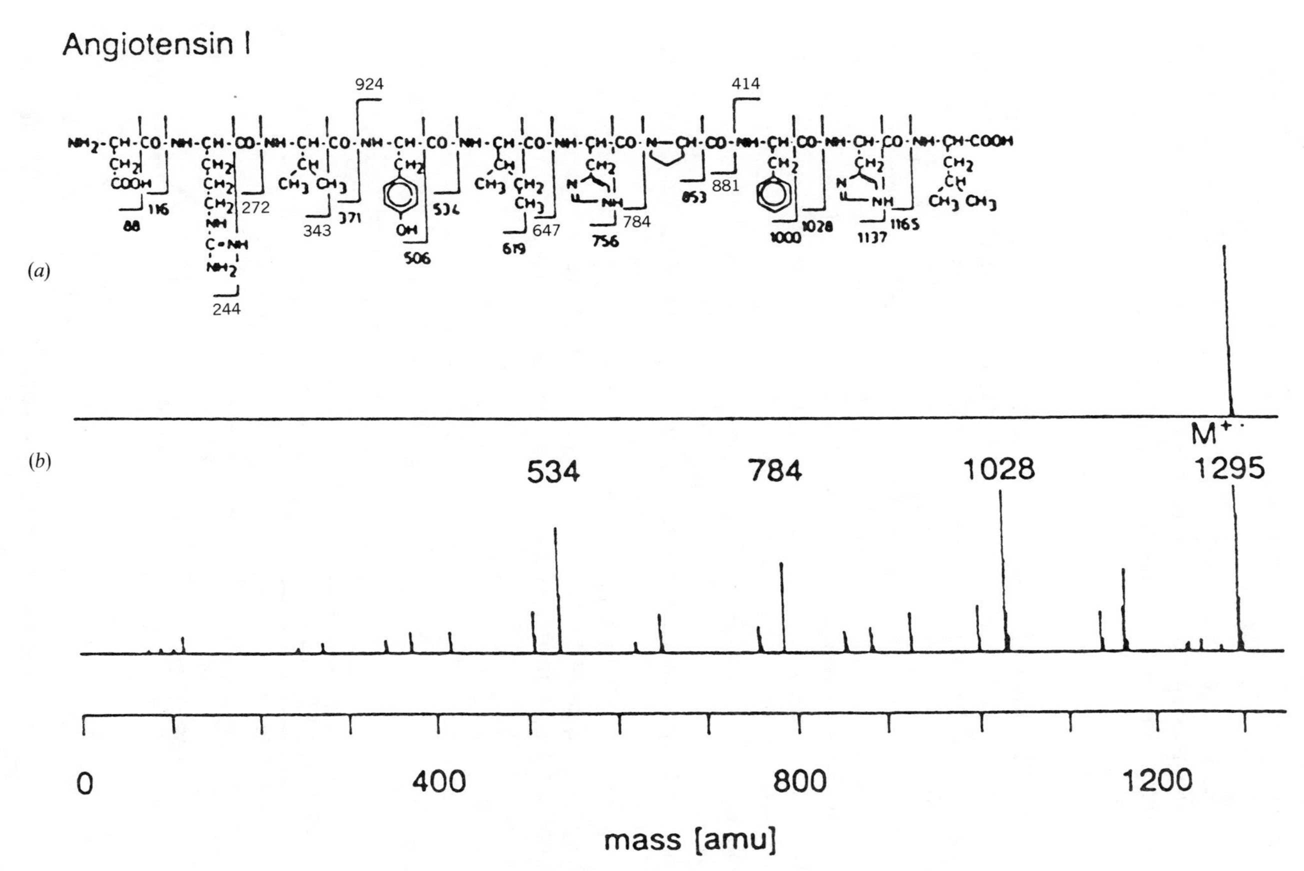

Figure 12. Mass spectra of angiotensin I under soft ionization (*a*) and hard ionization (*b*) conditions.

absorption lines it is possible to selectively ionize a given compound to eliminate many prepurification steps normally involved in sample preparation. Moreover, it is possible to adjust the degree of fragmentation by adjusting the laser wavelength (37). This is illustrated in Figure 12, which compares mass spectra of angiotensin I using a low (soft ionization)- and high (hard ionization)-output power of a dye laser.

2.2.2. *Ionization from Ionic Solution*

2.2.2.1. Thermospray Ionization. The technique of thermospray ionization has evolved from the efforts of Vestal (38, 39) to couple an HPLC, operated at mobile phase flow rates in excess of 1 mL/min, to a mass spectrometer. A schematic of the thermospray system (Fig. 13) illustrates the sequence of events associated with the ionization process. As the effluent emerges from the liquid chromatographic column, it is forced through a heated capillary tube and produces a supersonic jet of vapor engulfed in a mist of water droplets. These droplets become progressively smaller in size as the effluent proceeds in the direction of the mass spectrometer. Solute molecules present in the solution are ionized by interaction with electrolytes, such as the buffers used in HPLC, and are transferred as ions into the vapor phase upon breakup of the charged droplets. Ionic species that pass in front of the extraction plates of the MS ion source are drawn into the mass analyzer while the bulk of the neutral effluent is removed by a high-capacity pump placed in the line of flow of the vaporized liquid. The remarkable aspect of the thermospray ionization process is the production of ions without the use of electric fields or ionizing

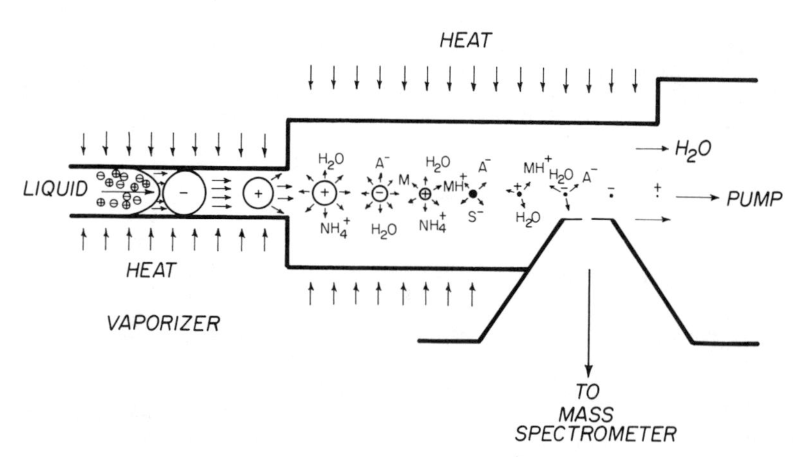

Figure 13. Schematic of thermospray HPLC–MS interface and ionization system. Reprinted from Ref. 39 with permission from the publisher.

beams; no electron guns are used for solute ionization, although a filament-equipped ion source is recommended to handle systems of low ionic strength.

2.2.2.2. Electrospray Ionization. In the thermospray process described above, ionic species are introduced into the mass spectrometer by atomization of a vaporized stream of liquid upon heating and expansion into a supersonic jet. By contrast, the electrospray ionization process (ESPI) may be viewed as a means of "producing atomization by charging" (39). In electrospray ionization, the sample liquid is introduced into a chamber containing an inert gas via a hypodermic needle. An electric field applied between the needle inlet (ground potential) and the cylindrical surrounding walls (-3500 V) promotes ionization of the emerging liquid and disperses it into droplets of gradually decreasing size eventually reaching the molecular level.

Ionic species are further extracted into the mass spectrometer as a result of a potential gradient established between the outer chamber and the rest of the inlet and the ion source system leading to the mass analyzer. The bath gas, whose flow direction is perpendicular to that of the liquid stream, is used to sweep away a portion of the neutral effluent. In addition to the mechanism of ion formation, a further difference between thermospray and electrospray is associated with the liquid flow rates. Optimum conditions for thermospray are found at flow rates of 1–2 mL/min while, presently, electrospray is operated at flow rates of about 10 μL/min or less corresponding to those of microbore or even capillary HPLC columns.

Dole and co-workers (40, 41) introduced the electrospray technique and obtained the mass spectra of macromolecules. Polystyrene ions of about 50,000 g/mol were observed as dimers and trimers, while larger macromolecules of 411,000 g/mol appeared as multiply charged species. Fenn and co-workers (42) resurrected the technique in conjunction with micro-LC–MS coupling. The propensity of larger molecules to form multiply charged ions in ESPI has been further exploited by Smith and co-workers (43, 44), as a means of extending the operable range of the mass spectrometer since mass analysis is based on the m/z ratio.

2.2.2.3. Electrohydrodynamic Ionization. The term eletrohydrodynamic (EHD) ionization was introduced by Evans and co-worker (45) to describe the ionization process that results from the interaction of a conducting liquid miniscus with a strong electric field. A liquid-metal alloy constituted a typical matrix, and metallic ions from a sample dissolved in it could be "extracted" by the field and analyzed mass spectrometrically. Initially, EHD ionization was viewed as a substitute for spark-source MS in inorganic analysis, but the wide ranging potential of the method and its applicability to organic MS were soon recognized. Glycerol has provided a convenient medium for the EHD–MS analysis of thermally labile organic compounds (46). To promote ion emission, addition of electrolytes into the solution matrix was necessary to increase conductivity. As a result, the EHD spectra of organic compounds are dominated by molecular–ion attachment peaks, $[M + Na]^+$, $[M + Li]^+$, and so on, similar to the situation observed in FAB ionization.

The interaction of alkali metals with poly(ethylene glycol) (PEG) dissolved in glycerol solvent has been the subject of recent studies using EHD–MS (47, 48). Poly(ethylene glycol) polymers were found to complex with more than one ion of, for example, Na^+, or K^+, depending on the length of the polymer chain. Large oligomers tended to attach multiple alkali metal ions and could be readily detected by the mass spectrometer owing to their reduced m/z value, even if their actual mass extended beyond the scanning mass range of the spectrometer. This feature is a means of extending the operating range of the mass spectrometer for polymer analysis by EHD–MS. In glycerol solutions containing more than one type of alkali metal ion, selectivity of complexation for one metal ion over the other was also observed. In addition to the apparent general utility of EHD–MS for polymer analysis, the studies of Cook and co-workers (47, 48) demonstrate that the technique may play a significant role towards understanding transport phenomena of metal ions through membranes or other biological processes involving ion–macromolecule interactions.

In reviewing the techniques, which we have classified as "methods of ionization from solution," a number of common features are apparent. It is obvious, for example, to cite the similarity between ESPI and thermospray since in both cases we have the atomization of charged droplets emerging from a vaporized liquid jet. However, it is also impressive to note the commonalities of EHD and ESPI (application of an external field). In fact, one can go even further and perhaps classify all these methods under the common category of *field desorption*. These commonalities notwithstanding, each of these ionization methods does have its own unique characteristics in terms of its practical applicability to chemical analysis.

3. TECHNIQUES FOR THE MEASUREMENTS OF ION MASS–CHARGE RATIOS

The power of the mass spectrometer is associated with its ability to discriminate between ions (positive or negative) of different mass–charge (m/z) ratios. This is accomplished by the action of magnetic and/or electric fields, either used in unison (for magnetic) or, frequently, in some tandem combination. In a broad sense one can classify the mass spectrometers in terms of continuous (scanning) and pulsed (nonscanning) analyzers. The more conventional single-focusing magnetic, the double-focusing tandem electrostatic–magnetic, and the quadrupole mass spectrometers can be viewed as continuous beam scanning systems. Time-of-flight and FT–ICR mass spectrometers would fall under the category of pulsed nonscanning systems. As indicated above, pulsed ionization methods are more compatible with pulsed nonscanning analyzers since an entire mass spectrum may be obtained with each ionization event.

3.1. Continuous Ion-Beam Analyzers

3.1.1. Ion Analysis by Deflection in Magnetic–Electric Fields

Conventional single-focusing mass spectrometers consist of a magnet that acts as a mass analyzer. An ion of mass m, charge z, and energy V, entering a magnetic field of intensity B will be bent along a path of radius r according to

$$\frac{m}{z} = \frac{B^2 r^2}{2V} \tag{2}$$

Several parameters influence the capability of the magnetic analyzer to separate and resolve ions of different m/z ratios. To obtain an ion ray as "thin" as possible and thus minimize the chances for overlap between ions of adjacent masses, it is important to introduce into the magnetic field an ion beam of a well-defined energy and as low an energy spread as possible. Placement of a radial electrostatic field of radius r before the magnet ensures the yield of an energetically monochromatic beam. As indicated from Eq. 2, the electric field acts strictly as an energy filter permitting only ions of a specific energy V to pass through the electrostatic analyzer along a path of radius r for a given value E of the field. Double-focusing mass spectrometers utilize a tandem combination of electrostatic (ESA) and magnetic analyzers for improved mass resolution

$$r = \frac{2V}{E} \tag{3}$$

where V = ion energy
r = field radius
E = field intensity

In the typical double-focusing MS, the electrostatic analyzer is placed before the magnet, which is then presented with an energetically monochromatic ion beam. Of particular interest in this kind of a system are ions that are formed in the field-free region between the ion source and the ESA.

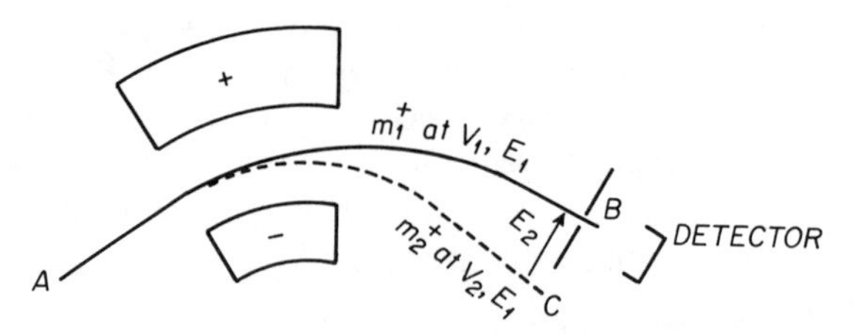

Figure 14. Ion separation in an electrostatic analyzer as a function of ion energy (V) and electric field (E).

Consider, for example, a molecular–ion of mass m_1 and energy V_1 entering the radial electrostatic field in Figure 14. This ion may decompose unimolecularly to m_2 or, alternatively, the same fragmentation may be induced by placing a collision cell containing an inert gas such as He or Ar before the ESA. The fragment m_2, or any other fragment derived from m_1 in this manner, will carry only a fraction of the original energy V_1. Thus, invoking the principles of conservation of momentum, a fragment ion of mass m_2 will have an energy of V_2 such that

$$V_2 = \frac{m_2 V_1}{m_1} \tag{4}$$

Under normal operating conditions, the electrostatic analyzer is adjusted at a field strength E_1 to transmit the ion beam of energy V_1 along path AB. As a result ions of lower energy, V_2, will be deflected by the ESA along the path indicated by line AC. The fragment ion can be brought into the line of the detector, however, by appropriately decreasing the intensity of the electric field according to the relationship

$$m_2 = \frac{E_2 m_1}{E_1} \tag{5}$$

Simultaneous scanning of the ESA and the magnetic field at a constant B/E ratio, permits the detection of the products related to a given precursor. Typically, the ESA is initially set at its normal operating field and the magnet adjusted to monitor the precursor of interest, m_1. The ESA field is then lowered to a ratio E_2/E_1 such that it will only transmit the metastable ion of interest, m_2^* (Eq. 5). At this point the magnetic field is lowered to a new value in order to detect m_2^*. Further decrease of both the ESA and the magnet by the appropriate amounts will bring to the detector another metastable m_3^* of the precursor m_1, and so on. Essentially, this scanning process produces a spectrum of the unimolecular decomposition (or of a collision induced decomposition) of any parent ion in terms of Eq. 5. If the rate of ion fragmentation is such that no unimolecular decomposition is occurring, "metastable" transitions may be induced by introducing an inert collision gas in the field-free region. A collision-induced dissociation (CID) spectrum is then obtained. Even though metastable ions may constitute less than 1% of the normal ions in the spectrum, this linked-scan mode provides a dramatic improvement in sensitivity as it removes interfering ions from the primary-ion beam thus enhancing the signal-to-noise (S/N) ratio. The next example illustrates this point.

The FAB spectrum of an epoxy resin obtained in a glycerol matrix is shown in Figure 15*a*. Certain features of this spectrum need to be emphasized. The large background from glycerol is a dominant feature that obscures recognition of the polymer peaks. Examination of the spectrum also reveals the presence of several oligomers, $n = 0$–2. The molecular ions of each occur at m/z 340 for the monomer, M; m/z 624 for the dimer, D; and m/z 908 for the trimer, T. Numerous less intense peaks due to the fragmentation of each oligmer are also present but difficult to discern. A B/E scan operation of the double-focusing mass spectrometer, however, permitted the acquisition of a "clean" spectrum for each oligomer in the mixture (49). The CID spectrum of the dimer (m/z 624) is shown in Figure 15*b*. Notable is the "orderly" fragmentation of the oligomer, which permits a clear definition of the fine structure of the dimer. This is similar to the CI example shown later in this chapter.

The CID spectrum in the preceding example was obtained using a tandem E-B mass spectrometer. The parent-ion selection process in this configuration is energy dependent and "leakage" of impurities reduces the specificity of the process. However, combination of electric and magnetic fields in three- and four-sector instruments introduces added power to the ion selection process and can provide considerable information in terms of the energetics of ionic decompositions, differentiation of isomeric compounds, analytical uses for trace level studies, and other related problems. For more details the reader is referred to Refs. 50 and 51.

3.1.2. Quadrupole Ion Analyzers

The quadrupole mass analyzer consists of four circular rods placed in a concentric arrangement with respect to each other. A +dc and a −dc potential are applied to each pair of diagonally opposed rods and a radio frequency (rf) field, superimposed to each of the dc voltages, is varied linearly forcing ions that enter the analyzer to oscillate in the space between the rods. Ions of a given m/z ratio will reach the collector only at specific values of the rf and dc fields.

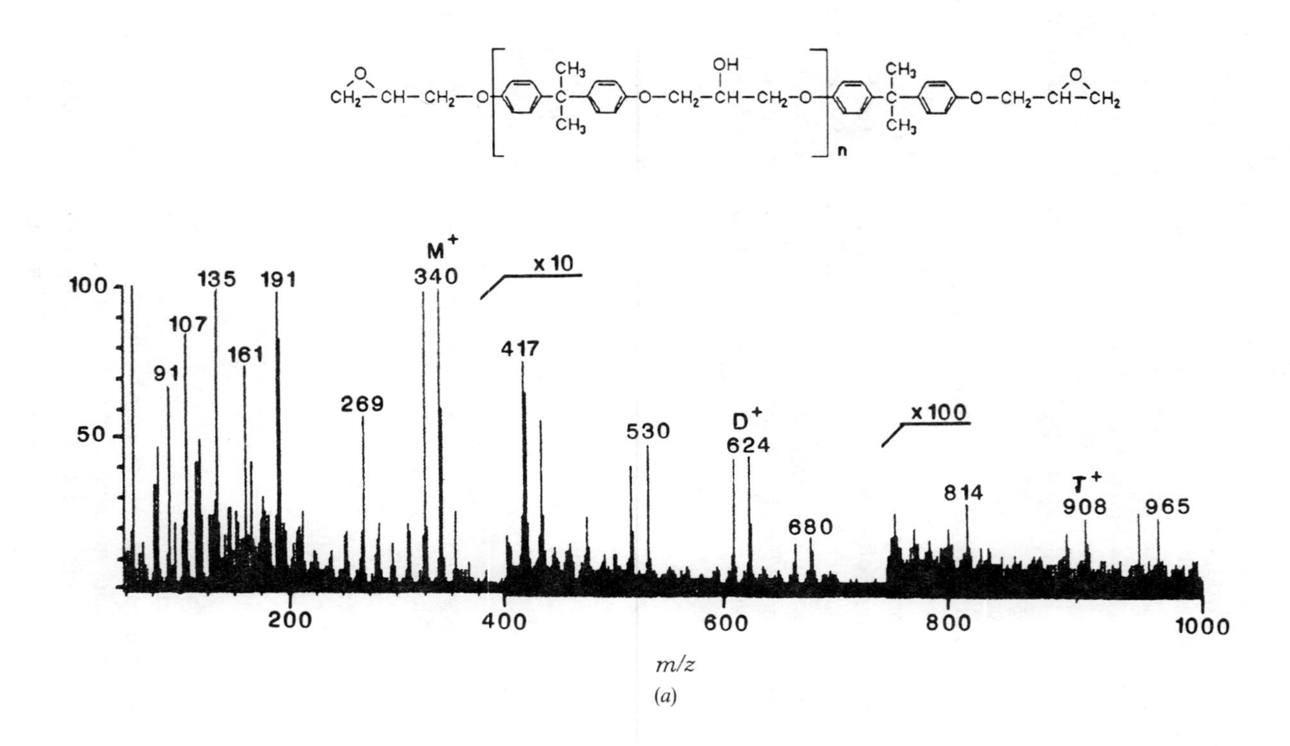

100
50
91
107
135
161
191
269
M+
340
x 10
417
530
D+
624
680
x 100
814
T+
908
965
200
400
600
800
1000
m/z

(a)

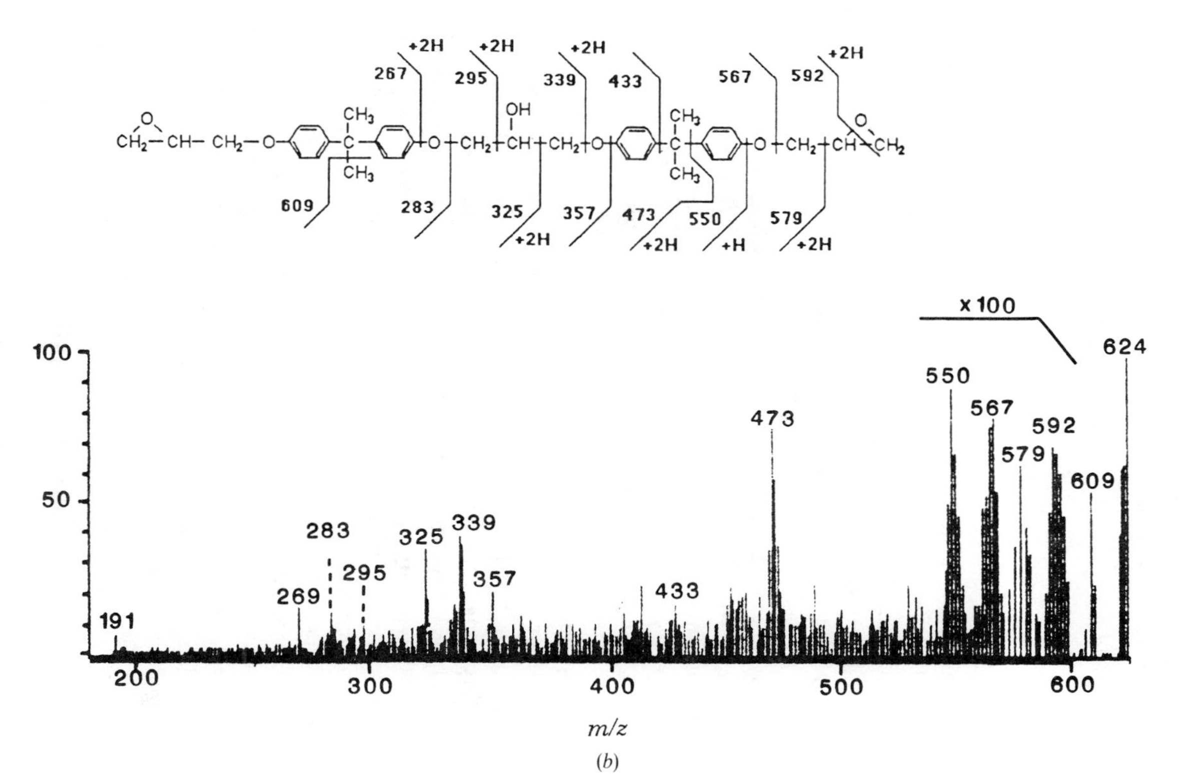

(*b*)

Figure 15. (*a*) FAB mass spectrum of epoxy resin in 3-nitro-benzyl alcohol. The molecular–ions of the monomer, dimer, and trimer are identified as M^+, D^+, T^+, respectively. (*b*) FAB–CID mass spectrum of dimer molecular–ion from epoxy resin. Reprinted from Ref. 49 with permission from the publisher.

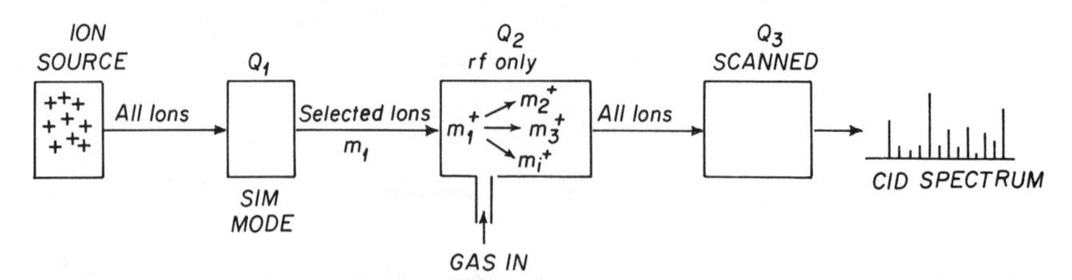

Figure 16. Schematic of triple-quadrupole mass spectrometer showing the operation of collision-induced processes.

Quadrupole mass spectrometers gained a major foothold in the field of MS during the early 1970s, partly because of their rapid scanning capability, which facilitated the coupling of capillary GC with MS, simpler interfacing to computers, and lower cost. These features are becoming less of a factor, however, with recent advances in magnet technology, which provide for almost equally as fast scanning rates, much higher mass ranges than quadrupole analyzers, and reliable data systems. For example, a mass range of 10,000 g/mol is well within the reach of presently manufactured magnets as opposed to an optimum of about 4000 g/mol in most quadrupoles.

Even with the recent improvements in other mass analyzers, quadrupole mass filters still have a significant role in modern MS. In addition to their well-defined and established role in GC–MS, quadrupoles are used in a variety of tandem combinations with magnetic mass spectrometers, in the so-called hybrid combinations (52). Perhaps the most common tandem use of quadrupole analyzers is in the triple-quadrupole mass spectrometer (53). As shown in Figure 16, three quadrupole analyzers are placed in a linear arrangement. The first quadrupole is used to select and transmit to the second quadrupole any ion of interest, typically a molecular species. The second quadrupole, Q_2, contains the collision gas and is operated in the rf mode only, which permits the transmittance of all fragment ions derived from the collision as well as any remaining m_1^+ species. Mass analysis of all ions exiting from Q_2 is conducted by Q_3, which is operated in the normal rf–dc mode of produce a CID spectrum. An important feature of the triple quadrupole mass spectrometer is its relative simplicity of operation. The inherent mass range limitation of the quadrupole analyzers may also be, at least partially, overcome by the judicious use of multiply charged species when used in conjunction with electrospray ionization (47, 48).

3.2. Pulsed Beam Analyzers

3.2.1. The TOF–MS

The principle of operation of the TOF–MS is based on the fact that ions of equal kinetic energy ($\frac{1}{2}\mathrm{mV}^2$), but different mass, deflected towards a collector in a field-free region, will travel with different speeds. Ions of lower mass will thus arrive at the collector ahead of ions of higher mass.

The collector voltage is synchronized with the pulse frequencies, which control ion production and "opens up" at selected intervals corresponding to the arrival times of ions of interest. The time t that it takes for an ion to travel a path of length l is related to the mass according to

$$t = k(m/z)^{1/2} \tag{6}$$

The resolution of a TOF–MS is a function of the length of the flight tube and, in addition to other parameters, the initial energy spread of the ions and their distribution in the ion-source volume. Ideally, it is desirable to produce all ions within a narrow plane so that, in effect, they will all leave towards the collector from the same starting point. In many ways this is equivalent to the need for a plug injection in chromatography. Nevertheless, until recently even with the most optimized conditions the resolution of a TOF–MS has been barely about $(M/\Delta M) = 500$. Recent advances, however, in the area of ion focusing have produced TOF analyzers with a resolution approaching 10,000 (54). The TOF–MS is ideal for the study of fast reactions, the detection of reaction intermediates, and has been the instrument of choice with pulsed ionization methods such as plasma desorption MS, as described earlier in this chapter. The spectra of biopolymers with MW approaching 200,000 g/mol have been reported using LD ionization and TOF analysis (33). Clearly, TOF–MS is destined to play a significant role in the analysis of high-molecular-weight compounds.

3.2.2. *The FT–ICR–MS*

Like all magnetic mass analyzers, the ICR–MS is based on the principle that an ion of mass m, charge z, and energy V, in a magnetic field of intensity B will be bent along a radius r as stated in Eq. 2. The ICR–MS differs in that the energy V is low enough that the ion is trapped in a circular motion in the magnetic field. The ions circle about the magnetic field at a frequency W given by

$$\frac{m}{z} = \frac{B}{W_c} \tag{7}$$

with a radius of curvature determined by their energy.

The ICR–MS is unusual among mass analyzers in that it does not separate the ions for detection. As shown in Figure 17, ion detection occurs by applying an rf electric field to the plates of the ICR cell perpendicular to the magnetic field. When the frequency of the rf field is equal to that of the oscillation of the

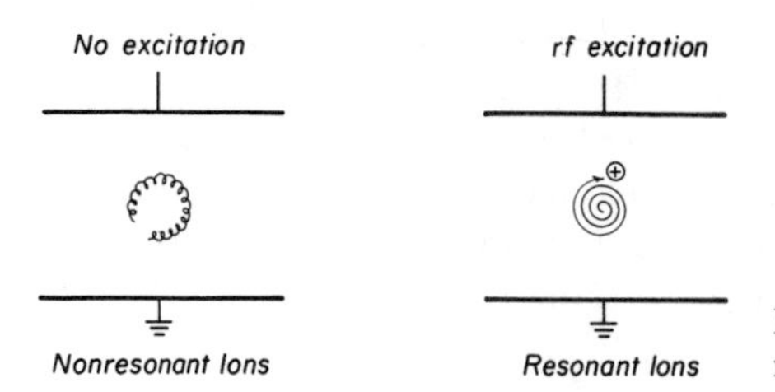

Figure 17. Schematic of nonresonant and resonant ions in an ICR cell.

ions, the ions are in resonance and absorb power from the magnetic field. The ICR mass spectrometers were originally scanning mass spectrometers in that either the magnetic field or the excitation frequency was scanned to acquire a mass spectrum while an excitation electric field was applied continuously while measuring the power absorbed by the ions (55). It did not take long for spectroscopists to realize that it would be a simple matter to use two or more excitation oscillators at the same time (56). This is equivalent to the double-resonance experiments performed in NMR but results in what is probably the simplest MS–MS experiment. This MS–MS feature has been used extensively to establish ion–molecule reaction mechanisms. In fact, ICR–MS is considered the method of choice for studying gas phase ion–molecule reactions that are the basis for CI(54).

As evidenced by the double-resonance experiment, the advantage of ICR–MS is that it is an rf technique. As such, it is possible to apply FT techniques to ICR–MS (57). The highly successful application of FT techniques to IR and NMR spectroscopy has only recently been applied to the ICR–MS. In FT–ICR the ions are formed via some type of pulsed ionization technique and then trapped in the ICR cell. In an analogous manner to FT–NMR, rf electric field containing the cyclotron frequencies of all the ions of interest is then applied. The ions absorb power and continue to oscillate in a coherent fashion even after the rf electric field is removed. The ICR cell is simply a parallel-plate capacitor and the coherent motion of the ions is an alternating charge in a capacitor. The alternating charge in the capacitor induces a time-domain voltage across the capacitor. If this time-domain signal is digitized an FT of the time domain signal will yield a mass spectrum in the form of ion frequencies and frequency intensities. The ability to use FT methods to excite and detect the entire mass spectrum at once is just beginning to make it a viable analytical technique. The ability of FT–ICR–MS to acquire an entire mass spectrum at once makes it ideal for pulsed ionization techniques.

Unfortunately, unlike the NMR experiment, the ICR experiment covers a much wider range of frequencies. Every decade of mass measured results in another decade of frequency to be measured. Even though NMR frequencies are higher, the actual range of frequencies covered by the NMR experiment is small. In FT–NMR it is possible to convert the entire spectrum to a lower frequency by "mixing" the time domain signal with a frequency close to the actual NMR resonance before the time domain signal is digitized. In the ICR experiment the wider range of frequencies makes it impossible to "mix" down to a lower frequency for the entire spectrum. As such, it is not only necessary to digitize at a much higher rate but the digitized time domain signal is also larger. Only recently has the necessary computer hardware become available at an economical price. The recent explosion in computer technology and its applications to rf techniques in radar and video imaging is helping to fuel the development of FT–ICR–MS.

The ICR may be considered a form of TOF technique in that the period of oscillation (the inverse of the frequency) for the ion is measured. In TOF resolution is a function of the length of the flight tube. In ICR the flight path consists of the radius of curvature and the number of times the ion oscillates. Since the ion is trapped in its circular motion by the magnetic field and the ions are not destroyed during detection, it is possible to select any path length or resolution desired as long as the ion motion is not disturbed. A large ion path length does not require a large spectrometer. The ions are actually confined to a volume on the order of a few cubic centimeters. The resolution will increase the longer the ion signal is measured. Of course, the longer the signal is measured the better the S/N ratio. In most mass spectrometers sensitivity must be sacrificed to increase resolution; in the ICR both increase at the same time.

To achieve the goal of high resolution and sensitivity, it is important not to disturb the motion of the ions. Background gas in the vacuum system results in ion-neutral collisions that interfere with the ion motion (58). Since the magnetic field is not scanned and the actual volume of the magnetic field need not be large, the technique lends itself to the use of superconducting magnets (59). With the high fields available from superconducting magnets there appears to be almost no limit to the mass range of the spectrometer. It is this seemingly unlimited mass range that makes FT–ICR–MS of particular interest to the field of polymer analysis.

As a pulsed method, FT–ICR–MS is ideally suited for use with laser ionization techniques. Their compatibility is further enhanced by the capability of laser desorption to desorb–ionize large molecules in their intact form (see above). Furthermore, the overall reduction in mass discrimination makes LD–FTMS ideal for the determination of the correct weight-average and number-average molecular weight of polymers in the range of 10,0000 g/mol. The spectra are dominated by peaks corresponding to the attachment of alkali metal ions (Na^+ or K^+) to the oligomers present and interpretation is further simplified by the general lack of fragmentation. Wilkins and co-workers (60) used LD–FTMS for MW characterization of commercial polystyrenes, poly(ethylene glycols), and polyethyleneimines in the range from 2000 to

10,000 g/mol. In comparing the technique to the other principal pulsed mass-analysis methods (TOF–MS), the authors pointed out that neither analyzer exhibits inherent mass discriminating properties. However, the use of electron multiplier detectors in TOF–MS introduces some mass discrimination and even some discrimination based on ion structure, which may tend to decrease the accuracy of molecular weight measurements in polymer analysis. By comparison, no such problem exists in FT–MS.

4. POLYMERIZATION MECHANISMS BY MS–MS

It was mentioned in Section 2.1.1 that tandem mass spectrometric techniques in conjunction with collision-induced dissociations, offer the opportunity to elucidate the mode of formation of ionic species and their structural characteristics. When the ion source of the mass spectrometer is used as a reaction chamber, considerably more mechanistic detail can thus be obtained with the employment of MS–MS techniques as opposed to the simple EI or CI spectra described in some of the preceding examples. While the mechanisms and processes observed in the vapor phase may not necessarily reflect the situations occurring in the condensed phase, such experiments can, at least, be considerably thought provoking. Such studies are facilitated by the use of tandem MS techniques, which might involve instruments with *B–E* or other combinations, or an FT–ICR system. In the former case, the ion source can be kept at a relatively high pressure, for example, 1 torr, while in ICR the polymerization can be conducted at a high vacuum using selected ion–molecule reactions. Two related examples are given below.

4.1. Polymerization of Hexachlorocyclophosphazene

Polydichlorophosphazene, $(NPCl_2)_n$, is an inorganic polymer prepared by thermal polymerization of its cyclic trimer, hexachlorocyclophosphazene, $(NPCl_2)_3$, **VI**, heated in an evacuated sealed tube at 250 °C for several hours (61). A cationic mechanism, involving the cleavage of a phosphorus–chlorine bond of **VI** to give a P^+Cl^- ion pair is proposed as an initiation reaction (Scheme 2). The polymerization is then propagated by electrophilic attack of the phosphonium center at P^+ on the free electron pair of the nitrogen of a second hexachlorocyclophosphazene **VI** molecule. Ring opening of the second molecule is postulated providing a new phosphonium cationic center for further reaction and subsequent polymerization (Scheme 3). The polymerization process is apparently irreproducible and influenced by additives such as metals, moisture, ether or ketones, and even the walls of the container. The occurrence of cationic intermediates, both in the initiation and

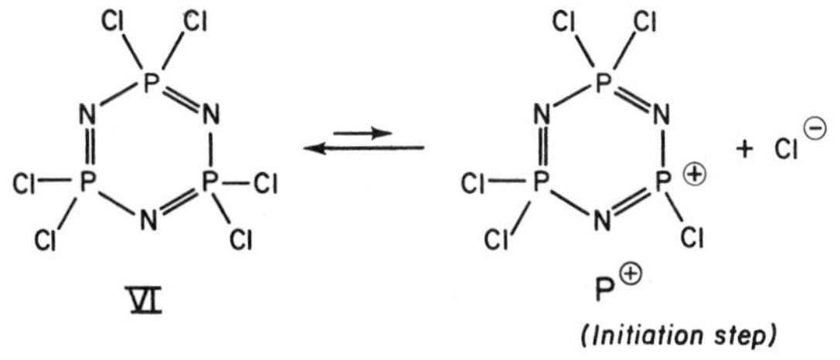

(Initiation step)

Scheme 2

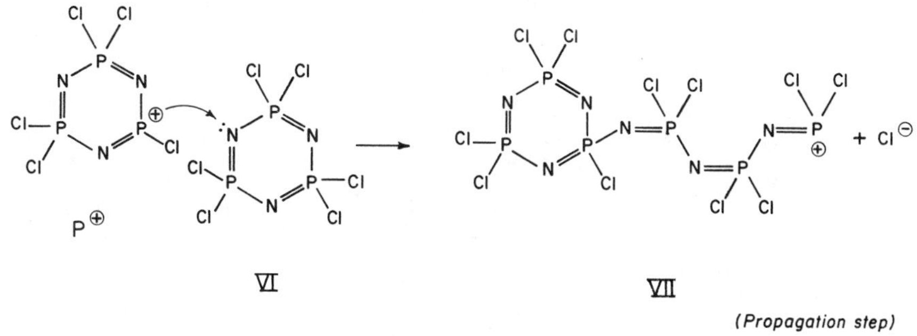

(Propagation step)

Scheme 3

the propagation steps, makes this system an ideal candidate for examination *in situ* by mass spectrometry and tandem MS–MS techniques. Several aspects of the formation of this polymer were examined in this study (61). We will discuss here the application of MS–MS towards the elucidation of the structures of the ionic species P^+ and **VII**.

The following series of experiments was conducted by introducing purified hexachlorocyclophosphazine, **VI**, into the ion source.

1. An EI (70 eV) mass spectrum of **VI** yielded a molecular ion, $M^{+\cdot}$ of *m/z* 345, $[M—Cl]^+$ of *m/z* 310, and a series of less abundant ions identified as $[M—2Cl]^+$, $[NP_2Cl_2]^+$, $[NP_2Cl]^{+\cdot}$, $[PCl_2]^+$, and $[PCl]^{+\cdot}$. The $[M—Cl]^+$ ion of *m/z* 310 is analogous to the P^+ ions referred to earlier as being involved in the initiation step.

2. The molecular-ion $(NP_2Cl_2)_3^+$ (*m/z* 345) was examined by mass analyzed ion kinetic energy spectroscopy (MIKES), and the MIKES spectrum gave only one fragment ion, that of $[M—Cl]^+$. This confirmed that the primary ionization process from **VI** involves the loss of Cl. In turn, MIKES of *m/z* 345 following CID gave a spectrum with $[M—Cl]^+$ as the predominant fragment.

Significantly, no $(NPCl_2)_n$ ionic species were observed under either EI or CID–MIKES conditions, supporting a cyclic structure for $(NPCl_2)_3^{+\cdot}$, the molecular ion of the trimer, **VI.**

3. The $[M—Cl]^+$ (m/z 310) ion from the EI spectrum was next investigated by MIKES and MIKES–CID. Again, no significant ions of the $(NPCl_2)_n$ variety, or losses thereof, were found, thus supporting retention of the cyclic structure for $[M—Cl]^+$, as well.

Collectively, the results from 1 to 3 suggest that in the ion source of the mass spectrometer it is possible to form the same species, which are involved in the *initiation step* of the polymerization reaction. In view of these encouraging results, it was then attempted to induce polymerization in the ion source, under increased pressure (0.1–1.0 torr). Indeed, the mass spectrum of the mixture showed peaks corresponding to polymers $(NPCl_2)_n$ with values as high as $n = 21$. Significantly, the most abundant ionic species in this spectrum has a mass of 655, which corresponded to that of the $[(NPCl_2)_6—Cl]^+$ ion, that is, **VII**, the substance involved in the propagation step of the polymerization. This ion was then examined by MIKES and MIKES–CID in order to ascertain its structure.

4. From the MIKES spectrum alone, it was found that m/z 655 underwent only one major primary fragmentation via the loss of $(NPCl_2)_3$ to give an ion of m/z 310.

5. The MIKES–CID spectrum of m/z 655 gave a series of intense fragment ions corresponding to sequential losses of $(NPCl_2)$ units. This mode of fragmentation is consistent with a linear trimeric $NPCl_2$ chain as shown for **VII** in Scheme 3. The final ion of m/z 310 in the series is, of course, the cyclic $[M—Cl]^+$ ion of the trimer $(NPCl_2)_3$, and, as expected, it does not fragment any further.

This series of experiments (1–5) provide strong evidence that the species formed in the ion source of the mass spectrometer have the same structure as those encountered in the initiation and propagation steps of the polymerization of hexachlorocyclophosphazene. In additional linked-scan experiments, the authors studied the effects of medium acidity by carrying out polymerizations in the ion-source chamber under CI conditions using acid–base combinations such as CH_4/CH_5^+ and NH_3/NH_4^+. In summary, the use of tandem MS–MS techniques has allowed the elucidation of a polymerization mechanism by inducing the polymerization reaction in the ion-source volume itself. This work may provide the basis for examination of other model systems by *in situ* techniques.

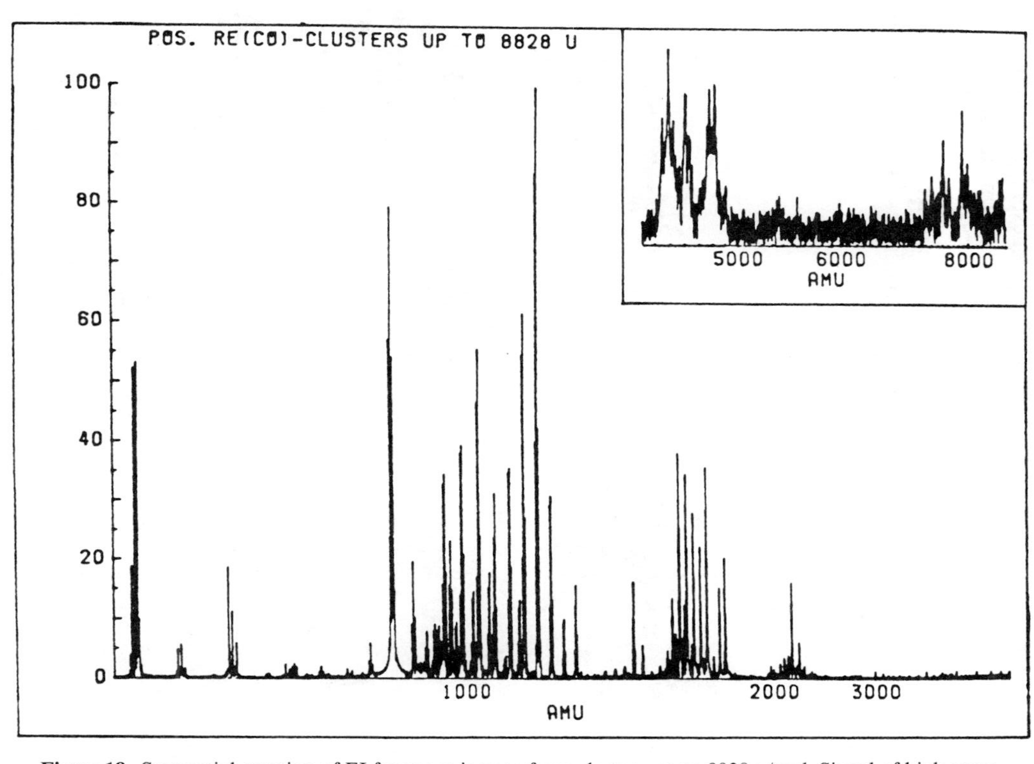

Figure 18. Sequential reaction of EI fragment ions to form clusters up to 8828 g/mol. Signal of high-mass ions shown in inset.

4.2. Time-Resolved Polymerization Experiments Using FT–MS

The polymerization of hexachlorocyclophosphazene in the ion source of the mass spectrometer may also be considered as a sequential ion–molecule reaction. The FT–ICR–MS has the ability to observe sequential ion–molecule reactions under time resolved conditions. This suggests that by combining the (MS–MS)$_n$ capabilities with time resolved experiments should allow one to vary the extent of polymerization, determine the structures of reaction intermediates, measure the absolute reaction rates for individual polymerization steps, and even introduce side reactions into the polymerization chain.

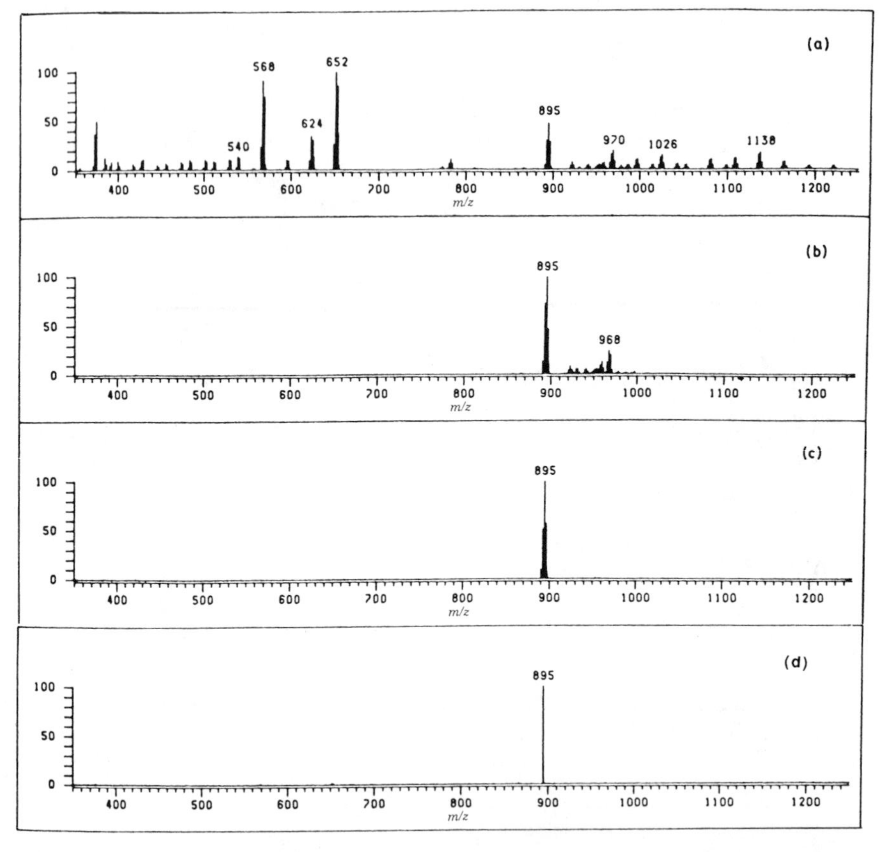

Figure 19. Isolation of an intermediate ion in a sequential reaction. (*a*) A 3-s self CI of 70-eV electron impact fragments of $Re_2(CO)_{10}$. (*b*) Hard sweeps from 814 to 180 and from 995 to 2400 amu. (*c*) Hard sweeps and soft sweeps from 882 to 814 and from 909 to 995 amu. (*d*) Hard sweeps and soft sweeps from 892.8 to 814 and from 896.8 to 995 amu.

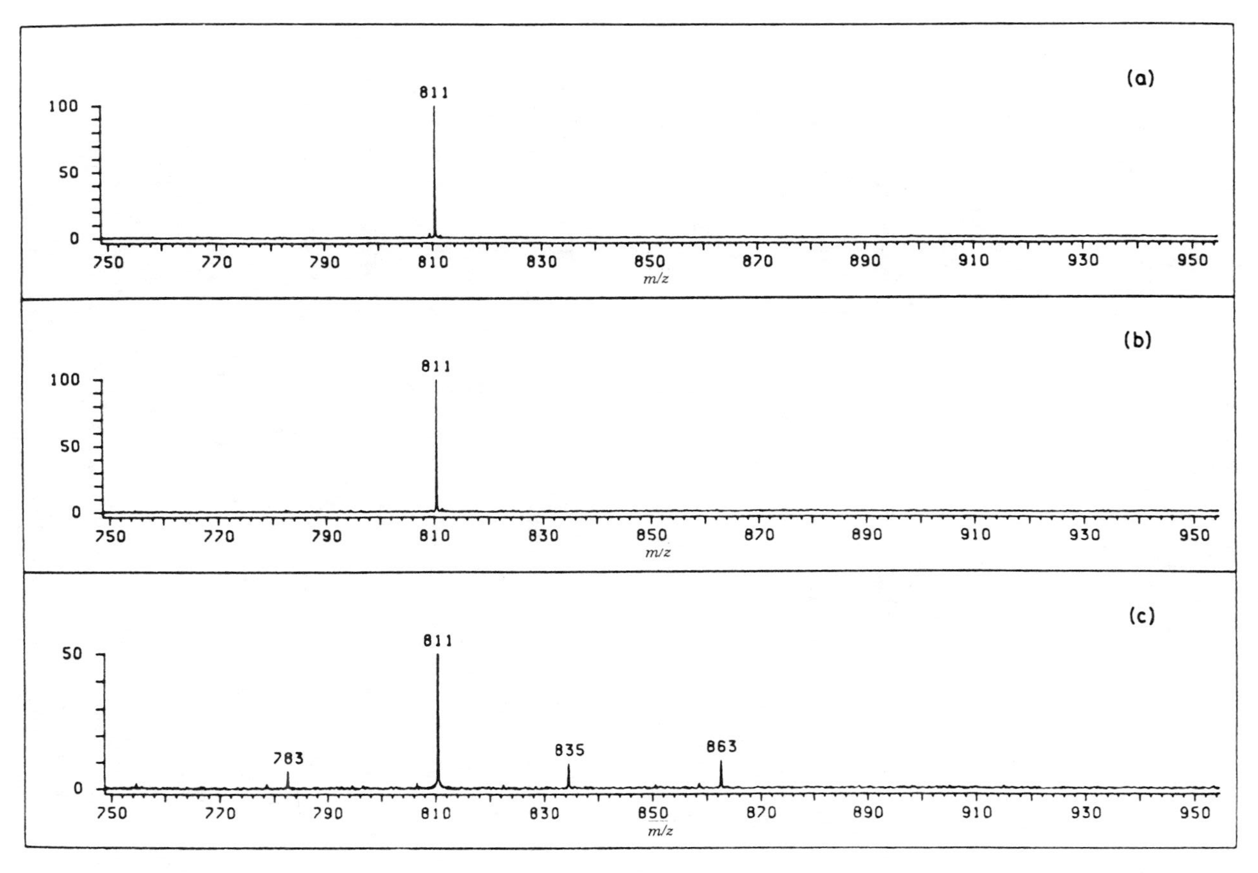

Figure 20. Reaction of m/z 811, an intermediate ion in the sequential reaction with cyclohexane, introduced via a pulsed valve. (*a*) Isolation of $(^{185}Re)_1\ (^{187}Re)_2\ (CO)_9^+$. (*b*) A 500-ms blank for $Re_3\ (CO)_9^+$. (*c*) Reaction of $Re_3\ (CO)_9^+$ with cyclohexane.

Wronka and co-workers (62) used metal carbonyls as model compounds to demonstrate a comprehesive methodology for the study of sequential reaction using FT–ICR–MS. Figure 18 is an analogous experiment to the polymerization of hexachlorocyclophosphazene in a high-pressure ion source. In this experiment, $Re_2(CO)_{10}$ is ionized with EI to form fragment ions. These fragment ions are allowed to react with the neutral metal carbonyl. Each step of reaction adds another $Re_2(CO)_{10}$ to the ion and eliminates either one or two CO neutral fragments to form large metal clusters. The highest mass product at 8828 g/mol represents 15 steps of sequential reaction.

Figure 19 represents a time-resolved experiment where the polymerization reaction is allowed to proceed to the first step, stopped, and one ion isolated for further experiments. Finally, in Figure 20, an isolated product ion (m/z 811) is reacted with cyclohexane to demonstrate that it is possible to stop the chain reaction, isolate an intermediate in the chain, and react this intermediate with another neutral to study side reactions. The authors demonstrated five steps of such MS–MS experiments. These experiments suggest that a wealth of information concerning ionic polymerization reactions is available utilizing FT–ICR–MS.

5. MASS ASSIGNMENT AT HIGH MASS: ISOTOPIC DISTRIBUTIONS

The recent advances in ionization techniques and the increasing capability of mass analyzers to transmit ions of high mass, for example, 10,000 g/mol and beyond at unit resolution, has brought closer to realization the expectation of acquiring the complete spectrum of at least a small polymer in a single scan. An interesting issue stemming from these developments, however, is the mass ion peak representative of a molecular or fragment ion when its m/z value is well in excess of 1000 g/mol? This issue can be appreciated if we consider the isotopic contributions and the mass defects associated with the individual elements comprising a given polyatomic ion.

Consider the example of a polystyrene polymer carrying a terminal butyl group, $C_4H_9(C_8H_8)_nH$. In a typical mass spectrum, we are interested in the nominal mass of the molecular ion, its monoisotopic mass, and the most abundant mass in the molecular–ion cluster. With larger molecules, however, the average mass becomes an additional important factor (63). For the polystyrene monomer ($n = 1$) of composition $C_{12}H_{18}$, the nominal mass is calculated on the basis of $^1H = 1$, $^{12}C = 12$, $^{13}C = 13$, and so on, while the monoisotopic mass takes into account the mass defects, that is, $^1H = 1.007825$, $^{12}C = 12.00000$, $^{13}C = 13.00335$, $^2H = 2.01410$, and so on. Accordingly, the nominal mass of the molecular–ion is calculated as 162 and its monoisotopic mass corresponds to 162.14085. The most intense peak in the molecular–ion cluster is, of course, also the monoisotopic mass. However, the average mass can be calculated from the weighted mass averages of the isotope combinations in the molecule. From isotopic contributions tables (64), and assuming that the contribution from deuterium is negligible, the molecular–ion pattern for a C-18 compound would be

$$[M]\ (100\%) \quad [M+1]\ (20\%) \quad [M+2]\ (1.9\%) \quad [M+3]\ (0.1\%)$$

If we add 1.00335 g/mol to each mass above M (162.14085) for every C-13 isotope, the average mass will be the summation:

$$(162.14085)(0.81967) + (163.1442)(0.16393) + (164.14755)(0.01557) + (165.1509)(0.0008197) = 162.3374$$

This is reasonably close to the monoisotopic mass and is effectively treated as the nominal mass in conventional low-resolution MS.

A radically different picture emerges when larger polymers of polystyrene are considered. The molecular–ion patterns for the decamer ($n = 10$) and the hecatomer ($n = 100$) are reproduced in Figure 21 (63). The nominal, monoisotopic, most abundant, and average masses are indicated on the bar

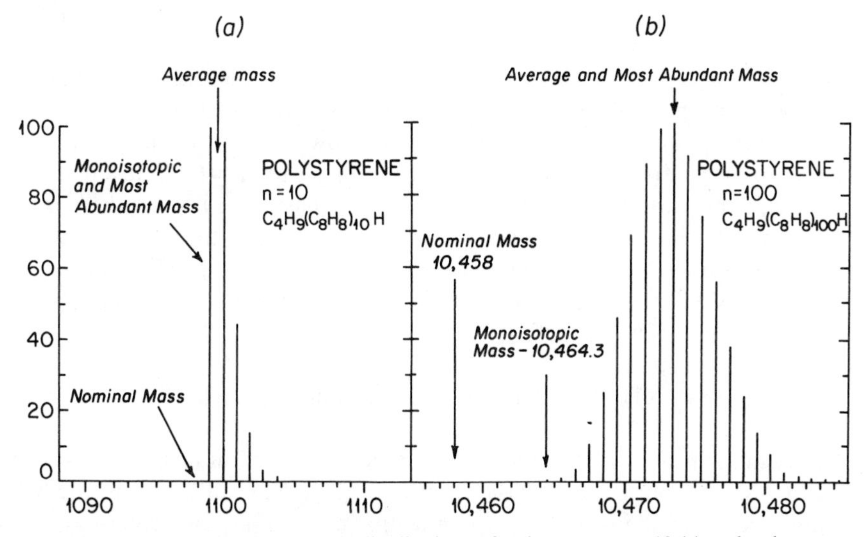

Figure 21. Theoretical molecular–ion distributions of polystyrene, $n = 10$ (*a*) and polystyrene, $n = 100$ (*b*). Reprinted with permission from J. Yergey, D. Heller, G. Hansen, R. J. Cotter, and C. Fenselau, "Isotopic Distributions in Mass Spectra of Large Molecules," *Anal. Chem.*, **55**, 353 (1983). Copyright © 1983 American Chemical Society.

graphs. It is obvious that with increasing molecular size the classical mode of mass assignment is no longer applicable. The nominal and monoisotopic masses are meaningless and, interestingly, a nearly Gaussian pattern of isotopic distributions is seen for $n = 100$. These examples demonstrate the need for new approaches when we consider the mass spectra of high-molecular-weight compounds. Invariably, mass-ion peaks will appear as clusters of ions reflecting the distribution of the various isotopes; the summed mass defects, and their natural abundances. It has been suggested (63) that the use of average masses may provide a more accurate method for making mass assignments of such ionic species.

6. ONLINE HPLC–MASS SPECTROMETRY

From the very early stages of development of modern mass spectrometry in the 1950s, the value of its combination with chromatography was quickly recognized. The coupling of GC with MS was a natural evolution since they are both vapor phase techniques. The technological problems associated with the interfacing of two methods, which operate at radically different pressures, were overcome relatively readily. Since the mid-1960s, GC–MS has been accepted as a standard component of the organic analytical laboratory. Historically, this milestone coincided with the beginnings of modern HPLC, but it has taken considerably longer to achieve a satisfactory and all-purpose mode of HPLC–MS coupling. It is easy to appreciate the difficulties associated with this task if we consider that vaporization of 1 mL/min of liquid translates into a vapor flow rate of approximately 500–1000 mL/min depending on the mobile phase. This compares with gaseous flow rates of 30–50 mL/min in standard (packed) column GC and 1–2 mL/min in capillary GC. Despite these seemingly insurmountable problems, during the past several years there have been several breakthroughs in this area, which have led to the development of commercial interfaces and a generally optimistic outlook regarding the future of HPLC–MS. Three of the most widely used modes of HPLC–MS coupling will be discussed in this section.

6.1. Direct Liquid Introduction Approach

The basic principle behind the direct liquid introduction (DLI) method of HPLC–MS coupling involves the transfer of the HPLC effluent into the MS ion source operated in the chemical ionization mode. The mobile phase is used as the CI reagent gas. In the initial work by McLafferty and co-workers (65), approximately 1% of the HPLC effluent was split into the MS in order to minimize “overloading” of the vacuum system. The subsequent development of narrow bore (0.5–1.0 mm) HPLC columns, which can be operated with optimum chromatographic efficiency at flow rates in the range of 10–50 μL/min provided a more viable alternative to this mode of HPLC–MS coupling. The DLI interface developed by Henion (66, 67) utilizes a diaphragm with a 5-μm pinhole through which the mobile phase is sprayed into the ion source as it is vaporized. A cooling jacket with circulating water prevents premature vaporization of the HPLC effluent before the pinhole and accumulation of dissolved salts or other additives, which may plug the transfer line and/or the orifice. The system can tolerate the use of volatile buffers and other modifiers and, significantly, relatively nonvolatile or thermally labile analytes such as, for example, corticosteroids, can be analyzed without derivatization.

It is important to recognize that with the DLI HPLC–MS interface the mass spectrometer can only be operated in the CI mode. To obtain more detailed structural information its use with tandem MS is recommended and has been used successfully (68). Gradient HPLC is possible, but one should beware of possible changes in the CI characteristics of the reagent gas (mobile phase) particularly at the extremes of the gradient. Some of the salient features of microbore HPLC–MS, with particular emphasis on the DLI interface, have been discussed in recent reviews by Lee and Henion (69). As indicated, the system has found numerous applications in equine toxicology and pesticide analysis. In a variation of the Henion DLI interface, Alborn and Stanhagen (70) recently reported on the use of fused silica packed HPLC columns (0.22-mm inside diameter) coupled directly to a magnetic mass spectrometer. The low flow rates (1–5 μL/min) are nearly equivalent to those of capillary GC–MS and, as a consequence, permit operation of the MS in the conventional EI mode. Moreover, chemical ionization can also be carried out with the reagent gas of choice without being necessarily limited to the mobile phase itself. A recent and promising innovation has been the development of what is essentially a DLI interface coupled to FAB ionization (71). Glycerol or other suitable matrix can be added **via** a concentric tube system (72).

6.2. Thermospray HPLC–MS

The principle of thermospray ionization was discussed in Section 1.2.1. The most dramatic feature of thermospray ionization is the transfer of ions directly from an aqueous solution into the vapor phase. In fact, attempts to couple HPLC with MS promoted the discovery of the thermospray ionization phenomenon. In general, thermospray HPLC–MS is the most widely used mode of interfacing the two techniques. Since thermospray ionization requires the presence of already ionized analytes in the condensed phase, it is best applicable to problems that involve the use of aqueous mobile phases containing ionic modifiers. In view of the fact that reversed-phase HPLC encompasses more than 80% of all HPLC applications, this is by no means a serious limitation. Nevertheless, the result is that while applications to the analysis of biopolymers such as peptides abound, its use in the commercial polymer area has been minimal.

6.3. Moving-Belt Interface

A mechanical transport system, the moving-belt interface (73) for HPLC–MS, was adopted after the moving-wire detector originally used in HPLC. As originally conceived, the moving-belt interface provides for the use of the mass spectrometer in its more conventional EI or CI modes. The HPLC effluent is deposited onto the belt surface and the volatile mobile phase selectively vaporized as it passes in front of an IR heater. The less volatile analytes are left on the belt surface and subsequently introduced into the MS ion source as they pass in front of the flash vaporizer. Sample deposition **via** the use of nebulization was shown to facilitate removal of the mobile phase and help maintain chromatographic efficiency (74, 75). An obvious limitation of the moving-belt interface **vis-a-vis** the thermospray and DLI interfaces is the requirement for sample vaporization thus rendering it inapplicable to the analysis of thermally labile or nonvolatile samples. Still, the ability to obtain EI spectra may often be advantageous. Moreover, this mode of HPLC–MS coupling is amenable to the use of surface ionization techniques (76).

Of necessity the moving-belt interface is limited to the LC–MS analysis of compounds of, at least, some moderate volatility. The analysis of selected polymer additives or small oligomers presented in the next examples is consistent with this requirement.

6.3.1. HPLC–MS of Antioxidants

A tandem HPLC–UV–MS system has been reported for the analysis of polymer additives (77). Since UV detection is nondestructive, comparisons can be made on the same sample in a single determination. The HPLC–UV and HPLC–MS chromatograms of a mixture of nine antioxidants are compared in Figures 22*a* and 22*b*. Comparable chromatographic efficiencies were calculated for both chromatograms, indicative of minimal contribution to band broadening by the moving belt HPLC–MS interface used in this example. Quantitative variations are the result of different responses of the UV and MS detectors to the analytes. It may be noted that component 9, distearyl thiodipropionate (DSTDP), was not detectable by UV at 280 nm, but was readily found in the HPLC–MS analysis. In this work, the mass spectrometer was used in the scanning mode under CI (CH_4) conditions. In general, the

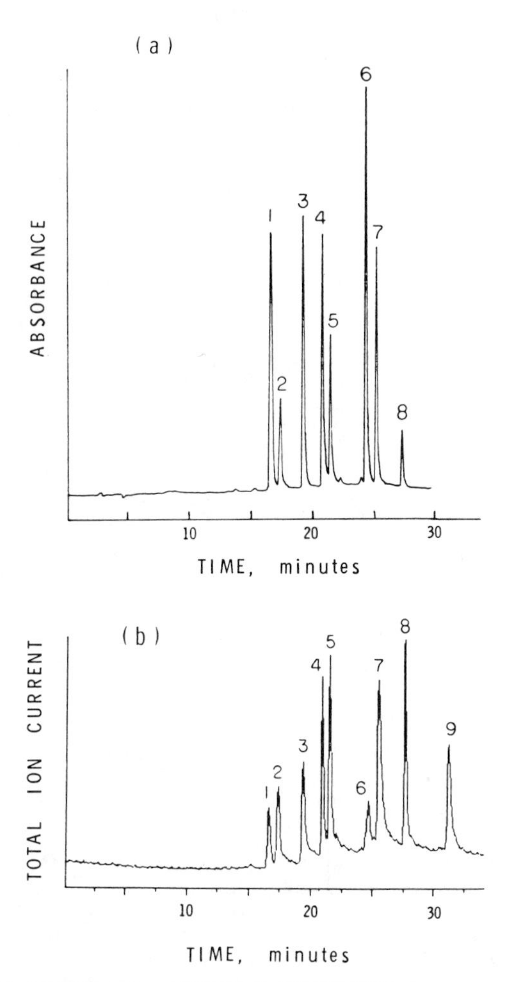

Figure 22. (*a*) HPLC–UV (280 nm) and (*b*) HPLC–MS (CI, CH_4) chromatograms of antioxidants and UV light stabilizers: (1) 2,6-di-*t*-butyl-4-methylphenol (BHT), (2) 4,4′-butylidenebis(3-methyl-6-*t*-butylphenol), (3) 1,1,3-tris(2-methyl-4-hydroxy-*t*-butylphenyl)butane, (Topanol CA), (4) 2-(2-hydroxy-5-octylphenyl)benzotriazole, (5) 2-hydroxy-4-(*n*-octyloxy)benzophenone, (6) pentaerythritoltetra-3-(3,5-di-*t*-butyl-4-hydroxyphenyl)propionate, (7) 1,3,5-trimethyl-2,4,6-tris(3,5-di-*t*-butyl-4-hydroxybenzyl)benzene, (8) octadecyl 3-(3,5-di-*t*-butyl-4-hydroxyphenyl) propionate, (9) distearyl thiodipropionate (DSTDP). Reprinted from J. D. Vargo and K. L. Olson, "Identification of Antioxidant and Ultraviolet Light Stabilizing Additives in Plastics by Liquid Chromatography/Mass Spectroscopy," *Anal. Chem.*, **57**, 672 (1985). Copyright © 1985 American Chemical Society.

authors report comparable sensitivities between UV detection (280 nm) and reconstructed chromatograms of selected ions (78). It is presumed that HPLC–MS in the selected-ion monitoring mode could improve the sensitivity into the picogram range. The CI (CH_4) mass spectrum of topanol (peak 3 in Fig. 22), showing structually informative fragment ions, is given in Figure 23. Typically, additives from plastics were extracted from the polymers overnight at ambient temperature with acetonitrile in a sealed vial. The extract solutions were filtered prior to HPLC–MS analysis.

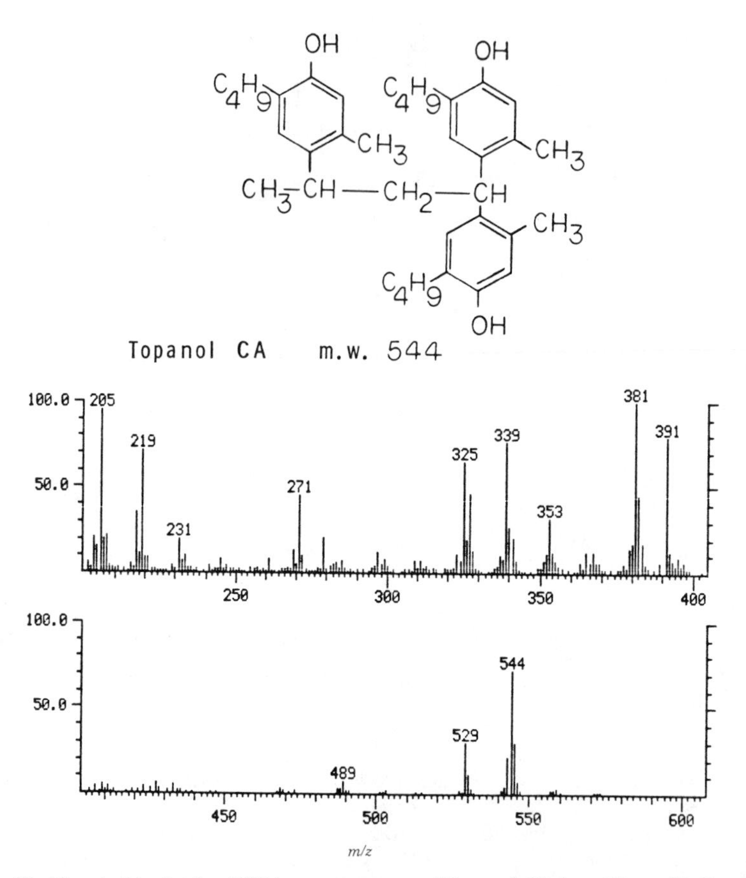

Figure 23. Chemical ionization (CH_4) mass spectrum of Topanol CA from Figure 22. Reprinted from J. D. Vargo and K. L. Olson, "Characterization of Additives in Plastics by Liquid Chromatography–Mass Spectrometry," *J. Chromatogr.*, **353**, 215 (1986). Reproduced with permission from Elsevier Science Publishers, Amsterdam.

6.3.2. *Analysis of Epoxide Oligomers by LC–MS*

High-resolution HPLC columns may even permit the MS characterization of individual polymers in complex mixtures. A mixture of oligomers of epoxy cresol novalac-9 (ECN-9) resin was examined by HPLC–MS (79). In this specific case the analysis was conducted with a moving-belt interface, which did not pass through the ion-source chamber, thus limiting the molecular size that could be vaporized without pyrolysis. Nevertheless, oligomers were well separated and mass spectra (CI, CH_4–NH_3 mixture) were obtained with clear definition of the oligomer sequence. In this context, it is interesting to consider the CI spectrum of the trimer ($n = 2$), Figure 24, from the ECN-9 resin that was analyzed under the above conditions (79). Notable is the intense molecular adduct ion $[M + NH_4]^+$ at m/z 710. Equally as important, however, is the fragmentation mode of the quasimolecular ion that yields distinct peaks at m/z

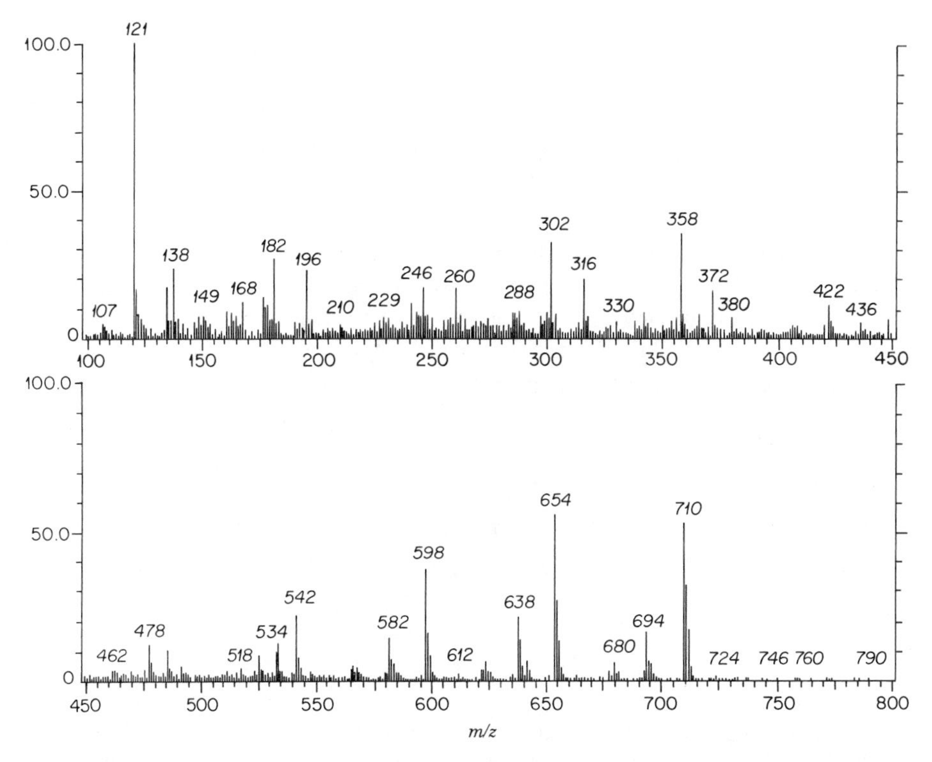

Figure 24. Chemical ionization (CH_4–NH_3) mass spectrum of the trimer from epoxy cresol novalac-9 resin obtained during online HPLC–MS. Reprinted from Ref. 79 with permission from the author.

values corresponding to the individual units of the oligomers as shown on ion structure **q**. In each of the indicated cleavages, retention of the charge can be postulated to occur on either side of the dotted line and is accompanied by a hydrogen transfer and addition of NH_4^+.

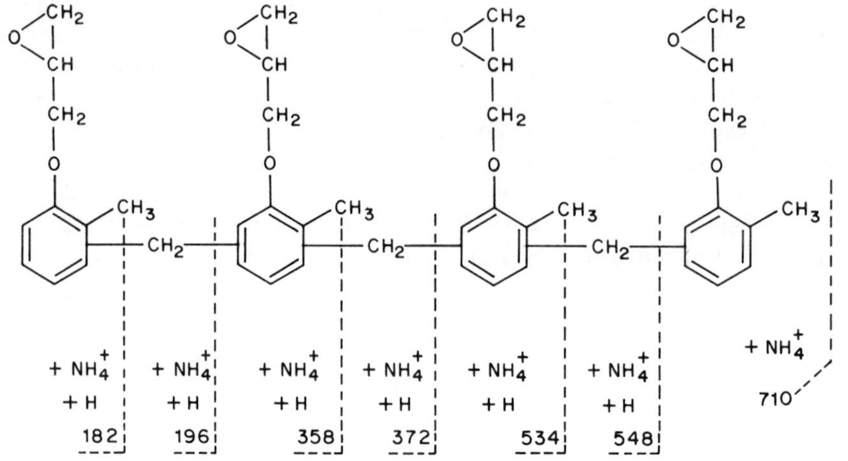

q

Fragmentation of trimer from ECN—9 resin

Remarkably, this is very similar to the CI fragmentation of peptides (80, 81) and, in principle, this type of oligomer analysis may permit the recognition of any variant residues in the structure. Further fine structural detail is provided by fragmentation of the terminal epoxides that leads to a series consistent with sequential elimination of $(CH_2)_2Co$.

$$m/z\ 710 \xrightarrow[-56]{} m/z\ 654 \xrightarrow[-56]{} m/z\ 598 \xrightarrow[-56]{} m/z\ 542 \xrightarrow[-56]{} m/z\ 486$$

Of significance here is the well-defined manner in which the oligomer fragments by chemical ionization, which permits elucidation of detailed structural features.

7. CONCLUSIONS

For many years MS has been utilized as a vapor phase technique and despite this inherent limitation, it has found numerous applications in the polymer field. Pyrolysis–MS can be carried out with good reproducibility and, as illustrated in the examples presented in this chapter, spectral patterns characteristic of a given polymer can be readily obtained. Chemical ionization procedures permit selective determination of molecular species in mixtures and finer structural details are obtainable by EI ionization. The resulting spectral patterns, however, are often prohibitively complex especially when the spectra are obtained from pyrolyzable and/or unknown materials. The examples of polyurethanes illustrate these difficulties. Additional data from IR, NMR, UV, and other physical methods are essential before proceeding with mass spectrometric studies, if sample quantities permit their use.

Most recently, we have witnessed an explosive growth in the field of MS. New ionization techniques such as laser desorption, electrospray, thermospray, and FAB are being introduced to the polymer field and high-mass range–high-resolution mass spectrometers will play an increasingly greater role in the characterization of polymeric substances. The use of FAB or LD to ionize selectively the additives present in a polymer shows the potential for carrying out specific analyses by appropriate chemical manipulation of a sample. Furthermore, it is expected that new online chromatographic–mass spectrometric techniques such as LC–MS will find their way into the polymer field in the not-too-distant future. Without a doubt, we can look forward to a more effective use of MS towards polymer characterization in the coming years.

ABBREVIATIONS

BE	Refers to tandem arrangement of magnetic and electric fields
CI	Chemical ionization
CID	Collision induced dissociation
DCI	Direct chemical ionization
DLI	Direct liquid introduction
EB	Refers to tandem arrangement of electric and magnetic fields
EHD	Electrohydrodynamic
EI	Electron impact
ESA	Electrostatic analyzer
ESPI	Electrospray ionization
FAB	Fast atom bombardment
FD	Field desorption
FI	Field ionization
FT–ICR	Fourier transform-ion cyclotron resonance
GC	Gas chromatography
HPLC	High performance liquid chromatography
ICR	Ion cyclotron resonance
LD	Laser desorption
MS	Mass spectrometry

MIKES	Mass analyzed ion kinetic energy spectrometry
NMR	Nuclear magnetic resonance
SIMS	Secondary ion mass spectrometry
TOF	Time-of-flight
VPO	Vapor pressure osmometry

1. H.-R. Schulten and R. P. Lattimer, "Applications of Mass Spectrometry to Polymers," *Mass Spectrom. Rev.*, **3**, 231 (1984).
2. F. Bekewitz and H. Heusinger, "Mass Spectrometric Characterization of Oligostyrol," *Angew. Makromol. Chem.*, **46**, 143 (1975).
3. J. Luderwald, "Mass Spectrometry of Synthetic Polymers," *Pure Appl. Chem.*, **54**, 255 (1982).
4. S. Foti, P. Mavavigina, and G. Montaudo, "Mechanisms of Thermal Decomposition in Totally Aromatic Polyurethanes," *J. Polym. Sci., Polym. Chem. Ed.*, **19**, 1679 (1981).
5. A. Ballistreri, D. Garozzo, and G. Mantaudo, 'Mass Spectral Characterization and Thermal Decomposition Mechanism of Poly(dimethyl siloxane)," *Macromolecules*, **17**, 1313 (1984).
6. A. Ballistreri, G. Montaudo, and R. W. Lenz, "Mass Spectral Characterization and Thermal Decomposition Mechanism of Alternating Silarylene–Siloxane Products," *Macromolecules*, **17**, 1848 (1984).
7. A. G. Harrison, *Chemical Ionization Mass Spectrometry*, CRC, Boca Raton, FL, 1983.
8. Y. Shimizu and B. Munson, "Pyrolysis/Chemical Ionization Mass Spectrometry of Polymers," *J. Polym. Sci., Polym. Chem. Ed.*, **17**, 1991 (1979).
9. R. J. Cotter, "Mass Spectrometry of Non-Volatile Compounds. Desorption from Extended Probes," *Anal. Chem.*, **52**, 1589A (1980).
10. V. N. Reinhold and S. A. Carr, "Direct Chemical Ionization Mass Spectrometry with Polyimide-Coated Wires," *Anal. Chem.*, **54**, 499 (1982).
11. D. O. Hummel, H. J. Dussel, and K. Rubenacker, "Field Ionization and Electron Impact Mass Spectrometry of Polymers and Copolymers," *Makromol. Chem.*, **145**, 267 (1971).
12. H. R. Schulten, N. Simmleit, and R. Muller, "High-Temperature, High Sensitivity Pyrolysis Field Ionization Mass Spectrometry," *Anal. Chem.*, **59**, 2903 (1987).
13. H. D. Beckey, "Field Desorption Mass Spectrometry: A Technique for the Study of Thermally Unstable Substances of Low Volatility," *Intern. J. Mass Spectrom. Ion Phys.*, **2**, 500 (1969).

14. R. P. Lattimer, D. J. Harmon, and K. R. Welch, "Characterization of Low Molecular Weight Polymers by Liquid Chromatography and Field Desorption Mass Spectrometry," *Anal. Chem.*, **51**, 1293 (1979).

15. R. F. K. Herzog, H. J. Liebl, W. P. Poschenrieder, and A. E. Barrington, "Solids Mass Spectrometer," *Sci. Tech. Aerospace Rept.*, **3**, 2190 (1965).

16. C. A. Evans, Jr., "Secondary Ion Mass Analysis: A Technique for Three-Dimensional Characterization," *Anal. Chem.*, **44**, 67A (1972).

17. W. Abath, K. M. Straub, and A. L. Burlingame, "Secondary Ion Mass Spectrometry with Cesium Ion Primary Beam and Liquid Target Matrix for Analysis of Bioorganic Compounds," *Anal. Chem.*, **54**, 2029 (1982).

18. I. M. Bletsos, D. M. Hercules, D. Greifendorf, and A. Benninghofen, "Time-of-Flight Mass Spectrometry of Nylons: Detection of High Mass Fragments," *Anal. Chem.*, **57**, 2384 (1985).

19. I. M. Bletsos, D. M. Hercules, J. H. Magill, E. Niehuis, and A. Benninghoven, "Time-of-Flight Secondary Ion Mass Spectrometric Detection of Fragments from Thick Polymer Films in the Range $m/z < 4{,}500$," *Anal. Chem.*, **60**, 938 (1988).

20. M. Barber, R. S. Bordoli, R. D. Sedgwick, and A. N. Tyler, "Fast Atom Bombardment of Solids (FAB): A New Source for Mass Spectrometry," *J. Chem. Soc., Chem. Commun.*, **1981**, 325.

21. S. A. Martin, C. E. Costello, and K. Biemann, "Optimization of Experimental Procedures for Fast Atom Bombardment Mass Spectrometry," *Anal. Chem.*, **54**, 2362 (1982).

22. M. Barber, R. S. Bordoli, G. J. Elliot, R. G. Sedgwick, and A. N. Tyler, "Fast Atom Bombardment Mass Spectrometry," *Anal. Chem.*, **54**, 645A (1982).

23. R. M. Caprioli, "Fast Atom Bombardment Mass Spectrometry for the Determination of Dissociative Constants of Weak Acids in Solution," *Anal. Chem.*, **55**, 2387 (1983).

24. J. A. Sunner, A. Morales, and P. Kebarle, "Dominance of Gas-Phase Basicities in the Competition for Protons in Fast Atom Bombardment Mass Spectrometry," *Anal. Chem.*, **59**, 1378 (1987).

25. R. L. Cochran, "Fast Atom Bombardment Mass Spectrometry (FAB/MS) and Its Industrial Applications," *Appl. Spectrosc. Rev.*, **22**, 137 (1986).

26. M. Youssefi, "FAB Analysis of Polymers by an Optimized Quadrupole Mass Spectrometer," ASMS, Proc. 30th Annu. Conf. Mass Spectrometry and Allied Topics, Boston, MA, 1983, p. 550.

27. A. Ballistreri, D. Garozzo, M. Giuffrida, and G. Montaudo, "Identification of the Ions Produced by FAB–MS in Some Polyesters and Polyamides," *Anal. Chem.*, **59**, 2024 (1987).

28. M. Youssefi, Eastern Analytical Symposium, New York, NY, (1985).

29. R. D. MacFarlane, "Particle-Induced Desorption Mass Spectrometry of Large Involatile Biomolecules: Surface Chemistry in the High-Energy-Short-Time Domain," *Acc. Chem. Res.*, **15**, 268 (1982).

30. R. D. MacFarlane, "Californium-252 Plasma Desorption Mass Spectrometry," *Anal. Chem.*, **55**, 1246A (1983).

31. E. Denoyer, R. V. Grieken, F. Adams, and D. F. S. Natusch, "Laser Microprobe Mass Spectrometry 1: Basic Principles and Performance Characteristics," *Anal. Chem.*, **54**, 26A (1982).

32. D. M. Hercules, R. J. Day, K. Baulesanmugan, T. A. Deng, and C. P. Li, "Laser Microprobe Mass Spectrometry 2. Applications to Structural Analysis," *Anal. Chem.*, **54**, 280A (1982).

33. M. Karas and F. Hillenkamp, "Ultraviolet Laser Desorption of Ions Above 10 kDa," 11th International Mass Spectrometry Conf., Bordeaux, France, August 29–September 2, 1988.

34. A. T. Hau and A. G. Marshall, "Identification of Dyes in Solid Poly(methyl methacrylate) by Means of Laser Desorption Fourier Transform Ion Cyclotron Resonance Mass Spectrometry," *Anal. Chem.*, **80**, 932 (1988).

35. (a) F. M. Kimock, J. P. Baxter, D. L. Pappas, P. H. Kobrin, and N. Winograd, "Solids Analysis using Energetic Ion Bombardment and Multiphoton Resonance Ionization with Time-of-Flight Detection," *Anal. Chem.*, **56**, 2782 (1984). (b) C. Becker and K. T. Gillen, "Surface Analysis by Nonresonant Multiphoton Ionization of Desorbed or Sputtered Species," *Anal. Chem.*, **56**, 1671 (1984).

36. H. V. Weissenhoff, H. Selzle, and E. W. Schlag, "Laser-desorbed Large Molecules in a Supersonic Jet," *Z. Naturforsch*, **40a**, 674 (1985).

37. J. Grotemeyer, U. Bosel, K. Walter, and E. W. Schlag, "Biomolecules in the Gas Phase. II. Multiphoton Ionization Mass Spectrometry of Angiotensin," *Org. Mass Spectrom.*, **21**, 595 (1986).

38. M. L. Vestal, "High Performance Liquid Chromatography—Mass Spectrometry," *Science*, **226**, 275 (1984).

39. M. L. Vestal, "Ionization Techniques for Non-Volatile Molecules," *Mass Spectrom. Rev.*, **2**, 447 (1983).

40. M. Dole, L. L. Mack, R. L. Hines, R. C. Mobley, L. D. Ferguson, and M. B. Alice, "Molecular Beams of Macroions," *J. Chem. Phys.*, **49**, 2240 (1968).

41. L. L. Mack, P. Kralik, A. Rheude, and M. L. Dole, "Molecular Beams of Macroions. II," *J. Chem. Phys.*, **52**, 4977 (1970).

42. C. M. Whitehouse, R. N. Dreyer, M. Yamashita, and J. B. Fenn, "Electrospray Interface for Liquid Chromatographs and Mass Spectrometers," *Anal. Chem.*, **56**, 675 (1985).

43. J. A. Loo, H. R. Udseth, and R. D. Smith, "Collisional Effects on the Charge

Distribution of Ions from Large Molecules Formed by Electrospray-Ionization Mass Spectrometry," *Rapid Commun. Mass Spectrom.*, **2**, 207 (1988).

44. J. A. Loo, H. R. Udseth, and R. D. Smith, "Solvent Effects on the Charge Distribution Observed with Electrospray Ionization Mass Spectrometry of Large Molecules," *Biomed. Environ. Mass Spectrom.*, **17**, 411 (1988).
45. B. N. Colby and C. A. Evans, "Electrohydrodynamic Ionization Mass Spectrometry," *Anal. Chem.*, **45**, 1884 (1973).
46. B. P. Stimpson, D. S. Simone, and C. A. Evans, Jr., "Mass Spectrometry of Solvated Ions Generated Directly from the Liquid Phase by Electrohydrodynamic Ionization," *J. Phys. Chem.*, **82**, 660 (1978).
47. K. W. S. Chan and K. D. Cook, "Mass Spectrometric Study of Interactions between Poly(ethylene glycols) and Alkali Metals in Solution," *Macromolecules*, **16**, 1736 (1983).
48. K. W. S. Chan and K. D. Cook, "Extended Mass Range by Multiple Charge: Sampling Quadruply Charged Quasimolecular Ions of Poly(ethylene glycol) 4000," *Org. Mass Spectrom.*, **18**, 424 (1983).
49. J. N. Kyranos and P. Vouros, "The Role of Mass Spectrometry in Polymer Science," *J. Appl. Polym. Sci., Appl. Polym. Symp.*, **43**, 211 (1989).
50. F. W. McLafferty (Ed.), *Tandem Mass Spectrometry*, Wiley, New York, 1983.
51. R. A. Yost and D. D. Fetterolf, "Tandem Mass Spectrometry (MS/MS) Instrumentation," *Mass Spectrom. Rev.*, **2**, 1 (1983).
52. J. D. Cinpek, S. Verma, A. E. Schoen, and R. G. Cooks, "Hybrid Mass Spectrometers," *Spectra.*, **9**, 7 (1983).
53. R. A. Yost and C. G. Enke, "Triple Quadrupole Mass Spectrometry for Direct Mixture Analysis and Structure Elucidation," *Anal. Chem.*, **51**, 1251A (1979).
54. B. A. Mamyrin, V. I. Karataev, D. V. Shmikk, and V. A. Zagulin, "The Mass Reflectron: A New Non-magnetic Time-of-Flight Mass Spectrometer," *Sov. Phys. JETP*, **37**, 45 (1973).
55. (a) T. A. Lehman, and M. M. Bursey, *Ion Cyclotron Resonance Spectrometry*, Wiley, New York, 1971. (b) J. L. Beauchamp, "Ion Cyclotron Resonance Spectroscopy," *Annu. Rev. Phys. Chem.*, **22**, 527 (1971).
56. L. R. Anders, J. L. Beauchamp, R. C. Dunbar, and J. D. Baldeschweiler, "Ion-Cyclotron Double Resonance (E/T)", *J. Chem. Phys.*, **45**, 1062 (1966).
57. M. B. Comisarow and A. G. Marshall, "Fourier transform Ion Cyclotron Resonance Spectroscopy," *Chem. Phys. Lett.*, **25**, 282 (1974); "Frequency Sweep Fourier Transform Ion Cyclotron Resonance Spectroscopy," *Chem. Phys. Lett.*, **26**, 489 (1974).
58. R. T. McIver, Jr., and A. D. Baranyi, "High Resolution Ion Cyclotron Resonance Spectroscopy," *Int. J. Mass Spectrom. Ion Phys.*, **14**, 449 (1974).

59. M. Allemann, H. P. Kellerhals, and K. P. Wanczak, "A New Fourier Transform Mass Spectrometer with A Superconducting Magnet," *Chem. Phys. Lett.*, **75**, 328 (1980).

60. R. S. Brown, D. A. Weil, and C. L. Wilkins, "Laser Desorption–Fourier Transform Mass Spectrometry for the Characterization of Polymers," *Macromolecules*, **19**, 1255 (1986).

61. M. Gloria, G. Audisio, S. Daolio, P. Traldi, and E. Vecchi, "Mass Spectrometric Studies on Cyclo- and Polyphosphazenes. 1. Polymerization of Hexachlorocyclophosphazene," *Macromolecules*, **17**, 1230 (1984).

62. (a) R. A. Forbes, F. H. Laukien, and J. Wronka, "A Comprehensive Methodology for the Study of Sequential Ion Molecule Reactions Using Fourier Transform-Ion Cyclotron Resonance: The Formation, Isolation and Reactions of Very Large Metal Clusters," *Int. J. Mass Spectrom. Ion Phys.*, **83**, 23 (1988). (b) R. A. Forbes, F. H. Laukien, J. Wronka, and D. P. Ridge, "Effect of Ligation and Cluster Size on the Reactivity of Metal Clusters: Reactions of $Re_n(CO)_m^+$ with Cyclohexane," *J. Phys. Chem.*, **91**, 6450 (1987).

63. J. Yergey, D. Heller, G. Hansen, R. J. Cotter, and C. Fenselau, "Isotopic Distributions in Mass Spectra of Large Molecules," *Anal. Chem.*, **55**, 353 (1983).

64. F. W. McLafferty, *Interpretation of Mass Spectra*, University Science Books, Mill Valley, CA, 1980.

65. P. J. Arpino, B. G. Dawkins, and F. W. McLafferty, "A Liquid Chromatography/Mass Spectrometry System Providing Continuous Monitoring with Nanogram Sensitivity," *J. Chromatogr. Sci.*, **12**, 574 (1974).

66. J. D. Henion, "Continuous Monitoring of Total Micro LC Eluent by Direct Liquid Introduction LC/MS," *J. Chromatogr. Sci.*, **19**, 57 (1981).

67. J. D. Henion, *Micro LC/MS Coupling In: Microcolumn HPLC*, P. Kucero (Ed.), Journal of Chromatography Library, Vol. 28, Elsevier, Amsterdam, 1984.

68. J. D. Henion, "Determination of Sulfa Drugs in Biological Fluids by Liquid Chromatography/Mass Spectrometry/Mass Spectrometry," *Anal. Chem.*, **54**, 451 (1982).

69. E. D. Lee and J. D. Henion, "Micro-Liquid Chromatography/Mass Spectrometry with Direct Liquid Introduction," *J. Chromatogr. Sci.*, **23**, 253 (1985).

70. H. Alborn and G. Stenhagen, "Direct Coupling of Packed Fused-Silica Liquid Chromatographic Columns to a Magnetic Sector Mass Spectrometer and Application to Polar Thermolabile Compounds," *J. Chromatogr.*, **323**, 47 (1985).

71. R. M. Caprioli, T. Fan, and J. S. Cotrell, "Continuous-Flow Sample Probe for Fast Atom Bombardment Mass Spectrometry," *Anal. Chem.*, **58**, 2949 (1986).

72. M. A. Moseley, L. J. Deterding, J. S. M. de Wit, K. B. Tomer, R. T. Kennedy, N. Bragg, and J. W. Jorgenson, "Optimization of a Coaxial Continuous Flow Fast Atom Bombardment Interface Between Capillary Liquid Chromatography and

Magnetic Sector Mass Spectrometry for the Analysis of Biomolecules," *Anal. Chem.*, **61**, 1577 (1989).

73. W. H. McFadden, H. L. Schwarz, and S. J. Evans, "Direct Analysis of Liquid Chromatographic Effluents," *J. Chromatogr.*, **122**, 389 (1976).
74. R. D. Smith and A. L. Johnson, "Deposition Method for Moving Ribbon Liquid Chromatograph-Mass Spectrometer Interfaces," *Anal. Chem.*, **53**, 739 (1981).
75. M. J. Hayes, E. P. Lankmayr, P. Vouros, B. L. Karger, and J. M. McGuire, "Moving Belt Interface for High Performance Liquid Chromatography-Mass Spectrometry," *Anal. Chem.*, **55**, 1745 (1983).
76. J. S. Stroh, J. C. Cook, R. M. Milberg, L. Braybon, T. Kihera, Z. Huang, K. L. Rinehart, Jr., and J. A. S. Lewis, "On-Line Liquid Chromatography/Fast Atom Bombardment Mass Spectrometry," *Anal. Chem.*, **57**, 985 (1985).
77. J. D. Vargo and K. L. Olson, "Identification of Antioxidant and Ultraviolet Light Stabilizing Additives in Plastics by Liquid Chromatography/Mass Spectrometry," *Anal. Chem.*, **57**, 672 (1985).
78. J. D. Vargo and K. L. Olson, "Characterization of Additives in Plastics by Liquid Chromatography–Mass Spectrometry," *J. Chromatogr.*, **353**, 215 (1986).
79. B. Petersen, "HPLC-MS Analysis of Epoxy Resins," Final Report to AAMRC, Watertown, MA from Battelle Columbus Laboratories, April 28, 1980.
80. M. Mudget, J. A. Sogn, D. V. Bowen, and F. H. Field, "Peptide Sequencing by Chemical Ionization Mass Spectrometry," *Adv. Mass Spectrom.*, **7**, 1506 (1978).
81. T. J. Yu, H. Sshwartz, R. W. Giese, B. L. Karger, and P. Vouros, "Analysis of *N*-Acetyl-*N*, O, *S*-Permethylated Peptides by Combined Liquid Chromatography–Mass Spectrometry," *J. Chromatogr.*, **218**, 519 (1981).

INDEX

(*continued from front*)

Vol. 63. **Applied Electron Spectroscopy for Chemical Analysis.** Edited by Hassan Windawi and Floyd Ho

Vol. 64. **Analytical Aspects of Environmental Chemistry.** Edited by David F. S. Natusch and Philip K. Hopke

Vol. 65. **The Interpretation of Analytical Chemical Data by the Use of Cluster Analysis.** By D. Luc Massart and Leonard Kaufman

Vol. 66. **Solid Phase Biochemistry: Analytical and Synthetic Aspects.** Edited by William H. Scouten

Vol. 67. **An Introduction to Photoelectron Spectroscopy.** By Pradip K. Ghosh

Vol. 68. **Room Temperature Phosphorimetry for Chemical Analysis.** By Tuan Vo-Dinh

Vol. 69. **Potentiometry and Potentiometric Titrations.** By E. P. Serjeant

Vol. 70. **Design and Application of Process Analyzer Systems.** By Paul E. Mix

Vol. 71. **Analysis of Organic and Biological Surfaces.** Edited by Patrick Echlin

Vol. 72. **Small Bore Liquid Chromatography Columns: Their Properties and Uses.** Edited by Raymond P. W. Scott

Vol. 73. **Modern Methods of Particle Size Analysis.** Edited by Howard G. Barth

Vol. 74. **Auger Electron Spectroscopy.** By Michael Thompson, M. D. Baker, Alec Christie, and J. F. Tyson

Vol. 75. **Spot Test Analysis: Clinical, Environmental, Forensic and Geochemical Applications.** By Ervin Jungreis

Vol. 76. **Receptor Modeling in Environmental Chemistry.** By Philip K. Hopke

Vol. 77. **Molecular Luminescence Spectroscopy: Methods and Applications** (*in two parts*). Edited by Stephen G. Schulman

Vol. 78. **Inorganic Chromatographic Analysis.** Edited by John C. MacDonald

Vol. 79. **Analytical Solution Calorimetry.** Edited by J. K. Grime

Vol. 80. **Selected Methods of Trace Metal Analysis: Biological and Environmental Samples.** By Jon C. VanLoon

Vol. 81. **The Analysis of Extraterrestrial Materials.** By Isidore Adler

Vol. 82. **Chemometrics.** By Muhammad A. Sharaf, Deborah L. Illman, and Bruce R. Kowalski

Vol. 83. **Fourier Transform Infrared Spectrometry.** By Peter R. Griffiths and James A. de Haseth

Vol. 84. **Trace Analysis: Spectroscopic Methods for Molecules.** Edited by Gary Christian and James B. Callis

Vol. 85. **Ultratrace Analysis of Pharmaceuticals and Other Compounds of Interest.** Edited by S. Ahuja

Vol. 86. **Secondary Ion Mass Spectrometry: Basic Concepts, Instrumental Aspects, Applications and Trends.** By A. Benninghoven, F. G. Rüdenauer, and H. W. Werner

Vol. 87. **Analytical Applications of Lasers.** Edited by Edward H. Piepmeier

Vol. 88. **Applied Geochemical Analysis.** by C. O. Ingamells and F. F. Pitard

Vol. 89. **Detectors for Liquid Chromatography.** Edited by Edward S. Yeung

Vol. 90. **Inductively Coupled Plasma Emission Spectroscopy: Part I: Methodology, Instrumentation, and Performance; Part II: Applications and Fundamentals.** Edited by J. M. Boumans

Vol. 91. **Applications of New Mass Spectrometry Techniques in Pesticide Chemistry.** Edited by Joseph Rosen